CONSTRUCTION CONTRACTS

THIRD EDITION

Jimmie Hinze

Professor of the M. E. Rinker, Sr. School of Building Construction
University of Florida

The **McGraw·Hill** Companies

CONSTRUCTION CONTRACTS, THIRD EDITION

Published by McGraw-Hill, a business unit of The McGraw-Hill Companies, Inc., 1221 Avenue of the Americas, New York, NY 10020. Copyright © 2011 by The McGraw-Hill Companies, Inc. All rights reserved. Previous editions © 2001 and 1993. No part of this publication may be reproduced or distributed in any form or by any means, or stored in a database or retrieval system, without the prior written consent of The McGraw-Hill Companies, Inc., including, but not limited to, in any network or other electronic storage or transmission, or broadcast for distance learning.

Some ancillaries, including electronic and print components, may not be available to customers outside the United States.

This book is printed on acid-free paper

1 2 3 4 5 6 7 8 9 0 DOC/DOC 1 0 9 8 7 6 5 4 3 2 1 0

ISBN: 978-0-07-339785-6
MHID: 0-07-339785-7

Vice President & Editor-in-Chief: *Marty Lange*
Vice President, EDP/Central Publishing Services: *Kimberly Meriwether David*
Publisher: *Raghu Srinivasan*
Sponsoring Editor: *Peter Massar*
Marketing Manager: *Curt Reynolds*
Development Editor: *Lorraine Buczek*
Project Manager: *Melissa M. Leick*
Buyer: *Laura Fuller*
Design Coordinator: *Brenda A. Rolwes*
Cover Designer: *Studio Montage, St. Louis, Missouri*
Cover Image Credits: *Bridge Construction:* © *Donovan Reese/Getty Images/RF; Contract and Pen:*
       © *Stockbyte/Getty Images/RF; Construction Site at Twilight:* ©*2007 Getty*
       Images/RF; Silhouetted Girders with Worker and Crane: © *Getty Images/RF, and*
       House under Construction: © *Getty Images/Somos/RF*
Media Project Manager: *Yeswini Devdutt*
Compositor: *S4Carlisle Publishing Services*
Typeface: *10/12 Times Roman/Palatine Display*
Printer: *R.R. Donnelley*

Library of Congress Cataloging-in-Publication Data

Hinze, Jimmie
    Construction contracts / Jimmie Hinze.—3rd ed
    p. cm.—(McGraw-Hill series in construction engineering and project management)
    Includes index.
    Includes bibliographical references and index.
    ISBN-13: 978-0-07-339785-6 (acid-free paper)
    ISBN-10: 0-07-339785-7 (acid-free paper)
    1. Construction contracts—United States. I. Title.
KF902.H56   2010
343.73'07869—dc22

                                     2010018909

www.mhhe.com

ABOUT THE AUTHOR

JIMMIE HINZE, a professor in the M.E. Rinker, Sr. School of Building Construction, has been at the University of Florida since 1996. In the prior twelve years, he was a professor of Civil Engineering in the University of Washington's graduate program in construction engineering and management. He had previously served on the civil engineering faculty at the University of Missouri-Columbia for seven years. He received a B.S. and M.S. in architectural engineering from the University of Texas and a Ph.D. from Stanford University. He has been involved in various types of projects including residential, commercial, highway/heavy, and industrial construction. His experience stems from representing general contractors, design-build companies, and specialty contractors. He has been consulted on a variety of construction claims. His primary research interest included construction safety, worker productivity, and construction contracts. He is active in construction organizations such as ASCE, AIC, CIB, ASC, ASSE, CII, NCCER, and various contractor associations.

To Maxine, Jacob, and Justin

CONTENTS

PREFACE

THE CONSTRUCTION OF various types of facilities often represents the culmination of the efforts of several designers. In fact, most projects undertaken by designers are ultimately constructed. Most construction projects begin with the award of a construction contract. This contract becomes the underlying foundation for the relationship that will exist between the various parties involved in the project. Thus, knowledge of contracts is beneficial to virtually all parties involved in the construction process. This book will provide information of general interest to anyone working with construction contracts.

This book was written to serve as a learning tool and a reference guide on construction contracts. The fundamentals of contract law are presented, along with an in-depth treatment of the construction topics that most frequently result in litigation. In addition, an overview is provided of other important construction-related topics, including the procurement process for construction contracts, methods of dispute resolution, surety bonds, construction insurance, construction safety, and construction labor laws. This third edition incorporates some of the changes that have occurred during the past decade.

In comparison with other books on contracts, two distinguishing features of this text should become apparent. First, this text includes summaries of a large number of legal cases involving construction, and discusses many topics that are germane to contract disputes. Well over 100 cases are described to help illustrate key points. These cases will also give the reader a greater understanding of the role of the judiciary in the construction industry. The reader will also appreciate the frustration that some contracting parties have had with decisions made by the courts. In some cases different interpretations merely reflect differences in the courts, while in other cases they reflect differences in statutes. In spite of some differences, most court decisions tend to be reasonably consistent. An understanding of how judicial decisions are made will give the reader insight into how the facts of a particular situation may be interpreted in a court of law.

The second feature not found in most texts is that many contract provisions are isolated for the reader. These provisions help the reader recognize the importance of the particular wording that is used in contract documents. Some of these provisions are presented to illustrate provisions that are in common usage, while others are presented to show exceptional provisions, particularly those that shift responsibility or risk. While the provisions that are presented primarily use terms such as *owner* and *contractor,* in actual usage, many terms are used in lieu of *owner,* including *owner's representative, architect, contract officer, agency, department, division, city, county,* and *district.*

Chapter 1 provides a description of the construction industry, including its size and importance to the U.S. economy. Chapter 2 describes the different contracting arrangements most often encountered in construction. Chapters 3 through 6 provide background information on the fundamentals of contracts, the role of real property laws in construction, the difference between agents and independent contractors, and the significance of different forms of organizations. Chapter 7 introduces the topic of torts. Chapter 8 describes bid bonds, performance bonds, and payment bonds. Chapter 9 describes how construction contracts are generally awarded. Various contract documents are discussed in Chapter 10. The methods of payment for construction contracts are discussed in Chapter 11. Chapters 12 through 16 cover topics that are common to many construction disputes, including changes, changed conditions, delays, payments, and warranties. Chapter 17 describes the various types of construction insurance. Chapter 18 discusses the role of subcontractors. Chapter 19 presents an overview of some of the major issues commonly encountered in international markets. Chapter 20 discusses the resolution of disputes by methods such as negotiation, partnering, arbitration, mediation, dispute review boards, and minitrials. Chapter 21 presents basic issues involving ethics that may be faced by construction professionals. Chapter 22 describes safety in the construction industry and related legislation. Chapter 23 presents terms related to labor relations, and introduces the more important laws that have a direct impact on the construction industry.

New for the third edition.

Added new material on:
• Day Labor Agencies
• Independent Contractors
• Statutory Employees
• Reverse Auction Bidding
• Multiple Bid Packages or Phased Approach
• Job Order Contracting
• Patent and Latent Defects
• Electronic Bidding
• OSHA Fines and Penalties
• Compliance with the Contracts Documents

Updated sections on:
• Performance and Payment Bonds
• Job Order Contracting

- Cardinal Changes
- The Miller Act
- Differing Site Conditions
- Termination
- Comprehensive General Liability
- Contractual Liability
- Codes of Ethics
- Record-Keeping Requirements
- Subcontractor Progress Payments
- Indemnification
- Arbitration
- Home Office Overhead
- Eminent Domain
- Differing Site Conditions
- AIA Documents A101 and A201
- ConsensusDocs 200 and 750

Many individuals contributed to the successful completion of the first edition of this book. Particular appreciation and gratitude are expressed to Bryce Coleman, William Shirk, Jennifer Tada, Paul Prost, and Kyle Hansen. Gratitude is also expressed to Richard O'Cull, James Franken, Randy Zuke, and Julie Dickens for their assistance. I continue to offer my sincerest thanks to Neal Benjamin for his many years of guidance and counsel. The second edition was similarly the result of guidance and input from several individuals, including Steve Auld, Michael Wozney, Jim Dunn, Michael Cook, James Milward, Debra Bosma, John Gambatese, Andrea Johnson, Sherwood Kelly, Leon Wetherington, Norma Andersen, and Ken Andersen. I would like to thank the reviewers for the third edition: Keith D. Berndt, North Dakota State University; Robert F. Brehm, Drexel University/Goodwin College of Professional Studies; John Gambatese, Oregon State University; Michael G. Headrick, University of Minnesota; Kelly C. Strong, Iowa State University; Marion R. Tuttle, New Jersey Institute of Technology; Edwin C. Weaver, North Carolina State University. I am particularly grateful to Kevin Bowen and Joseph Roesler for their assistance.

Jimmie Hinze

Text website: www.mhhe.com/hinze
For Instructors only, a Solutions Manual and Lecture PowerPoints
are available.

1

DESCRIPTION OF THE CONSTRUCTION INDUSTRY

CONSTRUCTION INCLUDES ALL immobile structures, such as buildings, tunnels, pipelines, dams, canals, airports, power plants, railroads, docks, bridges, sewage treatment plants, and factories. Most of the reshaping of the earth's surface can be attributed to the construction industry. The only exception is agriculture, which is responsible for clearing a significant amount of land for farming. The mining industry has played a minor role in reshaping the land. The forestry industry has also played a role in changing the earth's surface, but this can also be attributed directly to the construction industry.

THE SIZE OF THE CONSTRUCTION INDUSTRY

How large is the construction industry? It accounts for nearly two-thirds of $1 trillion in expenditures per year in new construction alone (see figure 1.1) and constitutes about 5 percent of the gross domestic product (GDP). Approximately 6 percent (7.5 million workers) of the industrial workforce is employed directly in the construction industry (see figure 1.2). These numbers do not include more than 2.5 million establishments with no payroll, or the more than 1.5 million nonconstruction (white-collar) employees required to keep the industry viable. A smaller percentage of workers is employed in agriculture (4 percent), the steel industry (1 percent), and the auto industry (1 percent). Furthermore, 15 percent of the industrial workforce is directly or indirectly involved in construction. This involvement includes the production, transportation, and distribution of construction materials and equipment. In the support industries, it is estimated that 42 percent of the volume is contributed by four industries: lumber products, stone and clay products, iron and steel, and heating and plumbing materials.

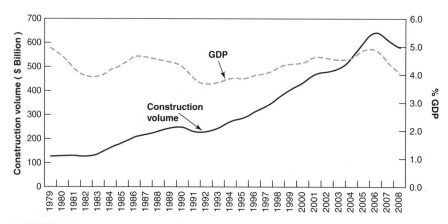

FIGURE 1.1
Annual construction volume and the percent of the gross domestic product attributed to the construction Industry. *(Source: Bureau of Economic Analysis.)*

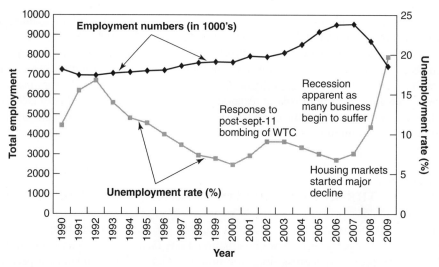

FIGURE 1.2
Number of workers employed in the construction industry and the associated unemployment rate. *(Source: Bureau of Labor Statistics.)*

Construction volume consists primarily of new construction, with the remaining portions being devoted either to maintenance and repair or to renovation and remodeling (figure 1.3). Expenditures for maintenance and repair are generally fairly consistent from year to year. Trade-offs are often made between expenditures for new construction and those for renovation and remodeling. Expenditures for new construction are usually highest when the economy is strong. In a weak economy and during recessionary periods, a greater percentage of the total expenditures will be devoted to renovation and remodeling; however, the total expenditures for construction will

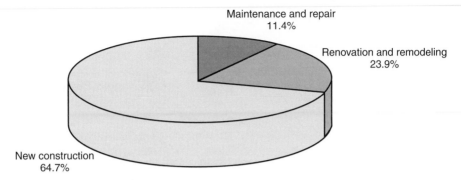

FIGURE 1.3
Relative distribution of construction expenditures by type of construction work.

obviously decline. Renovation and remodeling are more desirable than new construction during recessionary periods, as fewer funds are generally available at those times.

The construction industry represents the largest single production activity in the American economy. Despite this distinction, the construction industry has suffered in recent decades in terms of low productivity compared with other industries. This stems in part from the conservative nature of the industry. For example, the amount spent on research and development in the construction industry is estimated to be about $1.50 per construction worker per year. Of this relatively small sum, it is estimated that contractors contribute about 4 percent. Most research funding is derived from manufacturers (69 percent) and government agencies (18 percent) and is largely spent on the development of improved materials, while productivity receives little attention. Other U.S. industries spend nearly 2 percent of annual revenues on research and development. Even this percentage is well below the more than 2.5 percent spent on construction research in countries such as Japan and Germany. It is interesting to note the tremendous strides that have been made in the health care industry, which is heavily committed to research and development; the health care industry is comparable in size to the construction industry.

MANUFACTURING VERSUS CONSTRUCTION

The construction industry differs from typical manufacturing. The construction industry is characterized by custom-built projects, whereas standardized methods (mass production) are common in manufacturing. As a result of standardization, the control of quality is more easily assured in manufacturing. Standardization in construction may occur to some degree as a result of the prefabrication of different project components. This is limited, however, as the location changes with each project; manufactured products are produced at one plant location, while each construction project has a unique plant location.

Construction projects differ from most manufactured products in another sense as well. Construction projects are relatively complex and generally are completed through the combined efforts of different crafts. Differing percentages

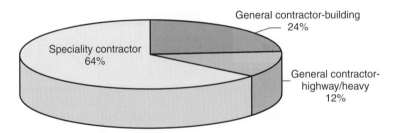

FIGURE 1.4
Distribution of the different types of construction firms with employees.
(Source: Bureau of Labor Statistics.)

of the work are subcontracted to specialty contractors. On some projects all the fieldwork is done by subcontractors. Subcontracting some of the work on a project is commonplace and is often required so that the necessary skills to complete the project can be obtained. The complexity associated with projects and the diverse numbers of trades required for their completion (typically about 10 to 15 different crafts on a single building project) result in inefficiencies not typically encountered in manufacturing. Most building construction projects will have over half (commonly 80 to 90 percent) of the work performed by specialty contractors, while on heavy construction projects about 15 percent is subcontracted. These subcontractors or specialty contractors provide skills in various trades, with each subcontractor typically specializing in one trade (figure 1.4). While many firms may be involved in the construction of one project, the composition of those firms is rarely repeated on subsequent projects. For this reason, there is little or no economy of scale when several different construction projects are undertaken.

Since the location is unique to each project, unique demands are made on construction workers. Workers who make a career in the construction industry must be willing to transfer to the location of the next project. An added disadvantage for these workers is that construction work, particularly in some areas, is seasonal in nature. This is especially true of highway work and heavy construction.

Recent estimates show that the total annual expenditures for the payroll of construction workers exceeds $160 billion. On the average, each privately employed construction worker accounts for about $160,000 worth of construction put in place each year. In the past, the average annual pay of construction workers was more than 10 percent above the pay of workers in all U.S. industries. In recent years, the wages of workers in some industrial sectors have surpassed those in construction, that is, wages of construction workers have not kept pace with some industrial workers. The average wages of skilled construction workers will vary by geographic region but tend to range from $13 to $17 per hour.

The cyclical nature of construction work also means that construction workers have limited employment opportunities during certain time periods; unemployment rates are frequently double the rates of other industrial workers. It is common for construction employment to fall more than 25 percent from the peak construction month of August to the February trough. The fluctuation in employment in the seasons is readily apparent in figure 1.5. The seasonal decline in employment can

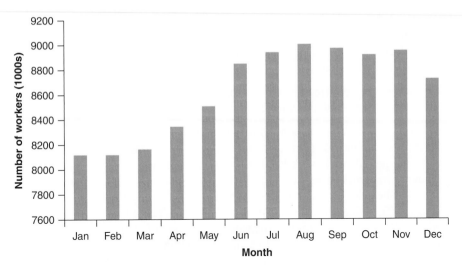

FIGURE 1.5
Variation of construction worker employment by month of the year. *(Source: Bureau of Labor Statistics.)*

be much higher in highway and heavy construction. Although the average number of hours worked per week by construction workers is about 40, this does not accurately reflect a steady state employment, as 25 percent of these hours represent overtime. This employment fluctuation results in 20 percent of the U.S. construction workers being employed less than 1700 hours per year. The only industry that is affected to a greater degree by the seasons is agriculture. These factors, in addition to the unique skills required of skilled construction workers, are largely responsible for the high wages in this industry.

THE ECONOMY AND THE CONSTRUCTION INDUSTRY

Because of the cyclical nature of the construction industry and its quick response to changes in the economy, entry into the industry must be facilitated. As the need for construction services quickly rises during periods of a strong economy, or "up cycle," the industry, with its inherent flexibility, is able to react with relative ease. Although prices tend to rise as competition is decreased, adjustments invariably take place to address the changing needs. This is possible because the construction industry is an easy entry industry. In fact, more than one in every eight business starts occurs in construction. This easy entry is made possible and is necessary for the following reasons:

- High growth rate in construction (the industry responds quickly to economic shifts).
- Low capital requirements (large investment required only for larger equipment).
- Little absolute cost or profit advantage for established firms (their most valued assets are their employees rather than their equipment and materials).
- Most states have no rigid licensing requirements or fees.

Just as the industry needs to be able to respond to a growing demand for construction services, it must also be able to recoil or adjust easily when the demand for its services declines. This is also easily accommodated within the construction industry for the following reasons:

- A company can be formed just to construct a single project (as in joint ventures).
- Firms are seldom sold as a unit (continuity is not assured or guaranteed).

Some firms exit the construction industry without the benefit of choice; that is, they fail. This does not include firms that simply stop doing business. During recessionary periods and periods of a weak local economy, competition increases, causing profit margins on construction services to decrease. This reaction of the industry often results in business losses that can be severe and result in the failure of construction companies. The failure rate for construction firms is quite high, with business failures in construction accounting for approximately 12 percent of all business failures. It has also been estimated that 20 percent of all construction-related businesses eventually fail. This does not include firms that simply stop doing business. Failures result from many factors, including overextension of resources, subcontractor default, inadequate insurance to cover major losses, errors in estimating, undertaking projects out of the area of expertise of the firm, death or departure of a key company officer, inadequate labor, acts of God, managerial inexperience, and other economic causes. Some otherwise successful firms fail because they cannot collect on their receivables. Incompetence results in some failures including the inability to focus on business issues due to family problems, business conflicts, and poor work habits. Although it is often said that new firms are the ones most vulnerable to failure (14 percent fail in the first three years of existence), nearly 40 percent of the business failures occur in firms that have been in operation for more than 10 years.

The construction industry is fragmented. There are over 3 million construction firms, but their influence is not uniformly distributed. For example, the 10 largest construction firms construct from 15 to 20 percent of the nonresidential construction projects. Approximately two-thirds of construction firms have no employees. The 650,000 establishments with employees (one-third of all construction establishments) account for 93 percent of the total value of work done (see figure 1.6). Approximately 100,000 firms have annual business volumes exceeding $1 million, while fewer than 10,000 firms have annual business volumes exceeding $10 million. In addition, approximately 80 firms engage in international construction projects totaling nearly $30 billion per year. There are many small specialized firms. The numbers of the different types of firms is characterized in figure 1.4.

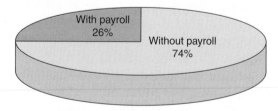

FIGURE 1.6
Distribution of construction firms by payroll status.
(Source: The Construction Chart Book.)

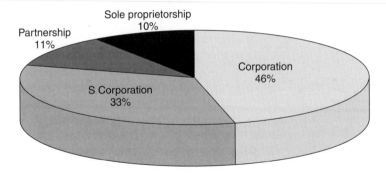

FIGURE 1.7
Distribution of construction firms by type of ownership.
(Source: The Construction Chart Book.)

Most construction firms can be characterized as being small. Over 80 percent of the firms with employees employ less than 10 workers (65 percent of the firms employ 1 to 4 workers) and the average number of employees per firm is 9.3 workers. There are nearly 2 million construction organizations. Most of these firms (2 million) consist of proprietors and working partners who have no employees. Almost all large construction firms are corporations and most of the small firms, especially those without employees, are sole proprietorships (figure 1.7). As was noted earlier, a large percentage of the construction work put in place was performed by a relatively small percentage of the firms.

Contractors are resource managers. The resources include labor, materials, equipment, money, and time. The effective management of all these resources is essential to the success of a business concern. Perhaps the most important resource is labor, which is often responsible for the greatest fluctuations in total anticipated costs. While the other resources are generally regarded as more controllable, their costs also can be altered significantly by factors over which contractors have little control. For example, large increases in the prime interest rate may cause some projects to be canceled. Large increases in the price of crude oil, such as that experienced during the oil embargo of the 1970s, can seriously alter the price of petroleum-dependent products such as cement, asphalt, and roofing materials. Business failures can be expected to increase as a direct result of such price increases.

The construction industry is an economic barometer of the country. Statistics on housing starts, the number of new houses placed under construction in a stated period of time (generally one month), are often used to indicate how well the industry is responding to the economy. Prior to 2006, single family housing starts had numbered in excess of 1.6 million per year, representing an annual volume of approximately $300 billion (see figure 1.8). A less often used measure is the value of the building permits that are issued. In times of prosperity, construction activity is sparked largely by extensive private construction, as evidenced by the total value of building permits issued. By contrast, in periods of recession, one of the first activities of the government is to stimulate the economy through the expansion of publicly financed construction.

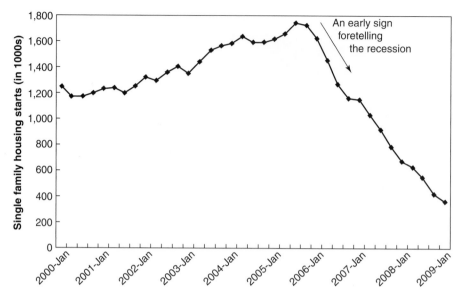

FIGURE 1.8
Annual single family housing starts 2000–2009. *(Source: Census and HUD.)*

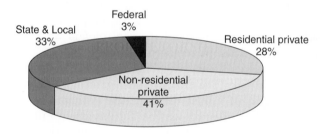

FIGURE 1.9
Distribution of construction funds by type of owner.
(Source: U.S. Census Bureau.)

PUBLIC VERSUS PRIVATE CONSTRUCTION

Public works projects are financed by government agencies in municipalities, counties, states, and the federal government (figure 1.9). Most highways, bridges, sewers, water supply projects, land reclamation projects, and public buildings are public works projects.

Privately funded projects account for an average of 68 percent of total construction volume. These projects include most buildings, railroads, and utilities. Such projects are generally owner-let, built through self-performance efforts, or built on speculation.

Before construction activities take place, the design work must be either completed or well under way. Before performing the design work, the owner of the project must ascertain that sufficient funds exist or can be acquired to finance the design and construction effort. The owner, whether private or public, will pursue different avenues to acquire the necessary funds. If adequate funds already are available, the issue of financing is greatly simplified. When sufficient funds are not immediately available, various means and sources can be pursued. Private owners use approaches different from those of their public counterparts.

Private projects are funded through the following means:

- Expenditure of existing capital.
- Direct loans from outside creditors.
- Sale of fixed assets.
- Issuance of additional shares of stock.
- Issuance of corporate bonds.
- Endowments.

Public projects are funded through the following means:

- Appropriations from annual operating budgets (obtained through general taxation).
- Special taxation assessments for specific purposes.
- Bond issues for specific purposes.
- Endowments.

A large portion of the private-sector projects are included in the residential community. Also included are many commercial buildings, such as retail stores, office buildings, hotels, etc. (see figure 1.10). The owners who finance these projects will range from the individual homeowners to the corporate giants. Thus, the

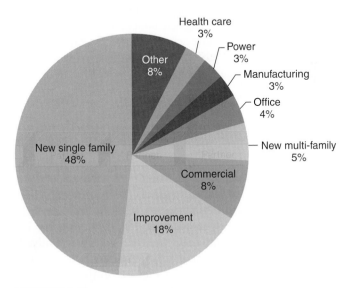

FIGURE 1.10
Allocation of private construction funds.
(Source: The Construction Chart Book.)

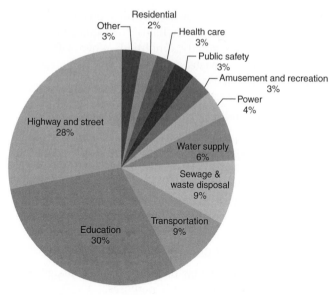

FIGURE 1.11
Allocation of public construction funds.
(Source: The Construction Chart Book.)

budgets of these owners may consist of only a few hundred dollars on a simple re-modeling project to as much as a billion dollars or more on a large facility.

Public works projects include a greater number of civil works projects. On a much smaller scale than in the private sector, residential projects will also be encountered in the public sector. While buildings are built in the public sector, they tend to be more institutional in nature, including schools, hospitals, fire stations, courthouses, and other similar structures. Civil projects tend to be built in the public domain, as are most roadways, dams, airports, canals, and similar structures (see figure 1.11).

CONSTRUCTION CATEGORIES

Construction projects can be categorized in several ways. The four broad categories described below include housing construction, nonresidential building construction, engineering construction, and industrial construction. These are general groupings. Some statistics include engineering and industrial construction projects in the same category. In more detailed groupings, even housing is divided into single-family detached, single-family attached, and multiple-family. The use made of these statistics will dictate the level of detail required in the groupings.

Housing construction consists of residential units (single-family homes, multiple-family dwellings, low-rise apartments, and high-rise apartments) constructed by speculative builders and contractors under contract with the owner. This is a major economic stabilizer of the U.S. economy, as it responds strongly and quickly to national monetary policy. Although some public housing projects are constructed, most housing projects are done by the private sector.

Building (nonresidential) construction includes institutional, educational, light industrial, commercial, social, religious, farm, amusement, and recreational projects, both public and private.

Engineering construction includes all structures in which most of the planning and design is performed by engineers. These tend to be structures that are nonarchitectural and may include the use of large amounts of earth, rock, steel, asphalt, concrete, timber, or piping. Engineering construction projects are often referred to as civil works projects because they are frequently designed by civil engineers. These are primarily public projects and include sewage and water treatment plants, water mains, canals, levees, pipelines and pole lines, reclamation projects, marine structures, tunnels, large bridges, streets, highways, airport runways, mass transit, and railroads.

Industrial construction consists of projects associated with the manufacture and production of a commercial product or service. The construction of complex industrial projects is usually undertaken by large specialized firms. Typical industrial projects include paper mills, petroleum refineries, steel mills, chemical plants, smelters, and electric power generating stations. In the United States these projects are usually privately financed.

The general mix of the different types of construction projects varies somewhat from year to year as legislation shifts emphasis to certain selected types of projects (see figure 1.12). Change in the demographics of the population also alters the needs for certain types of projects in different regions of the country.

While housing starts are monitored closely to provide an index of the strength of the economy, these shifts in strength are reflected in varying degrees in all types of construction. Projects that are characterized by long lead times or that require a long construction duration are affected less by short-term swings in the economy. As was

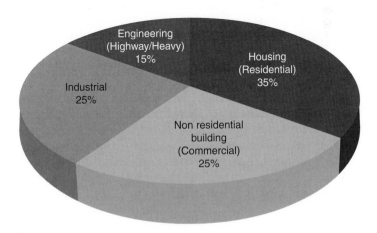

FIGURE 1.12
Distribution of construction funds by construction sector.
(Source: Bureau of Labor Statistics.)

mentioned before, the number of publicly financed projects tends to increase during recessionary periods, along with a decline in the number of privately funded projects.

The construction industry is large in size and significant in the role it plays in the economy. The nature of construction projects makes the industry unique in that the manufacturing facility or plant must move to the construction site. When the product is completed, the plant is shifted to another manufacturing or work site. This is very unlike typical manufacturing operations. Construction projects are made more complex by the changes that invariably occur among the contracting parties and in the labor requirements.

THE CONSTRUCTION EMPLOYERS

Employers in the construction industry constitute a large number of diverse entities. The Current Population Survey shows that there are nearly 2 million self-employed firms in construction. Of these, 2 million are unincorporated firms, most of them with no employees and no payroll. The actual number is probably smaller, but a more accurate estimate is difficult to establish. The reason that the number is probably less is that some workers may be fraudulently included among the self-employed. Some employers misclassify their employees as independent contractors (unincorporated), a ploy that is allegedly used by many firms as a means of avoiding the payment of Social Security, Medicare, workers' compensation, and other taxes. As noted earlier, many firms have no employees. In addition, most of those firms with payroll tend to have only a few employees (figure 1.13).

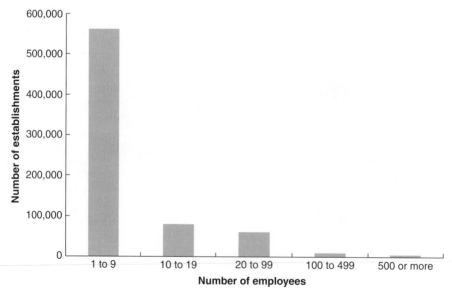

FIGURE 1.13
Number of firms by number of employees. *(Source: The Construction Chart Book.)*

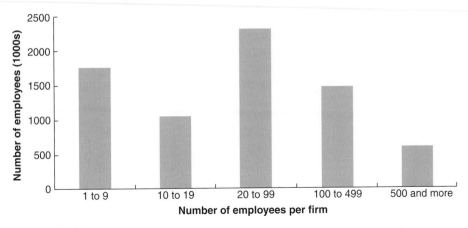

FIGURE 1.14
Number of employees by number of employees in the firm. *(Source: The Construction Chart Book.)*

Eighty percent of the construction firms employ less than 10 employees and employ 24 percent of all construction workers. Two percent of the firms employ more than 100 employees and employ 21 percent of all construction workers. Eight percent of the workers are employed by 0.1 percent of the firms with more than 500 employees (see figure 1.14).

THE CONSTRUCTION WORKFORCE

Just as the construction industry is unique in many ways, so too, is the construction workforce. There are many different craft jurisdictions represented on most construction projects. These different occupations include carpenters, laborers, painters, electrical workers, plumbers, operating engineers, roofers, bricklayers, truck drivers, heating and air-conditioning mechanics, drywallers, sheet metal workers, carpet layers, concrete finishers, welders, insulation workers, ironworkers, tile setters, glaziers, boilermakers, and others. Each craft represents unique skills.

The construction workforce has been a subject of much discussion in recent decades. Principal among the concerns has been the availability of skilled workers. In the mid-1990s, this issue became particularly acute, primarily because of the booming construction phenomenon that was being realized in almost all parts of the United States. While the shortage of skilled workers has been widespread, it has been particularly pronounced with regard to welders and bricklayers. It is not uncommon to see construction projects with large billboards advertising for carpenters and other workers.

The Construction Chart Book (2007) compiled interesting statistics on the construction workforce. The average age of construction workers is 39 years. Ten percent of the construction workers are women, with 74 percent of these being in managerial or support staff positions. The number of women working in the trades is disproportionately small when compared to other industries and the percentage of women in the general population, but their numbers in the construction trades are increasing.

In the construction workforce, slightly over 10 percent are racial minorities, including black, American Indian, Aleut, Eskimo, Asian or Pacific Islander, and others. Some Hispanics (Cuban, Mexican, Puerto Rican, Central American, South American, or other Hispanic origin or descent) are included among the white population and others are included among the minorities. Hispanics make up about 30 percent of the construction workforce (see figure 1.15).

While construction was booming until the start of the recession in 2006, construction workers generally had not experienced the rewards of their labors through greater purchasing power. For most construction workers, the wages have not kept pace with inflation during the past 30 years. In fact, when adjusted for inflation, wages have steadily declined, approximately a 30 percent decline from 1973 to 1996. In addition, it has been noted that during the 1990s construction worker wages fell below those of many workers in manufacturing.

A major change occurred in the construction workforce between the early 1970s and the 1990s. In the early 1970s, approximately 80 percent of the construction work was performed by a unionized construction workforce. By the early 1990s, only about 20 percent of the construction work was performed under negotiated labor agreements. While there still remain several union strongholds in construction, especially in some large metropolitan areas, much of the construction

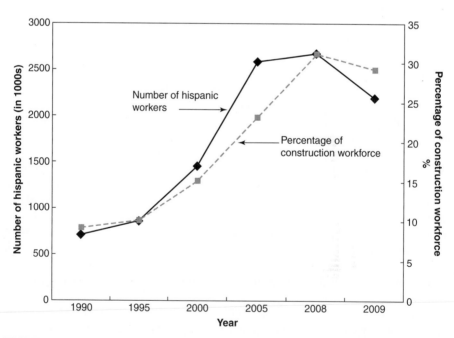

FIGURE 1.15

Number of Hispanic construction workers 1990–2009. *(Source: Bureau of Labor Statistics.)*

work (approximately 86 percent) is now performed under the open shop, an arrangement in which union affiliation has no bearing on the management-labor relationship. This is probably one of the primary reasons that wages have not kept pace with inflation.

In the public sector, it is common for wages to be established by law. Federal projects are governed by the Davis Bacon Act that mandates that the local prevailing wage be paid on the project. This has often been interpreted as being the local union scale. Thus, the union contractors and the open shop contractors must often pay the same wages. Since the open shop contractors must often pay the union scale wages, there is no wage advantage for the open shop contractor. It is not surprising that union contractors and a unionized workforce perform about 40 percent of the construction work in the public sector.

PROJECT DEVELOPMENT

This chapter has presented many different statistics related to the construction industry and the construction work force. It must be realized that construction projects do not just suddenly present themselves. Projects evolve through the considerable efforts of various parties. The roles of these parties will be described conceptually, but it must be realized that these efforts can be considerable, depending on the specific project.

The *owners* are the parties with whom the project begins. An owner might recognize a business opportunity that entails the construction of a specific facility, such as a restaurant, office building, retail store, and so on. The owner must then explore various factors, including financing, desired location, zoning restrictions, clientele, market analysis, projected sales, and so forth. At the end of this analysis, in which many variables are considered, the owner must make the "go or no-go" decision concerning the business venture. The expected rate of return on the investment and the inherent risks in the venture will surely play major roles in this decision.

If the owner decides to go forward with the business venture, the next stage will be to acquire the necessary *financing,* if the owner does not have adequate funds to finance the project. If financing will be needed, the owner will generally prepare a business plan to present to a financial institution or potential investor. If this effort is successful, the owner can continue with the project development.

Once the financing is obtained, the business venture begins to materialize. The first issue to be resolved is the specific location of the project. If the owner does not already own the desired property, it will be necessary to locate the ideal location for the project and to find an available parcel of land. This is when the *real estate* personnel will become engaged in the project. The realty personnel generally know the parcels of land that are on the market, and they also have fairly good insight about the fair market value of different parcels of land. A good real estate agent will be able to show the owner a number of sites that meet the owner's site criteria.

After the site is identified and acquired, the owner can then expend energy in getting the project designed. The *designer* of the project, typically an architect, will need to know exactly what the owner wants. This is generally communicated to the designer via a report that describes the program requirements. If the owner has identified a suitable designer, the program can then be translated into a construction document that shows all of the project features. The designer will generally communicate with the owner as the design is evolving. It is in these early stages that changes can easily be made with minimal impact on the construction costs.

REVIEW QUESTIONS

1. List several measures that describe the size of the construction industry.
2. What construction-related measures can be used to describe the strength of the U.S. economy?
3. Under what conditions might the funding of public construction projects be particularly high?
4. Why is the construction industry referred to as an easy entry and easy exit industry?
5. What are the ramifications of the cyclic tendencies experienced in the construction industry?
6. Contrast the financing of public versus private construction projects.
7. Give examples of publicly financed building construction projects. Give examples of publicly financed nonbuilding construction projects.
8. Contrast typical construction projects with the production of manufactured goods.
9. What would be some key indicators in the construction industry that conditions are improving during an economic slump?
10. Prior to an economic recession, the timing of the visible changes is different for the different construction sectors (residential, commercial, industrial, etc.). The same is true when economic conditions improve significantly. Explain why the different sectors respond differently in terms of timing.

2

CONSTRUCTION CONTRACTING METHODS

MOST CONSTRUCTION PROJECTS involve the participation of owners, designers, and constructors. The *owner* is the party that determines when a particular project is needed. This decision may be made after an extensive study of various alternatives. Once the decision is made to have a project built, it is necessary to obtain *designer* services. The design, closely adhering to the owner's stated project objectives, will serve as the guidance document for the *constructor* who will build the project. Depending on the nature and size of the project, the contractual arrangements between these parties may change. In some cases one party may play two roles or even all these roles. These different roles should be clearly understood and carefully evaluated to determine the contractual relationship that should lead to the most effective delivery of the project. These decisions lie primarily with the owner, with input frequently being provided by informed counsel.

There are essentially five different types of contracting procedures in the construction industry. Although relatively "pure" forms of each type do exist, modifications are frequently made by the owners. These pure forms will be discussed, along with construction management at risk, an arrangement that represents an emerging new approach.

GENERAL CONTRACT METHOD

The *general contract method* consists of a contract drawn up between the owner and a general contractor. The owner is usually represented by the firm that was responsible for drawing up the contract documents. On building-type projects the representative is usually an architectural design firm. On engineering projects the representative is usually an engineering firm. In either case, the owner will enter into two separate contracts, one with the designer and one with the constructor.

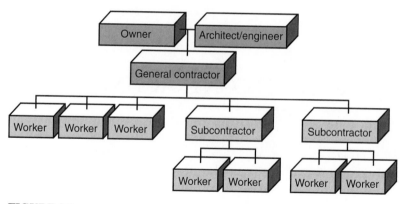

FIGURE 2.1

Typical organization for a general contract agreement.

On public works projects the role of each party is clearly defined. Rigid or formal procedures are usually followed in forming the general contract. This begins with a public advertisement which advises all interested parties of the particulars of the project and the upcoming bid date. At the designated time, sealed bids are submitted at the location and in the manner specified. These bids are opened and read to all those in attendance so that all the bidders will know how they stand at the end of the bid opening. The contract award is given to the lowest bidder, although the other (usually the second and third lowest) bids remain open until the contract has actually been signed. This procedure has legal controls designed to prevent fraud and collusion. The purpose of the guidelines is to establish a competitive spirit in which every bidder has an equal opportunity to be awarded the contract.

In the private sector a similar procedure is used, although the criteria may not be followed as rigidly. For example, the bid opening may be "closed," the owner may elect to award the contract to a contractor other than the lowest bidder, or the owner may try to negotiate a price lower than those stated in the bids. Such practices are not governed by law.

Whether public or private, the contractual arrangements are similar (figure 2.1). The general contractor usually has a specialization in one of the major components of the project (concrete, steel, etc.). However, the general contractor will often have no expertise concerning certain aspects of the project. When this occurs, the general contractor simply subcontracts those portions to firms that can perform the work. The subcontractor will then be responsible for providing the necessary tools, labor, materials, and supervision. On building projects, subcontractors are generally obtained for such work items as electrical, mechanical, roofing, masonry, ceramic tile, carpeting, wallpaper, insulation, drywall, suspended ceilings, millwork, mirrors, ornamental metals, and resilient flooring. It is common on even simple buildings to have 20 to 30 subcontractors; on more specialized buildings, such as hospitals, as many as 70 or 80 subcontracts may be let. By contrast, simple heavy construction projects generally require very few subcontractors, as the general contractors usually have the in-house capability of performing most of the work. For example, on a bridge, the general contractor may sub out only the guardrail and the seeding.

The general contract approach is often referred to as the design-bid-build approach in that the three key phases (design, estimating/bidding, construction) are undertaken in succession without overlap. This can be depicted very simply as follows:

An extreme situation occurs when the general contractor subcontracts all the work on a project. This is called *brokerage* and is generally not regarded as being beneficial to the owner. Since the general contractor's bid includes profit for both the subcontractor's work and the general contractor's work, the owner will expect additional effort from the general contractor to compensate for the additional markup. This added effort is generally expected to be expended through the coordination efforts of the general contractor. On typical projects, with perhaps 75 percent of the work subcontracted, it is the general contractor's responsibility to adequately plan, organize, supervise, and coordinate the work efforts. If 25 percent of the work is performed by the general contractor, there is a natural incentive for this to occur, because the general contractor's profit will be directly affected by the efforts expended in controlling the job. If the project is brokered, however, the general contractor has a diminished incentive to do this. In fact, the incentive is for the general contractor to minimize costs. In such a situation the subcontractors will be frustrated, as they will have to coordinate their own efforts. This will be difficult, as their contracts are solely with the general contractor, not with the owner or the other subcontractors. As a result, many owners place contractual limits on the amount of work that can be subcontracted, or stipulate that a certain amount of work, such as 15 to 20 percent, must be performed by the general contractor's own workers. A provision may state, "The Contractor shall perform with his or her own organization not less than 25 percent of the work." On some public works projects these limits may be required by law.

When is the general contract form advisable? It is generally assumed that the general contractor has unique skills that should reduce the costs of construction to the owner. These skills include the administration of construction operations, efficient procurement of materials, effective management of the workforce, and thorough planning and coordination of the construction process. This efficiency is generally attributed to the fact that the general contractor maintains a staff of trained supervisors, has available trained mechanics and workers, and owns the equipment needed to perform the required job tasks. If the owner had the requisite skills of management in construction, this method of contracting would probably not be preferred by the owner. This is the only method that gives the owner a firm idea of the final cost of the total project prior to the construction phase. The other methods may, at best, give an estimate during the design or early construction phase. For the owner, the general contract approach results in clearly defined roles for each of the contracting parties. The owner also minimizes the contractual liability for cost overruns and late project delivery.

Regarding disadvantages, the owner must be aware that the design-bid-build approach often extends the project duration. Another disadvantage with the general contract approach is that the owner does not have an agent or "friendly" party involved in the contractual arrangements. Some cynics have stated that design-bid-build is not desirable because this method leads to tight bids and small profit margins. This creates an incentive for general contractors to "beat up" on their subcontractors, cut corners on performance, and to look for loopholes in the contract that might bolster profits. The nature of the contracts creates inherent adversarial relationships between the different parties. The inherent inflexibility of this approach also exposes the owner to a greater probability of claims. This fact, perhaps more than any others, has given rise to the emergence and use of other contracting approaches.

SEPARATE CONTRACTS METHOD (OWNER AS GENERAL CONTRACTOR)

The *separate contracts method,* also known as the multiple prime contracts method, is an arrangement by which the owner lets contracts directly to specialty contractors for the various portions of the work (figure 2.2). The individual contractors may subcontract portions of their work. This is essentially a general contract method without the general contractor. This means that the owner must take charge of the management of the project, assuming the managerial functions ordinarily performed by the general contractor.

This method may be appropriate if the owner has the necessary in-house capabilities to manage a construction project. The benefit to the owner is that the profit that would have been earned by the general contractor is kept by the owner. A variation of this method may be exercised when the owner does not have the requisite in-house managerial capabilities. In this case the owner can let a separate contract to a firm to perform the management functions. The role of this management firm will have to be clearly outlined, and the other contractors will be bound by the coordination efforts of this firm. It is obvious that legal entanglements can arise from this arrangement. In general, the power of the management firm will not be as great as that of the general contractor because of the parties involved in the contract. As a result, it is advisable for the owner to retain managerial control or to award a general contract.

One complication that can arise under these types of contracts is related to obtaining a permit for projects in the private sector. It is a common requirement in

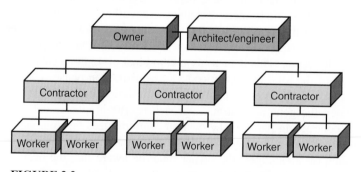

FIGURE 2.2
Typical organization for separate contract arrangements.

many states for the party obtaining the building permit to have a general contractor's license. If the owner does not have such a license, a considerable delay in the project can occur.

When is the separate contracts method advisable? A prerequisite for employing this method is the availability of a competent construction manager or construction engineer to administer the various contracts. If such personnel are employed by the owner, this method is particularly appropriate on projects where the required specialty work is restricted to a few types of construction. It is less advisable when many work items are involved. It must be borne in mind that this method forces the owner to assume a greater risk than is assumed with the general contract method. This method essentially eliminates the general contractor's profit from the cost of construction to the owner. It must also be remembered that the profit earned by the general contractor is generally very small (commonly less than 3 percent). This must be weighed by the owner and apparently is one of the reasons why this contracting procedure is not widely used.

SELF-PERFORMANCE METHOD

An owner may decide to *self-perform* some construction work or essentially do the work "in-house." Under this approach, also known as force account work, no contracts are written for a construction project. The owner's own workers or employees are solely assigned the task of performing the construction work (figure 2.3). The owner provides the necessary materials, labor, equipment, and supervision. The owner plays the role of the manager. On most projects employing this method, the designer plays a minor role, with the design function also often being performed in-house.

Since no contracts are let, the owner benefits by eliminating the expense of following through with formal contracting procedures. Time is also saved. In addition, this method eliminates the profit that would be earned by the general contractor and subcontractors. It is also alleged that a cost reduction is realized in regard to engineering and inspection. This is probably true on projects where the plans need not be elaborate.

When is it advisable for an owner to self-perform work? This type of work is particularly appropriate when the project is small in scope, simple in character,

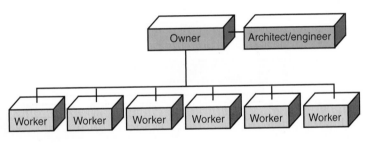

FIGURE 2.3
Typical organization for a self-performance arrangement.

and ongoing in nature. Of course, it is essential that the owner have within the organization a trained and skilled construction force. It is common to self-perform work on maintenance projects. For example, the maintenance of county roads is usually done in-house (self-performed), while the construction of new roads is done by the general contract method. Self-performance is common for the maintenance of railroad tracks, while the installation of new rail lines is usually contracted. Grounds maintenance on most campuses is done in-house.

Self-performed or force account work has been criticized by various contractors when public owners decide to do projects by self-performance. There have been instances in which the public owner submitted a bid on competitively bid projects. The owner is frequently the low bidder in such instances. In other cases, the owner does not even permit private contractors to submit bids. The contractors contend that they are placed at an unfair disadvantage. First, they allege that the owners do not charge properly for the costs of owning and operating heavy equipment. This may be the case when county maintenance equipment is owned by the public agency and is then priced at a reduced rate in bidding on new construction. Second, the contractors on such projects are forced to pay higher wage rates by the local Little Davis Bacon Law, while the public agency does not have to comply with the same regulations. There may be some merit in these allegations, as the trend to do work in-house has not materialized in the profit-oriented private sector. In fact, many private owners have begun a trend of outsourcing (contracting with others for performance) more of their construction and maintenance work. Some owners find that the only time they do self-perform work is when the contractor defaults on the contract and the owner must step in to finish the project. Even then, many owners will give serious consideration to contracting with yet another firm.

DESIGN-BUILD METHOD

In the *design-build method,* the owner lets a single contract for both the design and the construction of a project (figure 2.4). This is also known as design-construct or turnkey construction. This method utilizes the construction firm's experience in the design phase. As a result, the final project should have a higher degree of constructability. In a sense, it is like the general contract method, except that the

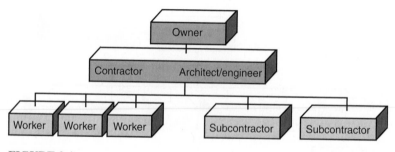

FIGURE 2.4
Typical organization of design-build projects.

contractor is also responsible for the design of the project. This single point of responsibility for design and construction services is very appealing to many owners. This approach has been very popular in constructing large, industrial-type projects such as petrochemical plants. In recent years, this approach has gained tremendous popularity in the construction community, especially in the private sector. Many public entities are also beginning to utilize this approach. Because there is no extensive design on which to compute a bid (perhaps there is only a narrative description of the project), contracts are awarded in a variety of ways. If the scope is reasonably well defined, a competitive approach can still be utilized to some extent, but the contract is generally then established as a target cost or guaranteed maximum price (GMP) that is not to be exceeded. It is common for these to be written as cost plus a fixed fee with a GMP. In the public sector, the owners prefer to keep all the savings when the cost is below the GMP, while on private projects different incentive schemes (sharing the savings) might be explored to motivate the design-build firm to control costs for the owner.

Since the design evolves with constructor input, it is understandable that fewer changes will arise during construction due to designer error. On conventional general contract projects, such changes can be very costly. For the owner, the design-build approach offers a considerable advantage over other approaches in that the potential for the owner being embroiled in disputes arising between the design firm and construction firm are essentially eliminated. In the ideal contract, the design and construction expertise exist within a single firm. This is often the case, but some design-build teams are created by a joint venture of a design firm and a construction firm. This teaming up of firms may make it possible for firms to pursue projects that they otherwise could not consider. In such partnership arrangements, it is common for either the design firm or the construction firm to play the lead role. While disputes can still arise between the design firm and the construction firm in such an arrangement, the owner is not a part of the dispute. For example, the construction firm may claim that costs on the GMP for the project were exceeded because of errors in the design. This type of dispute is internal to the design-build team. The owner's contract will be with the joint venture and any disputes between the design firm and the construction firm generally must be resolved without the owner's involvement.

When is the design-build method advisable? Since this method integrates the design and construction functions within one firm, it is possible for construction to begin before completion of the design for the project. This is accomplished by first designing the foundation and then developing the design as construction commences. This overlap of design and construction is referred to as fast-tracking, as it is meant to deliver the project to the owner earlier than would occur if the design had to be complete before the start of construction. In periods of high inflation, this approach has increased viability. The design-build approach is particularly attractive when projects are large and technically complex. Perhaps the owner's best contribution to the success of a design-build project is to clearly define the scope of the project prior to entering into a formal design-build contract. The primary concern of those contemplating the use of a design-build approach is that there are fewer checks and balances built into the process, there is less control by the owner, and, on public projects, there are laws and regulations that may place serious restrictions on the process.

Not only has the use of the design-build approach become quite popular in recent years, but also several variations of the process have evolved. These options that have emerged are intended to make the process more marketable to owners. For example, some firms offer a design-build-finance option; others offer the design-build-operate capability. The attractiveness to different owners will vary with the owner's needs. Still other projects may be completed as design-build projects with construction manager (CM) oversight. Regardless of the choices made, advocates of the design-build approach state that design-build projects can be delivered at the lowest cost and that the design-build method delivers projects in the shortest time, often claimed to be 20 to 30 percent faster.

When compared to design-bid-build contracts, advocates of design-build contracts are quick to point out that awarding contracts to the lowest bidder may result in a contract with an incompetent contractor. The competitive bidding process entails many tactics by general contractors and subcontractors to ensure that they provide the lowest bids. Some of these practices are often regarded as unethical. During construction, the concern for reducing incurred costs will be paramount in order to ensure a profit. This may lead to disputes with designers, between subcontractors, and between the general contractor and the subcontractors. Such disputes are rare on design-build projects.

PROFESSIONAL CONSTRUCTION MANAGEMENT METHOD

With the *professional construction management method,* the owner hires a firm with construction expertise to perform construction management services on the owner's behalf (figure 2.5). The professional construction management firm (CM) is generally hired by the owner before any substantial design work is done and before any construction work has begun. In the purest form, the CM may even be instrumental in selecting the design firm. While the design is being developed, the CM periodically reviews the project design to see how the cost and time of completion for the project can be reduced. Value engineering is perhaps most cost-effective under such circumstances when performed during the design phase, before any monetary commitments have been made to construction.

The compensation of the CM is arranged between the owner and the CM. The payment for the CM's services may be based on a flat fee, an incentive payment method in which cost savings are shared by the owner and the CM, or a cost plus a percentage fee. Thus, the CM is in fact working for the owner's benefit, representing the owner during the design and the construction phases. Designers are often paid in direct relation to the cost of the construction project; in other words, a more expensive project generates a larger design fee. The CM is hired in part to see that the owner actually receives the most economical project that satisfies the owner's needs. As a further incentive, the CM often gives the owner a GMP that the project cost will not exceed. As a further incentive for the CM, an arrangement may be made by which the owner and the CM share (at a predetermined rate) any savings below the stated maximum price.

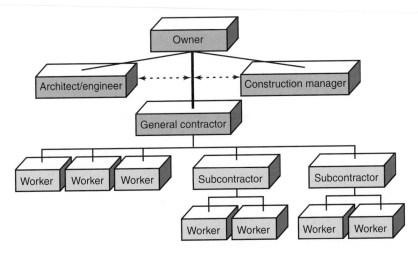

FIGURE 2.5
Typical organization for the professional construction management method.

As the owner's agent, the CM, is not responsible for the means and methods of construction and the CM does not guarantee the construction cost, time, or quality of the completed facility. Other firms (a general contractor or multiple contractors) who have direct contracts with the owner typically perform the construction work. These are the parties that provide the guarantees of performance.

Once the design is partially complete, construction work can begin. For example, the CM may let a contract for sitework as soon as that portion of the design is complete (fast-tracking is possible). Note that the CM does not perform any of the construction work with its own forces. At most, the CM may provide a skeleton workforce for general overhead work, such as cleanup. If the design is completed early in the construction phase, the CM can let a general contract for the remaining work. Whether this is done by general contract or by separate contracts, the role of the CM is to ensure that the project is delivered to the owner according to the plans and specifications.

When is the construction management approach advisable? This method is advisable on large or complex projects when construction expertise is needed during the design phase. The CM approach also permits considerable flexibility for changing the project as the design evolves. Hospitals are often constructed by this method. A prerequisite is that the owner have confidence in the ability and integrity of the CM. Projects to be delivered quickly (fast-tracking) are good candidates for this construction approach. Naturally, the owner must be able to specify and identify the professional qualifications of the ideal party to serve as the CM.

With the professional construction management approach, it is common to have the contract established as a cost plus a fixed fee. Under this arrangement, the costs are the actual costs of personnel to perform the CM services, as contractually agreed, with a stated multiplier to cover other items such as travel expenses, training, inspections, and so forth. For example, the CM may be reimbursed for the salaries of all personnel assigned to a particular project multiplied by 2.5 (or some similar agreed factor). The reimbursement for direct construction labor will vary considerably. As mentioned, in the purest form of a CM contract, the CM will perform no direct labor. However, the CM often has the in-house ability to perform some construction tasks, and this may alter the nature of the CM's role. For example, the CM may have the in-house ability to perform masonry work and the owner may be convinced that it is in the best interest of the project for the CM to perform this work. The CM's role obviously changes when more field labor is performed directly, but this may be deemed to be most desirable for the owner. When the CM firm does perform direct construction services, it is important that the owner recognize the potential for conflicts of interest for the CM. Only a CM that is well trusted should be allowed to self-perform significant portions of the work.

CONSTRUCTION MANAGEMENT AT RISK (CM AT RISK)

Changes have been made to the traditional approach utilized to obtain construction management services. While the traditional CM approach established the CM as an agent of the owner, the new approach, with particular popularity in the public sector, establishes the CM as an independent contractor (figure 2.6). As an independent contractor, the CM is "at risk." That is, the CM is responsible to the owner to complete the project by the established substantial completion date and within the agreed budget. The CM must compensate the owner when the construction put in place does not satisfy the established standards of performance for the project. The CM at risk approach clearly has some of the elements of the general contract approach in that the role of the CM in the "at risk" contract is similar to that of the general contractor. While "CM at risk" is the most common term used for this method, it has also been referred to as the GC/CM or the CM/GC approach. Similar to the general contractor, the CM at risk firm will be responsible for hiring all the subcontractors (perhaps prequalified in order to guarantee quality of performance to the owner) and for coordinating the activities involved with completing the project. A distinct difference from the GC approach is that the CM at risk firm enters the contract prior to design completion. Thus, the construction expertise of the CM can be effectively utilized in the early phases of a project. In addition to the CM, prospective subcontractors are also often involved in making value engineering suggestions to help control costs. Even during the subcontract bidding phase, value engineering conditions within the bids are given serious consideration.

With the construction management at risk approach, it is common to have the contract established as a cost plus a fixed fee with a GMP. The fee could also be established as being a stated percentage of the construction costs. Under this arrangement, it is generally assumed that the actual costs will be less than the GMP. If the price exceeds the GMP, the CM at risk firm will be required to absorb those costs, unless the scope of the project can be shown to have been changed such that the GMP should be modified. Conversely, if the actual costs of construction are less than the GMP, the owner (especially on public works projects) is inclined to keep the savings and not share these funds with the CM. In the private sector, many

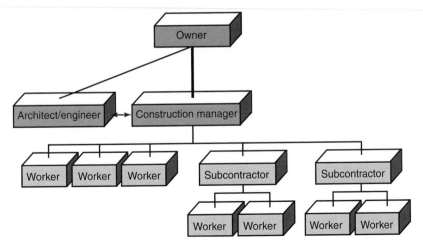

FIGURE 2.6
Typical organization for the CM at risk approach.

variations of these arrangements can be found. Some CM at risk firms give all savings to the owner to ensure that the interests of the owner are kept in focus. The accounting books may also be made available to the owner to give added assurances about the integrity of the CM at risk firm. Many variations can certainly be found on how CM at risk is implemented. Future work with an owner is most definitely impacted by the experiences on past projects, so honesty, integrity, and character are important attributes to communicate to the owner.

REVIEW QUESTIONS

1. What types of projects are most appropriate for the general contract approach?
2. What are some advantages and disadvantages of the general contract approach?
3. Give an example of a project for which separate contracts might be successfully used.
4. What conditions must exist for separate contracts to be viable?
5. Discuss the criticisms that have been made of public agencies performing construction services by the self-performance or force account method.
6. What type of construction work is particularly well suited for force account?
7. What type of project might be ideally suited for the design-build or design-construct method?
8. What are the advantages of using the professional construction management approach?
9. Which types of contractual arrangements lend themselves to fast-tracking?
10. Describe how the different contracting approaches differ in terms of allocating risk.

3

THE NATURE OF CONTRACTS

A *CONTRACT* IS an agreement, usually between two parties, that is enforceable by law. In some instances there may be a third-party agreement in which the benefit of the contract goes to a third party. An example would be an insurance policy, particularly a life insurance policy, in which a third party is named as the beneficiary. However, most construction agreements are drawn up between two parties for their mutual benefit. There are several ways in which contracts can be described.

LITIGATION

Once a contract has been written, issues may arise which the parties to the agreement cannot satisfactorily resolve between themselves. If the parties cannot amicably resolve the dispute, it is common for one of the parties to file a lawsuit against the other party. This is the beginning of formal litigation in which a court decision may ultimately have to be made that is binding on both parties.

When dispute resolution is pursued in court, the court will seek guidance in making its decision. Such guidance comes primarily from the court's interpretation of statute law, the Constitution, and common-law principles. Common-law principles, as interpreted in the United States, have been largely adapted from traditional usage in Great Britain. Common law is based largely on tradition or accepted practice over an extended course of time. Rather than written law in the form of statutes, common law typically is used to define what is construed as right or wrong. Common-law principles have not been formally adapted as a unified compilation of information; however, through numerous court decisions, there is considerable written documentation of common law. Thus, past court cases provide clear guidance in many situations for which no formal laws exist.

As a general rule, if a party does not comply with the terms of a contract, that party has breached the contract. The extent of the noncompliance may determine if the courts will regard noncompliance as a breach. The case of *Jacob & Youngs, Inc. v. Kent* (129 N.E. 889) is a good example. This 1921 case related to nonpayment of a contractor, Jacob & Youngs, Inc., who had constructed a country residence for George E. Kent. Work on the home ended in June, 1914, and Kent began to occupy the dwelling. Apparently there was no concern over any noncompliance or over defective performance until March, 1915. The contentious issue concerned the plumbing work that had been done by a subcontractor. The specifications stated "all wrought iron pipe must be well galvanized, lap welded pipe of the grade known as 'standard' pipe of Reading manufacture." There was apparently no obvious defect in the materials that were provided, but it was later determined that the materials were not a Reading product, but rather were Cohoes brand. The architect failed to notice that the pipe was not marked with the Reading manufacturer's stamp at periodic intervals along the pipe length. The owner asked that the pipe be replaced so the plumbing would be in compliance with the specifications. The contractor refused to replace the pipe, as considerable demolition would be required. The contractor contended that the installed pipe was of the same quality, appearance, and cost as that which was specified. The court determined that the failure to provide the Reading pipe was both trivial and innocent. Also, it concluded that the omission of the prescribed pipe was neither fraudulent nor willful. Much of the court's decision was based on the high cost of replacing the pipe versus the actual difference in value between the two types of pipe which was determined to be nominal or nothing. The court decided that the contract had not been breached.

DESCRIPTORS OF CONTRACTS

A contract is an agreement that can be executed or can be executory. A contract is *executed* when both parties to the agreement have fully performed in accordance with the contract's terms. A construction contract is fully executed only after the contractor has satisfactorily completed the construction work in accordance with the contract documents, and the owner has paid the contractor for this work. Normally the phrase "execute a contract" is used to mean that the contract has been signed, with the parties then being bound by it. It is another matter to execute the obligations created under the contract.

A contract is *executory* when some portion of the agreement remains to be done. It may be executory on the part of both parties, or it may be executory on the part of only one party. A construction contract is still executory if the owner has not paid for the work. A contract that is entirely executory on the part of both parties (neither party has performed the obligations of the contract) is easier to cancel than is one in which one or both of the parties have performed at least a portion of their obligations.

A contract can be bilateral or unilateral. A *bilateral contract* consists of an agreement created by mutual promises made by the contracting parties. In this type of contract, each party plays two roles: promisor and promisee. In other words, promises are exchanged. Most construction contracts are bilateral in that the contractor promises to perform the construction work as specified, and the owner promises to pay a stated amount for this work.

A *unilateral contract* is a one-sided contract in that only one of the contracting parties makes a promise, while the other party exchanges something other than a promise, most commonly some stated performance. The actions of the other party may substitute for a promise. These contracts are not common in construction, but examples do exist. A landscaping subcontractor might say to the contractor, "I'll sell you 100 landscaping railroad ties for $6 each." If the contractor sends the subcontractor a check for $600 for those ties, a contract was formed unilaterally. Note that the contractor was in sole control over whether a contract would be formed.

A contract can be express or implied. An *express contract* is one in which the terms of the agreement, whether verbal or written, are clear, concise, explicit, and definite. Most construction contracts are express. In fact, virtually all written agreements could be classified as being express.

An *implied contract* is one in which the terms of the agreement are not clearly stated, but are established through inference and deduction. The facts and circumstances surrounding a contract must be evaluated before the mutual intent of that contract can be determined. In other words, the terms of the contract must be implied from the actions of the contracting parties. Suppose a driver in the downtown area pulls into a pay parking lot, hands the attendant $2, and parks the car in the lot. One could reasonably infer that the $2 was paid in exchange for the privilege of parking in the lot. Since no words were spoken, this would constitute an implied contract.

Under some circumstances an express contract may be altered by an implied contract. This occurred in *John Eaton v. Engelke Manufacturing, Inc.* (681 P.2d 1312). Engelke asked Eaton to design an electronic video game. Eaton estimated the cost of the effort to be $1,500 and estimated that the work could be completed in three months. Engelke requested that Eaton begin the work. After Eaton had begun to work on the video game, Engelke asked that several changes be made. Eaton and Engelke had several conversations regarding the impact of those changes on the original contracted amount and on the time for completing the effort. After 11 months, the project was 90 percent finished and Eaton was fired without being paid. Eaton then filed suit to obtain $7,800, his estimate of the value of his efforts. Engelke stated that the express oral agreement was for Eaton to be paid when the design was completed, a condition that was never satisfied. Furthermore, Engelke contended that nothing of value was ever received because the design was never completed. The court decision was that Engelke had to pay Eaton on the grounds of a quasicontract to prevent the unjust enrichment of Engelke. The original terms of the oral contract were not crucial to the decision, since Engelke had ordered significant changes to be made. In effect, a new implied contract was formed which nullified the original oral or express agreement.

Sometimes a contract is made between one party and a number of individuals. These individuals can be contracting in a joint, several, or joint and several manner. These distinctions are important, as they establish the degree of liability assumed by each party.

A *joint arrangement* is one in which the individuals are "joined," in a legal and liability sense, as one party in the action. They are united and undivided and will be treated as such. If one of these individuals is released from an obligation, this has the effect of releasing all the other persons as well. Suppose several landowners decide to sell their property. Their properties are all adjacent, so that when they are combined, a large, attractive piece is created. Suppose they enlist the services of an individual to locate a buyer for their combined properties. The intent was to pay a percentage fee to the agent for locating such a buyer. If an altercation later developed concerning the payment of the agent for these services, the individuals would be sued jointly, as all mutually benefited from the agent's efforts. If one individual satisfied the debt, that individual could then sue the other landowners for the prorated amount due.

If the parties to a contract are considered to be *several, separate, or severed,* each party has a liability that is separate from that of any of the other parties. The 1894 decision in *O'Connor v. Hooper* (36 P. 939) is a good example. Several property owners in a neighborhood agreed in a single contract to hire a construction company to make improvements along their properties facing the street. Each property owner was contracting severally and promised to pay the contractor a predetermined prorated amount. The court ruled that this had the same effect as if each property owner had actually written a separate contract. The concept here is that the improvements on each property could easily be distinguished from those on the adjacent property. All the property owners could clearly identify the work performed for the benefit of their respective parcels of property.

A combination of these systems can occur and is known as *joint and several.* Contracts are frequently written so that a number of individuals are bound both jointly and severally. This can be expressly stated; for example, "We, Thomas Davies, Joan Walters, and Harold Herman, jointly and severally promise to pay for work to be performed by . . ." This has the effect of binding the individuals as a unit (joint) and also of having each individual accept separate (several) accountability. If the individuals defaulted, the suit would then be either joint or several, as it could not be both joint and several. Others might be more willing to enter an agreement when the other parties enter the contract as joint and several. Thus, such explicit wording is often for the benefit of the "other" party entering the agreement, rather than those who are actually entering the contract as joint and several.

ELEMENTS OF A CONTRACT

In order to be valid, all contracts must meet certain criteria. These criteria include an offer and acceptance, a meeting of the minds, consideration, lawful subject matter, and competent parties. While formal contracts are carefully drafted to ensure that each of these elements are fully incorporated in the contract documents, they may have no more validity than contracts that are drafted on the "back of an envelope." If the basic elements are fully addressed in the agreement, the contract is valid and will be binding on the parties to that contract.

Offer and Acceptance

An *offer* is considered to be made when one person signifies to another person a willingness to enter into a binding contract on certain specified terms. This party *(offerer)* confers on the second party *(offeree)* the power to create a binding contract by accepting the stated terms. The offer may be express or implied, as may the acceptance. Once an offer is made, it is revocable as long as it has not been accepted. *Acceptance* creates the contract, provided that it is made in the manner and at the time specified in the offer. If an offer is made through the mail, a mailed revocation of that offer is not considered to have been made until the revocation has been received by the other party. However, if an acceptance is sent through the mail, it is considered made at the date of the postmark.

The acceptance must be definite, unqualified, and unconditional or it will constitute a *counteroffer.* Once a party has made a counteroffer, the acceptance of the original offer is no longer possible without the specific approval of the person who originally made the offer. Thus, a party who rejects an offer and counters with a different offer cannot unilaterally accept the original offer if the counteroffer is rejected.

In construction contracts, the offer is the bid submitted to the owner. The advertisement of the project to be bid is simply a request, by the owner, for offers or bids.

The issue of offer and acceptance is not as clearly defined for subcontractors who submit bids to general contractors. This was shown in *Milone and Tucci, Inc. v. Bona Fida Builders* (301 P.2d 759). Milone and Tucci, specializing in underground utility installations, submitted a bid of $20,700 to Bona Fida, a general contractor, for four federal office buildings. Bona Fida was declared the low bidder and was awarded the contract. Bona Fida then began to closely examine all the bid quotations and noticed that another subcontractor, Ray N. Erickson, Inc., had a lower price on the sewer and water installation. This had not been noticed earlier, because Erickson had also included mechanical work in the bid. Bona Fida awarded the mechanical work, including the sewer and water work, to Erickson. Realizing that its quotation had been used in Bona Fida's bid, Milone felt that it should have been awarded the contract, and filed suit to enforce the implied contract. Milone contended that an implied contract existed since its bid was an offer and Bona Fida had accepted Milone's offer when it used that same bid in its offer to the owner. The court agreed that Milone's bid did constitute an offer to do the work, but that Bona Fida's use of Milone's bid amount, in and of itself, did not constitute acceptance. Formal acceptance is required.

In the case of *A & W Sheet Metal v. Berg Mechanical* (653 S.2d 158), the nature of the acceptance of the offer was central to the court's decision. A & W Sheet Metal was requested to submit a bid to Berg Mechanical, a subcontractor. After the bids were opened, Berg contacted A & W, saying that A & W was the low bidder, Berg had used its bid of $514,000, and that A & W was to make prompt preparations for the fast-track project. Later, Berg permitted River City Sheet Metal to submit a late bid. A week or so later, A & W contacted Berg to check on the progress of the job. A & W was told that it would be contacted as soon as the project team was formed. Later that month, Berg negotiated a contract with River City for $414,000. Berg then contacted A & W and informed the firm that the contract was awarded to River City because of their lower bid. Berg offered A & W $15,000 to cover their estimating expenses, but A & W refused. A & W filed suit for breach of contract and unfair trade practices. The court had to decide on the issue of acceptance. It ruled that by its actions, Berg had accepted A & W's offer. The court was unwilling to decide on whether using a bid constituted acceptance. In this case, however, Berg had told A & W to get ready for the project. This was regarded as sufficient evidence for A & W to construe that the offer had been accepted. Offering $15,000 to cover estimating expenses may also have been regarded as an admission of guilt by Berg, as this is not a normal practice in the industry.

Meeting of the Minds

The parties to a contract or agreement must have a *meeting of the minds*. This means essentially that the contracting parties agree on the basic meaning and legal implications of the contract. This is usually considered to be the underlying purpose of the contract. If a meeting of the minds is not achieved, the parties simply do not enter into a contract. However, circumstances can be such that it is not known until after a formal contract is made that there was in fact no meeting of the minds. This is known as *unreality of consent* and provides just cause for nullifying a contract. Note, however, that a contract that is executory on the part of only one party can lead to legal complications concerning nullification. The particular circumstances must be taken into consideration. At any rate, it is assumed that a contract is an agreement reached on the basis of fact. When this is not the case, nullification of the contract can be expected. These mistakes of fact can fall into the following categories:

- Unilateral mistake or mutual error (unintentional).
- The parties do not have the same perception of the identity of the subject of the agreement.
- The subject of the agreement does not exist as a result of death, destruction by fire, etc.
- Misrepresentation (innocent misrepresentation of fact).
- Fraud (false representation of fact with intent to deceive).
- Fraud (deliberate failure to provide relevant information that is vital to an agreement).
- Duress (threats forcing consent to an agreement).

While an agreement between two parties may satisfy the meeting of the minds, there are some contracts that are clearly "one-sided" with one party having an obvious advantage over the other. Such a contract is one of "adhesion" and is formed when one party exercises the strength of its bargaining position to get the second party to agree to the lopsided terms. This occurs often in subcontract agreements between general contractors and subcontractors where the subcontractors are forced to accept the terms of the general contractor. If the subcontractor would object to the terms, the general contractor may threaten to negotiate with another subcontractor. Adhesion contracts are generally binding.

The accuracy of the plans is primarily the responsibility of the owner, as was shown in *Enrico v. Overson* (576 P.2d 75). Enrico presented a set of house plans to Overson for estimating. Overson assumed that the plans were accurate with the exception of revisions Enrico had specifically marked on the building plans. These plans showed the finished and unfinished square footage calculations for the home. The square footage figures were used by Overson to prepare his bid. Enrico accepted Overson's bid and entered into a written agreement with him. To obtain a building permit, the plans had to be redrafted to "clean up" Enrico's markings. Enrico and Overson received the new plans, which showed the same square footage figures, with neither party noting any discrepancies. Within two weeks, Overson had taken the opportunity to check the permit set in greater detail, and had discovered that the new plans contained an additional 480 square feet of finished floor space that was not reflected in the figures provided. Overson called Enrico to tell him about the need to adjust the price to reflect the additional area. When Overson had completed his revised estimate, he informed Enrico of the price change, but was told that Enrico had awarded the contract to another contractor. Enrico had also filed suit against Overson for breach of contract for failing to build the house at the contracted price. Enrico claimed that Overson had made a unilateral mistake and should have verified the accuracy of the plans. Since Overson's bid was comparable to the others received, Enrico stated that poor judgment had been used by Overson in pricing the house. Overson said he had relied on the accuracy of the plans and had based his bid on the information provided. He contended that the error was one of fact and that the contract was not binding.

The court decided that Overson had been misled by the square footage calculations provided by Enrico. The error was attributable to both parties. Overson had acted in good faith. Gross negligence was not considered applicable since Overson had given timely notification of the errors found. Thus, without a meeting of the minds (errors of fact), the contract was void.

Consideration

In a broad sense, *consideration* is something of value. It is the primary reason or main cause for a person to enter a contract. It is something of value received by one of the parties in exchange for another item or action that is of value. It is not regarded as consideration unless it is so regarded by both parties. Both parties to a contract must obtain consideration or the contract is not valid. This may seem clear-cut, but the courts tend *not* to interpret the relative value of what is received by the contracting parties. If consideration exists on the part of both contracting parties, regardless of the value, the courts will probably consider it sufficient.

Something can be regarded as consideration even though there is no actual benefit for a party, merely a detriment to the person furnishing consideration (the offerer may receive no apparent benefit from the consideration being given). The courts have ruled that the surrender of one's legal rights can fulfill the requirement of consideration. For example, an offer to pay someone not to get married can lead to a binding contract upon acceptance.

The importance of consideration was shown in *Northern State Construction Company v. Bernard Robbins* (457 P.2d 187). Bernard Robbins represented Diners Incorporated and intended to build a concession building on land owned by two other corporations controlled by Dave Beck, Sr. After plans were completed, Robbins awarded a construction contract to Northern State Construction Company. At a later date, after the contract had been signed, additional documents were signed which stated that Diners guaranteed the payment of Robbins. After the building was completed, Robbins was unable to pay all his debts. Northern State then brought suit against Diners for default on the guarantee. Northern State contended that Diners had promised to fulfill the construction contract in the event of default by Robbins. In court, Robbins showed that the construction contract was signed on February 27 and that the guarantee by Diners was signed on March 1. Since the guarantee was signed after the contract award, there was no consideration for the guarantee. That is, since the guarantee was not part of the original contract, the guarantee constituted a separate contract. Since no consideration existed for this guarantee, it was not a valid contract and was not enforceable.

In the case of *Central Ceilings, Inc. v. National Amusements, Inc.* (873 N.E. 2d 754), the subject of a verbal promise was examined. Central Ceilings entered into a subcontract with Old Colony Construction Co. for the construction of a movie theater. The project completion date was extended from June 28 to September 3, 2000 (an aggressive deadline), due to poor groundwater problems. The problems intensified when Old Colony failed to pay any of its subcontractors for the theater project and prior projects. Without assurance of payment, Central Ceilings was reluctant to continue to work on the project. National Amusements then verbally agreed to guarantee payment of all Old Colony's debts to Central, provided that the project would be completed before Labor Day. Central accelerated its work schedule and completed the project in late August, so the theater could open prior to a competing theater complex. A partial payment was made to Central, but National Amusements refused to pay the balance of $593,237 because the promise about payment was not in writing. Central filed a suit for the breach of an oral agreement. The theater argued that there was not consideration, but the court did not agree as the theater received the benefit of early completion at the expense of acceleration by Central. The court determined that this was adequate consideration and ordered full payment to be made to Central.

Consideration may consist of any benefit received by the promisor or any detriment incurred by the party to whom the promise is made. This was tested in *Sylvan Crest Sand & Gravel Co. v. United States* (150 F.2d 642). Sylvan had a contract to deliver trap rock to Mollison Airport as required. After the contract was awarded, no materials were requested or accepted. Being deprived of anticipated profits, Sylvan filed suit. The U.S. argument was simply that it had decided to exercise its option of canceling the contract at any time. Since the term *at any time* is unlimited, the government's logic was that consideration was not given with this option. Sylvan contended that cancellation could occur only within a reasonable amount of time. The court viewed the government's argument as meaning that it could request a delivery of materials and still refuse to accept it by canceling the contract. The court did not agree with this definition. It construed the agreement to mean that the government would take delivery of the rock and pay the price, or give notice of cancellation within a reasonable time. By this definition, the court found sufficient consideration to support the contract.

Lawful Subject Matter

Another requirement of contracts is that they constitute *lawful subject matter;* that is, the subject must be definite and clearly defined. The subject matter cannot violate any fundamental dictates of common law. In addition, the subject cannot be contrary to public policy. The topic of public policy is particularly relevant on contracts for public works projects. The contracts on such projects cannot constitute a restraint of competition (e.g., collusion in bidding).

Richmond Company, Inc. v. Rock-A-Way, Inc. (404 So.2d 121) provides a good example of a contract opposed to public policy. In early 1979 Richmond and Rock-A-Way were independently planning to submit bids to the South Florida Water Management District for a construction project. Both contractors had the ability to perform the required work. Rather than compete for the work, however, Richmond orally agreed with Rock-A-Way not to submit a bid. In return, Rock-A-Way agreed to enter into a subcontract agreement with Richmond if it was awarded the contract. Rock-A-Way submitted the low bid for the project and was awarded the construction contract, but did not enter into a subcontract agreement with Richmond. Richmond filed suit to force the establishment of a subcontract. It was suggested that the arrangement between Richmond and Rock-A-Way had been made to eliminate competition. Richmond replied that this was not the case, as it had made similar arrangements with other contractors who otherwise would not have been able to undertake the work. Thus, they contended, more contractors were able to bid as a result of Richmond's failure to bid.

The court did not agree with Richmond and stated, "Oral agreements between the companies whereby one company agreed to employ and pay second company as a subcontractor provided second company refrain from submitting a bid as a prime general contractor was void as against public policy in that tendencies of the agreement were to extinguish competition between the partes [*sic*] as bidders for the primary contract and to eliminate any competition for the subcontract bid." Essentially, the court found no merit in Richmond's argument since the issue concerned a matter that was opposed to public policy.

In the case of *X.L.O. Concrete Corp. v. Rivergate Corporation* (597 N.Y.S.2d 302), the issue of an illegality was construed by the court as not being sufficiently compelling to void the contract. In this case, X.L.O. Concrete, prior to entering the subcontract with Rivergate, agreed to make payments to an organized crime organization in order to avoid retaliation and physical harm from the group. Rivergate, the general contractor, was not part of the extortion, but was aware of it. X.L.O. performed work under its subcontract, but was paid only a portion of the contract amount. X.L.O. sued Rivergate for breach of contract. Rivergate contended that the extortion arrangement was illegal and that this voided the contract. Since Rivergate knew of the extortion that was taking place, and since Rivergate had paid some of the funds to X.L.O., the court ruled that the contract was valid. The illegal aspects of the contract, by themselves, were not sufficient to cancel the validity of the contract. The court stated, "The contract and its performance can be proved without reference to the illegal arrangement."

A related topic concerns impossible promises. Contracts that call for payment for performing physically impossible tasks are not enforceable.

Competent Parties

Anyone, with a few exceptions, acting in good faith may enter into a binding contract. Those specifically excluded are minors, persons determined to be insane, and drunken persons. In a broad sense, the exceptions are persons who are, in a legal sense, infants or not mentally competent. If one of the two contracting parties is judged to be incompetent, the contract can be nullified. It should be noted that until the recent past, married women were not granted contracting ability equal to that of their spouses. However, women are now considered equal under the law in this regard.

ESTOPPEL

Estoppel is a principle by which a contract becomes binding in spite of the fact that no formal agreement was made between the parties concerned. Estoppel is essentially the result of a court action asserting that an agreement or contract exists, based largely on the behavior or actions of the parties involved. This matter arises when there is an implied agreement. In other words, a contract may be created by what a party does or says, without a written document, and that party is then "estopped" from denying that a contract exists. For example, if party A leads party B to believe that an agreement exists, party A cannot later claim that one of the basic ingredients of the contract is lacking. The courts will rule that party B relied on the actions of party A, causing party B to assume that a contract existed. Party A cannot claim that a contract does not exist after party B has suffered a detriment as a result of this reliance.

A charitable-subscription agreement is a case where estoppel applies. Suppose you go to a movie on campus that is sponsored by an honorary student group and a donation of $1 is requested at the door. You have essentially placed a great deal of reliance on the actions of the student group. You feel confident that you will see a movie after entering the auditorium. By estoppel, you are entitled to see the movie. If the movie is not shown, you are entitled to a refund of your donation.

Promissory estoppel is frequently encountered in the construction industry. *Loranger Construction Corporation v. E. F. Hauserman Company* (384 N.E.2d 176) is a typical example. In this case Loranger, a general contractor, submitted a bid for the construction of a building for the Cape Cod Community College. The bid, which was submitted on May 20, 1968, included a price of $15,900 for furnishing and installing metal partitions. This price had been provided by E. F. Hauserman, the low bidder for that work item. The price of $15,900 had been provided to Loranger by means of a telephone call from Hauserman. Hauserman's bid had actually been prepared about two weeks before the bid opening date, but to avoid the chance of bid shopping, the bid was not submitted to Loranger until just before the bid opening. Loranger was declared the low bidder and was awarded the construction contract. Loranger then sent a subcontract agreement to Hauserman for a signature, and Hauserman refused to sign. By this time Hauserman had determined that the work could not be performed for the price originally quoted, and refused to enter into an agreement with Loranger. Loranger then subcontracted the metal partition work to another firm for $23,000 and filed suit against Hauserman for the difference, $7,100. The Massachusetts Supreme Court decision rendered on October 4, 1978, stated that the estimate by Hauserman constituted an offer and was not an invitation to further negotiations, as contended by Hauserman. In addition, Loranger successfully argued the case on the basis of promissory estoppel. Either condition was sufficient for Loranger to win the suit.

The validity of subcontractor bids to general contractors is often the subject of disputes. The court decisions tend to be fairly consistent in upholding the subcontractor's bid as binding on the subcontractor if the general contractor relied on the quoted price. The case of *Branco Enterprises, Inc. v. Delta Roofing, Inc.* (886 S.W.2d 157) is typical. Branco, the general contractor, received a bid from Delta to install a newly built roof using a specified product. Branco's president confirmed the quote given to him by an estimator from Delta. Branco then used that price in the computation of its bid to the owner. Branco was awarded the contract, but Delta refused to enter into a subcontract agreement with Branco. The court ruled that it was foreseeable that Branco would rely on the Delta price. The court stated, "The doctrine of promissory estoppel applies." In the 1958 case of *Drennan v. Star Paving Co.* (333 P.2d 757), the subcontractor made a mistake in its bid, but was still bound to honor its commitment to the general contractor under the principle of promissory estoppel.

There are limits to the extent to which the courts will interpret the application of promissory estoppel. In *Lahr Construction Corp.* (as LeCesse Construction Co.) *v. J. Kozel & Son, Inc.* (640 N.Y.S. 2d 957), this is clearly demonstrated. LeCesse had requested and received bids from several subcontractors. LeCesse contacted one of the subcontractors, Kozel, and notified them that they wanted to have a meeting to discuss a possible "deal." The terms that LeCesse proposed to Kozel were different from those in their bid proposal. Kozel refused to enter into a subcontract agreement with the new terms. LeCesse filed suit against Kozel for breach of contract under the principle of promissory estoppel. When the terms of the agreement were altered, the court concluded LeCesse was essentially making a counteroffer. LeCesse had essentially rejected Kozel's bid. When LeCesse delayed acceptance of Kozel's offer in hopes of getting a better deal, it could no longer make a claim of reliance on the original offer.

FORM OF A CONTRACT

Not all contracts need to be in writing to be binding. Implied contracts, however legal, are not in written form. To be binding under the law, only a few types of agreements must be in writing, namely, those related to real estate. A good general rule is that all agreements should be in writing. This is one way to clarify the scope of the services that will be provided, the consideration that will be given for performance, and the time in which performance is to take place. These terms should be clearly outlined above the signatures of the contracting parties.

CONTRACT INTENT

In some instances, courts will consider the actual wording embodied in a contract and also the intent of the contract. The intent of the contract is often a vague concept, but in general it is the presumption of what one party wanted to have done when entering the contract. It is often difficult to comply with the intent of the contract if that intent is not clearly embodied in the plans and specifications. Courts would prefer to have clear language in contracts so that the issue of intent need not be addressed as a separate matter.

The case of *Knier v. Azores Construction Co. and Everett S. M. Brunzell Corp.* (368 P.2d 673) examined the issue of intent on a project involving moving and re-modeling a motel. A subcontract was awarded to Knier to paint the motel. The scope of the contract stated, "It is the general intent of these specifications that the work involved entails the moving of the existing motel units to the designated site. These buildings to be assembled and/or re-erected in the manner to present a final completed, habitable, operating facility in its entirety. All repairs . . . to be completed in a workmanlike, acceptable manner presenting in effect a newly built structure similar to its appearance while it was being operated as The Stage Coach Motel on Highway 40." The painting specifications stated, "Doors and windows to be trimmed in white. All interiors to have touchup work where patching or other damage occurs." Knier fully painted the exterior of the motel units as stipulated in the specifications, but only did touchup painting on the interior walls. Azores and Brunzell felt that the interior should be fully painted in order to restore the motel to the quality that existed when it was "The Stage Coach Motel." When final payment was denied due to this disagreement, Knier filed suit. The court reviewed the intent provision and the painting specification describing the touchup painting. Azores and Brunzell claimed that the motel did not have the appearance of the motel when it was The Stage Coach Motel. Knier defended his position by stating that he had followed the painting specification, and argued that touching up is quite different from repainting. The Supreme Court of the State of Nevada agreed with Knier. The intent provision was general and did not specifically mention the painting requirements, but the painting specifications were quite specific. Since Knier had followed the painting specification, he had complied with the contract.

ASSIGNMENT OF CONTRACTS

Assignment refers to transfer. In contracts, assignment occurs when one party to an agreement transfers the rights or obligations of the agreement to another party who was not originally involved in the agreement, but became involved only after the assignment was made. Normally it is possible to transfer such rights or obligations to another party. However, this may be expressly prohibited by the terms of the agreement.

Such assignments cannot be made when the performance of personal services or the personal skills of one of the parties are part of the consideration. In the construction industry, this might include the specialized services of an engineering, architectural, or construction firm.

Most construction contracts, unless expressly prohibited, are assignable. A contractual right can be considered that of receiving payment in exchange for the performance of a specific duty outlined in the contract. Assignment is simply a transfer of that right from the party possessing it to a third party.

A general rule of assignment is that the second party to the contract cannot be placed in a worse position than would have been the case if the assignment had not been made. Once a legitimate assignment has been made, both of the original contracting parties are bound by the assignment.

EXAMPLE. As loan security, or to satisfy a creditor, a contractor may make an assignment of the funds that will become due when portions of the construction work are completed. When this occurs, the contractor still performs the work as stipulated in the contract, but the payment that would otherwise be made to the contractor will be made to the creditor.

EXAMPLE. A contractor on a project may encounter serious cash flow problems as a result of circumstances encountered on other projects. If it is apparent that a default is imminent on the project, the contractor may successfully assign the entire contract to another contractor. The new contractor will then be expected to perform as if the original agreement had been signed again. The owner will be compelled to treat the new contractor as a legitimate contract member. Naturally, such an assignment is a serious matter, and the owner should be fully apprised of the process of assignment as it is taking place. The contract assignment may require the owner's approval.

EXAMPLE. A subcontractor may have insufficient cash with which to purchase construction materials. Since the subcontractor is operating on a shoestring (limited financial resources), the supplier will be reluctant to finance the subcontractor. An arrangement may be made by which the funds that will be paid to the subcontractor will be assigned to the supplier. The subcontractor may also assign the funds to a bank that is willing to finance the subcontractor's work. In any case, the contractor must honor the assignment and make payments to the assignee.

SOVEREIGN IMMUNITY

When public entities are involved, there are some states in which the issue of *sovereign immunity* might arise. Sovereign immunity essentially means that the government entity cannot be sued without its consent. This applies to both the federal and state governments. State laws will vary, so there will be considerable differences between some states on the extent that sovereign immunity is claimed. Some states have abolished sovereign immunity. As a general rule, when there is a clear contract involving a governmental entity, courts have been reluctant to permit the public agency to claim sovereign immunity. The Florida courts have ruled this way based on the interpretation that by granting a governmental body the power to enter a contract, it can be interpreted that the intent is for both parties to be bound by the contract terms. In the case of *County of Brevard v. M. E. I.* (703 S.2d 1049), sovereign immunity was granted because the contract was not in writing. In *Southern Roadbuilders v. Lee County* (495 S.2d 189), sovereign immunity was granted to a state agency because the contractor failed to show or prove that the state agency had breached the contract.

REVIEW QUESTIONS

1. What general rule is followed when one party to a contract wants to cancel the contract, even though the second party has already performed a portion of or all the obligations under the contract?
2. Give a construction example of a unilateral contract. Give a construction example of a bilateral or mutual contract.
3. Give an example of a contract in which the parties to the contract are bound jointly. Give an example of a contract in which the parties are bound severally.
4. Discuss bilateral and unilateral contracts in relation to express and implied contracts.
5. A drywall subcontractor named Jack Munster performed work for Ace Contractors, a general contractor. The value of the work was $1,500. When he did not receive payment, Munster sent a letter threatening a lawsuit if prompt payment was not made. Soon thereafter, Ace Contractors sent a load of drywall to Munster's business address. This material, which was labeled as being in fulfillment of the debt, was accepted without protest. Six months later, Munster sued Ace for $600, the balance that he had determined remained on the debt. Discuss the strength of Munster's case from the standpoint of consideration.
6. Give an example of an implied contract in which estoppel will prevent one party from canceling the contract.
7. Give an example of a practice in bidding that may be contrary to public policy, and thus may constitute invalid grounds to enforce a contract.
8. Discuss the significance of a bid being an offer rather than an acceptance.

4

ISSUES CONCERNING REAL PROPERTY

REAL PROPERTY OR *real estate* consists of land and any attachments. Many laws have been enacted which relate to real property and improvements made to real property. With rare exceptions, construction projects are improvements to real property. Thus, real property laws, statutes, regulations, and codes govern various aspects of construction projects. These laws restrict the types of improvements that can be made on real property, give safeguards to those who finance such improvements, and guarantee payment to those who carry out the improvements. A wide variety of other restrictions and safeguards may apply to various parties involved in the improvement of real property. An understanding of some of the rules and laws concerning real property is important to many of the parties involved in improving real property.

TAX LIENS

A *lien* is a legal claim placed on property. It gives the party filing the lien the right to retain possession of the property until a debt payable is satisfied. Potential buyers are very reluctant to purchase land that has a lien on it. New landowners usually obtain the services of a lawyer to ensure that the property they are buying is free of liens.

A *tax lien* is the right of the government to retain possession of property until the tax on it has been paid. If the tax is not paid when the land is sold, the lien transfers with the land title to the new owner. This is not desirable from the standpoint of the new owner. If a tax debt remains unpaid, the government can force the sale of the property in order to collect the tax. Payment of the tax debt will remove the lien.

EMINENT DOMAIN

Eminent domain is the right of the federal government or a state or other public agency to take possession of private property and appropriate it for public use. Private citizens and landowners have some safeguards regarding the exercise of eminent domain. First, property can be taken only by due process of law. This means that a landowner must be given proper notice about the government's intentions and that the landowner must have an opportunity to make a case against the seizure of the property. Second, a landowner must receive fair compensation for land that is seized. The quantification of fair value must often be determined through judicial proceedings.

Although the practice of eminent domain may seem harsh to the individual landowner, the purpose of this governmental power is to benefit the public in general. As the country was developing, eminent domain was exercised numerous times to acquire land for public roadways and highways. Property for public schools is often obtained in the same manner. It can be argued that the government should compete for the property on the open market (private sector), but that would cause land prices to escalate. This would not be to the public good or in the public's best interest.

Condemnation is a word that has a harsh sound, but it means the same thing as eminent domain; that is, it is the exercise of eminent domain. Condemnation proceedings can be justified if the public is served. In fact, the power of eminent domain has been granted to some private citizens and firms when it was determined that the general public's best interests were being served. Thus, private firms, such as railroads and private utilities (water lines and sewer lines), have occasionally been given the power of eminent domain. There have even been cases in which the government used the power of eminent domain to seize private property and transfer the land to another private citizen. Native Americans are still fighting court battles over such transfers, which they allege were illegal. The use of eminent domain for the transfer of real property between private (not public) parties is rare and can legally occur only if the public good is served. Private firms that can obtain the power of eminent domain are often called quasipublic firms.

The use of land for a public good can be broadly defined. In *Parks and Recreation Commission v. Schluneger* (475 P.2d 916), the commission brought a suit to condemn an easement on private property for the construction of a pipeline. The pipeline was to carry water from the Methow River to Alta Lake for the purpose of stablizing the declining water level in the lake. This use was challenged by Leslie and Marjorie Schluneger and John and Veronica Schluneger. The commission stated that the water was needed because the water level had dropped approximately 12 feet in Lake Alta from 1960 to 1968. This decline had made swimming areas muddy, had made boat-launching areas inoperable, and had compromised the beauty of the lake. The water level could be restored if water was pumped from the nearby Methow River. The proposed pipeline to be used to pump this water crossed over property owned by the Schlunegers.

The Schlunegers argued that the lake was not part of the park but was instead controlled by the Washington State Department of Games; therefore, the pipeline was not for a public use. The Department of Games, they contended, did not have the power of eminent domain. The Schlunegers also stated that an alternative pipe route along a public road had been established by the commission. The court ruled that Alta Lake was used by the public in conjunction with the park. Therefore, the lake was an integral part of the park, and raising the water level had become a public function. Although another pipeline route was considered, the court ruled that the route on the Schluneger property was the most direct and economical.

Does the government's denial of land use for a particular purpose constitute public seizure of property? This was tested in *State of Washington v. Lake Lawrence Public Lands Protection Association* (601 P.2d 494). Lake Lawrence Public Lands, a development firm, proposed single-family lot developments for a specific Thurston County property. During the period of the preliminary plot approval review, it was revealed in an environmental impact statement that the land was a nesting ground for endangered bald eagles. The plot was denied as proposed. The developer resubmitted the proposal, asking for 22 lots with 3 lots set aside as a bald eagle preserve with a 75-foot buffer zone around the nesting site. The plan was again denied. It was returned to the developer with an indication that fewer lots and a 200-foot buffer zone would be approved. The developer filed suit, claiming that the denial of the plot constituted an unlawful taking of the land. The denial was keeping the developer from making a reasonable profit, and so the county, it was claimed, had in effect seized the land without paying for it. The court ruled that the denial of the plot did not constitute an unlawful taking of the land. Since the county had proposed alternatives, the developer could still develop a portion of the property.

When land is seized through eminent domain, a fair price must be paid for the property. As was shown in *State of Washington v. Lawrence M. Wilson and Colette Wilson* (493 P.2d 1252), disputes may arise over the price to be paid. Wilson owned property on which he had previously converted a house into a doctor's office and two apartments. When eminent domain actions were started to acquire the property, a dispute arose concerning the fair price to be paid. Wilson wanted the value to be based on the cost of building a similar structure on another lot. The state contended that depreciation should be charged against the 20-year-old structure. The question then was whether the state should pay for building a new structure? "Just compensation" is not specifically defined in the applicable statute. The court reviewed other cases and concluded that the value of a 20-year-old structure is not the same as that of a new structure. Thus, the court concluded that depreciation had to be charged against the structure in order to come up with its fair value.

Once eminent domain proceedings begin, the most viable means of stopping condemnation may be to show that the land seizure is not for the public good; but this is difficult to prove. Such a tactic was used in *State ex rel. Church et al. v. Superior Court for King County* (240 P.2d 1208). In this case, the city of Seattle needed to build a sewage disposal plant. Condemnation proceedings were started to acquire the desired site. The people who owned the property, including Roberta H. Church, challenged the proceedings, claiming that the use of land for sewage disposal is not a public use. The owners claimed that the disposal of sewage is not a governmental function; therefore, the site would not be used for the public. The court decided against the landowners, stating that the use of land for sewage disposal constitutes a public use.

A recent (2005) Supreme Court ruling is casting a new light on when eminent domain may be exercised. This related to a dispute in New London, Connecticut, in which the city was permitted to seize several privately owned homes for another private party to develop them. The city's motivation to seize the property was that the tax base would increase once the land was developed. In other words, the court essentially determined that the city could exercise its eminent domain rights on behalf of another private party, because the public good was served by the increased taxes that this would generate. In response to the decision, many states have enacted legislation that prevents the seizing of private property for the purpose of retail, office, commercial, industrial, or residential development.

HIGHWAYS

Highway property can be obtained from private landowners by several means, including the following:

1. Outright or direct purchase (mutual agreement).
2. Eminent domain or condemnation proceedings (hostile acquisition).
3. Prescription (hostile acquisition). This is the acquisition of property that has been used by the public for a prescribed period of time. This falls under state law, and so the specific requirements may differ from state to state. A typical rule or law would be that if the public has used a given private road (as if it were public) for a period of 7 years, proceedings can be followed to acquire the roadway as public property.
4. Dedication (mutual agreement). This is the granting by the owner of the use of private property for the public at large (highway, street, park, or school). If the public has been permitted to use private property in a certain manner for a given (long) period of time, the owner will not be able to deny continued use. If the public use of the property stops, however, the private owner has free use of the property once again. With dedication, the land remains under the ownership of the private citizen; that is, there is no transfer of land title. The dedication of property for public use may be done in writing or implied by actions.

In the state of Washington, prescription can be exercised after 10 years of adverse use. The rights of prescription are not absolute. For example, in *Peeples v. Port of Bellingham* (613 P.2d 1128), prescriptive rights were sought after the port made a mistake. In 1957, the Port of Bellingham constructed a breakwater on the north and south sides of a tract of tideland belonging to Peeples. Part of the tideland had to be dredged in order to construct the breakwater. In 1966 the port discovered that it did not own the land, but in 1970 it began constructing facilities on the property anyway. In 1972 the port offered to purchase the land from Peeples, but the offer was refused. In 1974 the port claimed the property by adverse possession. Peeples claimed that the port's use of the land was not continuous, but involved only dredging the channel in 1970, and that the possession of the property by the port was never absolute or total. Furthermore, the port did not know that it was not the owner until 1966, and so adverse possession did not exist for 10 years. The court decided in favor of Peeples.

RIGHT-OF-WAY

A *right-of-way* is a tract of land, usually consisting of a series of connected parcels of property, that is used for the operation of a highway or public utility. Because of the nature of projects such as highways, railroad tracks, and sewer lines, the general requirement is that all parcels must be adjacent and that no gaps may exist in the tract. Right-of-way property is owned by either public or private firms. The most common means of obtaining such property is through an outright purchase arrangement. If they are granted the power, private firms may be able to acquire private property for a right-of-way through eminent domain.

For railroads, the property for the main track is often obtained by eminent domain. However, industrial track (rail spurs) property is more frequently obtained by outright purchase, as eminent domain powers may be restricted to main track right-of-way.

The above comments refer to right-of-way property that consists of an actual transfer of land title. This is how many transactions involving real property are handled. Another means is available for private firms that want to obtain the right to limited use of a particular piece of property; in other words, full title to the property is not essential. This can be done by obtaining an easement which states the limited and specific use that can be made of a particular piece of property. An *easement* grants a specific right of use to nonowners (neighbors, utilities, governments, etc.). Easements are quite appropriate for the installation of sewer lines, water lines, telephone lines, power lines, and gas lines as they are less expensive than acquiring the land and only limited access will be required after the installation is completed. If the land is sold, the easement is generally transferred with the land; that is, the sale does not adversely affect the easement.

The case of *Dunbar v. Heinrich* (605 P.2d 1272) provides an example. Shane Dunbar purchased a parcel of land in Snohomish County, Washington, in 1963. To gain access to his property, he drove on a road he believed to be a public right-of-way. After 7 years of usage he learned that the right-of-way was actually on private property owned by Emerson Investment Co. Dunbar continued to use the road without asking permission, but with no direct prohibitions against such use. In 1977 Emerson Investment's property was sold to H. Heinrich. Dunbar then initiated proceedings against Heinrich to gain a permanent prescriptive easement for access to his property. Washington law (RCWA 4.16.020) states that a prescriptive easement can be established after 10 years of adverse use of property during which no permission is sought. Dunbar claimed that he had used the roadway for more than the prescribed 10-year period, while Heinrich contended that the adverse use had occurred for less than 10 years. Heinrich based his allegation on the fact that Dunbar did not know that the property was private property until several years after he had begun using the roadway, and that the prior use could not be regarded as adverse use. The court ruled in favor of Dunbar, stating that he met the basic criteria of (1) adverse use of the land, (2) continuous use of the land for the required time, and (3) knowledge by the owner that he was using the land. The court stated that Dunbar's belief that the land was on public property was not relevant and that his actions, not his beliefs, satisfied the element of adverse use. The easement was granted to Dunbar.

The public right to the use of easements is not absolute, as was shown in *James J. Keesling v. The City of Seattle* (324 P.2d 806). Keesling owned two parcels of land, with his residence located on one of them; the other parcel was vacant and offered a view of Puget Sound. The city of Seattle placed two transmission poles on Keesling's vacant lot, and Keesling's view of Puget Sound was obstructed by the poles and lines. When his requests to move the poles were ignored, Keesling filed suit. The city offered no denial of the claims by Keesling, and it appeared that the city was testing Keesling's resolve. The court ordered the poles removed, and damages were awarded for trespass (one pole extended beyond the easement). The use of an easement is not absolute. If land is devalued as a result of an easement, damages may be assessed against the easement user, the party benefitting from the easement.

In *Robertson v. Club Ephrata, Dungan, et al.* (351 P.2d 412), the court had to decide if the owner could stipulate the location of the right-of-way easement. Dungan, a former mayor of Ephrata, challenged the city's right to establish a 30-foot right-of-way across his property. The right-of-way easement was for the purpose of installing a connecting pipeline. Dungan did not challenge the city's right to an easement, but the specific location selected for the easement. Dungan stated that verbal arrangements had been established three years earlier with the former sewer and water superintendent. In that agreement, Dungan contended that they had agreed on an alternative easement location. The city engineer, however, determined that the alternative route was inadvisable, since exorbitant costs would be incurred. The city was simply claiming that it was exercising its right of eminent domain. The court considered Dungan's alternative route, but found no substance in his claim.

ZONING

Zoning is the division of real property, especially in larger cities, into classifications of use. Each area of the city will have a particular designation regarding the use of the property or land within it. Restrictions may be placed on the minimum size of a lot, the minimum distance a building can be placed near the property boundary or street, the types of buildings that can be built in the area, the height of buildings, the number of stories, the sizes of billboards, the area of a lot that can be occupied by a structure, and the density of population. The use that can be made of the land will also be noted—residential, commercial, industrial (unrestricted), rural, recreational. Zoning requirements are essentially the master plan of a city to regulate the use of the land in each area or community.

Zoning is used to ensure an orderly development of the community and maintain the quality of life (health and welfare) for a city's inhabitants. Zoning requirements place restrictions on the use of the property (limited property rights). For example, industrial facilities cannot be established in commercial areas, rental property restrictions are often placed on property in residential areas, and commercial properties cannot be developed in residential areas. In a few instances, special permits may be granted for exceptions.

Zoning restrictions may change and adversely affect an owner who finds that a planned use is suddenly in violation of zoning restrictions. Does an owner have a viable grievance against a municipality that initiates changes in its zoning restrictions? This was the topic of *Tekoa Construction, Inc. v. City of Seattle* (781 P.2d 1324). In 1982 the city of Seattle liberalized zoning requirements by enacting a new land-use code. To promote development within the city, the code granted the development of large numbers of lots which did not meet the previous minimum area requirements. In effect, the new code meant that substandard lots that had not previously met the code requirements could now become sites for new construction. In response to the new code, however, developers were demolishing houses on lots that were adjacent to the smaller lots, and replacing them with two or three new houses. This was objected to by the communities, which contended that this increased the density of neighborhoods and changed their character. To counter this action by developers, the land-use code was amended to prohibit the demolition of existing housing to develop multiple undersized lots.

Tekoa Construction sought an injunction to prevent the city from enforcing the amended ordinance, and sought compensation for the losses resulting from the new restriction. The city believed that the ordinance was within reason and that it had the power to enforce it. Tekoa stated that this was essentially the same as taking property without due process, and violated its right to develop its property. Tekoa felt that it could develop property under the zoning restrictions that existed when the land was acquired. Also, the intent of the original ordinance was to encourage the development of these substandard lots. The city countered by stating that the change in the ordinance was adopted in the public's interest and was valid. The court ruled that since Tekoa had not previously obtained building permits on any of the lots under consideration, the new zoning restriction should apply. It ruled against Tekoa's challenge of the new zoning restriction.

MECHANIC'S LIENS

A *mechanic's lien* is a right created by law that permits workers and materials suppliers who provide improvements to real property, to place a claim on that land if they are not paid. This claim initially consists of filing the formal papers which secure for the worker or materials supplier a right to the property until the debt is paid. This is a powerful law that helps workers who perform work and establishments that furnish materials to improve land. The lien is similar to a mortgage in that the lien is an attachment to the land. The purpose of the law is to permit a claim or lien to be placed on premises when a benefit has been received by the owner and the value or condition of the property has been increased by labor and materials that have not been paid for. Lien laws were enacted to prevent unjust enrichment or the rendering of service without pay.

Mechanic's liens provide protection to all parties who directly improve real estate, including workers (the archaic word is *mechanics*), materials suppliers, subcontractors, and general contractors. For general contractors to file a lien, a written contract is often required. The general contractor must have performed the obligations of the contract to a substantial degree, as approved by the engineer, to qualify for filing a mechanic's lien. A general contractor will usually opt for a civil suit for breach of contract instead.

The owner of a project will obviously know that the general contractor has lien rights. The same is true of the subcontractors. Unfortunately, the owner cannot readily identify the sub-subcontractors who might be performing work on the project, or the materials suppliers who are providing materials to the subcontractors. To ensure that the owner has knowledge of these "second tier" parties, many states require that some official notification be provided to the owner about the work or materials that are being provided by these parties. This notification is referred to as the "notice to owner" or "notice of intent to lien" or some variation of this basic principle. Once the owner receives notice of the second tier firms providing labor or materials, they have secured their right to file a mechanic's lien in the event they are not paid for their services (see figure 4.1).

As long as the formal papers still exist, the mechanic's lien itself is not a major worry to a landowner. It is the discharge of the lien that is most troublesome. Of course, the simplest way for the lien to be discharged (eliminated) is for the landowner to pay the debt. If the debt is not paid, the lien can be discharged through *foreclosure,* or the sale of the property to obtain recovery. The latter option is most worrisome to landholders with liens on their property. If a lien is created, a lien claimant who is left unpaid can demand judicial foreclosure (judicial sale of property) of the property and have the obligation satisfied out of the proceeds.

For workers who improve real property, a mechanic's lien serves as guaranteed reimbursement. In satisfying the lien claimants, the first parties paid are the workers and materials suppliers. The next priority is the contractor and subcontractor category. Subcontractors have varying priorities and rights, depending on the state. The laws are more ambiguous concerning the contributions of designers. Some states do not give designers the right to file mechanic's liens, while others do. Some states permit designers to file mechanic's liens only if the actual construction work has taken place.

Just as the *notice to owner* is an official document that puts the owner on notice that a lien can be filed if a particular party is not paid, the owner would similarly want to have a document that clearly states that a potential lien is no longer a viable threat. This is a document from subcontractors and suppliers that informs the owner of the extent to which payments have been made to them by the general contractor. This is known as a *partial release of lien* and assures the owner that at least the stated payments have been made and that lien rights for that portion of the work are officially released (see figure 4.2).

NOTICE TO OWNER

Warning to Owner: Under state law, your failure to make sure that we are paid for labor and materials may result in a lien against your property and that you may be required to pay twice for the same services or materials.

To avoid a lien, you must obtain a written release from us every time you pay the general contractor.

To: *Riverside Development Group (Project Owner)*   Copy to: *General Contractor*

Copy to: *Construction Lender*   Copy to: *Surety*

We, *The Tile Setter Company,* hereby notify you that we have furnished or will furnish services or materials as follows:

Tile materials and labor to install same in Office Building "C" at 4700 Circle Drive in Memorial Office Park.

Under state law, this notice restricts your right to make payments under your contract.

For your protection, you should know that those who work on your property, or provide materials, and are not paid have a right to enforce their claim for payment against your property. This claim is known as a construction lien or a mechanic's lien. If your general contractor fails to pay subcontractors or material suppliers or neglects to make other legally required payments, the people who are owed the money may look to your property for payment, **even if you have already paid your general contractor in full.**

You must recognize that this Notice to Owner may result in a lien against your property unless we have been paid.

Date: *February 23, 2011*   By: *Larry Jacobs*
                                            Authorized Agent of The Tile Setter Company

---

FIGURE 4.1
Example of a notice-to-owner form (not specific for any particular state).

Lien rights are rarely given on public works projects. Public property is usually not subject to liens. Who would buy a public building, anyway? However, other mechanisms can be used. Some states permit workers, suppliers, and subcontractors to file stop notices. A *stop notice* permits a worker, materials supplier, or subcontractor to notify the owner when the general contractor has failed to make payments for labor and/or materials. Upon receipt of this notice, the owner will make no further payments to the general contractor until the claimant has been paid. Essentially, this is a method by which unpaid parties in the construction process can put a lien on public funds due the general contractor.

Another mechanism commonly used on public works projects is the requirement that the general contractor furnish a payment bond as a condition of entering into a construction contract. A payment bond is an extremely effective means of

---

PARTIAL RELEASE OF LIEN

KNOW ALL MEN BY THESE PRESENTS:

The undersigned, for and in consideration of payment of $ *$16,329*, paid by or on behalf of *PDR Builders* ("Contractor"), the receipt and sufficiency of which is hereby acknowledged, hereby releases said Contractor, its employees, officers, agents, sureties, successors and assigns, and *Riverside Development Group* ("Owner") of the premises being improved, all liens, lien rights, claims, and demands of every kind whatsoever, including any claims for extended or additional job costs, overhead, and lost profits, due to any cause, including any claims and demands arising from any claimed delays, disruptions, or changes to the work, which the undersigned now has or might have on account of work performed through the date of *April 20, 2011,* on the job identified and legally described as follows: *Office Building "C" at 4700 Circle Drive in Memorial Office Park* and including any claim or lien for any and all work, labor, materials, supplies, services, equipment, and/or rental of equipment furnished and/or used by the undersigned, its subcontractors, and material suppliers for construction of said improvements upon the Property and/or job referenced hereinabove. The Undersigned represents that all of the foregoing so furnished or used through the date of *April 20, 2011* have been paid for in full and all taxes imposed by applicable law, including, but not limited to, sale and use taxes, have been paid and discharged.

The undersigned Authorized Officer of the Undersigned acknowledges that the facts and matters set forth in the foregoing **Partial Release of Lien** are true and correct.

Signed, sealed, and delivered in the Presence of: _____(Seal)

By: *Larry Jacobs*_____   Title: *Owner of The Tile Setter Company*

Date: *May 9, 2011*

FIGURE 4.2
Example of a partial release of lien form (not specific for any particular state).

ensuring that the general contractor's workers, subcontractors, and direct materials suppliers will be paid.

For the foregoing reasons, mechanic's liens are used only in the private sector of the construction industry. Usually the owner is under sufficient pressure wrought by the mechanic's lien that the bills are paid outright. When there is a general contract, one of the conditions of the owner's obligation to pay for work performed is that the contractor keep the project free of liens. Theoretically, the liens against a property can exceed the contract price between the owner and the contractor; however, some states limit the amount of liens to the contract price.

The *notice to owner* must be submitted to the owner in a stipulated time frame from prior to commencing supplying of materials or labor but no later than a stipulated time period after beginning to provide such services. The specific time periods may vary considerably between states, so the individual requirements of individual states must be examined. The general procedure is outlined in figure 4.3.

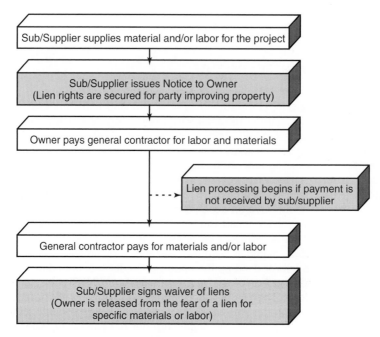

FIGURE 4.3
Process of obtaining and releasing lien rights.

Several cases help illustrate the power and complexity of mechanic's lien laws. The 1928 case of *Coast Trucking Co. v. West Seattle Dairy* (268 P. 598) addressed the issue of how to place a lien. The case concerned excavation work that was performed by Coast Trucking. In the original contract, West Seattle Dairy contracted with William Wilds to excavate several Seattle lots in an area designated Block 35. The hauling of the excavated material was performed by a subcontractor, Coast Trucking Co. The material was hauled to property (designated Block 58) owned by West Seattle Dairy. West Seattle Dairy paid Wilds for excavating the material and hauling it to Block 58; however, Wilds did not pay Coast Trucking. Coast Trucking then filed a lien on the lots in Block 58. This lien was based on RCW 60.04.040, which states in part, "Any person who, at the request of the owner of a real property . . . fills in or otherwise improves the same . . . has a lien upon such real property for the labor performed." Defending West Seattle Dairy, Wilds claimed that the material could have been hauled anywhere and that the contract was specifically for excavating the material in Block 35; that is, there was no contract concerning Block 58. The court ruled in favor of West Seattle Dairy. The decision was based on the fact that the contract was for the excavation of material from a lot and that the lien had been placed on the material that was dumped on Block 58. It was implied that Coast Trucking had a right to file a lien, but that the lien should have been filed on the lots that were excavated, not on the lots to which the material had been hauled.

In *Robert Burns Concrete Contractors, Inc. v. Norman et ux.* (561 S.W.2d 614), filing irregularities prevented the lien from being honored. Robert Burns Concrete Contractors entered a subcontract agreement to construct the foundation for a house. When the subcontractor was not paid by the general contractor, Burns contended that he had perfected a mechanic's lien on the property by giving notice to the general contractor. The court decided that since the general contractor was not the owner of record of the property, proper notification had not been given to the owner. The owner had to be specifically notified.

Various court decisions have dealt with the rights of architects to file mechanic's liens. In *Lamoreaux et al. v. Andersch et al.* (150 N.W. 908), Lamoreaux, the architect, provided the plans and specifications for a building to be constructed on a tract of land owned by Andersch. The final design documents were accepted by Andersch in accordance with the original contract between Lamoreaux and Andersch. Subsequently, Andersch abandoned the project and decided not to have the structure built. When no payment was made for the architectural services, Lamoreaux filed a mechanic's lien. The court recognized that Minnesota law states that liens do not apply "prior to the actual and visible beginning of the improvement on the ground," but that previous cases had granted such liens for materials furnished to a project but never used. The court rationalized that if prefabrication work that was never incorporated in the project qualified for a lien–*Howes et al. v. Reliance Wire-Works Co.* (48 N.W. 448)–the efforts of an architect should similarly be granted the protection of lien laws.

Lien rights and obligations are contract specific. This was the court's assessment in *Elton Lee v. All Florida Construction Company* (662 So.2d 367). All Florida entered a contract to build an addition to the home of Elton Lee. The contract contained a provision that called for mandatory arbitration to resolve disputes. After the contract was signed, the home sustained damage from Hurricane Andrew. Another contract was entered for All Florida to repair the hurricane damage. This second contract did not contain a mandatory arbitration provision. All Florida completed the addition and repaired the hurricane damage, but was not paid for all of the work because there was a disagreement concerning the quality of the work. All Florida then filed a lien and also demanded that arbitration begin. Lee claimed that there were two contracts and they should be handled separately. All Florida contended that the work pertained to a single structure and that one lien filing was sufficient and that the arbitration provision should apply to both contracts. The court was clear in its findings that there were indeed two contracts. This mandated that two separate liens must be filed and that mandatory arbitration applied to only one contract.

The time of filing liens and notices must be adhered to strictly. This timing can be confusing and was clarified in *Stunkel v. Gazebo Landscaping Design, Inc.* The Stunkels were having a new home constructed that included extensive landscaping to be performed under a subcontract agreement by Gazebo Landscaping. Gazebo was to plant trees to be selected by the Stunkels. On November 7, 1990, the Stunkels accompanied a Gazebo representative to Tampa to select and mark the trees they wanted. On December 5, holes were dug on the Stunkel property and on December 7 the trees were delivered and planted at the Stunkel home. When Gazebo was not paid for its services, it posted a notice to owner on the Stunkel's gate on January 18, 1991, after failing in other attempts to contact the Stunkels. Gazebo then filed suit on February 11 to foreclose on the lien. The underlying question that arose was whether the lien was filed within the required 45 days of providing the services. To answer this question, the court had to decide if the 45 days should be counted from November 7 (when the trees were selected by the Stunkels) or December 5 (when preparations were made to plant the trees). The Supreme Court of Florida decided that the 45 days began on December 7 when the trees were actually delivered to the site. Thus, the lien was valid, even though the notice to owner was not notarized, as the Stunkels acknowledged receiving the notice.

The intricacies of following the strict requirements of the applicable lien requirements were further demonstrated in *Gonas v. Home Electric of Dade County, Inc.* When Home Electric was not paid in full for services provided to Gonas, it filed a lien. Gonas responded to Home Electric with a certified letter demanding a written accounting to justify its claim. The statute required that the owner of property has a right to demand a written statement to justify the amount to be paid on account. Furthermore, failure to respond within 30 days "shall deprive the person so failing or refusing to furnish such statement of his lien." Gonas did not mention the 30-day requirement in its letter, and Home Electric, unaware of the statute, did not respond to the request within the 30-day period. The court decided that it was not the obligation of Gonas to inform Home Electric of the statute requirements. Strict compliance with the statute is required, and since the requirements were not followed, the lien rights were lost by Home Electric.

In the state of Washington, liens must be filed within 90 days "from the date the contributions to any type of employee benefit plan are due, of the cessation of the performance of such labor, the furnishing of such materials, or the supplying of such equipment." These terms were interpreted in *American Sheet Metal Works, Inc. v. Haynes* (407 P.2d 429). In 1961 Haynes was awarded a contract by Columbia Grain Growers, Inc., to construct a concrete grain elevator at Relief, Washington. The various metal appurtenances for the project were supplied and installed by a subcontractor, American Sheet Metal Works, Inc. American completed its work on July 27. Shortly thereafter, American was instructed to return to the project and install the remote controls for the aeration system. American performed the requested work on August 11. When American was not paid, a lien was filed on November 8, and foreclosure proceedings were initiated. The general contractor contended that the lien was invalid, as it had been filed more than 90 days after the July 27 completion date of American's work. American contended that its last work was performed on August 11, when the owner requested that remote controls be installed. The court stated that the lien was valid, since the general contractor requested that the subcontractor perform the work under the conditions of the original contract. Had the added work been performed under a separate contract, the lien would have applied only to the work performed under the second contract.

Accuracy of documentation is important in supporting a lien, as was demonstrated in the Wyoming case of *Gary Zitterkopf, Superior Woods Construction v.*

Basil C. Bradbury (783 P.2d 1142). Gary Zitterkopf, representing Superior Woods Construction, verbally agreed to do remodeling work on the home of Basil Bradbury. Cost was never discussed, but the costs were to be directly reimbursed by Bradbury. By mid-November Bradbury had received several billings that totaled $40,000. At that point Bradbury felt the project was too expensive and informed Zitterkopf that work was to cease and that he should contact his attorney. Zitterkopf worked an additional two weeks and then filed a lien on Bradbury's house. Although many issues were relevant to this case, some key points related to the question of Zitterkopf's integrity. Zitterkopf's failure to stop work was to his own detriment. In addition, it was shown that Zitterkopf had charged for his own time on Bradbury's house, when in fact he was working on another project. Furthermore, Zitterkopf had altered invoices before submitting them to Bradbury. While Zitterkopf testified that his markup on labor was approximately 15 to 17 percent, it was determined to be about 100 percent. Based on this evidence, the court could not find justification for the $40,000 Zitterkopf had charged Bradbury. The court found that no contract existed between the parties because there was no meeting of the minds. Obviously, the falsification of information to gain a larger payment was not helpful to the contractor's argument.

The case of *Ragsdale Bros. Roofing, Inc. and Corneau-Finley Masonry v. United Bank of Denver* (744 P.2d 750) shows the strength of a mechanic's lien. First Colorado Construction Company (general contractor and owner) began the construction of a building on a lot on which the First National Bank of Denver held the deed of trust. On April 10, 1980, the owner defaulted on the loan, and First National began foreclosure proceedings. Mechanic's liens were filed on April 16 by Corneau-Finley, a masonry contractor, and on April 23 by Ragsdale Roofing. The property was then sold on May 28 by United Bank of Denver at a public trustees' sale. The deed was signed on October 10. Foreclosure on the mechanic's lien began on June 30. The owner, United Bank, stated that the purchase meant that the property was free of liens and that the liens therefore were invalid. Essentially, United Bank was saying that Ragsdale and Corneau-Finley would have had to foreclose before May 28. The court ruled otherwise, concluding that the contractor's mechanic's lien was superior to the public trustees' deed. The trustees' deed was on the lot, while the contractor's lien was on the building which was attached to the land. An interesting point was raised in regard to some finished doors that Corneau-Finley had custom made for the project. Since the doors had not been delivered to the site, their value was not covered by the lien in spite of the fact that no other use could be made of the doors.

In *Abe Den Adel v. Edwin Blattman and Doris Blattman* (357 P.2d 159), the Blattmans had hired Abe Den Adel to construct a residence on their property in Seattle. The written contract stipulated payments to be made on the basis of the cost plus a fixed fee. After the residence was completed on June 11, 1957, the contractor signed a "receipt and lien waiver" along with an affidavit that all bills had been paid. In return, the contractor received the remaining balance of the loan the Blattmans had obtained from the bank. The payment received by the contractor was only for the direct reimbursement of costs incurred and did not include the fixed fee as stipulated in the contract. The contractor then filed a lien to recover this unpaid fee. The state supreme court decided in favor of the Blattmans. Since the contractor had signed a receipt and a waiver of lien, this could be interpreted only as meaning that he had waived his right to file a lien against the residence.

Liens filed erroneously with malicious intent can have an adverse result. This was shown in *Puget Sound Plywood v. Frank Mester* (542 P.2d 756). Frank Mester constructed speculative residential homes with financing provided by Coast Mortgage Company. Mester contracted with Puget Sound Plywood (PSP) for lumber on a per house basis. On April 23, 1973, Mester paid his account with PSP in full. In August PSP filed liens on three houses being constructed by Mester, claiming that he owed $3,172.53, $2,693.59, and $406.61, respectively, on those homes. PSP notified Coast Mortgage of the liens, in response to which Coast Mortgage was obligated to withhold any further loan funds from Mester. At the time the liens were filed, the true indebtedness was apparently about $600. Mester then filed a counterclaim, contending that the liens had been improperly filed and had greatly damaged his business reputation. The lien notices had been publicized in the local journals, and this had severely damaged his credit rating. With loan payments stopped, Mester could not undertake any additional work. The court found that PSP had wrongfully filed the liens with the intent to harm the reputation of Frank Mester, and to "lower him in the estimation of the community or to deter third persons from associating with him." At the time of the court decision, it was determined that Mester owed a total of $193.39 on his account with PSP. All liens were ordered released. In addition, PSP was required to pay Mester $20,000 for lost profits, $5,000 for damage to his credit rating, and $2,256.30 for his legal fees.

The case of *Pilch v. Hendrix* (591 P.2d 824) tested the credibility of a lien based on the poor quality of the work. David and Irelan Hendrix hired a contractor, Robert Pilch, to replace some concrete steps at their residence. The contract price was $850, with $200 paid in advance. When Pilch finished work on the steps, he requested payment on the balance owed. The Hendrixes said that the concrete steps were poorly built and refused to pay for the work. Pilch then filed a lien to force payment. The Hendrixes demonstrated to the court that Pilch had used poor workmanship in constructing the steps. To correct the flaws in the steps, it was concluded that the entire project would have to be redone. Thus, it was determined that Pilch had breached the contract. The court then removed the lien, stating that it had been improperly filed. Pilch was ordered to pay the Hendrixes $600 in damages and attorney fees.

The registration requirements for contractors also apply to mechanic's liens, but this is not an absolutely certain issue. One statute states that it is "a misdemeanor for any contractor having knowledge of the registration requirements of this chapter to offer to do work, submit a bid, or perform any work as contractor without being registered as required by this chapter." The purpose of the statute is to prevent unreliable, fraudulent, and incompetent contractors from victimizing the public. This was addressed in a contract involving the drilling of a well. The contractor was not registered with the state because he had not passed the exam for a well-water construction operator's license. Other than this license, the contractor satisfied all the requirements. The contractor then hired an employee who was properly licensed by the state. The employee carried out the well drilling without supervision by the contractor. The contractor submitted a bill to the owner. When full payment was not received, the contractor filed a mechanic's lien on the owner's property. The owner sought to get the lien dismissed on the grounds that the contractor was not personally licensed. The owner contended that RCW 18.27.080 states, "No person engaged in the business or acting in the capacity of a contractor may bring . . . any action in any court of this state for the collection of compensation for the performance of any work . . . of any contract for which registration is required." The court decision was based on the spirit of the registration statute. It was determined that the contractor had substantially complied with the registration requirements by employing a registered employee and had thus complied with the spirit of the law. The mechanic's lien was valid, and foreclosure could be exercised.

Mechanic's lien laws were established specifically to ensure that workers would receive compensation for their efforts in improving real property. While this ideal is generally meritorious, lien laws are not without shortcomings.

Mechanic's lien laws are not a perfect solution to a problem. Several strong criticisms have been levied. Some of the primary criticisms are as follows:

- Lien laws are complex. Notices have to be given, filings have to be made, and actions have to be taken, all within specified time limits.
- Lien laws are inconsistent between states. Since they are state laws, there is considerable variability from state to state; for example, design professionals are not covered by these laws in all states.
- Liens are no guarantee of payment and are often worthless. When mechanic's liens are filed, there are probably other parties who also want their money. Priority in collection is given first to parties who have taken out mortgages or who hold deeds of trust on the property. If this happens, claimants filing the mechanic's liens may not have their claims paid.
- Filing a lien is a severe means of collection if the debt is small.
- Among owners, a strong criticism is that payment may be made twice for the same work (first, to the contractor who did not pay a subcontractor or worker, and second, to satisfy the claim filed by an unpaid subcontractor or worker).

How can mechanic's liens be avoided or their impact minimized? While various responses can be given to this question, it should be apparent that some are more realistic than others.

1. Owners can post a notice of nonresponsibility within a specified time after the improvements on a property have been made. Any potential lien claimants should surface at this time. Note that improvements must generally exist for a lien to be valid. An exception might be in the design area. This has a further implication, however, if the project is destroyed or a project improvement does not commence; in this case the lien right may also be destroyed.
2. A no lien contract, in which the contractor agrees not to assert a lien, can be drawn up between the owner and the contractor in some states.
3. The owner can require the contractor to furnish an affidavit that all the bills related to a project have been paid. This method relies on having a trustworthy contractor. If a contractor fails to pay the workers, will he or she have any compelling moral obligation not to lie about it?

4. The owner can demand to see receipts or statements from workers, suppliers, and subcontractors that the contractor has paid them. The engineer or the owner must then be fully aware of the identities of all subcontractors. It is cumbersome to get such statements from all the workers on a project. There is one case on record in which a subcontractor was hired by the general contractor. The general contractor concocted a scheme in which the subcontractor would perform the work without the owner knowing about it. Later the subcontractor filed a mechanic's lien because the general contractor had not made any payments for the work. In this case the court ruled that the subcontractor was not entitled to file a claim, as the owner never had a chance to obtain a lien waiver from the subcontractor.

5. The owner can require the general contractor to pay the subcontractors and materials suppliers before the owner makes any payments. This is counter to standard policy in construction, as it places a substantial financial burden on the general contractor.

6. The owner can write joint checks to the general contractor and the subcontractors and suppliers so that they have to endorse the checks in order to get paid.

7. The owner can delay making final payment to the general contractor until the time of filing mechanic's liens passes. (This is called the "lean" period.)

8. The owner can keep a reasonable retainage out of the contractor's payments. This is a fairly common practice in the construction industry. The amount withheld will be sufficient only to cover small claims.

9. The owner can require a payment bond from the general contractor. The surety company is then liable for any failure in making payments to subcontractors, suppliers, and workers. The surety will pay the unpaid parties in the event of a contractor default. There is strong assurance with this method that no mechanic's liens will be filed.

10. Probably the best way for the owner to minimize the possibility of mechanic's liens is to obtain the services of a competent, reputable, and trustworthy contractor.

REVIEW QUESTIONS

1. What is the relationship between eminent domain and condemnation?
2. What is the essential difference between prescription and dedication?
3. Describe the major differences between a right-of-way and an easement.
4. Give examples of zoning restrictions that a municipality might enact to control or guide its growth and development.
5. Discuss what may happen to a municipality that does not enact any zoning restrictions.
6. Describe some of the conditions that must generally exist for a worker to successfully file a mechanic's lien.
7. What are some major criticisms that property owners have of mechanic's liens?
8. What are some major criticisms that workers have of mechanic's liens?
9. Speculate on the reasons why the procedures pertaining to mechanic's liens must be rigidly followed in order to be successful in filing a lien.

5

AGENTS

AGENCY AGREEMENTS ARE similar to contracts in that they must contain the same basic elements. An agency serves a major function in that it is a means by which a person or persons can let someone else do something for them. An agency agreement consists of a *principal* and an *agent*. In an agency, one party (the principal) authorizes another party (the agent) to represent the principal in certain specified business dealings with third parties (figure 5.1). As such, the agent must be a trusted individual; that is, there is a fiduciary relationship between the principal and the agent.

An *agency* is a consensual relationship; this means that the arrangement is mutual and that the agent cannot simply volunteer. An agency agreement can also be created by ratification of an unauthorized act (after the fact) and estoppel (implied by past actions). The agent's authority must come from the principal. The agent is appointed to act for the principal in transactions with third parties. The agent is authorized to do only what the principal wants. The agent can be placed in a position to exercise discretion, and can even enter into contracts that are binding on behalf of the principal. There are no limits on who can be an agent; for example, an employee can be an agent.

> **EXAMPLE.** A construction worker can be regularly employed to perform carpentry functions for his or her employer. This employee can also be granted the authority or power to purchase on the employer's account any tools or materials needed in the performance of the work.

A true agent is not self-motivated, at least not in terms of the agency agreement. The agent acts for the principal and not out of self-interest. In general, the agent can be empowered to do anything the principal can lawfully do. However, there may be some exceptions, including the following:

- Acts that are personal in nature or must be personally performed
- Acts that are illegal
- Acts that are immoral
- Acts that are opposed to public policy

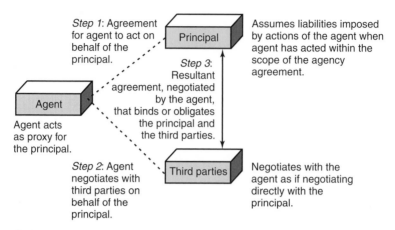

Step 1: Agreement for agent to act on behalf of the principal.

Assumes liabilities imposed by actions of the agent when agent has acted within the scope of the agency agreement.

Step 3: Resultant agreement, negotiated by the agent, that binds or obligates the principal and the third parties.

Agent acts as proxy for the principal.

Step 2: Agent negotiates with third parties on behalf of the principal.

Negotiates with the agent as if negotiating directly with the principal.

FIGURE 5.1
General relationship of the agent, the principal, and third parties.

THE PRINCIPAL

The principal is liable for all contracts made by the agent while the agent is acting within the scope of the stated and granted authority. In a like manner, the third party who is signatory to an agreement made with an agent is held liable to the principal for the contract. It is imperative that the agent act within the stated scope of authority. If an act of the agent is unauthorized, the principal will not be bound by it. However, if an act was unauthorized, it can become binding if the principal ratifies it.

The principal obviously should be very careful in selecting an agent. The principal is liable for acts of negligence committed by the agent if the agent operates within the scope of authority and in furtherance of the principal's business. This is true even if the agent acts in an irregular manner, or contrary to the explicit instructions of the principal.

The responsibilities of the principal to the agent should be clearly defined by the terms in the agency agreement. This agreement should include provisions concerning the reimbursement of the agent for services rendered, usually on a commission basis. Compensation is generally dependent on the success of the efforts of the agent. Greater success yields greater compensation; total failure generally means no compensation.

The principal is liable for the criminal acts of the agent if assent is given prior to their occurrence, or if they are part of the agency arrangement. Criminal acts, however, cannot be ratified.

THE AGENT

The agent must display complete loyalty and good faith, obey instructions to the letter, and not attempt to exceed the authority that has been granted. Furthermore, the agent is obligated to make a full disclosure of agency-related transactions to the principal whenever requested and when the agency is terminated. That is, all information acquired through the course of the agency should be communicated to the principal. All profits made from activities related to the agency agreement belong to the principal. Examples of agents include attorneys, brokers, auctioneers, and real estate agents.

The agent cannot compete with the principal, profit at the principal's expense, or do any self-dealing. For example, an agent who is authorized to make certain purchases for the principal cannot buy the items from himself or herself. This would constitute a conflict of interest or even fraud, and the principal could recover the full purchase price from the agent.

An employee of a company can be an agent for that company. An employee can also function as an independent contractor. The powers of the employee must be examined to determine if an agency arrangement exists. The decision in *Aisenberg v. C. F. Adams Co., Inc. et al.* (111 A. 591) provides some guidance about when an employee is effectively an agent. Mr. Aisenberg sold a variety of Adams products throughout Connecticut. In the course of his sales travels, Aisenberg was fatally injured. Aisenberg's widow sought workers' compensation benefits for his death. Her claim was based on the fact that he was an employee. She testified that the goods sold by her husband remained the property of Adams until they were purchased. The terms of the sales were also dictated by Adams, and Adams made subsequent transactions with the clients without Aisenberg's knowledge. Adams even provided Aisenberg with transportation in the form of a horse. She also claimed that Adams could have fired her husband at any time. Adams did not agree, claiming that Aisenberg was in fact an independent contractor. This argument was based on the fact that Aisenberg was in full control of determining where he would make sales, controlled his own work hours, was paid on a commission basis, and provided his own wagon. The court stated that the true test in deciding such a case is whether the employee has the "right to direct what should be done and when and how it should be done, the right to general control." In this case the court based its decision on the issue of general control. It stated that the terms of the sales and the proceeds of the sales remained in the control of the company. The method of payment was not relevant to the case, nor was the fact that Aisenberg used his own wagon. Thus, Aisenberg was determined to be an employee functioning as an agent for the company.

If an agent acts outside the scope of authority, the principal is not generally liable for those actions. However, if the principal ratifies such actions, the actions become authorized after the fact. It will then be as if the principal had originally authorized the actions. If ratification does not take place, the agent is personally liable for any actions taken outside the scope of authority.

If an agent is empowered to make purchases for the principal, the agent should make known the identity of the principal (the agent's client). Ignoring this identification can lead to legal problems.

In construction, agents are used to carry out various functions, frequently representing owners, contractors, suppliers, or subcontractors. As such, the role of the agent may be to design the project to be constructed; ensure that the performance requirements are met (the focus will be on quality, general supervision, monitoring time, monitoring costs, etc.); plan, direct, and manage construction activities (from the owner's point of view, e.g., CM work); and provide supplies for the project.

CREATION OF AN AGENCY

Ideally, an agency is created by an express contract. However, the contract can be implied. There are essentially three ways in which an agency can be created: by law or contract, by ratification, and by estoppel. As with other contracts, there must be a meeting of the minds, consideration, and the other requirements of a binding contract.

The Owner's Agent on a Construction Site

In construction, the owner's agent could be an architectural design firm, a professional construction manager, or a member of an engineering firm. On public works projects, such a person is referred to as the owner's resident engineer or site engineer. The engineer or architect acts as the agent for the owner; he or she has been selected as the owner's agent because of the specific skills possessed. As such, this professional is expected to be judicious in the exercise of those skills. If damage results through the negligence or lack of diligence of the engineer, the engineer will be liable to the principal (owner). For example, failure to inspect an important operation, or to enforce construction contract specifications, may cause the engineer to become liable.

If there is misrepresentation of the skills possessed when creating the agency, the engineer may be liable to the owner for breach of contract or fraud.

The engineer or other agents can be held criminally liable if injuries occur through negligence. The agent is not considered negligent for errors in judgment in most cases; however, this has not been true in several recent cases. Likewise, architects and engineers may be held liable for errors in the plans and specifications. Consequently, agents (engineers in particular) should have expert legal counsel and adequate professional liability insurance.

Generally, the owner exerts little control over the preparation of plans and specifications. This places the engineer or architect in the position of being an independent contractor. Thus, greater liability can be placed on the engineer's or architect's shoulders.

The Role of the Contractor

Owners of construction projects are generally aware of the implications of agency agreements. Consequently, they are careful to ensure that their contractors are not in positions of being agents, but rather that the contractors are independent. If the contractors are to be agents, their scope of authority must be clearly established.

In an agency, the principal controls and directs the methods and acts of the agent. The principal is responsible for acts and torts (civil breaches) of the agent that are within the scope of authority. If the principal specifies only the results to be obtained by the firm but has no control or directive role of the methods by which they are accomplished, the construction firm is an independent contractor. The owner is not obligated by the acts of an independent contractor. In construction, owners carefully draft contracts to make construction firms independent contractors. This means that the owner cannot take control away from the construction firm. The owner can specify only the final result, and cannot interfere with the methods and personnel (as with hiring) practices of the contractor.

Construction documents typically will be reasonably explicit in establishing the contractor as an independent contractor, as opposed to being an agent of the owner. One set of construction documents that was developed, adopted and finally released in 2007 by 22 leading construction associations, is known as Consensus-DOCS. The use of these documents has grown rapidly in the construction industry. While numerous documents have been issued, only a sample is included in the Appendix. For example, ConsensusDOCS 200 Owner/Contractor Standard Agreement & General Conditions (Lump Sum) clearly states that the contractor is in control and has responsibility for the construction means, methods, and so on. This provision states the following:

> §3.1.2 Contractor is responsible for construction means, methods, techniques, sequences and procedures. When Contract Documents give specific instruction for means and methods, the Contractor shall not be liable to the Owner for damages resulting from compliance with such instruction unless the Contractor recognized and failed to timely report to the Owner any error, inconsistency, omission or unsafe practice that it discovered in the specified construction means, methods, techniques, sequences or procedures.

The contractual arrangements must be examined to determine whether an agency exists. For example, it is sometimes assumed that a cost-plus contract automatically establishes an agency arrangement. This is not true in all cases, and the specifics of each case must be examined. In *Lytle v. McAlpin* (220 S.W.2d 216), a cost-plus arrangement had been established on a house remodeling project. The contract was between J. W. Lytle, the homeowner, and E. M. Carey, the contractor. Carey then engaged McAlpin to do some of the concrete and brick work for $535. After McAlpin was finished with the work, he submitted a bill for payment to Lytle. It was Lytle's intention that Carey would pay McAlpin for his work. Lytle identified Carey as having the directive to "engage and hire subcontractors and to furnish labor and materials necessary for and incidental to said construction." McAlpin filed suit when Lytle refused to pay. The suit was filed primarily to establish Carey as an agent so that a mechanic's lien could be levied against the house. McAlpin's argument was based on the fact that the cost-plus contract, established Carey as an agent, thereby obligating Lytle to pay for the work performed. Carey testified that there was a cost-plus contract, but that he was in complete charge of the work and Lytle never interfered with the process of construction. He was under Lytle's orders and could be dismissed at a moment's notice. The court decided in favor of Lytle, stating that the cost-plus contract did not automatically establish Carey as an agent. It also determined that although the owner reserved the right of general supervision and control over materials to be incorporated in the structure, this did not establish the contractor as an agent. Thus, McAlpin was barred from filing a mechanic's lien against the property.

The owner can exercise control over the quality of materials and workmanship in the finished project. This is accomplished through a general staff that does not supervise. The owner's agents must clearly focus on product quality, not on methods. Supervision is clearly excluded from the scope of authority of the owner's agent on most construction contracts.

The actions of the principal may also be considered in determining the existence of an agency. In such circumstances, an individual may be regarded as an agent in accordance with the apparent authority vested in the position held. Apparent authority often exists if another party is led to believe that the proper authority exists. This is not always a clear issue. For example, the architect often represents the owner on construction projects. In fact, the power of the architect is often considerable. For example, the architect may authorize payments to the general contractor from the owner amounting to millions of dollars in a single month. The contractor might then infer that the architect has almost absolute power on the project. The contractor might then regard a $10,000 change order authorized only by the architect as a valid change order. The contract language might state otherwise, however, as the contract might stipulate that all change orders are valid only if they constitute "agreement by the Owner." The contract must be read carefully in order that the scope of the architect's power as the owner's agent is fully understood.

In *Robert Payne Company v. J. W. Hill Construction* (387 S.W.2d 92), the issue of apparent authority was defined. Robert Payne, as general contractor, was constructing an apartment building in San Antonio, Texas. The superintendent in charge was S. W. Shelton, and he entered into a subcontract with J. W. Hill Construction to install certain driveways and designated parking areas for $2,397.50. The driveways and parking areas were work items included in Payne's general contract agreement with the owner. After the paving work was completed by Hill, it was discovered that water was draining onto an adjacent neighbor's property. Hill returned to the project and corrected the drainage problem at his own expense. Payne paid Hill $2,000, but refused to pay any additional sums requested. Hill Construction filed suit for $626.50, the unpaid portion of its original subcontract agreement plus expenses for the extra work performed. Payne essentially claimed that Shelton did not have the authority to enter into contracts on behalf of the company, and that the subcontract agreement was void. The pivotal issue concerned the authority of Shelton. In this case, the court determined that although there was no evidence of actual authority by Shelton to contract on behalf of Payne, Shelton did appear to have such authority. This was based on the fact that Shelton was the only superintendent Payne employed on the project who was constantly managing the project. Robert Payne had also testified that he had seen the subcontract. Since Payne knew of the subcontract and did not object to it, it was essentially ratified by Payne's inaction. Although not specifically addressed in the court decision, the fact that partial payment was made to Hill also signified that the subcontract was recognized as a binding agreement.

In *Ronald Reber v. Chandler High School District No. 202* (479 P.2d 852), the liability of the designer was tested. Reber was an employee of Verdex Steel and Construction Company, which had a contract for the construction of a gymnasium. The plans called for the construction of six "three hinged" steel arches for the roof support. Verdex fabricated and erected the steel arches, which were bolted together on the site before being placed on the support columns. Reber, a structural iron worker, was installing steel cross-bracing when the arches collapsed. He was seriously injured and brought suit against the school district. He argued that the plans and specifications were defective, and that there was negligent supervision of the steel erection by the architect, who was the agent of the school district. The school district argued that the plans and specifications did not specify a method or sequence of steel erection, and essentially implied that the sequence and method of erection were left to the sole discretion of the contractor. The court ruled in favor of the school district, stating that insufficient evidence existed to prove that the plans and specifications were defective. In addition, no evidence existed to show that the architect had seized any supervising control of the project.

TERMINATION OF AN AGENCY

It is quite easy to terminate an agency agreement. If an agency agreement is terminated, it is important that this be promptly communicated to all parties that have engaged in business with the principal through a particular agent. At this point, the agent must also communicate, by full disclosure to the principal, all the obligations that have been created by the agent. An agency agreement can be terminated by any of the following means:

- Death of the principal or agent.
- Destruction of the subject matter for which the agency was formed.
- Occurrence of a specified event.
- Fulfillment of the particular purpose of the agency.
- Bankruptcy of the agent or principal.
- Expiration of a time period set in the agency agreement.
- A development which makes the subject matter illegal.
- Mutual consent of the principal and agent.
- Unilateral termination by either party.

Unilateral termination is valid since the agency is a voluntary relationship and can be terminated by either party without breaching the contract. However, unilateral termination can lead to damages if the termination was made without a justifiable cause. If specific performance was part of the contract, the innocent party to a termination is entitled to damages, but the performance of the services cannot be forced.

CONTINGENT LIABILITY

Under the rules of *contingent liability,* an injured third party (not an employed worker) is not or should not be affected by a contract between two other parties. On a construction site, an injured third party can sue the owner under the premise that the owner is jointly or wholly liable. This may be the preferred route for the third party, as appropriate redress may not be obtainable or feasible from the general contractor. Of course, the owner will probably try to recover from the general contractor, or the owner may have a contract with the general contractor stating that the general contractor will cover such losses for the owner. Nevertheless, the third party essentially remains removed from the contract arrangement. In a like manner, the owner may be held liable, or have contingent liability, for damages sustained by a third party, but due in part to the actions of a subcontractor.

Contingent liability can place the general contractor in a vulnerable position. For example, a third party may be injured as a result of the work of a subcontractor. Although the subcontractor is independent and not an agent, the third party might sue the general contractor; that is, the general contractor can have contingent liability for damages caused by a subcontractor. Note that in some states contingent liability will be more narrowly defined than in others. In any case, the general contractor will probably seek redress from the subcontractor after the suit has been resolved with the injured party.

Essentially all parties on a construction site are considered to be independent. However, through contingent liability, the general or prime contractor may be responsible to the owner and third parties for the torts (civil crimes) of a subcontractor. Subcontractors are responsible to the prime contractor. For example, a subcontractor may perform work that damages the property of an innocent third party. That third party might go directly to the prime contractor for recovery of losses. The owner similarly may have contingent liability for injuries sustained by third parties, even if caused by the general contractor or a subcontractor.

If engineers or architects are judged to be independent, third parties can seek recovery directly from them. This can occur if actual supervision was performed on the job site. In fact, the third party may even be an employee of the contractor or the owner. The allegation of independence may be claimed successfully, even if little or nothing is said by the owner's representative about the method of work. As a result, the word *supervision* is a very sensitive one, and was initially replaced in standard form agreements by the word *inspection*. In more recent years, the trend has been to soften the role of the engineer and architect even more by using the word *observe* instead of *inspect*. Note that if the relationship between the owner and general contractor becomes one of an agency relationship, the owner has direct liability. The owner will try to avoid having this situation arise. To ensure this, many contracts specifically state that the general contractor has sole control of the construction activities, and that the general contractor alone will direct the "means, methods, techniques, sequences, and procedures" of the construction process.

DAY LABOR AGENCIES

The construction industry is large and needs the services of millions of workers. There are many aspects of construction work that make it less attractive to many individuals, including seasonal work, work performed in the elements, changes in the work place, and so on. During economic boom periods, when the demand for construction workers is high, construction firms frequently have difficulty hiring sufficient numbers of workers. Day labor agencies are one source of workers that has emerged over the years that is being increasingly used by construction firms. These agencies provide workers for all types of tasks in addition to construction, but construction is a major beneficiary of the services of these agencies. Day labor agencies are used by general contractors and specialty contractors to fulfill short-term labor needs, especially where lower skill levels are needed.

If a contractor is in need of a few workers for one or two days, it is often difficult to quickly locate the required number of workers. Even if such workers were located and hired, the contactor would be reluctant to dismiss them, knowing that the company might need their services again in a short while. This dilemma of locating workers to satisfy short-term needs is addressed by day labor agencies. An agreement with a day labor agency is a convenient way for a contractor to fill short-term labor needs.

Once an agreement with a labor agency is established, the contractor essentially has a line of credit with the labor agency. This is required before any workers will be dispatched to a contractor's job site. Once the line of credit is established, a contractor who needs one or more workers can request workers from the day labor agency with a simple telephone call. At this time, the contractor must give information on the number of workers needed and their anticipated tasks. The rate paid to the workers is determined by the day labor agency, and this will be influenced by such factors as whether the workers will be performing manual labor, or they will be working with tools. The hourly cost of workers to the contractor will be determined at the time the request for workers is made, and this will generally be confirmed with a fax message to the contractor. In addition to the wages, the labor agency will pay for the workers' compensation coverage, taxes, insurance, and other benefits. The cost of the workers to the contractor will vary with the skill level and the workers' compensation rate for the work classification, but this is taken care of with a single payment (typically weekly) to the day labor agency.

Day labor agencies serve as a labor pool for contractors. Workers who have signed up with a day labor agency may be asked to show up at the day labor agency early in the morning (ideally between 5:00 and 6:00 A.M.) and wait for a job opportunity for that day. When a contractor calls for a worker or workers, the terms are arranged with the contractor and the workers are dispatched. The workers may be transported to the construction site by the labor agency, or they may drive themselves if they have other means of transportation. The agency typically provides each worker with a hard hat, gloves, protective eyewear, hearing protection, dust mask, safety vest, and training (workers provide their own lunch). After workers are assigned to a project, the labor agency may make "spot checks" to verify the type of work being performed by the workers. The workers are typically paid at the end of each workday, after the contractor has confirmed the number of hours that have been worked. For the contractor, this arrangement is simple in that the workers can be hired at will, with minimal paperwork, and workers can be laid off whenever they are not needed.

Once the workers arrive at the construction site, the contractor is expected to provide them with instructions about what is to be done. This can present somewhat of a dilemma for the day labor worker. For example, who is the employer of the day labor workers? The day labor worker is paid by the labor agency, and the day labor agency also provides training and safety equipment to the worker; so the labor agency might be regarded as the employer. At the same time, the worker is given instructions about the work to be done by the contractor, and this might also be interpreted as a task performed by an employer. Normally, it would be inferred in the industry that the day labor agency is the employer, but this may not be clear to every worker (see figure 5.2).

There are instances when a worker might be viewed as being particularly productive and motivated. Contractors may wish to permanently hire such a

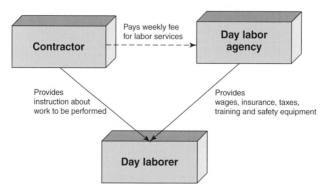

FIGURE 5.2
Relationship of day laborer to contractor and day labor
agency.

worker. This is possible, but the arrangement with the day labor agency will generally influence the cost of this hire to the contractor. One labor agency has a policy whereby workers can be hired permanently by the contractor, but the fee to the labor agency must be based on at least four hours of work. Other labor agencies may charge a much larger sum (possibly $2,000 or more) as a finder's fee. The specific agreement will dictate the cost to the contractor.

INDEPENDENT CONTRACTORS

Independent contractors, unlike agents, are held solely accountable for their actions. It is typical for the general contractor to be an independent contractor. This is often assured contractually by statements such as "the prime contractor shall be in sole control of the methods, means, techniques, sequences, procedures and safety precautions." Similarly, subcontractors are generally also established contractually as being independent contractors. Under an independent contractor arrangement, the liability for work performance lies with the independent contractor. Typically, with an independent contractor, there is no withholding of income taxes, there is no payment of the employer's portion of FICA, there are no workers' compensation payments, there are no requirements for overtime payments, there are no requirements for supervision, there are no requirements to maintain personnel records, and so on. These remain the responsibilities of the independent contractors. The benefits of employing an independent contractor are at risk if a worker has been misclassified as an independent contractor. There are no clear-cut distinctions that establish absolutely when a worker is an independent contractor and not an employee. When a contractor exhibits greater control over a worker, there is an increased probability that the worker is an employee and not an independent contractor.

When an arrangement is made with a general contractor or subcontractor, the contracting parties generally are quite careful so as not to undermine the independent status of the general contractor or subcontractor. It is commonly understood that the independent contractor status could be changed quite readily if the owner assumed a supervisory role with the workers of the general contractor, or if the general contractor started to supervise the employees of the subcontractor. To supervise the employees of an independent contractor transfers the liability for the work from the previously independent contractor to the supervising entity. Thus, it is quite acceptable to inform an independent contractor that a facility component was improperly installed, but it is not appropriate to direct the independent contractor on how the work should be done to correct the problem.

In many contracts with owners, the term "supervision" has been replaced by "observation" to help ensure the independent contractor status of the general contractor. When a party is involved in supervision, it is implied that there is a level of control over the construction task being performed and that this involves directing the process. This directly impacts the outcome of the effort. Observation, on the other hand, excludes this direct involvement with the work effort and suggests that the owner or owner's representative is merely acquiring information about the work being performed.

There are allegations that some contractors employ undocumented or unlawful workers as independent contractors. The allegations continue that these workers are often exploited, as they are not protected by workers' compensation, they are not guaranteed minimum wages, and in general they have no viable recourse to address grievances. Many states have begun to explore enacting legislation to penalize contractors who employ undocumented workers, whether as employees or independent contractors.

In a recent New Mexico case, it was also affirmed that the general contractor must establish that the independent contractors they hire are qualified to perform the work. This related to a sewer line installation in which Phillip R. Tafoya Bobcat and Dump Truck Service, an unlicensed contractor, was subcontracted to Chuby's Construction Inc. (193 P.3d 551). While making a tie-in for the sewer line, a trench cave-in killed Mr. Tafoya. His widow sued Chuby's Construction because they knowingly awarded the contract to an unlicensed contractor in order to get a lower price. The court acknowledged that Tafoya willingly entered into the contract and performed the work for which there was no license, making Tafoya a possible contributory factor in his own death. The court found that Chuby's was also liable as it sought out the unlicensed contractor. Chuby's had a duty to hire a qualified subcontractor or independent contractor.

STATUTORY EMPLOYEES

Statutory employee is a designation of employees that places them somewhere between regular employees and independent contractors. This is a designation that might be imposed on a worker when certain conditions are evident, or in certain occupations. There are four typical occupations where individuals are considered to be statutory employees, including full-time life insurance salespersons, agent drivers and commission drivers, traveling or city salespersons, and homeworkers. When a worker is declared to be a statutory employee, the employer must with-hold FICA taxes on their income, but is not required to withhold income taxes.

Workers that may be declared as statutory employees might have been initially regarded as independent contractors, or they may also have been employees of other firms. Statutory employees are employees of other firms who may become classified as employees of a firm for whom specific services are provided. This may occur when the work being performed pertains to the business, trade, or occupation of the firm. A subcontractor who merely supplies materials to a project for the general contractor is not likely to be regarded a statutory employee. The distinction of being a statutory employee is made on the basis of other tasks that might be performed in addition to the delivery of materials. In the case of *Bosley v. Shepherd,* a decision was made regarding the nature of the work performed. The subcontractor's employee, Bosley, was using a crane under the direction of the general contractor to distribute sheetrock into a building that was under construction. The court ruled that Bosley was not a statutory employee, as the use of the crane did not appreciably change his task of making material deliveries.

A statutory employee classification can arise when another firm has a high degree of control over certain workers, especially the employees of other firms. Day laborers provided by a day labor agency may become statutory employees under certain conditions. In essence the firm that exercises considerable control over the employees of others may become their common-law employer.

The designation of a worker as an employee, instead of an independent contractor, has serious implications in terms of workers' compensation and the regulations of the Occupational Safety and Health Administration (OSHA). It is especially where temporary employees or day laborers are utilized that this distinction must be clear; it is especially true when training is necessary. If specialized training is required to qualify temporary workers for specific tasks, and if the day labor agency is not aware of this need for training, the contractor utilizing those services might be deemed to be responsible for the training. Failure to provide the training will then make the contractor liable, as the worker may be viewed as a statutory employee. When a high degree of control is exercised over temporary workers, the contractor may inadvertently become a common-law employer of the temporary workers.

A host employer may create a statutory employee by exerting considerable control over the activities of an independent contractor and his or her employees. There are no discrete rules for the creation of a statutory employee arrangement, but consideration will include such factors as the skill required to perform the work, the location of the work, the source of tools, the duration of the working arrangement, whether the hiring party can assign the hired individual to other projects, the discretion of the hours worked per day, the method of payment, whether employment benefits are provided, and similar factors. One characteristic that is associated with statutory employees is that the work of a firm, such as a subcontractor, essentially mirrors the trade, business, or occupation of the host employer.

To avoid the creation of a statutory employee, there should be a clear contract that establishes the nature of the relationship between the various parties before work is begun. This agreement should clearly identify the party that is responsible for safety and health compliance. In the absence of clear contract language, courts will base the determination of the statutory employee status on the level of control exhibited over the employees of others. There must be diligence in adhering to the terms of the contract, if there is one.

The establishment of a statutory employee can also be a benefit to a contractor in some instances. This was shown in the case of *Elliott v. Turner Construction Co.* (381 F.3d 995) where a bridge manufacturer's employee was sent to assist in the installation of a steel pedestrian walkway across the Platt River near Invesco Field in Denver. Turner contracted with Mabey Bridge and Shore to rent a temporary bridge. Mabey provided an employee, Eugene Elliott, to assist in the installation. As the bridge was being launched by B&C Steel, Elliott went out on the bridge to stop the operation for fear that the bridge might fall. After he was out on the bridge, the bridge shifted and Elliott fell and suffered severe injuries. The court had to determine whether Elliott was a statutory employee of Turner. It determined that the work being done by Elliott was part of the work that Turner would have performed as part of its regular business. In other words, this type of work was typically done by Turner's own employees. Since erecting bridges was part of Turner's business, it would have been forced to use its own employees to supervise the work. Thus, Turner was ruled to be Elliott's statutory employer and was thereby protected from suit under the Colorado workers' compensation law. The court ruled that B&C Steel was liable for Elliott's injuries, but that Turner was immune from suit by a statutory employee.

REVIEW QUESTIONS

1. Describe the circumstances in which an employee can act as an agent.
2. An employee for a particular firm has not formally received authority to act as an agent. Describe the circumstances in which an agency agreement might be established by ratification. By estoppel.
3. Among the contracting arrangements discussed in Chapter 2, in which arrangement is an agency agreement typically assumed to exist?
4. Is the principal bound by the actions of an agent if the agent exceeds the stated scope of authority?
5. What are the implications to the owner if the general contractor is considered to be an agent of the owner?
6. Describe the circumstances under which a general contractor can inadvertently lose independent contractor status and thereby become an agent of the owner? Similarly, how can a subcontractor become an agent of the general contractor?
7. If the general contractor is considered to be an independent contractor, is the owner shielded from any lawsuits resulting from the actions of the general contractor?
8. Examine the American Institute of Architects (AIA) documents and the ConsensusDOCS that are included in the Appendix and determine the similarities and differences in the provisions related to the contractor's role during the construction of the project. Specifically examine AIA provisions §3.3.1 and ConsensusDOCS §3.1.2.

6

FORMS OF ORGANIZATIONS

THERE ARE ESSENTIALLY three forms of business organizations: proprietorships, partnerships, and corporations. Each type is found in the construction industry.

PROPRIETORSHIPS

Proprietorships are firms owned by an individual (figure 6.1). The structure of the organization is whatever the owner wants it to be. No formal documents are required to establish a business of this type. Such a business also can be discontinued at will at any time. The distribution of profits is very simple: The owner receives the profits as if the firm and the individual were the same (in the eyes of the law, they are the same). The owner pays income tax on the company profits as personal income.

All the tax and other liabilities incurred by the company are the responsibility of the owner. The proprietor remains his or her own boss and is personally responsible for the liabilities incurred by the business. The individual and the proprietorship are not viewed as being separate under the law.

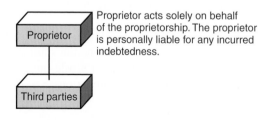 Proprietor acts solely on behalf of the proprietorship. The proprietor is personally liable for any incurred indebtedness.

FIGURE 6.1
Proprietorships and third parties.

PARTNERSHIPS

A partnership is an association of two or more persons to carry on a business (figure 6.2). Each person entering into a partnership usually has something unique and of value to contribute to the firm. Such contributions can include money, special skills or talents, equipment, land, facilities, or other assets, all of which will help the company achieve a common objective or course of action. In general, the partners will have joint control of the firm since their resources have been joined. This pooling of resources enables the partners, as a unit, to undertake projects of a larger scope and volume. The partners share in the management of the firm, often with the intent of utilizing each partner's strongest capabilities.

The profits of the company are divided among the partners in some way. Usually this apportionment is related to the value of the contributions of each partner, whether translated in terms of technical expertise, annual financial performance, or the value of the original contribution. It is simplest if each partner has contributed exactly the same amount as the others. If the distribution of profits is not established when the partnership is formed, it is usually assumed that the profits will be shared equally among all the partners.

A partnership pays no income tax. It is not considered a separate legal entity apart from the individual owners. At most, a firm that is a partnership must file an information form. The burden of income tax falls on the partners. Since it is not a separate legal entity, the partnership cannot file suit in its own name; the suit must be filed in the names of the individual owners or partners.

Suppose a partnership pays each partner a salary. In addition, some of the company profits are reinvested in the firm to permit greater growth potential. Each partner will then be required to pay income tax on the salary that was earned plus a prorated share of the profits that were left in the company. This is done because the partnership is not a separate legal entity. In fact, the partnership cannot own real property in its own name. However, the partnership can own other types of property.

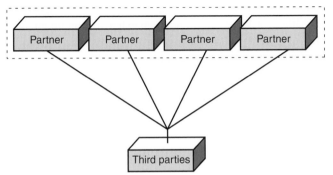

Each partner is an agent for the other partners. Partners are liable for any debts created by partners when acting for the partnership.

The partners remain as individuals under the law.

The partnership is not viewed as a separate legal entity apart from the individual partners.

FIGURE 6.2
Partnerships and third parties.

The Formation of a Partnership

A partnership can be formed by an oral agreement, but it is advisable to form it in a written fashion. Written articles should be used to set forth the basic provisions of the partnership, clearly stating the rights, responsibilities, and obligations of each partner. Each partner is an agent of the other partners and can cause the partners to bear considerable liability, as contracts entered by one of the partners will obligate each of the other partners as well. Each partner should be familiar with the other partners. There must be a great deal of trust and confidence in the other parties, or the partnership may not function as it was intended. This trust must be real. Each partner is entitled to the use of the partnership property for any partnership purpose. Thus, each partner is a fiduciary of the other partners. Conversely, each partner must account to the other partners for any personal profit realized through private use of the firm's assets.

In *Vance v. Ingram et ux.* (133 P.2d 938), the terms of a partnership were rigidly enforced. Kathryn Vance, a skating rink operator, entered into a contract with L. P. Ingram for the construction of a skating rink for $18,000. Informally, the parties agreed that Vance would provide $9,000, mostly through outside financing, and that Ingram would finance the remaining $9,000. It was agreed that Vance would make payments of $150, plus interest, to Ingram on a monthly basis. Before construction began, Vance was informed that no loan would be made to her unless Ingram was listed as a partner in the operation. Vance reluctantly entered into a partnership agreement with Ingram after he assured her that he had no interest in operating the rink. Basically, Ingram was included as an equal partner in exchange for providing the necessary funds for the labor and materials to complete the rink. The financing of these materials and labor was deemed to be consideration for transfer of the half interest. The partners were to share equally in the profits and losses of the business until the partnership was dissolved.

Ingram's interest in the partnership could be purchased at any time by Vance. The purchase price was stated as being $9,000 plus 6 percent interest annually from the date of completion of the building. The rink proved to be successful. Over a period of three years, Vance made payments to Ingram that totaled more than $4,000. Vance then decided to exercise her purchase option. She felt she owed the unpaid balance of the $9,000, while Ingram contended that the option required the full $9,000. Although Vance contended that her arrangement with Ingram was essentially that of a debtor and creditor, the court decided that a unilateral purchase option in the contract did not negate the partnership or the contract. The fact that payments were made to Ingram did not prove that they should be applied to the purchase price. The court ruled that Ingram's investment in labor and materials was sufficent to justify the rate of return being realized.

The Role of Each Partner

Each partner is an agent of the partnership and has the power to enter into binding contracts in the name of the partnership. Each partner is a general agent, one who is empowered to transact all the business of the principal. However, limitations concerning such powers can be placed on any or all partners. This must be stated in the partnership agreement. In summary, each partner is a recognized member of the firm, is active in management, and has unlimited liability to creditors.

Being a partner is not without a great deal of risk. Each partner assumes unlimited personal liability to third parties for the full amount of all debts of the partnership. This is true regardless of that partner's contribution to the formation of the partnership. The partners are jointly and severally liable for company debts. Furthermore, each partner is liable for any acts of fraud or misrepresentation perpetrated by another partner in the ordinary conduct of the partnership's business affairs. Each partner is a principal for himself or herself and an agent for the other partners for agreements made within that partner's actual or apparent authority.

Each partner can act for the entire partnership. If a partner withdraws from the partnership, that partner will be personally liable for the partnership obligations incurred up to the date of the withdrawal. This is another reason why partners should be carefully selected. A partner cannot sell, assign, or mortgage an interest in a functioning partnership without the express consent of the other partners. On the other hand, a new partner in a firm is not responsible for any indebtedness generated by the partnership prior to that partner joining the firm.

The case of *Carey v. Wilsey et al.* (185 P. 600) demonstrated the role of a partner. This case concerned the purchase of timberland. Carey and Wilsey formed a logging partnership. Carey, an experienced logger, would handle the logging, and Wilsey would handle the marketing of the logs. They set up their partnership in 1917 and purchased the timber on several parcels of land on Vashon Island. Wilsey visited many owners of timberland on the island. One landowner, Frasch, offered to sell his 10-acre tract for $100. Before payment was made to Frasch, the partnership had already sold about one-third of the timber harvested on Frasch's property. A few months later, but before the Frasch tract had been purchased, the partnership was dissolved by mutual agreement. Immediately thereafter, Carey paid Frasch $100 for the property and had the deed registered in his son's name, presumably to insulate the property from the defunct partnership. Carey promptly filed suit against Wilsey for $900, the alleged value of the timber harvested on the 10-acre tract. The court had to decide if the land had been purchased by Carey or by the partnership. The court ruled that the land purchase by the partnership was effective when harvesting began and was not predicated on the actual payment for the land. The land had been purchased under the partnership agreement in spite of the fact that Carey now held the receipt of payment for the land. Carey, in essence, had fulfilled the obligation of the partnership to pay for the land. This case further illustrates the need for partners to be trustworthy.

Limited Partners

The foregoing comments were primarily focused on general partners or general agents. There is another type of partnership that is commonly used: the *limited partner.* A limited partner generally contributes cash or property to the business (partnership) and shares in the profits and losses. However, a limited partner provides no services and has no vote (or voice) in matters of management.

The liability of a limited partner is indeed limited. This partner's liability for partnership debts is no greater than the amount of the investment made in the firm. This limitation occurs only because the limited partner is not an active member of the firm.

Since this is essentially a means of obtaining capital, the limited partner's interest in the partnership can be assigned. At the same time, the partnership does not dissolve if the limited partner dies. Of course, if there is only one general partner, the death of the only limited partner can make the organization a proprietorship.

Limited partners can be very useful in a firm. This is an excellent way to raise capital for the enterprise while not spreading around too much managerial strength. Too many general partners can lead to problems involving company control and decision making. In addition, it may be easier to borrow from a limited partner than to borrow from a lending institution (banks, finance companies, etc.). The intended limited partner should be careful before entering into such an arrangement. If the company is operating on a shoestring, it may be advisable for the outside person simply to make a direct loan to the partnership. The risk must be assessed before the decision can be made. A limited partner has a lower priority in terms of payment if the company is dissolved. However, the returns for the limited partner may potentially be much greater.

The establishment of a partnership alters the responsibilities of the various partners from that point in time; that is, each partner becomes an agent for the other partners. In *Dwinell's Central Neon v. Cosmopolitan Chinook Hotel* (587 P.2d 191), the court's decision was based on the actual point at which the partnership was formed. The Cosmopolitan Chinook Hotel, which was owned by a partnership, had contracted with Dwinell's Central Neon for the lease-sale of neon signs. The contract stated that Dwinell's would provide several signs and maintain them over a number of years. A unique payment provision was added which stated that the regular payments would be accelerated if the hotel began to have financial difficulties. When the contract was signed, the partners had been in the process of converting their association with the hotel to that of limited partners. Three months after the contract was signed, the hotel partners officially filed a certificate of limited partnership with the Yakima county clerk. After another eight months had passed, the hotel failed to make timely payments to Dwinell's, prompting a suit to enforce the accelerated payments provision of the original contract. The partners of the hotel felt that the indebtedness was not their responsibility since they had become limited partners. The Washington statute (RCW 25.08.020) requires that partnerships "file for record the certificate in the office of the county clerk of the county of the principal place of business." The court ruled in favor of Dwinell's on the grounds that the limited partnership was formed on the date of the filed certificate, regardless of its publicly understood intentions of becoming a limited partnership. Thus, the owners were each regarded as general partners of the hotel venture and were therefore each liable for the indebtedness to the sign company.

Silent Partners

A *silent partner* is a person who is a partner in a firm but remains unknown to the public. This partner obviously cannot be active in the management of the regular business affairs of the company. Individuals who are silent partners may have a variety of reasons for staying anonymous.

Dissolving a Partnership

A partnership can be dissolved by any of several means, including the following:

- The death of a partner constitutes automatic dissolution. This can be circumvented by making provisions that the business will continue and the surviving partners will purchase the decedent's interest or that the heirs will be active in the partnership.
- Bankruptcy.
- A duration provision in the articles of the partnership may stipulate a time period after which the partnership will automatically be dissolved.
- The mutual agreement of the partners that the partnership should cease to exist.
- The partnership can be dissolved if one of the partners is judged to be insane.
- The partnership can be dissolved if one of the partners decides to withdraw.
- A court decree may be issued that will dissolve the partnership.
- A change in the partnership will constitute a dissolution of the original partnership.
- Expulsion of a partner for just cause may terminate a partnership.

If everything went according to plan, every firm would be successful in its objectives. However, failures do occur, particularly in the construction industry. Since failure is always a possibility, it is important to recognize or understand the priority with which the debts of the partnership will be paid if or when dissolution occurs. This priority is as follows:

1. Outside creditors are paid first. In fact, if the assets of the partnership are insufficient to pay off the outside creditors, the partners must come up with the deficient amount out of their private or personal funds.
2. The next priority is the repayment of loans or advances to the partnership made by any of the partners above and beyond the capital contributions stated in the articles of partnership.
3. The next priority is to return each partner's capital investment. The partnership agreement may stipulate the priority that different partners have if insufficient funds exist to repay all the partners.
4. If anything is left, the profits are distributed according to the partnership agreement.

Joint Ventures

Joint ventures are very popular in the construction industry. They are a special form of temporary partnership. *Joint ventures* are essentially the combined efforts of two or more construction firms to build a project. It is a special-purpose partnership that can be between two or more proprietorships, partnerships, corporations, or even combinations of these types of organizations. Many joint ventures are established with one project in mind; that is, the joint venture dissolves when the project is completed. Such temporary mergers make it possible for companies to undertake larger or more complex projects. Construction firms consider joint ventures when projects place unique constraints on them. Some compelling benefits or reasons for joint venturing are as follows:

- Increase the bonding capacity for the job by combining the bonding capacities of two or more firms.
- Gain familiarity with the local labor market by enlisting the services of a local firm.
- Gain familiarity with a foreign government by combining efforts with a firm with the requisite experience.
- Increase capabilities by combining and sharing construction knowledge and expertise.
- Pool equipment to expand the capabilities to perform the required work.
- Gain familiarity with unique construction materials by joining forces with a knowledgeable firm.
- Compensate for the deficiencies of in-house personnel.
- Increase the available capital to undertake a project.

Joint venturing on a project can serve a company in many ways. With a joint venture, smaller companies may successfully compete against the giants in the industry. Larger firms may find advantages in sharing risks through a joint venture agreement. For some, the joint venture will permit a company to undertake a project for which it simply does not have adequate internal resources. The resources gained through the joint venture may include capital, bonding capacity, equipment, or some required knowledge. Some joint ventures may be established simply as a means for one company to penetrate a particular market. Such a company would utilize the joint venture as a means of learning from its partner and developing the required expertise. This expertise might be too costly or impossible to obtain if a company decided to pursue it on its own. After the successful completion of such a joint ventured project, the company might feel comfortable in singularly pursuing work on projects in a particular locale or projects of a particular type that would otherwise not have been considered. For some joint ventures, the motivation may be strictly financial. One company may be sought as a partner simply to gain access to capital or its bonding capacity. Such a partner would not be expected to play a strong role during construction. The company with the capital or the bonding capacity is simply paid for the use of its resources.

Some joint ventures have been created to minimize or eliminate competition. While this certainly may be a reason for establishing a joint venture, this is not a motivation that should be entertained lightly. Such practices may be considered unlawful if they constitute a "restraint of trade" and may result in serious legal entanglements for the participating firms and their principals.

Just as in the formation of any partnership, the joint venture partners should have a clear understanding of the duties, rights, and responsibilities of each of the partners. This should be formally established in writing. One of the partners must be clearly designated as the "sponsor" and spokesperson for the joint venture. Other issues must also be clearly established in the agreement. This agreement should include such matters as the managerial role to be played by each partner; the financial investments of each partner; the sharing of the profits (and possible losses); limitations of liability; ownership and use of equipment; means to be followed to resolve disputes; access of partners to proprietary systems or methods; and the conditions under which the joint venture will be terminated.

CORPORATIONS

A *corporation* is a legal entity (artificial tax-paying individual) created to act as an individual while protecting the owners or stockholders in the firm. A corporation is owned by one or more individuals who form an independent body or unit under a special or corporate name (figure 6.3). Corporations have certain privileges and duties which make them different from both partnerships and proprietorships. Corporations are authorized to do business, own and convey real property, enter into binding contracts, and incur debts in the name of the corporation. A corporation can bring suit and can be sued in its own corporate name. It is a private business organization that limits the liability of its owners (the stockholders); that is, the owners are at risk only to the extent of their investment in the corporation. It has perpetual life or perpetual succession. It can raise capital easily and can be owned by large numbers of people without jeopardizing business operations. The owners pay income tax only on the profits actually paid to them.

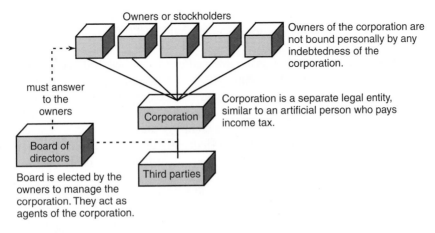

FIGURE 6.3
Corporations and third parties.

Formation of a Corporation

A corporation can be formed by a person or a group of persons who wish to conduct business as a corporation. Since corporations have unique privileges, this formation is more formal than that of the other business forms. The formation of a corporation (called incorporating) is governed by the articles of incorporation and the bylaws under which it will be operated. In addition, state laws place regulatory constraints on corporations. Such laws have to do with taxation, monopolies, and rights in general. It is not uncommon for a proprietorship or partnership to change to a corporation to take advantage of the unique features of corporations.

Most corporations are private, established to conduct business in the open market, usually a competitive market. Some corporations, however, have a more direct interface with the public, such as telephone companies, water companies, and power companies; they have characteristics of both private and public corporations. They almost, and sometimes literally, function as monopolies. Such companies are referred to as quasipublic companies and are subject to strict regulation because of their direct impact on the public. Although they conduct business for private gain, they frequently are granted the power of eminent domain.

Private corporations may or may not have capital stock. Those without capital stock are generally charitable, social, and educational organizations. Most, however, are stock corporations that issue shares of stock and distribute dividends (profits) to the owners (stockholders).

Stockholders

The owners of a corporation are called *stockholders* or *shareholders.* Individually, these owners are not agents of the corporation. The actions of the owners are essentially limited to voting at company meetings. The only exception occurs when an owner is also an employee. The stockholders elect their agents, often called directors. Because they own shares, the owners share in the ownership of the company. They do not own the corporate assets, which belong to the corporation through its own legal title.

Shares of stock are freely transferable and can be owned 100 percent by one person. In the transfer of stock, a requirement may be added in the bylaws of the company that states that before any transfer of stock can occur, the stockholder must first offer the shares to the other shareholders. Various types of constraints can be added. For example, a construction company may issue shares of stock for distribution among the employees. This is frequently done to entice the workers to stay with the company and to give them more of a feeling of being part of the business. Such a company will very likely place a constraint on the ownership of the stock so that the stock will have to be sold back to the company if termination of employment occurs.

There are various types of shares. *Common stock* is typically the type referred to when shareholders are discussed. Common stock entitles the owner to one vote per share. The value of common stock is perhaps the best measure of the strength of a corporation, as its value rises and declines on a daily basis. If profits are high for a particular year, the firm may pay large dividends to the stockholders. The long-term health of a company is sometimes measured by the number of consecutive years in which the stockholders received dividends. Typically, dividends are not paid if no profits were earned.

In addition to common stock, a company may issue preferred stock. *Preferred stock* is a more stable type of stock in that it does not generally have wide variations in pricing, as is the case with common stock, because the rate of return is fixed and is guaranteed as long as a profit is made. An added security for preferred stockholders is that the dividend is paid prior to common stock dividends. If there are no profits with which to pay preferred stock dividends, the indebtedness to the stockholders will be carried to the next year. If the subsequent year is more profitable, the stockholders will receive dividends for the no-profit year and for the profit year. This is often a nonvoting type of stock.

Another way of generating capital is for a firm to sell bonds. Bonds are generally sold in $1,000 denominations with a specified interest rate and a stipulated maturity date, often 20 or 30 years in the future. Bond interest is paid before the payment of any dividends to the stockholders. Thus, bonds tend to be fairly low in risk provided that there is considerable capitalization based on the sale of stocks. Of course, a business with high inherent risks will not be a guarantee of security regardless of the ratio of the value of bonds to that of stock. Bonds are often rated on the basis of the risk they present to investors.

Essentially all stockholders have certain rights and privileges. These are as follows:

- They have the power to enact bylaws.
- They have a right to share ratably in declared dividends; that is, they are entitled to their due proportion as determined by the number of shares owned.
- They have a preemptive right that states that if more shares are authorized for issuance, they will have the first opportunity to purchase them. They are guaranteed the right to maintain their relative proportion of ownership in the company.
- They have a right to participate in the distribution of the assets of the company if liquidation occurs.
- They have immunity from personal liability for corporate debts.
- They have a right of a reasonable inspection of the corporate records.
- They have a right to file justifiable claims against the corporation.

Corporate Profits

Profits of corporations are taxed by the U.S. government. In the eyes of the government, a corporation is an artificial tax-paying person. The tax structure for corporations is not the same as it is for individuals, however. It is more simple than individual rates. After the tax has been paid, the directors must decide what to do with the remaining funds (after tax profits). One option is to reinvest the funds in the company; another option is to give the funds to the owners in the form of stock dividends. Generally, a combination of the two is preferred.

Management of Corporations

The business activities of a corporation and the management of the company's property are controlled by its board of directors. The members of the board need not be shareholders. This board must operate within the firm's bylaws. The directors are agents of the company, and their actions can be ratified by the owners (by votes cast in person or through a proxy) if those actions are beyond the scope of their authority. As a rule, the directors act as a unit through their board meetings.

The directors serve the company. They are not supposed to use their positions in the company for personal gain. Consequently, a contract drawn between the company and a director can be voided even if the deal seems fair.

Corporate Powers

A corporation has legal powers conferred to it in the certificate of incorporation. In addition, a corporation has certain implied powers, including the following:

- It can buy and sell real or personal property.
- It can sue and be sued in its own name.
- It can make contractual commitments.
- It can invest its funds.
- It can lend or borrow money for business purposes (loans by the corporation to its directors or officers are usually prohibited).
- It can make bylaws.
- It can appoint officers and agents.
- It can distribute profits to its owners (dividend payments to stockholders are prohibited if this will make the company insolvent).

If a corporation is dissolved, the outside creditors are paid off first, then the bondholders, and lastly the stockholders. The dissolution of a corporation is accomplished by the surrender or expiration of its charter. The corporation simply ceases to exist. This can be done with the approval of the stockholders. It can also occur through a natural dissolution that was established in the charter, as in reaching an expiration date. The worst case is an involuntary dissolution, which occurs through the directive of a court order or through bankruptcy.

Ultra Vires Activity

Corporate contracts that go beyond the scope of a corporation's implied or expressed powers are referred to as *ultra vires contracts*. Ultra vires activities are those that go beyond the powers granted to the corporation by its charter or articles of incorporation. Ultra vires actions may occur inadvertently if a company principal mistakenly enters an agreement for which proper authorization has not been granted. In some cases, the principal may become overzealous when negotiating for the company and make commitments on behalf of the company that the principal is not empowered to make. Essentially, ultra vires activities are those in which the agents of a corporation are acting outside of their authority. For example, a firm may have been incorporated as a nonprofit organization. If the firm enters into an agreement with the intent of making a profit, that would be regarded as an ultra vires contract.

If an ultra vires contract is counter to statute or public policy, it will probably be unenforceable. If it is not illegal but merely "unauthorized," the question is more complex. Some precedent for the state will indicate the likely posture of the courts in certain situations. If the corporation has exceeded its authority in making a contract, but both parties have fully performed, the courts generally will not interfere. If the contract is entirely executory on each party's side, neither party can enforce the contract. If it is only partially executory (partly executed), the party that received the benefit cannot claim ultra vires as a defense when pressed for counterperformance.

Holding Companies

A *holding company* is a "supercorporation" in that it is created to hold such a dominant interest in one or more other companies that it can prescribe, through its voting power, the management policies of those other companies. This controlling interest is gained through stock purchases of the company to be controlled. The company that is being controlled is known as a *subsidiary* of the holding company.

A holding company is essentially a big stockholder. As such, it has the normal rights and privileges of individual stockholders. The voting for the holding company is done by proxy. It is not essential that the holding company have over 50 percent of the stock in order to control the dictation of management policy; the firm needs only to own a controlling interest. A controlling interest is usually much less than 50 percent because a smaller percentage, such as 20 percent, is often sufficient to achieve the holding company's objectives.

Obviously, holding companies are powerful. As a result, governments scrutinize their actions very carefully. A holding company cannot violate antitrust laws. Large firms are frequently accused of operating on the fringes of such laws.

Of course, a holding company can be established quite by accident or without intention. For example, a corporation may decide to invest money in another firm. If the investment looks lucrative, the company may buy enough stock that it obtains a controlling interest. However, an investment company has its primary purpose of holding securities in other companies that look profitable. This is done for investment, not control, purposes.

Subchapter S Corporations

Another form of corporation is the *subchapter S corporation* or *S corporation,* which offers some unique advantages for firms fitting certain criteria. The advantages often apply well to closely held construction businesses. In many ways, the S corporation is like a typical corporation in that there is limited liability for shareholders; but the big difference is that it generally does not pay income taxes as a separate legal entity. S corporations have been referred to as incorporated partnerships. All company profits are taxed directly to the shareholders. Since the S corporation earnings are not taxed prior to distribution to the shareholders, there is no double taxation as is common for typical corporations. A new firm or an existing corporation can become an S corporation provided that certain requirements are met. These include that there be no more than 35 shareholders, that the shareholders be U.S. citizens or naturalized citizens, that the shareholders be individuals (not corporations or partnerships), and that there is only one class of stock. These are the major criteria, but others may also be applicable. The nuances of advantages of an S corporation may also vary from state to state, especially in states with a state income tax. Naturally, expert counsel should be sought prior to becoming an S corporation.

While a regular corporation can avoid some double taxation by taking earnings and applying them as compensation to the "owners," this practice may attract the attention of the Internal Revenue Service if they fluctuate considerably from year to year. This dilemma is avoided with the S corporation. It is common for S corporations to distribute more of the earnings than is typical in a regular corporation. This practice may be a concern to a surety when a request for payment and performance bonds is made. Creditors may also be concerned with the distribution of earnings practice. The concerns of the surety and creditors are that the firm may have insufficient funds to satisfy a possible indebtedness. These concerns may be reduced and essentially eliminated if the S corporation makes a commitment through a formal agreement to retain a stipulated level of cash and other liquid assets in the S corporation. If assurances are provided that debts can be serviced, the S corporation should be able to function well with the surety and creditors. There are some unique advantages for establishing an S corporation and these should be carefully evaluated.

REVIEW QUESTIONS

1. What are the essential differences between a proprietorship and a partnership?
2. Give reasons why a joint venture might be established for the construction of a particular project.
3. What are some of the advantages of being a limited partner instead of a general partner? What are some of the disadvantages?
4. Discuss the pros and cons of being a limited partner instead of being simply an outside creditor.
5. Who are the agents in a typical corporation?
6. Give an example of an ultra vires contract.
7. What is the essential difference between common stock and preferred stock?
8. Discuss the advantages of a partnership compared with a corporation.
9. Discuss the disadvantages of a partnership compared with a corporation.
10. What is an advantage of an S corporation when compared with a typical corporation?

7

CONTRACT DISPUTES AND TORTS

COURT CASES INVOLVE either criminal or civil suits. Criminal suits relate to violations of the law, and civil suits relate to disputes concerning contract matters and torts. On construction projects, civil suits are more common.

Disputes can be generated in virtually any environment; indeed, conflict or an argument can occur whenever two individuals try to work together. A conflict or argument has been defined by Leonard Neubauer as "a discussion which has two sides and no end," as reported in *The Dictionary of Quotable Definitions,* edited by Eugene E. Brussell, Prentice-Hall, Englewood Cliffs, NJ, 1970. As this definition implies, conflicts are not necessarily clear-cut issues of right and wrong. Two individuals may simply have differing interpretations of an issue. Contract disputes and torts involve matters of conflict for which no laws have been specifically enacted.

CONTRACT DISPUTES

On construction projects, particularly projects in the private sector, matters of statutory law are seldom the root causes of contract disputes. Since contracting procedures in the public sector are subject to compliance with more laws, issues of law do arise in this area. However, these disputes are often outside the contract, involving the procedures related to the award of the contracts.

While disputes about the interpretation of almost any provision in a contract can arise, major disputes that ultimately are resolved in the courts tend to relate to recurring issues. These disputes often concern topics such as changes, differing site conditions, delays, and payments. When standard provisions are used in contract documents, there is greater consistency in the way the courts view such provisions. However, when provisions have not been "court-tested," the disputing parties have fewer court precedents to rely on to provide assistance in resolving the matter.

How do courts interpret provisions on which the contracting parties cannot agree? In many cases the courts will look at previous decisions involving similar conflicts. In some instances past courts have ruled differently on seemingly identical or similar disputes. These different rulings may stem from the fact that the courts were in different states or jurisdictions, or the fact that the decisions were made in a different time frame. When differing past decisions are presented, the court may decide which cases are most relevant to the current conflict. The court may also discount prior decisions and base its decision solely on the merits of the current case.

What is the general philosophy of the courts when a dispute relates to different interpretations of the same provision? Several approaches can be taken. One approach is for the court to decide on the obvious intent of the provision in question. The intent may be based on the wording of other provisions within the contract documents, or may be inferred from the actions of the contracting parties. Fairness is not the overriding objective of the court; rather, its intent is to give an accurate face-value assessment of the meaning of the provision in dispute.

Is the obvious intent the first issue the court tries to resolve? The nature of the dispute and the nature of the provision may be crucial to the approach taken by the court, particularly if the provision is exculpatory in nature. An *exculpatory provision* is one in which one party, typically the contractor, is asked to assume liability that would not otherwise be assumed. That is, exculpatory provisions are typically ones in which the owner contractually shifts liability or responsibility to the contractor. Such provisions are valid in many instances; however, the courts tend to give such provisions greater scrutiny. In general, the courts try to interpret exculpatory provisions very narrowly or literally. In some cases the strict interpretation of exculpatory provisions has rendered them virtually ineffective. *Strict interpretation,* as used here, means that the court tries to interpret the provision, as much as it can within reason, against the party that seeks protection under the provision or against the party which drafted the provision. Thus, while exculpatory provisions are often found to be valid, if they are not drafted with care, the courts may find weaknesses or loopholes in them.

A strict interpretation of contract language was demonstrated in *U.S. Industries, Inc. v. Blake Construction Co., Inc.* (671 F.2d 539). Blake was the general contractor on a construction project at Walter Reed General Hospital in Washington, DC. The steel work was the responsibility of Blake, and U.S. Industries (USI) had a subcontract for the mechanical work. The subcontract agreement contained a provision that stated that the "Contractor shall not be liable for any damages that may occur from delays or other causes on the part of other contractors or subcontractors involved in the work, or the furnishing of materials." The work to be performed by USI could not begin until the steel work was completed. The steel delivery was delayed, causing a considerable delay in the mechanical work. USI sought to recover delay damages from Blake. When Blake refused to pay, USI filed suit. Blake sought protection under the subcontract provision, contending that the delay was caused by the steel supplier, who was essentially a contractor. The court made a narrow interpretation of the provision and concluded that the steel supplier was not a contractor but a supplier. Thus, the court found for USI, concluding that a supplier is not the same as a contractor. If delays caused by suppliers were to be covered, the provision should have specifically included suppliers.

The case of *John F. Miller Company, Inc. v. George Fichera Construction* (388 N.E.2d 1201) gives some indication of the strength of contract provisions. This case involved the plumbing work on a project, subcontracted to Miller, that was designed in such a way that it did not comply with state plumbing codes. Miller's bid was submitted with the assumption that a different piping system could be substituted that would be in compliance with the codes. When construction work started, Miller refused to install the plumbing system as originally specified. This delayed construction, and Fichera sought damages. Fichera claimed that Miller should have mentioned the nonconforming specification before bidding. Even if the subcontractor was planning to use a system in accordance with code requirements, it still had to provide a system that was "equal" to the one specified. Fichera further argued that materials that can generally be substituted refer to any "article, assembly, system, or any component thereof," but Miller's substitute was more extensive, involving changes in the size, number, and location of fittings, pipes, and vents. In addition, the supplies were made of materials that were not specified, and the subcontractor's design was never approved by the owner. Miller contended that there was no obligation to perform work not conforming to code. When the substitute was mentioned to the architect, the architect refused to approve the system without the submittal of shop drawings. Miller felt that this was work an architect should do.

The court based its decision on the wording in the contract. It ruled against Miller, stating that Miller could have satisfied the code by making a few simple changes. Even if Miller had submitted a superior design, he had no right to ignore the general conditions or specifications of the contract. In other words, a party to a contract is not above complying with the contract no matter how just his or her cause may seem. The contract clearly stipulated the procedure to be used for making modifications, and this procedure had to be followed. Thus, the decision was not based on the quality of the design submitted by Miller, but on the fact that Miller failed to follow the clearly outlined procedures for making substitutions.

TORTS

Torts are disputes that relate to matters not addressed by statutory law or contract obligations. Torts are wrongs committed against others that do not involve contracts. Since these wrongs are not addressed by law or by contract, common-law interpretation is often required. These wrongs or breaches of duty may stem from injury or damages incurred by one party as a result of the action or inaction of another party who had a duty to prevent the injury or damage. A tortuous act is often one that violates a social norm. A tort can result from a specific action or can be caused by failure to act. In most cases a tort is an offense against a person that does not involve a crime or the violation of a law. An action can, however, be both a tort and a crime. No tort actions take place unless the injured party seeks redress. For a tort to occur, the following conditions must be met:

1. One party owes a duty to another party.
2. That party does not conform (breach in the performance of that duty) to the standard.
3. The second party is harmed by the act or failure to act.
4. There is a clear causal relationship between the act and the harm that results.

The breach of duty could result from failure to act properly in the performance of a specific duty, or it could involve failure to act when there is a duty to act. The party to whom a duty is owed must be damaged. Damages may include physical injury, destruction of property, and defamation of character. It must then be shown that the damage is a direct consequence of the breach of the duty. Examples of torts include the following:

- Defamation of character through libel or slander.
- Unlawful entry onto another's premises.
- Unwarranted seizure, alteration, or destruction of another's property.
- Unauthorized use of another's patents, trademarks, or copyrights.
- Violation of another's freedom through nuisance and negligence.
- Failure to exercise care in the exercise of one's duty to another.

Torts can arise from damage or injury caused by failure to act with the proper standard of care. *Standard of care* is broadly interpreted as conduct that is expected of someone acting in a given capacity. For engineers, the standard of care is essentially the conduct that can reasonably be expected of other engineers in a similar situation. The issue of *negligence* also arises in many tort cases. The definition of a tort is often applied to negligence suits. Negligence arises when a legally protected interest is overtly invaded or violated in some way.

The responsibility of the designer was defined in *Rosos Litho Supply Corp., et al. v. Richard T. Hansen* (462 NE.2d 566). Hansen, an architect, was to design an addition to a structure for Rosos Litho Supply Corp. (Rosos), a firm that acted as the general contractor. The construction sequence was such that the fill was to be completed and then the structural frame was to be constructed. After the roof was in place, the concrete slab was to be completed. The fill had been exposed to the elements for three months and snow was even shoveled onto the fill in places prior to slab placement. Despite this, Hansen did not request soil tests on the fill, prior to slab placement, as stated in the contract. After the concrete slab was completed, it began to crack and some displacement occurred. Hansen was sued by Rosos on the principle that he did not exercise due care in regard to the fill. Hansen argued that as a professional he did not guarantee perfect plans or satisfactory results. The court ruled that the negligence of architects was recoverable since they "hold themselves out and offer services to the public as experts in their line of endeavor." They become liable for professional negligence when they fail to exercise the level of skill that is required. Hansen was liable for the damages sustained in the concrete slab.

Many injuries in the construction industry are sustained by workers as a result of the negligence of their employers. These workers are generally barred from suing their employers because of the protection provided employers by the workers' compensation laws. There are some nuances that still offer a means of recovery to injured workers outside of the workers' compensation benefits. The case of *Hawkins v. Cordy* (642 So.2d 1115) involved Hawkins, a painter, who was hired by W. J. Miranda Construction Company. He was instructed by his supervisor, Cordy, to paint a warehouse. Hawkins was doing some elevated work, and was injured and paralyzed when the scaffolding collapsed. This scaffolding had been provided by Cordy. Hawkins and other workers had previously told Cordy that the scaffolding was defective and Cordy had promised to fix the scaffolding. Assuming that the necessary repairs had been made, Hawkins was not aware of the unique hazard. Hawkins contended that since Cordy required him to use the scaffolding, Cordy played a major role in his injury. The court was asked to determine if Cordy was acting in a managerial or policymaking capacity when he provided the scaffolding for the work. Although Cordy was an agent of Miranda when acting as the supervisor, he was independent when acting as the scaffolding supplier. The court essentially viewed the supplying of scaffolding as being separate from the employment scenario. Cordy was not protected by workers' compensation immunity.

Trespass violations are torts. This issue becomes clouded on a construction site, which can be defined as an *attractive nuisance.* Construction sites can add considerable interest to a neighborhood and may capture the curiosity of children. Suppose a child walks onto a construction site during the weekend, crawls into an unlocked truck, releases the emergency brake, and incurs injuries when the truck strikes a corner of the building under construction. If the individual had been an adult, that individual would be regarded as a trespasser and the cost of the damage would be borne by the trespasser through a tort action. However, since the intruder is a child, the contractor may be forced to pay for any injury incurred under the attractive nuisance doctrine. Furthermore, the contractor will have no recourse against the child for the damage inflicted. Attractive nuisance is defined in state statutes. While these vary from state to state, the conditions of attractive nuisance are generally defined as applying when (1) the party controlling a piece of property should know that children are likely to trespass, (2) the party should realize that there is an unreasonable risk of death or serious injury on the site, (3) the children, because of their age, will probably not recognize the risk involved, and (4) the party could reduce the risk with a small effort by keeping the children out or by reducing the dangerous condition. If all four conditions are satisfied, courts tend to rule that the attractive nuisance doctrine applies.

Cases in which attractive nuisance has been found to apply include a variety of conditions and situations. A sampling of such cases will show the extent to which this doctrine has been found to apply. The case of *Helguera v. Cirone* (178 Cal App.2d 232) involved defective scaffolding. A 7-year-old boy wandered onto a construction site and ended up walking on scaffolding that had no handrail and a loose plank. He sustained injuries when he fell from the scaffolding. Cirone, the owner and builder, was held liable because the site was an attractive nuisance. In the case of *Ridgewood Groves, Inc. v. J. A. Dowel* (189 So.2d 188), a 7-year-old boy visited a neighborhood construction site that was being cleared for a housing development project. The boy's body was found the next day under a large pile of sand and debris that was 15 feet high and had steep sides. The construction firm knew children played on the site and clearly had the option to make smaller piles. The site with the piles of debris was determined to be an attractive nuisance. In *Carter v. Livesay Window Co., Inc., et. al.* (73 So.2d 411), the subcontractor placed windows in the rough openings of a building under construction over the weekend. Unfortunately, the windows were not fastened or secure. The frames were essentially resting on an 8-inch sill. One of the 325 pound precast concrete window frames fell on a 4-year-old boy who wandered onto the site. The subcontractor was deemed negligent in not providing safe conditions for the possibility of a child entering the site. In *Chase v. Luce* (58 NW.2d 565), a house under construction was visited by a 5-year-old girl. The girl managed to climb to the second floor even though the stairs were not complete. The flooring was not in place, so the girl walked on the joists that were spaced 16 inches on center. Unfortunately, the girl fell and was injured. The developer was found liable. The case of *Johnson v. Wood* (21 So.2d 353) involved lime that had been left on the job site. Several children ventured into the construction area. One of the children threw some of the contents of an unguarded mortar box in the face of Thelma Johnson, a 5-year-old. The lime in the contents resulted in loss of hearing and sight in one eye. The mortar box was viewed as an invitation, and satisfied the criteria for an attractive nuisance.

Despite the apparent consistency between the cases defining attractive nuisance, there are numerous cases in which at least one of the elements of an attractive nuisance was found to be lacking. In *Patterson v. Gibson* (287 SW.2d 853), a 5-year-old boy entered a construction project and was injured when he tripped on a plank that was used to cross an excavated area. In this case, the court determined that the excavation was not inherently dangerous and that the plank was not the source of attraction. *Latta v. Brooks* (169 SW.2d 7) involved an injury resulting from several boys entering a construction site. One of the boys threw lime in the face of 5-year-old Jimmy Latta, resulting in a permanent eye injury. The court decided that lime is not classified as a dangerous substance. It further determined that the injury was due to the actions of one of the boys throwing the lime. *Guelda v. Hays & Nicoulin* (267 SW.2d 935) was a case involving an injury sustained by a boy who fell in the basement of a building. The boy was crawling through a basement window and fell 2 feet, breaking his leg. The court ruled that a muddy basement floor, 2 feet from the ground level, does not constitute an unreasonable risk of death or serious injury to children. In *Concrete Construction, Inc. v. Petterson* (216 So.2d 221), an 11-year-old boy found .22 caliber cartridges on a construction site and took them home. The next day, the boy was playing with the cartridges when one exploded and injured him. The court did not recognize this as a case of attractive nuisance because the contractor had no control over the actions of the boy when he returned home. In *Witte et al. v. Stifel et al.* (28 SW891), a 7-year-old boy was injured on a construction site. The boy tried to look into an unfinished basement window by pulling himself up. He tried to hoist himself up by crawling onto a 600 pound stone that was placed across the window. The boy was crushed by the stone that was not secured. The court ruled that the boy was trespassing and that the contractor was not negligent. This was an 1894 case and may no longer reflect the interpretation of courts today.

Contributory negligence may also be a defense offered to avoid liability for an injury. This was examined in *Thomas Orrin Frevele v. Bernard McAloon and Lawrence Diebolt* (564 P.2d 508). McAloon, an employee of Diebolt Lumber and Supply, was attempting to unload an order of building materials at a residence by hoisting the bed of the truck. Thomas Frevele and two coworkers happened to be behind the truck when the materials began to slide off. Frevele felt that the coworkers were in immediate danger, and so he tried to hold back the materials. Because of his efforts, Frevele suffered injuries that left him unable to work for several months. He then filed suit against McAloon and Diebolt for compensation. As a defense, Diebolt argued that Frevele had willfully placed himself in danger by stepping in front of the moving material and thus had negligently contributed to his own injuries. Frevele said that McAloon was told not to raise the truck bed too high, but had continued to raise it even though he had been told to stop. The court ruled in favor of Frevele, stating that a person is not negligent when placing himself in danger to protect another person.

On construction projects, the types of torts that are perhaps of greatest concern are those that involve personal injuries. Injured parties may seek compensation from any number of individuals associated with the project. Their basic intent is to show that the personal injury was a direct result of the actions of a specifically named party, or that the injury occurred as a direct result of the failure of a particular party to act when there was a duty to act. Personal injury torts may result in large monetary settlements if gross negligence is proved. While the contractor has generally been shown to be the party with primary responsibility for the safety of employees, this may also be the responsibility of the architect, owner, or professional construction manager.

Several cases have involved the issue of the liability of professional construction managers for construction workers' safety. This duty may be interpreted from Occupational Safety and Health Administration standards which state that employers "shall do everything reasonably necessary to protect the life and safety of employees." In *Bechtel Power Corp. v. Secretary of Labor* (548 F.2d 249), an injured party successfully sued the construction manager on a project. The court ruled very broadly that Bechtel was an employer on site and that it did not matter that its employees worked only in managerial and supervisory capacities. In *Carollo v. Tishman Construction and Research* (109 Misc.2d 506), a subcontractor's employee was injured and successfully sued the construction manager for being passively negligent. In *Brown, Jr. v. MacPherson's, Inc.* (545 P.2d 13), the actions of the construction manager were used to determine the extent to which the manager owed a duty to the workers for safety. In that case, the construction manager had marked a stairway hazard leading into a job site trailer and the court then determined that this assumption of duty extended to the entire site. In *Parsons, Brinckerhoff, Quade & Douglas, Inc. v. Johnson* (288 S.E.2d 320), the employee of a subcontractor was injured by an electrical shock. When the injured worker sued the construction manager, the court decided that the worker's only recourse was against the employer, and that the limit of recovery was essentially the benefits derived from workers' compensation. The court did state that the construction manager had a contractual obligation to advise the general contractor on matters of safety. It also recognized that the construction manager had failed in the performance of that duty, but this Georgia court ruled that the construction manager was insulated from suit. In *Johnson v. Bechtel Assoc. Professional Corp.* (717 F.2d 574), several workers sued because of their exposure to high levels of dust and other industrial pollutants on a project. The court ruled that the construction manager was the agent of the owner and could not be sued.

Cases involving construction managers and the various ways in which court interpretations are made show that no general conclusions can be made about whether the construction manager is obligated to provide for work safety. The courts look at statutes, contract verbiage, job site actions, and common-law interpretations to make their judgments. The key questions relate to the duty that the construction manager had to the injured party, and whether the construction manager was negligent in the execution of that duty.

Torts are possible occurrences in virtually all walks of life. Construction projects are likely to be the sites of torts; however, a greater number of construction lawsuits relate to contractual disputes.

REVIEW QUESTIONS

1. How do torts differ from criminal acts?
2. What conditions must exist for an injured party to successfully sue another person in a tort action?
3. Explain the rationale for regarding attractive nuisances as torts.
4. Describe duty of care as it pertains to contractors.
5. What duty of care might subcontractors on a construction project have to other subcontractors on the site?
6. Contrast how a tort suit would generally differ from a contract dispute.

8

SURETY BONDS

THE EFFORT AND expense of undertaking a construction project entail a great deal of risk for the owner. Despite detailed market analyses and extensive feasibility studies, the viability of projects is often based on reliable estimates of construction costs. If the actual costs of a project are considerably above the estimated amounts, a project can fail. When lump sum contracts are used, the owner has some idea of the anticipated costs. If lump sum bids are received and are excessive, the owner can still cancel or terminate the contract. If they are within the budget, the project will go forward. Even at this stage, the ultimate costs of the project can be in excess of the budget if the contractor defaults on the project. This is a risk that the owner would like to avoid. The risk can be diminished by requiring the contractor to provide a surety bond. A *surety bond* is essentially a guarantee provided by a firm that states that the contractor will fulfill the terms of the contract (figure 8.1).

If the contractor defaults on the contract, the surety will then be obligated to satisfy the terms of the contract. This type of surety arrangement is common in construction. The arrangement includes the following three parties:

SURETY. The surety (the bond company) is obligated to perform or to pay a specified amount of money if the principal does not perform. The surety is the guarantor on the bond.

PRINCIPAL. The principal debtor, or principal, is the party (general contractor) whose performance is promised or guaranteed.

OBLIGEE. The obligee is the party (owner) to whom the promise of the principal's performance is made. The obligee is the beneficiary of the bond.

When considerable amounts of the work on a project are to be performed by subcontractors, the general contractor is placed in a position of risk if the subcontractors default on their work. To minimize this risk, the general contractor may require each major subcontractor to provide a surety bond. The role of the general

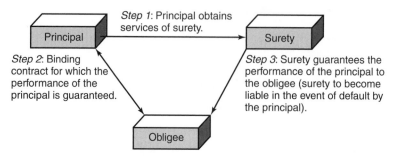

FIGURE 8.1
General relationship of the surety, principal, and obligee.

contractor is now changed, as the general contractor becomes the obligee and the subcontractor becomes the principal. This is summarized as follows:

The principal is the subcontractor.
The obligee is the general contractor.
The surety is the bonding company.

The principal is responsible for obtaining the bond from the surety and then, after signing it to establish the obligation of both the surety and the principal, deliver it to the obligee (owner). With this bond, the performance by the surety and the principal have been promised to the owner. The surety is immediately obligated in the event of contractor default. Note that the beneficiary of the surety agreement is the obligee, even though the person applying for and obtaining the bond is the principal. The surety guarantees the performance of the principal's obligations under the contract with the owner (obligee). The surety in essence lends its reputation and credit to the contractor.

The function of the surety is to assure the owner that the contractor is backed up by someone who is financially responsible, not to be confused with insurance. For the following reasons, a surety bond is not the same as insurance:

An insurance policy is a two-party instrument between the insurance company and the insured that protects the insured against a specified type of loss. In a surety agreement, the contractor is not providing the guarantee for himself or herself, but for the owner.

Insurance premiums are based on actuarial rates. Surety premiums are different in that the surety "presupposes no losses" and is generally structured on a set fee. It is more like a lending transaction at a bank. The premium is essentially a fee for the extension of credit, like the interest charged by banks for loans. If a loss is anticipated, no bond will be issued by the surety.

By presupposing no losses, the surety is more like a bank than an insurance company. The surety is indemnified by the principal, while this is specifically avoided in an insurance policy. Indemnification is a means by which the principal is obligated to the surety for any debts it incurs as a result of a default by the principal. Thus, if a contractor provides the owner with a performance bond and subsequently fails to complete the project, the surety will be obligated to finish the project, but the surety can seek reimbursement from the contractor because the contractor indemnified the surety. If the surety pays for losses, subrogation rights also exist for the surety against other parties. Under the principle of equitable subrogation, the surety does not use subrogation against its principal, but instead, by being subrogated to the principal's rights, the surety can in all respects seek compensation from other offending parties (owners, subcontractors, suppliers, etc.). That is, if the contractor default on a project can be shown to be due to a particular party's actions, the surety (through subrogation) has the right to seek compensation from that party. Essentially, the surety "steps into the shoes of the contractor" and can file suits that the contractor would have been able to file.

Insurance covers specific losses, while a surety bond is for losses of any kind, for those guarantees given, for example, performance and payment.

Insurance transfers risk, whereas a surety agreement does not.

The underwriter of an insurance policy often has the ability to cancel the policy during the policy period; this is not true of surety bonds. Once the surety bond is issued, the bond is regarded as irrevocable, even if the premium has not been paid.

A surety bond is analogous to a minor who must have an elder sign a contract to purchase an item on credit, so that the merchant is assured that the debt will be paid. If the minor fails to make the required payments, the elder cosigner becomes fully liable for the debt.

The contractor-surety relationship is based on trust. This relationship works best if the surety is fully aware of the contractor's work in progress and the other projects being bid. As in banking, the relationship is built on confidence.

SURETY UNDERWRITING

The risk of failure in the construction industry is high. A disproportionate number of construction firms fail each year. For example, over 25 percent of the construction contractors doing business in the United States between 2000 and 2002 failed for one reason or another. Information on these types of failures is reported regularly by Dun and Bradstreet. In economic "hard times," sureties will be more cautious about issuing surety bonds to construction firms. In recent years, surety loss ratios have been below 20 percent, considerably below the level that they had been just a few years ago (figure 8.2). These ratios can be expected to climb when contractor failures increase, primarily due to increased competition that is associated with lower margins. Increased competition causes some firms to pursue work in construction sectors in which they have no experience, and this is often associated with more business failures. During shaky economic times, general contractors are also more inclined to require surety bonds of their subcontractors.

Clearly, the surety company assumes a tremendous risk. As a result, it behooves the surety to make a thorough scrutiny and analysis of the contractor's operations and

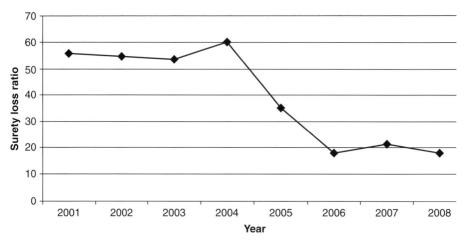

FIGURE 8.2
Surety loss ratios 2001–2008.
Source: The Surety and Fidelity Association of America.

solvency. This process involves determining the liabilities of the contractor and investigating the background, capabilities, and financial responsibility of the contractor. This will include the current bank line of credit, a broad list of references, résumés of stockholders and/or key personnel, corporate tax returns, information on project control systems in place, the contractor's experience, character, exposure and progress on other construction jobs, along with the size and nature of the proposed project in relation to the contractor's prior work experience and current backlog of work. This investigation will probably include the personal interests and activities of key personnel in the firm, particularly the owner or owners. The result of this research should help indicate if the construction firm will fulfill all the obligations placed upon it in the contract. This *underwriting* (rating the acceptability of risks being solicited) *procedure* is usually so carefully performed that many owners disregard the need for prequalifying the bidders, since the surety considers essentially the same variables, perhaps even more rigorously.

The underwriting is performed by a surety agent, who is an agent of the surety. Usually the surety is a firm with national standing. The contractor will deal directly with the surety agent concerning the bid bonds, payment bonds, and performance bonds. Contractors are wise to select their surety agents carefully. Qualified and experienced surety agents can be very valuable in ensuring the future success of a company. At times the bond company, acting through the surety agent, may limit the work that can be undertaken by denying a bond request. Denying a bond request on a low-risk project is not to the agent's or bond company's advantage, and so the contractor should heed the cautions given by the agent. The agent will not be paid unless a bond is issued; therefore, the denial of a bond request is something about which the contractor should be concerned.

A surety must know a contractor quite well in order to assume the risk of contractor default on a project. Thus, the contractor is probably getting good advice if the bond company is reluctant to authorize a bond. The surety and the contractor arrangement is similar to the relationship that a bank might have with a firm that wants to take out a loan. If the bank refuses to issue a loan, the bank must have a good reason for this assessment. Knowing that the bank cannot survive in the long run without approving loans, it should be evident that there is probably a good reason for refusing the loan. The same argument can be made when a surety refuses to issue a bond.

The surety underwrites against losses of any kind. The odds of incurring such losses are difficult to determine. Every possible loss must be considered. Specific problems to look for are overexpansion (more work taken on by the contractor than can be handled with the existing working capital), bids that are too low, subcontractor failures, unforeseen labor problems, unknown soil conditions, loss of key personnel, harsh contract provisions, and extraordinarily high wages. While requiring major subcontractors to be bonded and obtaining insurance for serious losses may provide some protection from default, it is impossible to anticipate all losses.

Other considerations, particularly for new contractors, include experience, character, reputation, financial standing, equipment owned, personal integrity of key personnel, personal habits, professional ability of the firm, and verification of bank credit. Once a relationship is established, the surety underwriter keeps current with the contractor's progress on projects currently under construction. Since the surety will then have an understanding of the contractor, the next bond request will not be examined as rigorously as the first. After that, each bond request will be judged individually. Each request will probably require an investigation of the following items of information:

Size, type, and nature of the project being bid.
The owner of the project and his or her ability to pay.
Uncompleted work of the contractor (bonded and not bonded, including work not yet awarded).
Adequacy of working capital and available credit.
Volume of work permitted to be undertaken by the contractor.
Money "left on the table."
Experience of the contractor in this type of work (diversification may be the sign of a possible challenge).
Contract terms: bonds required, how payments will be made, retainage (money earned by a firm on a construction project but withheld until project completion or until a specified date), time of completion, liquidated damages, amount subcontracted, and qualifications of subcontractors.

It is not in the best interest of either the contractor or the surety for the underwriter to grant excessive lines of credit, a process that can in itself lead to the financial failure of the contractor. This is a clear conflict since commissions of the surety agent depend solely on the sale of the bonds. The surety agent must balance the merits of issuing a bond with the importance of helping to ensure the long-term survival of the construction firm.

Bonds are a guarantee against unqualified and unscrupulous contractors. Surety companies are used for weeding out unworthy and irresponsible bidders. If a contractor has passed the investigation by the surety underwriter and can obtain a bond, that contractor is generally considered to be responsible. Conversely, if a contractor cannot get a bond, he or she is probably a poor risk. Incompetent or irresponsible contractors are generally screened out or culled by the surety underwriter.

Should the owner stipulate which surety the contractor should use, as is required in some contracts? In general, this practice is not advised. Research has shown that contractors do not shop around for sureties. Once a good relationship has been established, the contractors stay with "their" surety. If that surety denies a bond request, the contractor simply does not bid on the job. If the owner is concerned about the surety, the owner should make sure that the surety is qualified.

There are also financial limits on what a surety can do. The government adopted a rule that a surety cannot assume a single obligation that exceeds 10 percent of the surety's equity or surplus; that is, the surety's capital structure limits the size of the projects that can be underwritten. This limits the exposure of the surety and serves as a further guarantee to the owner that the surety, as well as the contractor, is sound. On very large projects, some surety firms do not have sufficient capital to issue a bond. In such instances several sureties will join together and issue the bond. There may be as many as five or six cosureties on a large project.

The premium rates for a bond are regulated but are usually based on a specified percentage of the total or face value of the bond. Regulations are set by states and by the U.S. Treasury.

Successful sureties are those that are careful about bond issuance (careful selection) and skillfully manage projects that go into default.

Lump sum projects are particularly vulnerable for the following reasons:

• Prices can increase.
• Labor difficulties can arise.
• Subsurface conditions may be different than expected.
• Government policy may change, affecting the ability to borrow.

Part of the success of a surety lies in its ability to carefully assess the stability of contractors, particularly when surety bonds are provided to contractors who have been in business for only a short time. A surety may be inclined to refuse to provide bonds to such firms. A common practice, applied to almost all contractors, is for sureties to require company owners to sign personal guarantees as a condition of providing bonds. In fact, the surety may issue bonds to some firms only after indemnity agreements have been signed by the corporate officers, major stockholders, and spouses in closely held firms. Through this mechanism, a surety effectively removes the protection offered by the corporate structure. Thus, if a contractor default occurs, the surety will be able to seek redress against the corporate assets as well as the personal assets of the company owners. Essentially, the indemnity agreement states that the principal (construction firm owners) will reimburse the surety for any payment (caused by contractor default) it must make under the bond obligation. Although exceptions exist for some very successful firms, sureties are very reluctant to provide a bond if the company owners lack the confidence to place their personal assets at risk.

THE MILLER ACT

In 1935 the Miller Act was enacted, stipulating that surety bonds are required of construction contractors on all federal and federally assisted projects. These bonds offer protection to the awarding agencies, laborers, subcontractors, and materials suppliers. The requirements of payment and performance bonds are stipulated for contracts that exceed $100,000 for the construction, alteration, or repair of any building or public work of the United States. On federal projects, it is a discretionary decision on the part of the owner if payment and performance bonds will be required on smaller projects, valued less than $100,000. On projects valued in excess of $100,000, a performance bond must be furnished in an amount that the contracting officer determines to be adequate for the protection of the federal government. To guarantee the contractor's performance on a project, a performance bond of 100 percent of the contract amount is generally required. The performance bond assures the owner that the project will be satisfactorily completed by the contractor. A separate payment bond is required for the protection of the suppliers of labor and materials. The principal amount of the payment bond should also be determined by the contracting officer, but shall not be less than the amount of the performance bond.

THE BID BOND

The *bid bond* is issued to give assurances that the contractor will enter into a binding construction contract and will provide the required payment and performance bonds if the contract is awarded to him or her (figure 8.3). If the contractor fails to do this (sign the contract or furnish the required bonds), the bond stipulates that a responsible party (the surety) will pay the damages. The damages may consist of the forfeiture of the face value of the bond, or may be limited (most commonly) to the difference between the amounts of the low bidder and the next low bidder up to the face value of the bond. The face value of the bond (penalty) is usually set at 5 to 20 percent (typically 5 percent) of the contract amount. A cash deposit, certified check, or cashier's check may be used instead of a bid bond on some jobs. The bidder would prefer not to offer cash or a check for security, as interest must be paid on any borrowed funds, or interest will be lost on the money that would otherwise be kept on deposit. A bid bond is preferred to a cash deposit by the owner, however, as the cash will not have the attached prequalification assurances obtained through the underwriting process. Generally, contractors also prefer bid bonds because no cash outlays are involved.

The bid bond can be viewed in more than one way. It can be a liquidation of damages (limitation of liability), a security device, or an unenforceable penalty. The distinctions can be clarified if the provisions address this issue very specifically. If forfeiture of the bid bond is regarded as liquidated damages, the owner will retain the full amount of the bid bond if the low bidder does not sign the contract and provide the required additional bonds. Some owners (particularly state agencies) simply regard the bid bond as a security device for which the difference

BID BOND

Principal: RST Constructors   Date Bond Executed: _Oct. 18, 2011_
               Charleston, SC

Surety: Round Top Insurance Company
               15 South Mountain Road
               New York, NY

Penal Sum of Bond: _10_ percent of bid price, not to exceed _7,500,000_ dollars,
Bid Date: _Oct. 18, 2011_

Know all persons by these presents, that we, the Principal and Surety, hereto, are firmly bound to _Naval Health Center_ as Obligee, hereinafter called the Obligee, in the above penal sum for the payment of which we bind ourselves, our heirs, execu-tors, administrators, successors and assigns, jointly and severally: provided, that Surety binds itself, jointly and severally with the Principal, for the payment of such sum only as is set forth above, but if no limit of liability is indicated, the limit of lia-bility shall be the full amount ot the penal sum.

The condition of this obligation is such, that whereas the Principal has submitted the bid identified above.

Now, therefore, if the Principal, upon acceptance by the Obligee of the bid identi-fied above, within the period specified therein for acceptance (sixty [60] days if no period is specified), shall execute such further contractual documents, if any, and give such bond(s) for the faithful performance of such Contract as may be required by the terms of the bid and the payment of debtors, as accepted within the time specified (ten [10] days if no period is specified) after receipt of the forms by the Principal, and if in the event of failure of the Principal so to execute such further contractual documents and give such bonds, the Principal shall pay the Obligee for any cost of procuring the work which exceeds the amount of the Principal's bid, but which does not exceed the penal sum stated above, the Principal shall pay to the Obligee the difference, not to exceed the penal sum, between the amount specified in said bid and such larger amount for which the Obligee may in good faith contract with another party to perform the Work covered by said bid, then this obligation shall be null and void, otherwise to remain in full force and effect.

Signed and sealed this _18th_ day of _October_, 20_11_.

                                        RST Constructors
_____   by _____Edward J. Billings_____ (Seal)
            Witness   Edward J. Billings, President

_____   _____President_____
            Witness   _   Title

                                        Round Top Insurance Company
                                        by _____
                                                    Attorney-in-Fact

FIGURE 8.3
Example of a bid bond.

between the low bidder and the second low bidder is lost if there is a forfeiture on a bid bond. Such a provision might state that the amount retained by the owner will be "the difference between the amount of the contract as awarded and the amount of the proposal of the next lowest responsible bidder but not to exceed the total amount of the proposal guaranty." Whether the bid bond is redeemed as a forfeiture of the face value of the bid bond or as the difference between the low bid and the second low bid will be determined by the actual wording of the bid bond. There is a tendency to view the bid bond as a demand instrument valued as the difference between the low and second low bids, up to the face value of the bid bond.

Even if the bidder makes a computational error, it is not certain that the bond should be returned. That is usually the case, although some courts have ruled otherwise. Although the contractor may be relieved of obligation, some courts have ruled that the bond should be kept by the owner because the promise to enter into a contract was breached. If the promise is breached, the bond can be kept and the owner cannot sue for additional damages. This is not true when the bond is not a limit of liability. Note that the courts have not made a clear distinction between these two interpretations; however, most courts state that the bid bond should be returned to the contractor when a bidding error consists of a mistake of fact. Remember, mistakes of fact are generally sufficient to have the bidder relieved of any obligation to enter a contract, but errors in judgment are not. Thus, bid bond claims are more frequently involved when bidders make judgment errors in their bids rather than mistakes of fact.

The amount of the proposal guaranty that is forfeited may be only the face value of the bond in the event of a default. Private owners may refuse to consider future bids from a contractor who has previously defaulted on a bid bond. Some public agencies may follow state laws that also penalize contractors who have defaulted. Low bidders who did not enter into a contract with the owner, regardless of whether the bid bond was forfeited, may be prohibited from doing any work on the project, such as in the capacity of a subcontractor or supplier. Furthermore, it is also common for a defaulting contractor to be barred from bidding on the same project if it is readvertised for letting.

After the bid opening, the apparent low bidder is forwarded a contract and bond forms that must be completed. The owner will generally state in the contract documents the period of time in which the owner must provide the low bidder with the contract for execution. Failure of the owner to send the low bidder the contract within the stated period will allow the low bidder to decide if the bid is to be withdrawn. The low bidder and the owner may also mutually agree to extend the time period for the contract award. The provisions should stipulate the period in which the contractor must complete the bonds and return them to the owner with the signed contract. The following provision is typical:

> The contract shall be executed by the bidder, and the contract bond shall be executed by the principal and the surety, and both shall be presented to the Owner within 15 days after the date of the notice of the award of the contract.

The time period for executing the contract documents and the bonds may vary depending on the type and size of project, but most owners require that forms be returned within one to four weeks, with 10 to 15 days being most common.

Failure to return the documents in the stated time period may cause forfeiture of the bid bond. This too is often specifically addressed in the contract documents. The following provision clarifies the definition and consequence of a default in entering into a contract or providing bonds:

> Failure on the part of the successful bidder to execute a contract and an acceptable contract bond within the stipulated time period will, at the discretion of the Owner, be just cause for the annulment of the award and the forfeiture of the proposal guaranty to the Owner, not as a penalty but in payment of liquidated damages sustained as a result of such failure.

The bid bond of the low bidder becomes null and void when the construction contract is signed and the payment and performance bonds are posted. When are the bid bonds of the unsuccessful bidders returned? This information is usually stated in the instructions to bidders. The bid bonds of the second and third lowest bidders may be held by the owner until the contract has been signed. The bid bonds of the other bidders are generally returned immediately after the bid opening or within a stipulated time period, such as 3 to 15 days, after bid opening. Some owners hold all bid bonds until the contract documents have been executed. The following is an example of a provision addressing the return of bid guaranties:

> Proposal guaranties in the form of bid bonds, certified checks, or cashier's checks will be returned immediately following the opening and checking of proposals, except that of the lowest qualified bidder; however, the owner may also retain the proposal guaranty of the second lowest qualified bidder at its discretion. Proposal guaranties that have been retained will be returned promptly upon contract award and after the owner has received satisfactory bond and contract forms executed by the award recipient.

The following is a provision that is more specific about when most unsuccessful bidders will receive their bid guaranties:

> The proposal guaranty of the three lowest bidders may be retained until after the contract has been awarded and executed and the required bonds have been provided. Proposal guaranties of all except the three lowest bidders will be returned within 72 hours after the bids have been opened.

The fee for issuing a bid bond is usually nominal; often there is no fee. If a fee is charged, it is not usually based on the value of the bid bond. The fee may be $100 or $200, an amount that will cover the administrative cost of preparing the bond. Surety firms generally issue bid bonds at no cost because they anticipate providing the payment and performance bonds in the event of contract awards to their clients. The surety understands that the award of the contract will result in the purchase of the payment and performance bonds, and so the costs of issuing bid bonds are generally recovered through the other bonds.

While a bid bond is generally issued at no cost, this can present a dilemma in some instances. For example, suppose a private owner requires a bid bond, which is provided by the low-bidding contractor. When the contractor is awarded the contract, the owner waives the requirements regarding the payment and performance bonds. This owner obviously has no guarantees against contractor default, but the owner has already benefited from the underwriting that has effectively prequalified the contractor. When such a waiver of the payment and performance bonds occurs, the surety will be inclined to charge an administration fee for the bid bond.

PERFORMANCE BONDS

A *performance bond* assures that a financially responsible party will stand behind the prime contractor if he or she does not perform properly (figure 8.4). These bonds usually state a specified dollar amount as a limit to the liability of the surety, more commonly referred to as the bond's penalty (figure 8.5). This is not an absolute guarantee that the project will be built as specified for the contract price. The surety simply states that it will back the contractor to the limit of the face amount of the bond. Surety bonds must be in writing. Performance (and payment) bonds are required on public projects and are required by some private owners who want strong assurances that their projects will be built with minimum risk. In some cases, financial institutions, such as banks, also require performance bonds in order to protect their "investments."

These bonds usually have a face value of 100 percent of the contract price. The actual bond premium does not change much for a lower bond percentage, so a 100 percent bond is generally the best buy. The premium for performance bonds, often combined with the payment bond premium, is generally in the vicinity of 1 percent of the contract amount on projects valued at more than $1 million. The premiums are closer to 1½ to 2 percent for projects less than $1 million and as great as 3 percent for projects valued less than $100,000. This is the fee for underwriting, not an amount to cover losses. To some extent, the fees may be reduced for a contractor who is considered to be a particularly good credit risk.

All states have some version of the Miller Act that requires payment and performance bonds on state projects. In one case (*Davidson Pipe Supply Co. v. Wyoming Co. Industrial Development Agency et al.*) in New York, an energy

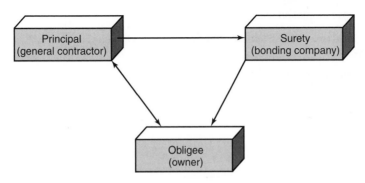

FIGURE 8.4
Parties involved in performance bonds provided by the general contractor.

PERFORMANCE BOND

Principal: RST Constructors   Date Bond Executed: *Jan. 10, 2011*
             Charleston, SC

Surety: Round Top Insurance Company
             15 South Mountain Road
             New York, NY

Penal Sum of Bond: *67,893,000* dollars   Contract No.: *ACW 77-291*

Know all men by these presents, that we, the Principal and Surety, hereto, are firmly bound to *Naval Health Center* as Obligee, hereinafter called the Obligee, in the above penal sum for the payment of which we bind ourselves, our heirs, executors, administrators, successors and assigns, jointly and severally: provided, that Surety binds itself, jointly and severally with the Principal, for the payment of such sum only as is set forth above, but if no limit of liability is indicated, the limit of liability shall be the full amount of the penal sum.

The condition of this obligation is such, that whereas the Principal entered into the contract identified above.

Now, therefore, if the Principal shall:

(a) Perform and fulfill all the undertakings, covenants, terms, conditions, and agreements of said contract during the original term of said contract and any extensions thereof that may be granted by the Obligee, with or without notice to the Surety, and during the life of any guaranty required under the contract, and shall also perform and fulfill all the undertakings, covenants, terms, conditions, and agreements of any and all duly authorized modifications of said contract that may hereafter be made, notice of which modification to the Surety being hereby waived; and

(b) If the said contract is subject to the Miller Act, as amended (40 U.S.C. 270a-270e), pay to the Obligee the full amount of the taxes imposed by the Government which are collected, deducted, or withheld from wages paid by the Principal in carrying out the construction contract with respect to which this bond is furnished; then the above obligation shall be void and of no effect.

In witness whereof, the Principal and Surety have executed this performance bond and have affixed their seals on the date set forth above.

Signed and sealed this _10th_ day of _January_, 20_11_.

                                              RST Constructors
_____C. W. Braun_____   by _Edward J. Billings, President_   (Seal)
        Surety   Principal

___Attorney-in-Fact___   ___President___   (Seal)
        Title   Title

FIGURE 8.5
Example of a performance bond.

cogeneration plant was being developed. In order to secure tax advantages, an agreement was made with Wyoming County Industrial Development Agency (WCIDA), a public-benefit corporation. According to the agreement, WCIDA would be the owner of the project during construction; the project would revert to private ownership at project completion. This agreement with the public-benefit corporation made the project a public improvement. During construction the steel supplier was not paid and subsequently filed suit. The suit was not filed against the steel erector, but against WCIDA. The state court found that, since WCIDA failed to obtain the required payment bond, it was liable for the payment of the steel supplier (595 N.Y.S. 2d 898).

What happens if the contractor makes a misrepresentation to the surety when applying for a bond? For example, suppose the contractor states that the firm has capital which does not exist. Is the bond valid if the contractor subsequently defaults on a project on which a performance bond was provided? In the case of misrepresentations on the part of the applicant, the bond is valid. The surety will then be expected to seek another means to remedy its claim against the contractor. If the owner knew that the contractor received the performance bond by offering fraudulent statements, the bond would be less likely to be enforced. Since the obligee knew about the misconduct in obtaining the bond, the bond was not properly obtained.

Contracts to build projects that are illegal cannot be enforced. If the legality is marginal, such as when a contractor has not been properly licensed or a building permit has not been obtained prior to contract award, the bond is usually considered valid.

The performance bond and the payment bond are considered valid for the life of a contract. This protection period extends through to final acceptance by the owner, and generally extends through the one-year warranty period.

Surety contracts should note whether the surety "waives notification of any alteration or extension of time made by the owner." A major change in the construction contract may release the surety if such an exclusion does not exist. Usually sureties waive the requirement to be notified about minor changes that fall within the scope of the original contract. The definition of *minor* should be clear. Generally, a change of 10 percent, or an aggregate sum of changes amounting to 10 percent of the contract amount, should be brought to the surety's attention.

Since the bonds are written for a specified limit, it is important to recognize the impact of change orders that increase the contract sum. The surety generally must be informed about any changes of a stipulated magnitude for a bond to cover the work performed for a change order. The surety was not fully informed of changes in *Union Pacific Casualty Insurance Co. v. Port of Everett* (46 P.2d 736). The port had a contract with James N. Main to furnish and install approximately 10,000 tons of riprap rock in the harbor of Everett. The contractor supplied a performance bond of $12,300. Shortly after the riprap work was completed, the port made a supplemental agreement with Main to deliver approximately 4000 tons of additional rock. Union Pacific agreed to be bound by the terms of the supplemental agreement. A short while later, the port recognized the need for an additional

6091 tons of rock. In spite of testimony indicating that the port had intended to obtain the consent of the surety, Union Pacific was never informed about the changes to the contract. The work performed by Main was completed and accepted. Within 30 days a suit was filed against the bond due to nonpayment by James N. Main. Union Pacific promptly sued the port. The surety argued that its liability extended only to the first 14,000 tons of rock included in the original contract and the supplemental agreement. The port contended that the surety's obligations extended to all the obligations of the contractor. The court ruled that the surety's liability included only the 10,000 tons of rock in the first contract and the 4000 tons in the supplemental agreement. The port effectively became surety for the extra work involving 6091 tons of rock when it failed to require a bond.

The case of *Ramada Development Co. v. U.S. Fidelity & Guaranty Co.* (626 F.2d 517) illustrates the extent of liability of a surety. Ramada Development Co. (Radco) was the general contractor for the construction of a 10-story Ramada Inn. The electrical work was subcontracted to Mays Electric, Inc. Mays was required to furnish a payment and performance bond in the subcontract agreement. Mays had bought materials from an electrical products supplier, Millen-Seldon Electric Company, but had failed to pay for the materials. Since Radco was confident that Mays would successfully complete the project, in spite of financial difficulties, Radco paid Millen-Seldon $17,358, the total amount owed by Mays. Radco also made monthly payments to Mays after the financial difficulties became apparent. Eventually, Mays defaulted and the electrical work was completed on a time and materials basis by Trans-Pac Construction Co. Radco then filed a claim against U.S. Fidelity. The major issue in the suit concerned the amount U.S. Fidelity was obligated to pay. Radco was awarded $223,851. This represented the sum that had been paid to Trans-Pac ($287,468), plus the amount paid to Miller-Seldon ($17,358), minus the amount Radco held in retainage ($48,710), and minus the two payments made to Mays ($32,265), for which the surety was not liable.

Performance bonds usually contain provisions that permit the surety to remedy the default and complete the construction contract itself, or to pay the owner to complete it, up to the limit of the bond. If the surety elects to take over the completion of the project (the option most commonly exercised), the surety must be diligent in its management of the completion of the project. While the surety may argue that the face value of the bond is the limit of liability of the surety, this may not be the case if the surety takes over the task of project completion. Despite this possible interpretation, sureties are inclined to actively participate in project completion. The initial step by sureties is generally to provide financial support to the contractor in hopes of salvaging the project and having it completed by that contractor. If this strategy fails, the next step is usually a more direct involvement of the surety in project completion.

Although payment and performance bonds are relatively simple instruments, specific guidelines generally must be followed. For example, the revised code of Washington contains a statute which states that any claims on the bond "shall be commenced by filing the complaint with the clerk of the appropriate superior court within one year from the date of expiration of the certificate of registration in force at the time the claimed labor was performed." Does this imply that a surety bond is not valid if the contractor is not registered when the bond is furnished? This was the issue in a case involving a defaulting mechanical contractor, *Joint Administrative Board of the Plumbing and Pipe-fitting Industry v. Fallon* (569 P.2d 1144). The surety, Empire Pacific Industries, claimed that the bond was invalid since the contractor had not been registered when the bond was furnished. The court ruled that the status of contractor registration has no bearing on the validity of a bond. The time period of the statute was simply interpreted as being a statute of limitations for processing claims. The bond became effective upon delivery.

Notification of the surety is required when the surety may be placed at risk. This issue was examined in *Lazelle v. Empire State Surety Company* (109 P. 195). Lazelle hired a contractor to construct a project, and the contractor obtained a performance bond from Empire. In December 1908, the owner noticed that the contractor had constructed the roof of the second story too high. The contractor was asked to correct it and agreed to make the necessary adjustments. The project was not completed by January 8, 1909; so, on January 21, Lazelle gave notification of the default to Empire. On February 11, after making several complaints to Empire, Lazelle asked Empire to take charge of the project. On February 16, a statement outlining the details of the contractor's default was sent to Empire. Lazelle had the project completed and subsequently sent a notice of the cost of completion to Empire on July 19. The surety denied liability for the debt since it had not been notified of the error in constructing the roof too high. Empire stated that this error in notifying the surety nullified the bond. However, Lazelle responded that damages were not being claimed for the error in the roof, but for the contractor's failure to complete the project. The court agreed with Lazelle. Since no damage claim was being filed for the roof error, Lazelle was under no obligation to notify Empire of the incident. The court stated that "mistakes of the contractor, which are corrected on calling his attention thereto, do not constitute defaults."

A performance bond provides protection from virtually all types of losses. This was tested in *Spokane and Idaho Lumber Company v. Boyd* (68 P. 337). The city of Spokane entered into an agreement with G. J. Loy in 1897 to furnish materials and construct the Oliver Street Bridge for $5,000. Idaho Lumber furnished the materials, which cost $1,200.33, but received only $717.83 from Loy. Before the project was completed, Loy assigned the contract to Boyd, who was acting as the surety. Loy had become "insane" and was unable to complete the project. Boyd completed the project at considerable expense. He failed to pay the balance owed to Idaho Lumber and also asked the city to pay for the additional expenses incurred on the project. In the ensuing court battle, Boyd claimed that the city had let the contract to Loy despite knowing that he was insane. This, he contended, nullified the surety agreement. The city argued that it was not a party to the surety bond agreement and thus could not be made liable. The city continued to state that failure to obtain a bond would have made it liable. The court agreed with the city and stated that the bond vouched for the good judgment and responsibility of the contractor; therefore, in a default, the surety was liable. The court stated that the insanity issue was irrelevant. It also stated that Idaho Lumber had to be paid. Had Boyd been able to complete the project with a profit, he would have kept the profit. Similarly, if the costs overrun, the surety must absorb those losses.

The issue of the time in which a claim must be made against the surety was addressed in *Comey v. United Surety Co.* (111 N.E. 832). In August 1908, J. Cadoza made a contract with the Pucci Contracting Company to excavate a plot of ground in New York City. The work was to be completed by April 15, 1909, for a price of $20,000. The contractor provided a bond for $7,500 conditioned on faithful performance of the contract. The contractor officially abandoned the job on March 1 and expressly refused to go on. The owner promptly made demand upon the surety, stating that Pucci was in default and that United Surety would be held responsible for damages. United Surety was also notified that the same contractor had submitted an offer stating the terms under which the work would be resumed and carried to completion. A new contract was made, with the surety's approval, for $3,300 above the original contract. The work was completed in February 1910, and in November 1910 a claim was sent to United Surety. United Surety's refusal to pay resulted in a suit. United claimed no liability based on the fact that the bond provision stated that claims against the surety had to begin "within six months after completion of the work specified in the said contract." Since the claim was made nine months after completion, United claimed that it was under no obligation to pay. The court did not agree, stating that the cause for action on the bond was assured when United was notified of the contractor's default in March 1909. Thus, the project did not have to be completed prior to obligating the surety.

In some cases the bond provisions are interpreted very narrowly. This occurred in *Southern Patrician Associates v. International Fidelity Insurance Company* (381 S.E.2d 99). Southern, as an owner, contracted with CM Systems to renovate the Roswell Mall in Georgia. CM subcontracted a portion of the work to R&M Mechanical Inc. To satisfy the terms of the subcontract, R&M provided a surety bond issued by International Fidelity. Before R&M had done any work, CM went bankrupt. CM's surety company assumed the obligation to complete the work and assigned all of CM's rights to Southern. Southern dismissed R&M for nonperformance and expected its surety, International, to ensure that R&M's work would be completed. International did not take action on the request, and so Southern had R&M's work completed, expecting International to cover any expenses in excess of the contract amount. A suit ensued when International refused to pay for the overrun costs. Southern argued on the issue of its contract with CM, which stated that it would bind "themselves, their heirs, executors, administrators, successors, and assignees." Since Southern was assigned CM's rights, Southern felt that it had a right to action where "successors" were named. International argued that the right to action clause in the surety bond, the only contract binding to the surety, stated that only the obligee, CM, could act on the bond with the exception of the "heirs, executors, administrators or successors of the obligee." The omission of the assignees from the surety bond provision was the basis of the court's decision. The court stated that since the assignees were not named, they did not have a right to claim, that is, the assignees had been omitted intentionally. Thus, International was not liable for R&M's nonperformance.

The manner of filing a claim for nonpayment was a central issue in *Delmar Davis and The City of Great Falls v. C. E. Mitchell and Sons* (511 P.2d 316). Davis had a contract with the city to improve a Great Falls golf course. Davis provided the required performance and payment bonds. Mitchell was employed on the project as one of several subcontractors. When Mitchell was not paid for its services, a statement requesting payment was sent to the city. Other subcontractors submitted similar requests. When the project was completed, most of the subcontractors were paid by the city; however, Mitchell remained unpaid. Mitchell then filed a formal claim against the city for the amount due. In the meantime, Davis declared bankruptcy. The city contended that it had no direct liability to laborers, subcontractors, or suppliers on city contracts on which a bond had been posted. Further, the city argued that it had no contractual relationship with Mitchell, thus precluding any recovery from the city. The court stated that the payment and performance bond was for the benefit and protection of the city. The city also had no obligation to subcontractors. The court stated that Mitchell's grievances were with Davis and that claims would have to be filed against either Davis or the surety company. If the claim had been filed against the surety, not the city, payment would probably have been assured.

PAYMENT BONDS

A *payment bond* gives protection to the owner if the subcontractors and suppliers are not paid by the prime contractor. Payment bonds prevent liens. The subcontractors are paid by the surety if the contractor fails to pay them. With the added assurance of being paid, the subcontractors are more inclined to bid and bid lower for their work. Payment bonds provide these assurances through such provisions as "every claimant as herein defined, who has not been paid in full before the expiration of a period of ninety (90) days after the date on which the last of such claimant's work or labor was done or performed, or materials were furnished by such claimant, may sue on this bond for the use of such claimant, prosecute the suit to the final judgment for such sum or sums as may be justly due claimant, and have execution thereon. The Owner (Obligee) shall not be liable for the payment of any costs or expenses of any such suit."

In private projects, the payment bond keeps a project free and clear of liens. Subcontractors and suppliers can file liens in most instances of nonpayment. Liens are permitted since the benefit is conferred to the owner by the inclusion of labor and materials in the project. Thus, without a payment bond, the owner may pay twice for work done: once to the prime contractor and once to the subcontractor or supplier. If the lien is perfected, the lien claimant has a right to a judicial foreclosure sale on the property and to have the claim satisfied out of the proceeds. The owner can, of course, sue the contractor if the contractor can be found. Under a payment bond, the surety pays the subcontractors and suppliers if they are not paid by the prime contractor. Another reason for a payment bond is to get lower subcontractor and supplier prices. Prices are lower if they are assured of being paid.

In public works projects, without a bond, subcontractors and suppliers would have to file a stop notice if they were not paid. A stop notice informs the owner that the subcontractors and suppliers have not been paid and that further payments are to be withheld by law from the contractor. Sometimes an unpaid party may present a claim against a public body through a request for special legislation for payment whether it be Congress, a state legislature, or a city council. Bonding of the contractor avoids these entanglements. On federal projects, the face value of the payment bond is to be at least equal to the performance bond.

The general contractor payment bonds provide assurance of payment by the general contractor, but this assurance is limited to certain parties. For example, the payment bond will provide protection to materials suppliers who provide materials directly to the general contractor, the suppliers of materials to the first-tier subcontractors, assurance of payment for the subcontractors (first-tier) of the prime or general contractor, and the second-tier subcontractors (sub-subcontractors) who have agreements with the first-tier subcontractors. Thus, the payment bonds do not grant any protection to subcontractors of or materials suppliers (second-tier suppliers) of suppliers. Neither do they provide protection to the suppliers of the sub-subcontractors or third-tier subcontractors (figure 8.6). If lower-tier subcontractors or suppliers are to be covered, a payment bond must be provided by the subcontractors.

In order to secure the right to file a suit on a payment bond, certain procedures must be followed. Subcontractors and suppliers should be fully aware of the conditions stated in the payment bond. For example, the payment bond may stipulate that

> No suit or action shall be commenced hereunder by any claimant: a) Unless claimant, other than one having a direct contract with the Principal, shall have given written notice to any two of the following: the Principal, the Owner, or the Surety, within ninety (90) days after such claimant did or performed the last of the work or labor, or furnished the last of the materials for which said claim is made, stating with substantial accuracy the amount claimed and the name of the party to whom the materials were furnished, or for whom the work or labor was done or performed. Such notice shall be served by mailing the same by registered mail or certified mail, postage prepaid, in an

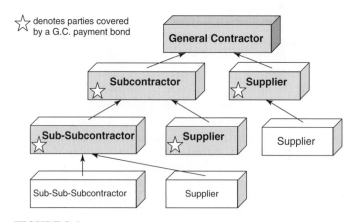

FIGURE 8.6
Parties typically covered by a payment bond furnished by a general contractor.

envelope addressed to the Principal, Owner, or Surety, at any place where an office is regularly maintained for the transaction of business, or served in any manner in which legal process may be served in the state in which the aforesaid project is located, save that such service need not be made by a public officer. b) After the expiration of one year following the date on which Principal ceased work on such contract, it being understood, however, that if any limitation embodied in this bond is prohibited by any law controlling the construction hereof such limitation shall be deemed to be amended so as to be equal to the minimum period of limitation permitted by such law. c) Other than in a state court of competent jurisdiction in and for the county or other political subdivision of the state in which the project, or any part thereof, is situated, or in the United States District Court for the district in which the project, or any part thereof, is situated, and not elsewhere.

Payment bonds may state that the subcontractor or supplier notify the surety if he or she has not been paid within a specified time from the completion of work, provide data about the amount claimed, and identify the party to whom the materials and labor were furnished. This time period should be such that the rights of lien claimants are no longer valid. A typical requirement of this type would be as follows:

> Give written notice to the surety of nonpayment within 90 days after the last day that labor or materials were furnished under the contract. If a sub-subcontractor or a materials supplier of a subcontractor has not received payment from a subcontractor for work performed or materials furnished, the written notice must also be provided to the general or prime contractor.
>
> Civil action on the payment bond must be brought no later than one year after the last day labor or materials were furnished under the contract (see figure 8.7).
>
> Civil action on a performance must be taken within two years on performance bonds.

The reporting requirements are not recommended procedures, but mandated procedures. Failure to fully comply can result in a party not being able to collect on a payment bond. This was the situation in *State Ex Rel. Martin Machinery v. Line One* (111 S.W.3d 924). In this case, Martin Machinery rented heavy equipment to Line One for the excavation and installation of a pipeline. When Martin was not paid by Line One, Martin presented a written claim against the payment bond to the surety. No written notice or request for payment was made to the general contractor or to the owner of the project. The failure to notify the general contractor or the owner was fatal to Martin's case. Martin's claim for payment was denied.

The payment bond guarantees payment to parties involved in the construction of a project. To what extent does this guarantee apply to parties that are removed from the direct contractual arrangement of the general contract? This was tested in *Layrite Concrete Products of Kennewick Inc. v. H. Halvorson, Inc.* (411 P.2d 405). Halvorson was a subcontractor on a nuclear power plant project who had entered into a sub-subcontract with Keystone Masonry for the masonry portion of the work. Layrite had an agreement with Keystone, the sub-subcontractor, to supply masonry materials. Before the masonry work was completed, but after Keystone

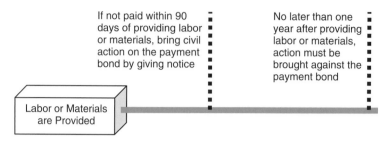

FIGURE 8.7
Timeline for filing a notice of non-payment and for bringing action on a payment bond.

had received some payment, Keystone went bankrupt, leaving Layrite unpaid. Layrite brought suit when Halvorson refused to pay for the materials. Layrite contended that Halvorson should comply with the obligations set forth by the payment bond, which guaranteed payment to all parties performing labor and furnishing materials. Halvorson argued that the bond was intended to protect the owner and prime contractor, not a separately contracted materials supplier. The bond language also was presented as evidence of this intent to protect the owner and prime contractor. Since they did not know about Layrite at the time of issuing the bond, Halvorson contended that there was no obligation to pay Layrite. The court found the bond language to be most compelling in that it guaranteed payment only to those supplying materials directly to Halvorson. The bond did not protect third parties who were not included in the contract made between the joint obligees and the subcontractors.

Unique statutory requirements may exist concerning the guidelines that must be followed to preserve a legal recourse in the event of nonpayment by the general contractor. This was shown in *Clyde Austin v. C. V. Wilder & Co., Inc.* (397 P.2d 1019). Wilder was a general contractor on a sewer project, and Austin was a supplier to Glen Eyerly, a subcontractor. Austin supplied materials to Eyerly, but the invoices were not paid. After Eyerly went bankrupt, Austin filed a suit against Wilder for the unpaid balance. Wilder's case rested on a Washington statute that states,

> Every person, firm or corporation furnishing materials, supplies or provisions . . . shall, no later than ten days after the date of the first delivery of such . . . to any subcontractor . . . deliver or mail to the contractor a notice in writing stating in substance and effect that such person, firm or corporation has commenced to deliver materials . . . and that such contractor and his bond will be held for the payment of the same, and no suit or action shall be maintained in any court against the contractor . . . unless the provisions of this section have been complied with.

The court interpreted the statute literally and found that Austin had lost any rights under the bond when he failed to provide the statutory notice within 10 days of the material delivery. The intent of the statute is presumably to provide information to the general contractor concerning materials delivered so that the general contractor is not required to pay twice for the same materials.

A technicality was addressed in *Eureka Sandstone Co. v. Long* (39 P. 446). Long, a general contractor for the construction of a courthouse, received $16,000 worth of materials from Eureka. When Eureka received only partial payment, a suit was filed against Long and Long's surety to recover the balance. In reviewing the various documents, it was discovered that Long had not signed the final agreement between himself and the surety. The surety contended that Long was solely liable for the indebtedness because he had not signed the agreement. Long argued that there were no problems during the time of construction and that both parties had assumed in good faith that they had a contract. The court ruled that the lack of the signature was significant. Its decision was based on the reasoning that if it ruled otherwise, any party could claim a contract with a surety without a signature. Thus, Long was responsible for the balance due for compensation to Eureka Sandstone Company.

An interesting technicality occurred in *Hoiness-LaBar Insurance v. Julien Construction Co.* (743 P.2d 1262). Park County, Wyoming, hired Julien to rebuild the county courthouse. The mechanical portion was subcontracted to Neilson Plumbing and Heating, Inc. The subcontract required Neilson to provide payment and performance bonds. The bonds were requested from Hoiness-LaBar, a broker with Allied Fidelity Insurance Company. A mistake was made when the bond was issued guaranteeing performance only. The error was not detected by Neilson when the bond was received. Neilson subsequently became unable to pay all its suppliers. The suppliers then requested payment from Allied. Allied refused to pay any claims because there was no payment bond and then refused to issue a payment bond as again requested by Hoiness-LaBar. Julien finished the project, paid Neilson's suppliers, and then sued Neilson, Allied, and Hoiness-LaBar. Regarding the Hoiness-LaBar suit (Allied went into receivership in the meantime), the court ruled that Julien was a third-party beneficiary of the contract between Hoiness-LaBar and Neilson. Hoiness-LaBar had violated the agreement to provide a payment bond, and thus Hoiness-LaBar had to pay Julien.

BONDING LIMITS FOR THE CONTRACTOR

Sureties generally stipulate a maximum value of uncompleted work that a contractor can undertake at one time. This is referred to as the contractor's *bonding capacity*. This level or limit is based on the surety's appraisal of the contractor's abilities and resources. This assessment is based on the "3 c's" of underwriting, namely, character, capacity, and capital.

The bonding capacity of a firm is vital in dictating the sizes of projects that a company can undertake. If a company wants to bid on a particular project that would cause the company to exceed its bonding capacity, the project is not generally worth pursuing. The bonding capacity separates the companies that are deemed capable of undertaking certain projects from those who are not.

A conservative rule is that the bonding capacity is about $10 of uncompleted work for each $1 of net working capital, depending on job size. With the strong construction market in recent years, the determination of bonding capacity has become more relaxed. There are some instances in which the bonding capacity has been established as being as much as 60 times the net working capital. Another rule is to set the bonding capacity at $10 to $15 of uncompleted work for each $1 of net worth or owner's equity, but this is not as widely used. Owner's equity should never be less than the working capital.

The difference between the bonding capacity and the current total of uncompleted work, both bonded and unbonded, is a measure of the amount of additional work for which the surety will issue a bond.

Another general practice is for the amount of a single bonded contract not to exceed 50 percent of the bonding capacity. Thus, a contractor with a bonding capacity of $4 million will be given bonds for projects up to $2 million in value.

CONTRACTOR DEFAULT

Although surety bonds are issued with the assumption that there will be no losses, contractors may fail to complete their construction obligations for a variety of reasons. These failures often result in bankruptcy. Once the contractor has been declared bankrupt, the contractor has no further obligation to continue work under the original construction contract. No claims by the owner against the defaulting contractor are likely to be satisfied. However, the surety is not released. In fact, it is for just this type of default that a performance bond is required by owners.

When a contractor defaults, the surety generally has three possible options to follow. These are as follows:

- Surety provides financial support to the defaulting contractor in order to expedite the completion of the project, with the surety to be reimbursed later (this is the most common option taken by some sureties).
- Surety solicits bids or quotes from other contractors to complete the project.
- Surety informs the owner to finish the project (this is not a common option).

Essentially, when a contractor defaults, the surety is asked to step in for the contractor to complete the work or pay the owner to do it. This does not mean that the surety is powerless. The surety must perform under the bond, but some defense is available to the surety. In fact, any defense that would be valid for the contractor can be claimed by the surety. Typical defenses include the following:

- Owner not making payment, even when the issuance of progress payment certification occurred.
- Unjustified interference by the owner in the contractor's work.
- Delayed approval of shop drawings.
- Discovery of unforeseen subsurface conditions.

Sureties try to recover from their losses in the following ways:

- They can recover from the contractor (not a likely option since contractor default on a project probably entails bankruptcy).
- They can sue the owner for any claims that the contractor could reasonably allege.
- They can try to get the retainage held by the owner. In this case, the surety must usually compete with other claimants, particularly if the contractor went bankrupt.

When taking over a project for a contractor, a surety will notify the owner to make all payments to the surety that would otherwise go to the contractor.

Private works projects are more complex, as they are not governed by law. If an owner requires a bond, it is not clear that an unpaid subcontractor can sue. The question here is, "Did the owner guarantee the benefit only to himself or herself?" If the answer is yes, the subcontractor cannot sue.

Note that if the surety takes the responsibility of completing the project, the actual cost to the surety may be greater than the face value of the bond. If the owner is permitted to complete the project, the limit of liability is the face value of the bond. In general, sureties assume that the cost to complete projects is less than the face value of the bond, and so the most common action by sureties in a default situation is to take over the responsibility for completing the troubled projects. If the surety does not take over this responsibility, the completion of construction will become the obligee's responsibility.

It is common practice for any major changes in the contract to be communicated to the surety. These added amounts constitute an added risk for the surety and warrant the payment of an additional fee to the surety.

The 1909 case of *Fransioli v. Thompson et al.* (104 P. 278) illustrated the need to keep the bond current with the project. This case involved a contract for street improvements that Thompson & Langford had with the city of Tacoma. The surety for Thompson & Langford was Empire State Surety Company. Thompson & Langford subcontracted a portion of the work to C. D. Elmore that involved the installation of cedar cribbing along a portion of the street. At the same time, a change was made in the original plans, substituting a concrete wall for the cedar cribbing. The cost was considerably higher as a result of this change, but the city did not take an additional bond on the work being changed. Elmore performed the work, obtaining the cement from Fransioli. When no payment was received for the cement, Fransioli notified the city, Thompson & Langford, and Empire State of the claim against Elmore and demanded payment from each. The three parties refused to pay, prompting a suit. The city claimed that the commissioner of public works had no authority to make changes to the plans, and so the claim was not valid. Fransioli argued that the city had accepted the completed work according to the change in the plans and therefore could not claim that the change was not valid. Empire State and Thompson & Langford claimed that the only liability they had to Fransioli was the liability created by the bond. The changes in the plans, they contended, were "radical," and therefore they should be relieved of any obligation to pay for the cement. The judge ruled that the change was indeed radical, relieving Empire State and Thompson & Langford of any liability. However, the city of Tacoma was obligated to pay for the cement.

SUBCONTRACTOR BONDS

The general contractor is responsible for the job performance of the subcontractors. The general contractor becomes liable if a subcontractor fails to pay for
materials, labor, or sub-subcontractors. Protection is afforded to the general
contractor by requiring the subcontractors to provide him or her with a payment
and performance bond (the contractor is the obligee as shown in figure 8.8).
Unfortunately, not all specialty contractors can obtain performance bonds. If
this occurs, the general contractor may elect to obtain performance bond coverage on behalf of the particular subcontractor and deduct the cost of the bond
from the payments made to the subcontractor. In some instances, the surety of
the general contractor may require performance and payment bonds of certain
subcontractors.

It should be noted that the use of subcontractor bonds does not obviate the
need for careful selection of subcontractors. A bad sub spells trouble. It is difficult to recover losses due to work stoppage, delays, and disruption of the work
routine.

It is becoming more common on large construction projects for the prime contractor to require the subs to obtain payment and performance bonds. If a sub does
not perform as obligated, the prime then has a financially responsible party to
back the sub. There is a cost incurred when a subcontractor is required to provide
payment and performance bonds. The subcontractor's bid to the general contractor
is simply increased by an amount that reflects the cost of the bonds. Of course, the
general contractor does have added confidence in a subcontractor who has gone
through a rigorous underwriting procedure, an excellent prequalification mechanism. Recognizing the prequalification that accompanies the issuance of bonds,
some general contractors have resorted to the practice of determining if a subcontractor can provide the bonds. If the surety writes a letter stating that the
subcontractor is bondable, the general contractor may simply use the letter as a
prequalification tool and save the cost of the bonds. This obviously does not eliminate risk to the extent that bonds do.

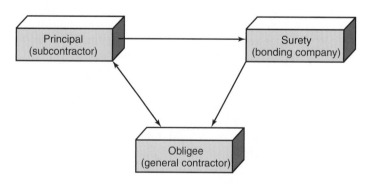

FIGURE 8.8
Parties involved in performance bonds provided by the subcontractor.

If the sub fails to pay his or her subs and suppliers, the prime is responsible since he or she is obligated to erect the project, free and clear of all liens. Thus, the prime should consider requiring payment bonds from the subcontractors.

On very large projects it is common for there to be a bonded prime contractor, bonded subcontractors, and even bonded sub-subcontractors. Litigation that might arise would then be in the hands of the sureties, with each trying to shift responsibility to another surety.

BONDS ON DESIGN-BUILD AND PROFESSIONAL CM PROJECTS

Payment and performance bonds are generally considered to be used on general contract projects. Traditionally, public works projects have been performed under general contract arrangements and as a result the Miller Act pertains especially well to those projects. What should occur on design-build projects or professional CM projects? The owner may still want to minimize risks, but must an alteration occur in the bonds that are written? It is incumbent for the owner to consider the type of bonds that are to be required. Is bonding different for a design-build project? If the design-build firm provides a performance bond, the bond may be viewed as extending the liability of the surety to the design of the project. The surety may be considered as applying to the design, especially if the professional liability insurance is inadequate. Since the owner acquires design and construction services in a design-build contract, it may be natural for the owner to assume that the bond also extends to the design effort. It should be noted that for the surety to cover the design work, it alters or changes the bond to a form of liability insurance. This is not the way in which surety bonds are underwritten. Also, designers typically have limited familiarity with surety bonds. Design-build firms would be wise to isolate the design and construction functions when the surety bond is written. For example, it is perhaps simplest for the bond to be written so that it explicitly does not cover negligence, errors or omissions in the design, or the warranty of the design. The design work would then be best covered by an appropriate professional liability insurance policy.

Consider what could occur on a CM at risk project. The owner could require the CM to provide 100 percent performance bonds for the entire guaranteed maximum price. The CM will, in turn, require 100 percent performance and payment bonds from the specialty contractors. Note that this will result in double bonding of much of the construction work. This can be avoided if the CM bond covers only the CM's fee, the general conditions, the insurance (builder's risk insurance) on the facility being constructed, and other items not normally covered by the bonds of specialty contractors. The specialty contractors would then provide 100 percent performance and payment bonds for their own work. This will reduce the cost of surety bonds, but not impact the extent of coverage.

REVIEW QUESTIONS

1. In what ways are surety bonds not like insurance policies?
2. How is it determined that a contracting firm has exceeded its bonding capacity?
3. What is the limit of liability of a surety if the contractor defaults?
4. What does the bid bond promise to the owner?
5. What rights does a surety have in the event of a contractor default?
6. What party or parties are given the most protection by a performance bond? By a payment bond?
7. Give an example in which a surety may not be forced to honor the bond in the event of a contractor default.

9

THE BIDDING PHASE

ONCE THE DESIGN phase has been completed, the owner must obtain the services of a construction firm to construct the project. In public works projects this procedure (the bidding procedure) is strictly outlined, while in the private sector the process is often varied. No laws govern the process by which private contracts are let.

ADVERTISEMENTS FOR BIDS ON PUBLIC WORKS PROJECTS

Specific procedures must be followed in awarding contracts on public projects. This process is governed by law. Here is a brief description of the procedure.

1. Notice must be given to interested and qualified members of the construction community in advance of the bidding. Notice of bids (advertisements) must be placed in newspapers, magazines, trade publications, and the like. The jurisdiction of the owner and the type of project will dictate the frequency with which advertisements must appear in publications and the length of time notices must be posted. This advertisement is commonly known as the notice to bidders. Information commonly placed in the notice to bidders includes the following:

   - Nature or type of project.
   - Location of the project.
   - Type of contract for construction.
   - Bonding requirements.
   - Dates in which to perform work.
   - Terms of payment.
   - Estimated construction cost (some specifically exclude this).
   - Time, manner, and place to submit bids.
   - Location to obtain bid documents.

- Deposit required on bid documents.
- Owner's right to reject any and all bids.
- Requirements regarding wage rates.

2. The invitation for bids must be posted in public places and distributed to the local construction community. This publicity usually includes the agency's (owner's) bid list of possible interested contractors.
3. All bidders must be treated alike and be afforded an opportunity to bid under the same terms and conditions.
4. Prequalification may be required.

Advertisements must be carefully written. They must convey the essential elements of the facilities to be constructed so the contractors specializing in such projects are encouraged to enter the bidding process. Examples of these advertisements, also called request for bids and invitation to bid, are shown in figure 9.1.

A Washington statute (RCW 47.28.050) on advertisements stipulates that "the department of transportation shall publish a call for bids for the construction of the highway according to the maps, plans, and specifications, once a week for at least two consecutive weeks, next preceding the day set for receiving and opening the bids, in not less than one trade paper of general circulation in the state."

An earlier statute stated that the "notice shall be published for at least two consecutive weeks previous to the date of letting, in one or more daily or weekly papers." This provision was the key issue in *Wyant v. Independent Asphalt Paving Co.* (203 P. 961). Bid advertisements for a road improvement project in Yakima County were published on Tuesday, January 25, and on Tuesday, February 1, 1922. Bid opening was on Monday, February 7. On February 11 Independent Asphalt was awarded the contract as the low bidder. Mr. Wyant, a landowner who wished to stop the highway improvement project, filed suit, claiming that the advertising statute had been violated. The first advertisement did not occur two full weeks before the bid opening, but rather appeared 13 days before the bid letting date. The county canceled the contract, and essentially no work was performed until the dispute was resolved. The technicality of 13 days not constituting two weeks was supported by the court.

ADVERTISEMENTS FOR BIDS IN THE PRIVATE SECTOR

There are no well-defined rules concerning the letting of contracts in the private sector. However, most owners, particularly those who make regular use of construction services, follow procedures not unlike those used in public works. The process options can be summarized as follows:

1. The owner may select a contractor by any means.
2. Public advertising is frequently used to obtain the advantages of open and free competition.
3. The owner may elect to negotiate a contract with a particular contractor. This is most common in residential construction and in industrial construction involving

---

REQUEST FOR BIDS

OVERLAND SERVICE CENTER ANNEX BLDG., Springfield, IL,
415 Industrial Way North

BIDS TO OWNER FEBRUARY 17 AT 2:00 P.M. (CST)
Owner - Overland Transport Company, 415 Industrial Way North, Springfield, IL

Architectural, Electrical, Mechanical, Structural by Owner

Building 1 story above grade and 0 stories below grade - 3,840 Total Square
Feet - Concrete Paving. Work consists of the erection of a 40′ × 96′ × 14′ pole
building - includes 6′ × 8′ poles set in concrete - metal roof framing and metal
decking - sheet rock walls but not ceiling - site work includes but is not
limited to installation of an 8″ waterline with fire hydrant - asphalt parking
area - approximately 5,000 SF asphalt paving and associated drainage
basin construction - associated electrical and mechanical work - Water
Distribution - Concrete Forms and Accessories - Metal Fabrications -
PreEngineered Structures - Basic Mechanical Materials and Methods - Basic
Electrical Materials and Methods - Site Clearing

Plans from owner @ $200 deposit - 100% refundable (mailing fee is not refundable)

INVITATION TO BID

VEHICLE MAINTENANCE FACILITY

Houston, Texas

GC BIDS TO OWNER APRIL 19 AT 1:00 P.M. (PST)

Owner/Architect - Harris County Police Department

Bid solicitation issued on electronic bid set only CD-ROM form only - disks
available on June 2 - refer to number HAVX0133 when ordering - requests
must be in writing or faxed to 800-555-7212 - plans and specs picked up Mon.
thru Fri. 8:00 A.M. to 4:00 P.M. - completion period is 450 days - work includes rock
removal - earthwork - construction of sand filter system - sedimentation chamber -
drying bed - equalization and water supply basins - oil water separator - associated
pumps - pump house - piping - lighting - asphalt paving and curbing - concrete joint
repair

Plans from architect @ $100 deposit - 100% refundable

5% Bid Bond

YES Payment Bond

FIGURE 9.1
Example of request for bids/invitation to bid.

highly technical work. The contractor would be selected early in the life of the project and then work constructively with the owner, architect, and other design professionals throughout the design phase of the project.

4. The most common approach is for the owner to select a few prime contractors who are reputable and capable of doing a good job. This list of contractors is called a select bidders list. These contractors are asked to bid in a process called invitational bidding. This process has the advantage of the competitive market while restricting bids to a selected group of contractors.

BID VERSUS NEGOTIATED CONTRACTS

The public bid process, which is adopted by many owners in the private sector, is often referred to as the design-bid-build process. Thus, the design must be completed before the bidding phase can occur, and the bidding is completed prior to the start of the construction phase. The advantages of this process include: (1) the owner benefits from the competitive marketplace; (2) the owner has the appearance of being impartial; (3) the process fully embraces the fundamentals of the free market system; (4) it may be the only viable method available for some governmental agencies. This method is not without its shortcomings, and these must be carefully weighed. The disadvantages include: (1) accurate costs cannot be known until the design is completed; (2) bids that exceed the owner's budget cannot be readily accommodated without jeopardizing the project; (3) the various parties tend to be adversarial under this process; (4) errors or omissions in the design may lead to costly change orders and the opportunity for the contractor to bolster profits after contract award. Negotiated contracts eliminate many of these disadvantages, but this often occurs by compromising some of the advantages associated with competitive bid projects. These factors should be carefully examined prior to deciding on the mode of contracting to be employed.

PREQUALIFICATION

Prequalification may be a requirement imposed by owners on some projects. Bidding will be restricted to firms that have been prequalified. Prequalification of bidders is not a common practice; however, it may be employed on both public and private projects. The prequalification process results in a select bidders list, or short list, which identifies firms that have demonstrated to the owner that they have the necessary abilities to perform the required work. Prequalification generally consists of submitting specific information about the types of projects successfully completed by the firm, the current work load, the personnel employed by the firm, the experience of the personnel to be assigned to the proposed project, the financial stability of the firm, and other information that the owner may deem germane to the successful completion of the project. This information may be requested via a list of specific questions, such as those contained in the AIA Document A350, Contractor's Qualification Statement. Prequalification averts construction problems by limiting bidding to contractors who have the experience and financial stability to complete the project. It is more desirable to eliminate a contractor before bidding rather than show that a contractor is not qualified after that firm has submitted the low bid.

It is the responsibility of the owner to show that a contractor is not qualified. This is true for prequalification as well as postqualification. The elimination of contractors through prequalification cannot be arbitrary or so restrictive as to eliminate competition. This was clearly illustrated in *Manson Construction and Engineering Co. v. State of Washington* (600 P.2d 643). In February 1979 an exceptionally strong storm destroyed half of the Hood Canal Bridge. The state wanted to quickly construct a temporary bridge as a means of relieving stress on the alternative routes. The Washington State Department of Transportation (WSDOT) decided to prequalify bidders for this project, particularly because this was a unique type of bridge, a temporary floating bridge. The advertisement for prequalifying bidders included an additional provision requiring interested parties to "provide evidence of previous successful use by the contractor of the proposed floating bridge configuration." A WSDOT bridge engineer visited several floating bridges that had been constructed by Acrow Corporation of America. Only Acrow met the prequalification conditions. Three firms interested in the project were rejected on the grounds that they could not satisfy the added provision. Suit was then filed against WSDOT on the grounds that the provision requiring prior experience with floating bridge construction was overly restrictive. The state defended the provision, as it was considered imperative that the replacement bridge be completed quickly. Essentially, the contractor would have the responsibility for a considerable amount of design work. The court was sympathetic toward the state's objective of expediting the bridge installation, but said that the WSDOT could not add the restrictive prequalification provision. The contract award was effectively canceled by the court decision.

The definition of the lowest responsible bidder is more complex than simply being the lowest bidder. This was demonstrated in *Crest Construction v. Shelby County Board of Education* (612 So. 2d 425) which involved bids received for a new school building. The prospective bidders were required to submit qualification statements on an AIA form that included financial statements, prior projects, and specific information on personnel to be assigned to the project. The low bid was submitted by Crest Construction Co., but the Board awarded the construction contract to the second lowest bidder. In making this award, the Board noted: Crest was a one-person operation with minimal equipment; it had no work in progress; it had not reported income for the year. The Board determined that there was a high risk in awarding the contract to Crest Construction. Crest filed suit when the second lowest bidder was awarded the contract, claiming that as a public entity, the Board was obligated to award the contract to the lowest responsible bidder. The court ruled that the Board had the discretion to determine the lowest responsible bidder as long as it did so in good faith.

REPORTING SERVICES

Plan service centers are another source of bidding information. These centers publish and distribute, on a regular basis, bulletins that describe all projects to be bid on in the near future in a locality. In addition, they provide services during the bidding stage. They keep copies of the bid documents on file for the use of general contractors, subcontractors, materials suppliers, and other subscribers. This service provides valuable information to a wide spectrum of the construction community. Without such a service, subcontractors and suppliers would have to obtain their own copies of the bidding documents. This could be very costly to small subcontractors, such as ceramic tile or wallpaper subcontractors, who often bid on many jobs. Without such services, general contractors are forced to assume a greater responsibility in providing bid documents to their key subcontractors and suppliers, usually by maintaining a plan room of their own. Any subscriber to the services of a plan service center can review the plans of a project and make a decision on the merits of bidding without paying a deposit for the documents.

From a general contractor's viewpoint, some valuable information can be gained through plan service centers. Two major questions to be answered are,

- Should the general contractor bid on the project?
- Which subcontractors and suppliers are bidding?

Plan service centers are typically used for detailed estimating only by subcontractors and materials suppliers. Suppliers and subcontractors may be able to conduct a complete quantity takeoff within an hour or so, without having to bear the burden of paying a deposit for access to the bid documents. The effort to complete the takeoff of quantities is too extensive for most general contractors to use these centers for this purpose. Consequently, general contractors usually obtain their own sets of the plans by paying the required deposit to the architect/engineer.

Subscribers to reporting services receive daily reports concerning jobs to be bid and all the known general contractors who are bidding. This helps contractors know the level of competition they are facing. If too many contractors appear to be bidding on a project, a firm may decide not to bid. Subcontractors and materials suppliers also benefit by knowing who is bidding. They will know which contractors to contact concerning their bids.

With the technology that is creeping into the construction arena, changes are taking place in how projects are bid. Subcontractors and suppliers are not always required to go to a contractor's plan room or to a plan service center to prepare their estimates. As more plans and specifications are becoming available electronically, these subcontractors and suppliers will be able to prepare their estimates in "electronic plan rooms" in their own offices. This will reduce the cost and time of preparing estimates. These Web-based plan rooms will give the users access to a large number of projects in a short period of time. Authorized bidders will have ready access to all public projects, and they will have access to those private projects for which they have received access authorization. The opportunity still exists for the estimators to use plotters to print out drawings of particular interest.

VALUE ENGINEERING

Value engineering refers to a specific procedure that is carried out to critically analyze the various aspects of contract documents in relation to the owner's objectives, to determine if alternative methods or materials might be more appropriate. A value engineering review on a project may result in a variety of changes in the contract documents that may reduce costs, improve or maintain project quality, and/or decrease the duration of construction. Such a review, conducted prior to releasing the contract documents for bidding, can save the owner considerable sums of money without compromising project quality. The term *value engineering* relates essentially to reviewing the contract documents with the owner's best interest in mind. *Value* includes elements of the delivered cost of a project, the costs of maintaining a completed project, the ease and duration of construction, the probability of disputes or litigation, and various other factors of interest to the owner.

Value engineering reviews can be conducted at two periods in the development of a project. The first is in the design phase. Naturally, the designer will try to focus on the owner's objectives as the contract documents are developed, but an independent review by an impartial third party may prove beneficial. This review incorporates the perspectives of others. Consultants who conduct such reviews should be familiar with the construction process, a variety of materials and their associated in-place costs, and the long-term life and maintainability of various materials. The second period in which a value engineering review may take place is during the construction phase. This review is conducted by the contractor who has been awarded the construction contract. The review by the contractor will be similar to the review conducted by the design-phase consultant. A primary difference is the manner in which compensation is made for the review. The consultant is employed on a fee basis, while the contractor will generally be compensated by sharing any savings with the owner. The construction-phase review is therefore more biased, in that the contractor will generally be more oriented to suggesting less expensive means of constructing a project, even though more expensive installation costs of some items may result in lower operating or maintenance costs.

A project that has gone through such a value engineering review will probably result in "cleaner" bidding documents. If the review is conducted by the contractor on a project, construction may take place as the value engineering suggestions are being evaluated by the designer and owner. This may delay some construction activities. The primary problem encountered with a construction-phase review is the determination of the actual savings. The owner may sense that the contractor discounted the original bid in anticipation of having some recommendations accepted by the owner. If this is the case, the owner will not realize the same cost reduction that might have been received from another contractor with the same proposal. Although this may appear to imply that an independent review is preferred, that is not necessarily true. Since the contractor has conducted a detailed estimate of the quantities of materials that constitute a project, the contractor has a much better knowledge of the project than would a consultant who is to consider plans never seen before. The contractor may already have given consideration to the costs of using specific methods and materials and the ease with which given materials can be installed.

The benefit of a construction-phase value engineering review is that the contractor's expertise is utilized in the design. To the extent that the design is used as a bidding document, fast-tracking is not always possible; however, the contractor's experience is used to advantage by the owner. This may result in a better project or a comparable project that costs less.

CONSTRUCTABILITY REVIEW

In recent years the term *constructability* has been widely used. As the word implies, constructability relates primarily to issues of construction. A constructability review is an assessment of the contract documents, prior to the bidding phase, to identify problem areas and suggest improvements. Problem areas consist of virtually any aspect that may present obstacles to the efficient construction of a project. Examples of such problems include the skill level or availability of the workforce, the cost or availability of specified materials, limitations of equipment typically used for the construction of similar projects, and unique environmental or social concerns. An element of practicality is inferred by constructability. The stated alignment tolerance for a concrete wall may be excessively strict when the specified face brick will effectively conceal any minor aesthetic flaws in the concrete wall. The spacing of reinforcing bars may be too close to accommodate the maximum size of aggregate that is specified. The routing of ductwork may be incompatible with piping runs as shown on the drawings. These types of problems may be costly, impractical, or impossible.

Constructability reviews should be conducted by the design team as the contract documents evolve. Specific construction expertise can be included in this review if a professional construction manager is used, or if the project is done under a design-build contract. On some projects, the contract documents may be subjected to a constructability review by an independent firm. Such an independent review can be cost-effective on complex projects. The parties conducting the constructability reviews will probably vary on projects of different sizes and complexity. A successful constructability review will invariably result in the smooth construction of a project.

THE DECISION TO BID

Once a contractor is informed about a project that is to be bid, a decision must be made about whether the company will be one of the bidders. This is not a trivial question, as a great cost is involved in preparing an estimate. It is said that on a small building project, the cost of estimating can run as much as 0.2 to 0.3 percent of the total bid amount. Various factors must be considered in deciding if the company will bid, including the following:

- Bonding capacity
- Location of the project
- The owner
- The owner's financial status
- The architect/engineer
- Nature and size of the project
- Probable competitors
- Labor conditions and supply
- Availability of in-house staff
- The company's need for work

PLAN DEPOSIT

If the contractor decides to submit a bid on a project, a deposit will be paid to the architect/engineer or the contracting authority. A deposit is generally required to cover the costs of production, or to guarantee the safe return of the bidding documents. The deposit amount typically ranges from about $100 to $200, but deposits of $1,000 may be required. The contractor must check in advance to see if the deposit is partially or totally refundable with the return of the bidding documents. If the project is large or complex, or if little time is made available to prepare the estimate, the contractor may obtain several sets of the bidding documents.

As technological advances are made, the use of traditional (paper) drawings will diminish. Some bid documents are already being made available electronically. The access to these documents is fast, and generally there are no fees that must be paid. This inexpensive access to plans will possibly result in more potential bidders, as contractors will be more inclined to have a quick look at any project that warrants consideration.

THE BIDDING (ESTIMATING) PERIOD

What is often referred to as the bidding period is actually the estimating period, which culminates with the bid (figure 9.2). The bidding period does not begin until the plans and the specifications are completed. It must be borne in mind that this estimating is done when contractors are busy with the actual construction of ongoing projects. If little time is available to prepare the estimate and the resulting bid, the contractor may make costly errors. Contractors are aware of this fact and include a larger markup when the bidding time is short or limited. Owners are well advised to be cognizant of this result: More time devoted to estimating means lower prices to the owner, since less uncertainty will exist in the contractor's bid.

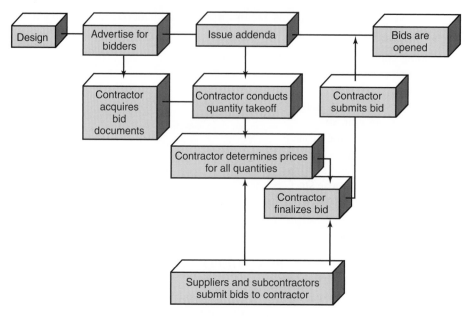

FIGURE 9.2
The estimating and prebid phase of competitive bids.

Unfortunately, owners are reluctant to provide long time periods for bidding, because this is seen as effectively delaying the start of construction.

Even when there is sufficient time in which to prepare a detailed estimate, the owner should consider very carefully the time and the day when bids are to be submitted. For example, it is not generally advisable to set bid dates on a Friday or a Monday. In general, mornings are not a good time to submit bids if many subcontractors are expected to submit prices to each general contractor. Naturally, the bid date should be carefully coordinated so that there is no conflict with bid lettings by other major owners.

ACCURACY OF THE BIDDING INFORMATION

Although the bid documents should be thorough and complete, it is a major mistake for a contractor to accept the information in the documents at face value. Usually assumptions are made in preparing the bid documents, and if an error is made in them, the contractor is expected to discover it prior to the bid date and allow for it in the bid. In particular, information about subsurface conditions is frequently a source of conflict. This includes information about soil borings, test piles, test pits, and so on, made by the owner for design purposes.

Who is responsible for the accuracy of this information? Unless there is a statement to the contrary, the owner is responsible for any errors in the bid documents.

However, statements are often specifically included which place direct responsibility for the information on the contractor. The responsibility is often placed on the contractor by a disclaimer which appears on the drawings or in the specifications. In addition, a site visit by the contractor is a requirement that is usually included in the general conditions or the instructions to bidders. Under some circumstances contractors may ask, Where is the design engineer? or What is the design engineer being paid for?

INSTRUCTIONS TO BIDDERS

The instructions to bidders, as was already mentioned, essentially are the rules by which the project will be bid. These rules concern the bid itself and the project during construction. These rules include the following:

1. Typical instructions relative to the procedure for writing and submitting the bid include the following issues:

   Bids must be submitted on the forms furnished.
   Erasures must be initialed by the signer of the bid proposal.
   All items in the bid schedule must be priced.
   Alternatives are not considered unless called for.
   Discrepancies between the unit price and the extended amount (how handled) are discussed.
   Are mailed bids accepted and considered?
   Can bid modifications be made (if or how)?
   The submission policy (sealed, marked, and addressed as directed) is specified.

2. The contractor may be required to submit an experience record to demonstrate the capability to perform. This is used for postqualification.
3. The instructions clearly list all the documents that are part of the bid documents (drawings, specifications, alternates, instructions to bidders, etc.).
4. The construction time period is carefully spelled out in the instructions. This includes the start date and the number of calendar or working days allotted for project completion, or it may permit the contractor to state the construction time required.
5. The instructions usually indicate who is responsible for the subsoil data, test borings, errors in the plans, and so on. Site visits by the bidders are usually required in the instructions.
6. The requirements of a bid guarantee are outlined.
7. The insurance to be provided by the contractor is stipulated.
8. The bonding requirements are given.
9. Conditions for handling bid irregularities are often stated.
10. Where and when to deliver the bids are stipulated.
11. Closed or public opening of the bids is indicated.
12. A prebid conference may be described.

ELECTRONIC BIDDING

The computer age is evident in many aspects of our lives, and in the past decade it has made inroads on the procurement process in construction. Computers are now being used by some owners for communications and for awarding contracts. One significant change is the adoption of electronic bidding, or e-bidding, on some construction projects. The software that makes e-bidding possible is generally user-friendly, stable, secure, and reliable. Although the paper-based process is still well entrenched, electronic bidding can be expected to expand with the increased use of technology in construction, and the realization that the exchange of information between the various parties involved in the bidding process is more efficient, faster, and more effective with computers. In fact, on some projects, the entire process of constructing facilities is done in a paperless mode.

In an e-bidding environment, the owner may require proposals to be submitted electronically. The submission is to comply with specific detailed instructions. The criteria will generally stipulate the time by which the submission is to be made and the format to be utilized (Word, WordPerfect, Excel, Lotus 1-2-3, or PDF formats). Although it might appear that the time of submission is more controlled in the electronic age, a proposal can still get delayed in cyberspace, so submittals should be made with sufficient lead times to allow for the delivery of electronic proposals. That is, a large attachment can cause a delay in the submittal of a proposal, just as a traffic jam can result in a delay in a bid submittal that is hand delivered.

Some projects let the contractor decide on the method of submitting an electronic bid, whether delivered in person on a floppy disk or CD-ROM, or submitted electronically via e-mail or "datafax." Some projects require bidders to submit their bids on paper and electronically. Some owners stipulate that bids for large projects (valued over $1 million) must be submitted electronically. On smaller projects, the bidders are given the opportunity to submit bids electronically or in person. Some owners offer a web-based electronic bidding process for their construction projects. The intent is to make the bidding process easier. At the owner's website, a prospective bidder can download the plans and specifications and also, through the same site, submit the bid that is prepared. Although the construction industry has generally been slow to embrace technology, electronic bidding will surely increase dramatically in the near future.

The software for submitting bids electronically offers bidders opportunities that they did not enjoy with the traditional hand-delivered procedures. This is the ability to modify a bid after it has been submitted. Suppose a bidder submits a bid at 10:30 A.M. on a project where bids must be submitted no later than 2:00 P.M. After the bid has been submitted, the contractor obtains a bid from a subcontractor that would constitute a significant reduction in the bid. With some of the bidding software, the bidder can withdraw the initial bid and submit a revised bid. This can be done up to the bidding deadline of 2:00 P.M.

ADDENDA

Although the design is generally fairly complete when the bid documents are advertised, a perfect set of plans and specifications is never a realistic goal. A number of events can occur after the bid documents are in the hands of potential bidders that should be communicated to all bidders. For example, the owner may review the specifications again and decide that a change is warranted in the way one item is specified. Upon additional review of the plans, the designer may identify a flaw in the design, or decide there is a better way to accomplish the same end result. A contractor may notice an error or incongruency in the plans as the quantity takeoff is being conducted. Whether the issue is an owner change, a designer change, or a clarification for a contractor, the owner wants to have the information communicated to all potential bidders. A clarification made to only one of the bidders would place the other bidders in an unequal competitive position. Thus, these types of changes or clarifications should be communicated in a formal manner to all the bidders through the issuance of an addendum or several addenda.

Addenda are formal changes or clarifications issued by the owner or owner's representative to all identified bidders during the bidding period. When modifications are not included in the original bid documents, the issuance of addenda is a process by which bidders can be updated on design changes and clarifications. If such changes or modifications were made after the contract award, these items of work have to be addressed as changes. The owner wants to avoid such postaward changes, as the associated costs will not be reflected in the bids. In addition, the costs of changes are more likely to be lower if the bidders are competing against each other. After the contract award, there is only one contractor with whom the costs of changes must be negotiated. This contractor is in a better bargaining or negotiating position if no other contractors are involved.

Addenda are issued during the bidding or estimating period. However, addenda should not be issued up to the time for submitting bids. It is prudent not to issue any addenda in the days immediately before the bidding deadline because some of the bidders may already have prepared their bids, or may not receive notice of an addendum if it is issued at a late date. The bidders will be asked to acknowledge the number of addenda that were received. If the low bidder does not acknowledge receipt of all the addenda, that bid will usually be rejected as being nonconforming. Thus, care must be exercised to assure that all bidders receive all addenda. Of course, a bidder is well advised to call the owner's representative shortly before the bid day to verify the number of addenda that have been issued.

ALTERNATES

Ideally, on a lump sum contract the low bidder will be determined as the party submitting the lowest bid. The determination and selection of the lowest bidder are made more complex when the project includes alternates. *Alternates* can be viewed as modifications to the base bid. They may consist of changes in the structure of a project, changes in the quality of the material to be furnished, the inclusion of additional items of work, the deletion of specified work items, and so on. Typically, the bidders are expected to submit a base bid that represents a sum of money required to construct the project without regard to any alternates. The bidders are then asked to state the amount by which the base bid would be changed for each alternate listed. Alternates can have the effect of increasing or lowering the base bid. The use of alternates gives the owner more flexibility in making decisions about changes to a planned project with full knowledge of the cost impact of those decisions. Thus, the owner can make award decisions on the basis of the limits imposed by available funds.

The awarding of contracts is made complex when alternates are included. For the owner, alternates pose a significant advantage, since their pricing can be used as a shopping list. Changes proposed after the contract award will typically be higher in cost, since competitive bidding will not be involved. Contractors, as a rule, do not see alternates as having strong advantages for them. The primary concern of contractors is that the owner might be able to manipulate the alternates that are accepted, so that a preferred contractor is awarded the contract. This is illustrated in the following bid tabulation:

Bid Component	Bidder "A"	Bidder "B"	Bidder "C"
Base Bid	$700,000	$725,000	$735,000
Alternate #1	$70,000	$40,000	$65,000
Alternate #2	$80,000	$70,000	$50,000
Alternate #3	$95,000	$75,000	$55,000

In this bid tabulation, several interesting aspects can be identified. When considering only the base bids, Bidder "A" is the obvious low bidder with a bid of $700,000. If the owner considered the base bid and Alternate #1, the low bidder would be Bidder "B" with a total bid of $765,000. If Alternate #2 is the sole consideration, the low bidder is Bidder "A" with a bid of $780,000. If only Alternate #3 is considered, the low bidder is Bidder "C" with a bid of $790,000. If all three alternates are considered, the low bidder is Bidder "C" with a bid of $905,000. There are also other combinations of utilizing the alternates. The observation of note is that each of the bidders could be deemed the low bidder, depending on the priority placed on the alternates to be accepted.

The use of alternates presents an interesting dilemma to a public agency that would like to know beforehand about the costs related to various aspects of a project. The owner also wants to present an image of being fair in dealing with contractors. How can this be done when alternates are involved, and the determination of the low bidder is highly dependent on the specific alternates selected by the owner? The owner should try to stipulate the priority of each alternate. This is difficult in some instances, since the priority may change in accordance with the amount of the base bid. The owner may want to use the base bid solely to determine the low bidder, but then the alternates will not necessarily be priced in a competitive fashion. Owners who wish to avoid accusations of favoritism are well advised to consider avoiding the use of alternates, or, prior to bid opening, clearly outlining the criteria to be utilized in determining the low bidder. Some federal government agencies have rules or policies that they follow when awarding contracts that contain alternates. For example, the agency might stipulate that alternates will be awarded in numerical order only. It is advisable that the procedures for awarding contracts based on bids with alternates be announced to the bidders after all bids have been submitted, but before any bids have been opened. Alternates are not as important if the owner or owner's representative has access to a good historical cost database. In general, alternates are to be avoided if possible. Failure to use alternates, however, will force the owner to rely more heavily on the integrity of the contractor if change orders (instead of alternates) are issued.

THE BID FORM

The bid documents usually include a *bid form* on which the bids are to be submitted. There are very compelling reasons to use a specified bid form for all bidders. This form will facilitate analysis and comparison of the bids so that irregularities can be detected quickly. For contractors, it ensures accuracy in providing the necessary information and prevents the possibility of having omissions in the bids.

On the bid form itself (figure 9.3), the following are common requirements:

- Price (lump sum or unit price).
- Time of completion (often given by the owner).
- Bid surety.
- Agreement to provide contract surety.
- Acknowledgment of having reviewed addenda.
- List of subcontractors used in the final bid.
- Experience record, financial statement, plant and equipment inventory.
- Declaration regarding fraud and collusion.
- Statement regarding site examination.
- Signature.

MODIFICATION AND WITHDRAWAL OF BIDS

The preparation and submission of a bid by a contractor is a complex procedure involving the processing of many price quotations from subcontractors and materials suppliers that are received shortly before the bid submittal. The use of electronic spreadsheets and customized estimating and bidding software for compiling bids have greatly reduced the probability of mathematical errors in the bid process. Despite recent advances, many unknowns are still unresolved until just before bid submittal, and human judgment remains a vital component in the process. Errors may occur if information is improperly recorded over the telephone. The chance of error has been reduced to some extent by the increased use of fax machines; however, the obvious shortcomings or errors in a quotation may go unnoticed in the rush to finalize a bid. Any major issues that

---

BID PROPOSAL FORM

To: Edward T. Crowley, Architect   Date: March 9, 2011
      7639 Elm Street, Suite 210
      Kansas City, MO

For: Clay County Civic Center
      3499 Stonewall Jackson Way
      Augusta, GA

Pursuant to and in compliance with the Invitation to Bid and the proposed Contract Documents, the undersigned, having become thoroughly familiar with the terms and conditions of the proposed Contract Documents and with the local conditions affecting the performance and costs of the Work at the place where the Work is to be completed, and having fully inspected the site in all particulars, hereby proposes and agrees to fully perform the Work within the time stated and in strict accordance with the proposed Contract Documents, including furnishing any and all labor and materials, and to do all the Work required to construct and complete said Work in accordance with the Contract Documents, for the following sum of money:

BASE BID *Nine Hundred Seventy Thousand Four Hundred Sixty*   Dollars, *$970,460*

ALTERNATES:
      #1 Demolish existing structure.............. *add $45,000*
      #2 Delete fountain in entry area............. *deduct $13,000*

UNIT PRICES:
For the purpose of adjusting the Contract Price in future changes, provide unit prices for the following, specified more fully in Section 01-026, Unit Prices:

Description	Units	Unit Price
1. Muck Removal	CY	$ *4.75* / CY
2. Concrete Sidewalk (5″ thick)	SY	$ *15.25* / CY
3. Underdrain Piping	LF	$ *2.10* / LF

All Allowances specified in the Contract Documents are included in the appropriate Base Bid.

We acknowledge receipt of Addendum #1 and 2 dated December 3, 2010 and February 4, 2011, respectively.

Respectfully submitted,
By: Edward J. Billings, President
      RST Constructors
      Charleston, SC

(followed by witness certification and seal)

FIGURE 9.3
Example of a bid proposal form.

are unresolved at bid finalization must be dealt with as contingent items. Obviously, the bid amount becomes more exact as the bid submittal time approaches. Once the bid is finalized, it can be submitted. This often occurs just minutes before the stated deadline.

In some cases bids may be submitted hours or even days before the deadline. Once a bid is submitted, it is common to permit bidders to withdraw or modify their bids, provided that the request is made prior to the bid opening. If no modifications can be made to a bid, or if no bids can be withdrawn after submission, contractors tend to submit their bids as late as possible. An uncommon lenient provision that will put a bidder at greater ease is one similar to the following:

> A bidder may withdraw or revise a proposal after it has been deposited with the Owner, provided the request for such withdrawal or revision is received by the Owner, in writing, before the time set for opening of proposals.

Some owners permit contractors to withdraw unopened proposals after the time set for final bid submittal if bids are being accepted for several projects at the same time. This will permit a contractor who is the apparent low bidder on one project to withdraw any unopened bids that have been submitted for other projects.

When estimating the various components of a project, it is frequently necessary for the estimator to make assumptions about the proper interpretation of the contract documents. If assumptions will have a significant impact on the final tabulation of the bid, it is prudent for the bidding contractor to request an interpretation that is typically made by the designer in the form of an addendum. If questions arise within a few days of the bid submittal date, there may be insufficient time to issue an addendum. Under such circumstances, the contractor must simply make an assumption so that the bid can be prepared. Naturally, the bid cannot be submitted in the public sector with conditions stated in the bid. If the contractor is the low bidder and is awarded the contract, the unresolved questions may subsequently become sources of conflict between the contractor and the owner's representative. The owner's representative may regard questions of document interpretation as a ploy to "mine the contract." These are clearly sources of some claims. One way disputes have been minimized by some owners is to require the low-bidding contractor to submit a sealed copy of the estimate shortly (within a day) after bid opening. This copy of the estimate is an *escrow estimate* and is not viewed by the owner unless it is necessary to resolve disputes. Since the escrow estimate is presented just after bid opening, there is little time to significantly alter the original estimate. Thus, the escrow estimate documents will include copies of worksheets that accompany the estimate that was prepared. The documents represent the bidding strategy of the contractor and embody assumptions that were made to prepare the bid. While escrow estimates can be used to demonstrate the strategy used to prepare an estimate, their use can be cumbersome; that is, the practice of requiring escrow estimates is not common.

THE AWARD

It is generally thought that the contract award is given to the lowest bidder. On public works projects, the stipulations are clear and are more inclusive. There are three components to the decision to make a contract award. The contract award on public works projects is awarded to the *lowest responsible* bidder submitting a *regular* bid.

On lump sum contracts and unit price contracts (to a lesser extent) it is usually easy to determine the lowest bidder.

How can a responsible bidder be assessed? This issue involves postqualification. Through this procedure, it is determined after the bid proposal has been submitted whether the contractor has the necessary qualifications to construct the project. Most public works owners require some form of qualification standards to be satisfied. The purpose is to eliminate incompetent, overextended, underfinanced, and inexperienced contractors from consideration. The endorsement of a contractor by a surety (bid bond, performance bond, etc.) is often deemed sufficient to satisfy prequalification criteria.

As was mentioned earlier, many states have laws that stipulate that contractors on public works projects should be adjudged qualified before they are permitted to submit a proposal. This prequalification restricts bidding to the "best" contractors. As with postqualification, the contractors may be asked to submit information about their experience on previous jobs, the capital structure of the company, the personnel available to construct the project, the machinery and equipment to be available for the project, and so forth. Licensing may be a requirement. If a project is complex, it may be a good idea to prequalify the bidders, since having too many bidders on a project may cause some of the better ones not to bid.

What constitutes a regular bid? This question is frequently asked in court. The answers are not always the same, even for seemingly identical circumstances. However, it is generally understood that a bid without a bid security (bid bond, check, or cash), or one with a bid security that is too low, will not be considered. The same is true for bids that fail to acknowledge an addendum. Late bids are usually not considered. Some minor irregularities that may still be considered include bids mailed but not received, bids not dated, bids submitted but not in the required number of copies, and bids with no signature. Bids without a signature are often considered, because of the obvious intent of the bidder. An unsigned bid proposal that is referenced in a signed bid bond may be assumed to be acceptable. As a general rule, however, failure to comply explicitly with the instructions to bidders will result in bid rejection.

There is a tendency among owners to allow some flexibility in regard to changes, corrections, or withdrawals of submitted bids if this is done before the bid opening. If there is a policy that is too liberal in this regard, however, the integrity of the owner will be hurt.

The ultimate decision about the contract award lies with the owner. The owner invariably has the right to reject any and all bids while obligating the award (if made) to the lowest bidder. The selection of the low bidder is made more complex if the project contains several alternates. This is where the owner can show favoritism by placing certain undisclosed priorities on the different alternates. An owner wishing to avoid questions of integrity, will establish, prior to bidding, the priority by which the various alternates will be considered.

After the low bidder is determined on the bid opening day, that bidder is informed of his or her apparent status. The owner also expresses the desire to enter into a contract with that bidder. The unsuccessful bidders should be informed of the status of their bids, but the second and third low bidders may also be told that their bids will be considered valid until a formal construction contract has actually been signed.

Most owners state in the instructions to bidders how long the submitted bids will remain open, or how long after the bid opening the owner has to award the contract. The contract time will start sometime after the contract award, with the total number of allotted construction days being stated in the bid solicitation. The time owners typically permit from bid opening to contract award ranges from 30 to 90 days, with 30 to 60 days being fairly common. It is also common for specific statements to be made that permit a contractor to withdraw a bid if the contract award is not made in that stated period. Since the contract award is rarely made immediately after bid opening, contractors must account for this time lag. Labor, material, and equipment costs can change with time, and so a prudent bidder will anticipate the increase in costs due to a delayed contract award and make a commensurate adjustment in the bid amount. The following is an example of a provision that specifies these time constraints:

> The award of contract will be made within 45 calendar days after the opening of proposals to the lowest responsible and qualified bidder whose proposal complies with all the requirements prescribed. The successful bidder will be notified by letter that his or her bid has been accepted and that he or she has been awarded the contract.
>
> If a contract is not awarded within 45 days after the opening of proposals, bidders may file a written request with the Owner for the withdrawal of their bid, and the Owner will permit such withdrawal.

Once the owner is fully satisfied with the bid of a particular contractor, a contract is issued. After the contractor has returned the signed contract, the owner may notify the contractor to proceed with the work by issuing a notice to proceed. In some cases the notice to proceed may accompany the contract forms. The notice to the contractor may be called either a notice of award or a notice to proceed, but the effect is the same. This is a means of notifying the contractor of the decision to award the contract and specifying the terms under which the contract time will start. The effective date of the notice to proceed may be when it is mailed, when it is received, or when it is signed (most commonly).

The contract start date is often stated specifically in the notice to proceed, or it may begin when the notice has been signed. The notice to proceed form prepared by the Engineers Joint Contract Documents Committee illustrates the simple means by which relevant and necessary contract information can be conveyed (figure 9.4). Some owners define the contract start date as being 10, 15, or 30 days after the effective date of the notice to proceed. Note that the contract may not be signed by the contractor when the notice to proceed is issued. It is possible

City of Arponia

NOTICE TO PROCEED

Date: _____

To: _____
                        (Contractor)

Address: _____

            _____

            _____

RE: **Arponia Project Name:** _____

Dear: _____
                  (Contact Person of Contractor)

Notice to Proceed with the referenced work is hereby granted effective _____,
20____. Enclosed is a completely executed copy of the Authorization for Construction for your files and the fully executed payment and performance bonds.

Sincerely,

Shirley H. Braxton, City Manager

ENCLOSURES:
(1) Authorization for Construction (Arponia Form #8A)
(2) Fully Executed Payment and Performance Bonds

COPY:
Arponia Construction Accounting Office
A/E
Project File G3.7

FIGURE 9.4
Example of the notice to proceed.

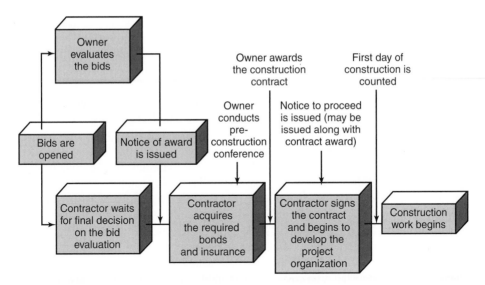

FIGURE 9.5
The contract award phase for competitive bids.

to have the contract "clock" started without the contract documents having been fully executed (figure 9.5). Some owners prefer to have the clock begin on the date the construction contract has been executed by all parties. The following examples are typical provisions:

> The contractor will be bound to the performance of the contract when given notice to proceed with the work on April 1 or no later than 45 calendar days after the date of execution of the contract by the contractor or the deposition of the performance contract bond and payment bond, whichever is last.

> The Notice to Proceed issued to the contractor by the Owner or his or her authorized representative will stipulate the date on which it is expected that the contractor will begin construction and from which date contract time will be charged. In no case, however, shall the contractor begin work prior to the date stipulated in the Notice to Proceed.

It is generally accepted that the notice to proceed obligates the owner to the contract. Only formalities remain in the contract formation after this notice is issued.

On some occasions, a contractor who is notified of being the low bidder may not want to enter into an agreement with the owner. This is often the case when the bidder is "low" by a large margin, suggesting a bidding error. A bidding error does not automatically justify a contractor in refusing to enter into a binding contract. For example, an error of judgment does not constitute valid grounds on which to withdraw a bid. To be released from the bid, the contractor must generally show that the error is one of fact, and it is a material error. Furthermore, the contractor must promptly communicate this information to the owner in written form.

There have been numerous cases in which the validity of bids was questioned, or the merits of a bid had to be assessed by the courts. Examples of bids that were considered irregular or nonresponsive include: a bid bond provided by a surety that was not included on the approved list of bonding companies, failure of the bidder to satisfy the participation goal for minority business enterprises among the subcontractors, submitting a bid two minutes late, failure to indicate that the general contractor would perform sufficient work with in-house (not subcontracted) forces, failure to submit a bid on all alternates, and failure to sign the bid. A decision by the owner that the low bid is irregular may be challenged in court. Conversely, the second low bidder may regard the low bid as irregular and may challenge the award by filing suit. The following cases illustrate the various types of disputes initiated in the bidding process concerning irregular or nonresponsive bids and the methods owners use to determine the recipient of the construction contract. Note that these cases arise only on public works projects.

The case of *Telephone Associates, Inc. et al. v. St. Louis County Board et al.* (364 N.W.2d 378) involved bids submitted on a new state office building in Duluth, Minnesota. This building was to include some office space for St. Louis County. As the project was nearing completion, bids were received for installing the telephone system. The bids were to be evaluated on the following basis: 50 percent on initial cost (including the total installation cost and the monthly maintenance cost), 20 percent on vendor support, 20 percent on system quality, and 10 percent on optional items. Telephone Associates submitted a bid of $544,681, and Norstan submitted a bid of $544,904, a difference of $223. Upon further evaluation, it was discovered that Norstan's bid did not include a fixed monthly rate to cover labor maintenance costs; instead, it included only a fixed monthly rate for materials. All the other bids were properly submitted. When the error was discovered in Norstan's bid, the county evaluators simply inserted the average of the high and low maintenance bids in order to make Norstan's bid complete. Telephone Associates was the low bidder, but on the advice of the county attorney and the county purchasing agent, the contract was awarded to Norstan on May 10, 1982. Telephone Associates unsuccessfully sought a restraining order to prevent the contract award. On August 24 Telephone Associates filed suit for monetary damages. The court determined that the bidding procedures of St. Louis County were indeed improper. The insertion of a dollar amount to complete Norstan's bid provided an opportunity for fraud and collusion. Although there was no evidence of fraud or collusion, the court stated that this type of action was not to be tolerated. At the time of the decision, the installation of the work was already complete, and so the court had to determine the appropriate relief for Telephone Associates. It was decided that Telephone Associates was to recover the costs of preparing the bid and the attorney fees incurred from the time the restraining order was sought. Loss of profit was not to be considered as an expense item.

The case of *Gostovich v. West Richland* (452 P.2d 737) concerned an irregularity in bidding. On a municipal sewer project for the city of West Richland, E. M. Gostovich was declared the apparent low bidder at the public bid opening that took place on August 18, 1961. Three days later, on August 21, the city received a bid from Pieler Construction of Seattle. The bid from Pieler was postmarked 5 P.M., August 17, more than 24 hours before the bid opening. When it was realized that the bid from Pieler was the low bid, the city decided to accept the bid, since the late arrival of the bid was an unintentional informality. When the contract award was given to Pieler, Gostovich sued the city.

Gostovich's argument was based on the city's contract award statute, which said that "the city or town shall let the contract to the lowest responsible bidder or shall have power by resolution to reject any or all bids and to make further calls for bids in the same manner as the original call." It was further stated that the invitation for bids was clear in that all bids were to be filed with the town clerk by 7:30 P.M. on August 18. Since the Pieler bid did not meet this stipulation, Gostovich contended that it was invalid. The city argued that it could waive minor details of the bidding process. The city stated that normally the bid would have been delivered on time. The late delivery of the bid was not the fault of Pieler; there was no fraud or deception by Pieler. The court ruled in favor of the city. It explained that the city should not accept a bid proposal with substantial irregularities, but that the tardiness of the Pieler bid could be waived as the city had done. Thus, the nature of the irregularity in a bid may determine whether it must be rejected.

The case of *Saturn Construction Co., Inc. v. Board of Chosen Freeholders, Middlesex County* (437 A.2d 914) concerned the way in which the owner evaluated bids. Middlesex County advertised for bids for the construction of a correctional facility. Bidders were requested to submit bids on the general contract and additional work and to list the fees to be charged for the administration of five other contracts to be let on the project. Saturn Construction Co. submitted the low bid based on the general contract and the additional work. The bid of M. Gordon Construction Co., Inc., was somewhat higher on those bid items, but its bid for administration was a 3 percent fee compared with Saturn's 5.5 percent fee. When the county calculated the total bids of each firm, including the administration fee, the total bid for Gordon was determined to be $8,596,000, compared with $8,667,000 for Saturn. When the contract award was given to Gordon, Saturn filed suit, claiming that the county should not have considered the administration fee when determining the low bid. The court did rule in the county's favor, but it did so on a technicality. It simply enforced a New Jersey statute that does not permit an unsuccessful bidder to challenge bid specifications after the bids are opened, so that Saturn had no basis for the suit. Essentially the court ruled on the merits of the bid specification rather than on the issue of challenging a contract award. Although this did not influence the decision, the court said that the county's interpretation of the specifications would have been upheld.

In *A & D Construction, Inc. v. Vineland* (418 A.2d 1319) the contract award was successfully challenged. The city of Vineland had advertised for bids on a construction project. Five bids were received, but confusion had existed in the specifications, so the city decided to reject all bids and rebid the project. When the project was rebid, only two bids were received (the low bid was submitted by A & D Construction, Inc.), and the lowest bid was higher than the low bid received at the initial bid opening (the low bid submitted by Commercial Concrete, Inc.). Commercial had not submitted a bid for the second bid opening. The city council rescinded its action of rejecting the bids received at the earlier bid opening and awarded the construction contract to Commercial. This prompted a suit by A & D to restrain the city from executing the contract documents. The court found that the city's reconsideration of the first set of bids that had been rejected was improper and an abuse of discretion. The only low bid that was before the council was the bid submitted by A & D.

Sometimes the apparent low bidder is not the true low bidder, as was illustrated in *Budd Construction Co., Inc. et al. v. City of Alexandria et al.* (401 S.2d 1070). In September 1980 the city council of Alexandria advertised for bids for the replacement of a bridge. Slocum Construction Co. submitted a bid of $97,600, while Budd Construction Co. submitted a bid of $98,600. Slocum was the apparent low bidder on the basis of the total bid price, but the unit prices on this unit price bid altered the positions of the bidders. Slocum had priced a unit item consisting of 567 units at "$12." However, Slocum had entered "eighteen and 00/100" for the unit price in writing. The city council had previously passed a resolution stipulating that written prices were to govern over prices entered in figures. Thus, the unit price to be used for this item was $18, increasing the total bid price by $3,400 to $101,000. The city, however, decided to award the contract to Slocum when Slocum informed the council that it intended to be bound by the $12 unit price. Budd filed suit to stop the city from awarding the contract to Slocum. The court ruled against the city, stating that the city was acting in an arbitrary manner by ignoring an established policy on bid discrepancies. Budd was the actual low bidder and should have been awarded the contract.

An unlicensed surety was involved in *George Harms Construction Co. v. Ocean County Sewerage Authority* (394 A.2d 360). The Ocean County Sewerage Authority had received bids for the construction of a sewer interceptor. The low bidder, Somerset Valley Construction Co., submitted a bid with a performance bond issued by Standard Indemnity Co., a surety that was not licensed in New Jersey. The award was conditionally awarded to Somerset, prompting the second lowest bidder (George Harms Construction Co.) to file suit to block the award. The court ruled against the authority, stating that the provision requiring sureties to be licensed in the state had to be enforced.

In *Ruck Construction Co., Inc. v. City of Tucson* (570 P.2d 220), the city of Tucson had included in its instructions to bidders a requirement that each bid had to be accompanied by a list naming the subcontractors who would be used to perform the work. It was also stated that only subcontractors whose names appeared on the list would be permitted on the project. When Ruck, the low bidder, submitted its bid, an error was made in naming one of the subcontractors. Ruck explained this clerical error to the city on the same day it received its award notice, and requested permission to substitute the originally intended subcontractor for the one who appeared on the list. The city verbally denied the request before the contract was signed, and formally denied it after the contract had been signed. Ruck then filed suit against the city for the difference between the price of the subcontractor who was listed and the price of the subcontractor it had intended to use. The court noted that the evaluation of the subcontractors had to be made by the city before the contract was signed, not after the award. For the award to be justified, the owner had to evaluate only the subcontractors listed. In addition, the court stated that the city was not at liberty to exercise any discretion after the contract was awarded, thereby changing the material terms of the bid. Since the contractor signed the contract after being told that no subcontractor substitutions would be permitted, this served as evidence that the contractor had confirmed the bid that was made.

The case of *George & Lynch, Inc. v. Division of Parks and Recreation, Department of Natural Resources and Environmental Control* (465 A.2d 345) also concerned the listing of subcontractors. George & Lynch submitted the low bid for the construction of a sewage treatment facility in which "none" was written in the space provided for listing the electrical subcontractor. The owner regarded this as indicating that the bidder would perform the electrical work, although George & Lynch was not licensed to do electrical work in Delaware. However, the bidder never intended to perform the electrical work, but was still negotiating with different subcontractors when its bid was submitted. The award was given to the low bidder, prompting an immediate protest from the second low bidder, who had submitted a bid that was $41,448.25 higher. The court ruled that there must be strict adherence to the subcontractor listing statute. It acknowledged that this might actually cost the taxpayers a greater amount, but it also inferred that adherence to the statute would prevent the sort of bid shopping that might actually have been the source of the nonconforming low bid. The listing requirement could not be waived, and the low bid from George & Lynch was determined to be nonresponsive.

A late change in a bid was at issue in *Harry Pepper & Associates, Inc., et al. v. City of Cape Coral et al.* (352 S.2d 1190). The information to bidders on a water treatment plant for the city of Cape Coral stipulated that unspecified equipment had to be submitted for approval before such equipment was incorporated in the bid price. Although this was a clear provision, Gulf Contracting, Inc., submitted a low bid that included an unspecified pump. After the bid opening, Gulf was informed that its pump was not specified and was not acceptable. Gulf then amended its bid, and the city accepted it. Harry Pepper & Associates, Inc., the second low bidder, filed suit to block the award. The court ruled that the city had violated the bidding statute and the rule of fair play. The action of the city was unfair and gave Gulf the option of staying with its original bid or changing it to satisfy the city. In either case, Gulf had an option not enjoyed by any other bidders.

A nonconforming bid also existed in *Charles N. White Construction Co., Inc. v. Department of Labor* (476 F. Supp. 862). On a Job Corps center project in Batesville, Mississippi, White submitted a second low bid of $2,934,900. The contract had been set aside for small business contractors, and so the low bid was rejected when the low bidder was determined to be unqualified. At that point the U.S. Labor Department rejected White's bid because White had failed to acknowledge on the bid form that it had received five addenda. It awarded the contract to the third low bidder. White filed a bid protest, claiming that its bid was responsive. White contended that the company president, Charles White, had orally acknowledged receipt of the addenda to the architect/engineer's secretary. But this was not substantiated. The court ruled that the U.S. Labor Department was correct in rejecting White's bid as nonresponsive.

Sometimes the parties may not fully agree on whether a bid has actually been accepted or not. This was tested in *Cal Wadsworth Const. v. City of St. George* (898 P.2d 1372). Cal Wadsworth Construction submitted the lowest bid on a project to expand the airport terminal in St. George, but its bid was still $100,000 higher than the city had budgeted. The city council awarded the contract to Cal Wadsworth Construction under the condition that it would work with the city to negotiate the work requirements and the price to satisfy the original budget. After the negotiations failed to result in a suitable price reduction, the city rejected all bids as being over budget and decided to rebid the project. Wadsworth sued the city, claiming that its bid had been accepted and that it was entitled to lost profits. It based its argument on the fact that the city had asked the bidders to keep their bids open for 45 days and to furnish bid bonds. The Supreme Court of Utah determined that the city had acted in a manner consistent with standard procedures on public projects, and that its actions did not constitute acceptance of Wadsworth's bid or an offer to Wadsworth to enter into a contract. Since the city's initial bid solicitation stated that the city could reject any and all bids, no party could claim that a contract had been formed until formal acceptance occurred.

California law states that a contractor who made a mistake on a bid cannot participate in future bidding on the same project. This was examined in *Columbo Construction Co., Inc. v. Panama Union School District, et al.* (186 Cal. Rptr. 463). On August 8, 1978, Columbo submitted a low bid of $1.8 million for the construction of a school building. Columbo notified the school board of a $100,000 clerical error and asked to be relieved of its bid if design changes to reduce costs were not made. The school board rejected Columbo's bid because of the error, and also rejected all the other bids since they exceeded the budget. Minor changes were made to the proposed school building, and the project was readvertised. When the bids were opened, Columbo was the low bidder, but the contract was awarded to the second low bidder. Columbo filed suit. The school board stated that it was simply enforcing the bidding statutes, which effectively barred Columbo from rebidding. Columbo contended that the changes that were made established a new project. The court ruled that these minor changes did not make a different project, as the essential features still existed. Columbo's second bid was disqualified.

In *Funderburg Builders, Inc. v. Abbeville County Memorial Hospital et al.* (467 F. Supp. 821), discretion in granting a contract was challenged. Funderburg, a Georgia corporation, submitted a low bid of $468,500 for an addition to Abbeville County Memorial Hospital in Abbeville County, South Carolina. The bid advertisement stated that the hospital reserved the right to reject any and all bids. Funderburg was notified that it was the low bidder, but that the hospital intended to award the contract to the second low bidder, a South Carolina firm. Funderburg filed suit to stop the contract award. The court ruled in favor of Funderburg on the basis that the state statute clearly stipulated that contract awards had to go to the lowest responsible bidder. The statement in the advertisement regarding the hospital's right to reject any and all bids did not give the hospital power to waive the statutes in order to reject Funderburg's bid.

A similar principle was illustrated in *Hilton Construction Co., Inc. v. Rockdale County Board of Education* (266 S.E.2d 157). By submitting a bid of $718,000, Hilton was the low bidder by $2,600 on a high school construction project. The bid forms requested the bidders to state the duration of the contract. The low bidder and the second low bidder, Cube Construction Co., stated that the duration would be 300 days. Since Cube was performing other projects in the county and was known by the school board members, Hilton's bid was rejected and the contract award was made to Cube. Hilton filed suit. The court ruled on the meaning of making awards to "the responsible bidder submitting the lowest acceptable bid." It ruled against the school board, stating that the fact that a contractor is unknown does not automatically satisfy the criteria of being not responsible.

The absence of a signature on bid documents was the primary issue in *George W. Kennedy Construction v. City of Chicago* (481 N.E.2d 913). Kennedy submitted the low bid on a water main project for the city of Chicago. The bid documents included several different pages to be completed by each bidder. Kennedy submitted all the requested forms, but the corporate signature page was not signed. Kennedy was notified of the omission, and the president of the company was permitted to sign the form five days after the bid opening. Soon thereafter the city notified Kennedy that its bid had been rejected as being nonresponsive. Kennedy filed suit. The court ruled in favor of the city, stating that the signature was an instrumental feature in the bid, and that without the signature, Kennedy could not be compelled to enter into an agreement. Thus, Kennedy was in a position not enjoyed by any other bidders. Without the signature, Kennedy would not be bound by the terms of the contract.

A similar case is represented by *Farmer Construction v. Washington State Department of General Administration* (656 P.2d 1086). On the bid form for a construction project at a reformatory, Farmer had no signature on the bid document; however, the president's name had been typed on the form. Farmer's bid was accompanied by the required bid bond. The state rejected Farmer's bid as being nonresponsive. Farmer filed suit on the premise that the lack of the signature did not invalidate the bid. The state contended that Farmer could claim that its own bid was invalid if it was to its advantage to do so. This was construed as placing Farmer in a unique position not enjoyed by the other bidders. The court reviewed the bid documents and found that the bid and the bond were in writing, and that the two instruments were connected by internal reference. The court ruled that Farmer's bid was a firm offer and that the absence of the written signature was not material.

The number of cases involving unqualified bidders is small in comparison to those involving nonresponsive bids. One such case is *Suburban Restoration Co. v. Jersey City Housing Authority* (432 A.2d 564), which concerned bids submitted on a contract to waterproof and caulk the windows of a building owned by the Jersey City Housing Authority. Suburban submitted the low bid for this work, but the contract award went to the next lowest bidder after Suburban was determined to be not responsible. Suburban challenged this action in court. The housing authority stated that it had knowledge of Suburban's failure to perform properly on a prior housing authority project that was very similar to the one being undertaken. Based on this experience, the court ruled that the housing authority was justified in regarding Suburban as not responsible.

MISTAKES IN BIDS

Mistakes in bids are often considered, at least by the party making the mistake, as grounds for nullifying a bid, since there is no meeting of the minds. The general rule is that mistakes of fact are grounds for relieving the bidder of any further obligations, while mistakes in judgment provide no basis for relief. The following cases demonstrate how this has been evaluated in the courts.

A mistake in the bid was the basis for *Puget Sound Painters, Inc. v. State of Washington* (278 P.2d 302). The Washington State Highway Commission asked for bids to clean and paint the two main towers of the Tacoma Narrows Bridge. A low bid of $45,354, accompanied by a bid bond of $2,267.70 (5 percent), was submitted by Puget Sound Painters, Inc. Puget's bid was considerably below the other bids submitted, and so the company principals rechecked their estimate. An error was quickly located, and prompt notice was given to the state. The following day the contract was awarded to Puget, but Puget refused to enter into the agreement and sought recovery of its bid bond. When the state refused to release the bid bond, Puget filed suit. The state then argued that Puget's mistake had been careless and showed "willful neglect." Puget contended that the mistake was not intentionally made; it was made when one of the principals used the estimated quantities prepared by another principal. The estimate was assumed to represent the entire surface area of the main towers, whereas in fact only half the surface area of each tower had been calculated. Thus, the surface area to be cleaned and painted was only half of what it should have been. Upon reviewing the information, the court decided that the bid bond should be returned to Puget. It stated essentially that the decision was based on the following: (1) Puget had acted in good faith, (2) there was no gross negligence, (3) prompt notification of the mistake was given, (4) Puget would suffer hardship for forfeiture of the bond, and (5) the state would not suffer damages by relinquishing the bid bond.

A bid error was the key issue in *Jensen & Reynolds Const. Co. v. State of Alaska, Department of Transportation & Public Facilities* (717 P.2d 844). Jensen & Reynolds submitted the apparent low bid at $1,327,000. Further examination of its unit price bid revealed that Jensen & Reynolds had written a unit price of "thirty-five thousand," but that the extensions showed that "3,500," the amount expressed in figures, had been used in determining the total bid price. Since 10 units were estimated upon, Jensen & Reynolds's bid for this item was $35,000. The state determined that since written values take precedence over figures, the unit price to use was $35,000 and that the bid item should have been $350,000. This adjustment meant that Jensen & Reynolds was no longer the low bidder. Jensen & Reynolds sued to stop the contract award from being made to the second low bidder. The court ruled that the lowest bid is preferred. It further stated that it was clear that, by the nature of the extensions, Jensen & Reynolds meant the unit price to be $3,500. In this case, the intent of the bidders was the primary concern addressed by the court. Jensen & Reynolds was the low bidder.

Mistakes in bidding made by a subcontractor may not allow relief to the subcontractor. In *Arango Construction Company v. Success Roofing, Inc., et al.* (730 P.2d 720), a subcontractor, Success Roofing, submitted a bid of $34,659 on December 6, 1983, over the telephone to Arango for roofing two buildings at Fort Lewis. The bid opening date was extended to December 21. Arango then asked Success to confirm its bid as being accurate. Success confirmed the quoted price of $34,659. Arango was the low bidder when bids were opened and was awarded the prime contract. On January 30, 1984, a standard subcontract agreement was sent to Success. Success then asked for a set of plans for the project. When Success reviewed the plans, it became apparent to them that they had unknowingly estimated their quantities from a set of plans that had been reduced 50 percent. Thus, their bid was only half what it should have been. On February 13 Success informed Arango of the mistake and stated that they would not perform the installation at the originally quoted price. Arango contracted for the roofing with another subcontractor and then brought suit against Success for breach of contract. Success contended that a meeting of minds had never occurred, and so all contract elements had not been satisfied. Success also contended that they expected the contract to be formed when all parties had signed a common instrument. The court determined that the verbal quote was irrevocable. When the bid opening was extended, Success had an opportunity to recheck its bid carefully, but apparently failed to do so. On the basis of promissory estoppel, the court determined that Success had breached the contract and that its offer was irrevocable.

A subcontractor's bid was also involved in *Universal Iron Works, Inc. v. Falgout Refrigeration, Inc.* (419 S.2d 1272). Falgout prepared its bid on a construction project and submitted it to Universal Iron Works. Falgout stated that its bid was for the air-conditioning work, while Universal contended that it had accepted the bid as also including the heating and ventilation work. Universal awarded the contract to Falgout, but Falgout refused to enter into the agreement. Universal then brought suit against Falgout for breach of contract. The court ruled that Falgout's misconception about the work being bid invalidated a fundamental prerequisite for the formation of a contract. Falgout could not be forced to enter into the contract.

The magnitude of the bid mistake may also be a material consideration as to whether a contract can or cannot be awarded. This was shown in a 1998 case in Ohio of *Smith & Johnson Construction v. Dept. of Transportation* (Court of Appeals of Ohio, June 30, 1998) (731 N.E.2d 720). R. F. Scurlock submitted a bid of $3.8 million for a road improvement project. This bid did not include the cost of site preparation. Scurlock's bid was $165,700 lower than the next lowest bidder, Smith and Johnson Construction Company and Robert Johnson (Smith-Johnson). The contract award was issued to Scurlock, after which Smith-Johnson sued to stop the award to Scurlock, claiming that Scurlock's bid was defective since it did not include a cost amount for site preparation. The court found that every deviation from the instructions to bidders did not invalidate the character of a bid. While a substantial variation would be grounds for invalidating a bid, the court ruled that the cost of site preparation, being less than $100,000 and constituting only 2.5 percent of the bid, was not a substantial portion of the bid. The Ohio Department of Transportation was justified in awarding the contract to Scurlock.

REVERSE AUCTION BIDDING

In recent years (beginning in the mid-1990s), a new variation of bidding was devised that was initially employed on some public works construction projects.

This is an Internet-based bidding process known as reverse auction bidding. In this process contractors are required to use a prescribed website to submit their bids. This is an open website in that all bidders can readily see other contractors' bids that have been submitted. This is the reverse of a typical auction in which competing bidders bid upward. Unlike a sealed bid procurement process, a reverse auction bidding process allows all bidders to see their competitors' total bid prices, generally without seeing the identity of the bidders. With this knowledge of the bids that have been submitted, each contractor knows the amount that must be bid to underbid the current low bidder.

Reverse auction bidding is not without criticism. This process has been described as unethical, a method of bid shopping, commodity shopping, and so on. Since it is a process that ends with the lowest bid that has been received at the time the auction duration expires, it elevates the importance of overall project cost, while de-emphasizing other important factors. Certain variables that may be critical to the sealed-bid process, such as a contractor's safety record, reliability, financial soundness, and the overall quality of the contractor's past work, are difficult to quantify and tend to be pushed by the wayside in this process that favors the contractor who can produce the overall lowest bid. It is possible to prequalify bidders, but agencies that promote reverse auction bidding tend to discourage short-listing for some reason.

A commodity product can be purchased via a reverse auction bidding process because an owner can expect a great degree of similarity between bidding options. Products such as concrete aggregate, lumber, and nails are examples of these kinds of commodities, where the products themselves are essentially identical and the largest difference between suppliers is price and availability. It has been argued that construction services are very different from commodity products; therefore, their procurement must also be different. This is because general contractors' services do not have this sort of across-the-board similarity. A great many variables and levels of risk-taking are included in a contractor's price to an owner. These variables include means and methods, equipment used, type of materials used, safety concerns, material delivery method, material removal method, production rates, and so forth, all of which are essential factors in choosing a contractor.

The results of reverse auction bidding have been compared with the results of the traditional sealed-bid process; no significant savings were found in the use of the reverse auction bidding process on publicly awarded projects. With this process, a contractor can submit more than one bid. It has been suggested that owners might not get the best prices as a contractor will not start out with the lowest bid, and a contractor might be declared the low bidder before the bidding gets to that contractor's lowest price. Despite the controversy surrounding reverse auction bidding, some private owners have also experimented with the use of this practice on their projects. With the widespread opposition to reverse auction bidding among contractors and contractor associations, the practice has not been widely adopted in the industry.

MULTIPLE BID PACKAGES OR PHASED APPROACH

It is common for projects to be awarded to a single prime or general contractor that subcontracts all or major portions of the work to specialty contractors. The overall responsibility for project completion then rests with a single general contractor. This is a standard approach that has been implemented on many projects. Occasionally, an owner may decide to break up a project into multiple contracts that will be undertaken by several general contractors. There may be a number of reasons for doing this. For example, a project may be of such magnitude that many contractors would not consider bidding on the project because of their bonding capacity limits. Apportioning the work via several general contracts also makes it possible to construct the project in phases, allowing some portions of the overall project to be started while others are still in the design stages.

A project may not be huge by many standards, but the size might preclude many smaller, local contractors from the opportunity of participating in the project. For example, a municipality may wish to have a $250 million museum built with local contractor participation. This participation may be realistic only if the work packages are small. This will require that the work be organized or repackaged so that many smaller firms can assist in the construction effort. The bonding capacity of the smaller firms is also accommodated by this repackaging of the work. With the repackaging of the work, there will be multiple prime contractors on the project. The various contractors will need to coordinate their efforts to minimize conflicts. This cooperation between contractors can be addressed contractually. To assist in the smooth undertaking of the construction effort by multiple prime contractors, the municipality will have a construction manager to represent its interests. The construction manager will help in the coordination of the project so that the many parties work smoothly together to construct a successful project.

In *Martin Engineering v. Lexington County School District and Sharp Construction Company* (615 S.E.2d 110), Sharp Construction of Sumter, Inc., submitted the low bid of $16.3 million on a school project in South Carolina. The second low bid was for $17.375 million that was submitted by Martin Engineering. Immediately after the bids were opened, Sharp notified the school district that it had made a significant error by not including the cost of roofing in its bid. Sharp requested permission to modify its bid by including the cost of roofing. The district allowed the modification and continued with the process of awarding the contract to Sharp, as its bid was still the lowest bid at $16,913,500. Martin sued to stop the contract award, arguing that the bid modification should not have been allowed. The court concluded that the contract could be awarded to Sharp, and that the modification was justified as the relative position of Sharp's bid remained unchanged.

While it is customary to award contracts to the lowest responsible contractors submitting a regular bid, the definition of what constitutes a regular bid is not uniformly defined. This was shown in the Georgia case of *R. D. Brown Contractors, Inc. v. Board of Education of Columbia County* (626 S.E.2d 471). McKnight Construction Co. submitted a low bid of $11,259,000 on a school project that stipulated that the bidders were to include the list of subcontractors with their bids. The bid was submitted without this list. On the day following the bid opening, McKnight provided the list of subcontractors, and the school board proceeded to enter into a contract with McKnight. Brown, the second lowest bidder, promptly filed suit to stop the contract award, claiming that the failure to provide the list of subcontractors was a material irregularity. Deviating from many other similar court decisions, the court upheld the school district's decision to award the contract to McKnight. The court stated that the school board could waive certain bidding requirements. This is not consistent with many other court decisions with similar circumstances.

CONTRACTOR LICENSING

Most states have statutes regarding licensing for contractors. The purpose of licensing is to ensure that the public is protected from unscrupulous contractors. This does not necessarily mean that contractors must be licensed in order to perform construction work, although some owners and most public agencies stipulate this as a specific requirement. Most licensing laws simply stipulate that an unlicensed contractor cannot use the court system as a means of redress for any construction-related disputes. This indicates to contractors that it behooves them to be licensed. Licensing laws vary between states, but generally they can be satisfied by paying a licensing fee of a few hundred dollars. In some cases, the process includes an examination that must be passed before a license will be granted.

In the state of Washington, the statute related to licensing (RCW 18.27.080) states the following:

> No person engaged in the business or acting in the capacity of a contractor may bring or maintain any action in any court of this state for the collection of compensation for the performance of any work or for breach of any contract for which registration is required under this chapter without alleging and proving that he was duly registered and held a current and valid certificate of registration at the time he contracted for the performance of such work or entered into such contract.

The case of *Bremmeyer v. Peter Kiewit Sons Co.* (585 P.2d 1174) illustrates the strength of this statute. Peter Kiewit Sons Company was the prime contractor for the construction of several miles of Interstate 90; it subcontracted with Bill Bremmeyer of Bremmeyer Logging Company for Bremmeyer to clear timber from the right-of-way. Bremmeyer agreed to pay $35,000 for the timber (to be sold to a lumber mill), furnish payment and performance bonds for $50,000, obtain liability insurance, and remove the timber. Bremmeyer made the required payment and furnished the required bonds and insurance, but the timber was never removed. Although work on the project had begun, the project was interrupted because of environmental concerns and was eventually terminated. Kiewit recovered $1,729,000 from the state for cancellation costs, but paid Bremmeyer only $38.73 for termination of the subcontract. Bremmeyer promptly filed suit against Kiewit to recover the value of the unharvested timber. Bremmeyer felt that he had complied substantially with the statute in providing the required bonds and insurance. Bremmeyer also felt that he should not be regarded as a contractor and therefore not be bound by the statute. The court stated that the law requiring registration is designed to protect the public from abuses by contractors, but it concluded that this legislation is not meant to protect prime contractors from actions initiated by unregistered subcontractors. Thus, the court ruled in favor of Bremmeyer on the basis of the intent of the law, rather than a strict interpretation of it.

It is clear that the registration provisions are designed to protect the public from unscrupulous individuals, but they may also bar seemingly innocent individuals from seeking compensation through the courts. This was exemplified in *Stewart v. Hammond et al.* (471 P.2d 90). Archie Stewart, a carpenter, was engaged by Mr. and Mrs. Hammond of Clarkston, Washington, to make minor repairs and alterations to their home. It was verbally agreed that Stewart, who was not a registered contractor, would furnish all the labor and procure the needed materials. He was to be reimbursed for his costs and paid an hourly wage for his own work. Eventually the project escalated in size to the point where Stewart hired four helpers and several specialty contractors. The Hammonds made periodic payments to Stewart, but refused to pay the final amount requested because they felt that the request was excessive. Stewart then filed suit for the unpaid balance. Stewart stated to the court that he was essentially an employee of the Hammonds, and that the registration requirement did not apply to him. The court did not agree. Although circumstances led Stewart from the role of an employee to that of a contractor, the court ruled that he was in noncompliance with the registration requirements.

Cameron v. State of Washington (548 P.2d 555) further illustrates the need for registration. William Cameron submitted a bid of $41,443.60 for the construction of a parking lot on a college campus in Skagit County. The bid was submitted with a $3,413.10 cash bid bond. Cameron was the low bidder and was awarded the construction contract. Cameron proceeded with construction of the parking lot and was near completion when the state notified him to stop work because he was in noncompliance with RCW 18.27.20, the registration statute. The state used the parking lot, but refused to pay for the work. Cameron claimed that since the state was able to use the parking lot, it should pay a reasonable value for the benefits it had received. The state contended that without registration, the contractor had no basis to make a claim. The court simply stated that it could not nullify the registration statute in spite of the fact that the state had been enriched by Cameron's work. It also addressed the issue of the bid bond. Since the statute related to compensation for breach of contract, the court ruled that Cameron was entitled to the cash bid bond, since this was not covered by the statute.

Vedder v. Spellman (480 P.2d 207) was similar in that Vedder was an unregistered contractor. In this case Spellman was a homeowner for whom Vedder performed some alteration and repair work. Spellman gave a check to Vedder for $2,500 to compensate Vedder for most of the work performed. Before Vedder deposited the check, Spellman halted payment. Vedder filed a suit for the money owed, but the court barred any recovery. The fact that the check was written did not change the circumstances. Vedder was an unlicensed contractor, and the courts could not be used to seek compensation.

Not all decisions are based on the letter of the law. In *H. O. Meyer Drilling Co., Inc. v. Alton V. Phillips Co.* (486 P.2d 1071), Meyer entered into a contract with Phillips. At the time of the contract agreement, Meyer was not a licensed contractor because he had not renewed his registration. Meyer had filed a surety bond, obtained the required insurance, and essentially satisfied all other requirements, short of being licensed. After Meyer had performed the work, Phillips refused to pay on the grounds that Meyer was not licensed. The court first stated that the statute was unambiguous and should be followed to the letter. However, it also decided to consider the intent of the legislation. The prime issue here was that contractors should be responsible. The court ruled that the bonds and insurance that were provided by Meyer showed that substantial compliance had taken place, and that Meyer could use the courts to receive payment.

The case of *Northwest Cascade Construction Inc., et al. v. Custom Component Structures, Inc.* (519 P.2d 1) also showed that the courts do not view these provisions in an absolute fashion. The dispute concerned a contract, dated October 1968, in which Northwest agreed to frame the walls of 12 Seattle apartment buildings for Custom. Northwest had completed four buildings when it began to work on another project in Olympia. At that point Northwest made a verbal agreement with Custom that two employees, Dave Kuipers and Russell Mowry, would take over the project and finish the remaining eight buildings. Since Kuipers and Mowry were not registered contractors, Northwest permitted them to use Northwest's bond and file required reports under the name of Northwest. Custom was informed of this arrangement and raised no objections. During construction, Mowry and Kuipers used the name M-K Construction Co. and made billings under that name. As construction progressed, Northwest equipment continued to be used on the project, and Paul Box, the vice president of Northwest, maintained almost daily contact with Custom concerning progress. Box also received a salary from the account of Mowry and Kuipers. The buildings, plus some extra work, were completed by April 1969. Custom then refused to pay for the work, since Kuipers and Mowry were not registered under the state registration requirements. Northwest filed suit to obtain payment. The court ruled in favor of Northwest, as the completion of the eight buildings and the extra work were considered extensions of the original contract. Kuipers and Mowry were not regarded as being separate from Northwest, but as functioning as employees. There was substantial compliance with the registration statute.

REVIEW QUESTIONS

1. What information should be included in an advertisement that does not relate directly to the nature or scope of a project to be constructed?
2. What is the purpose of a short list?
3. What type of information might be requested of a contractor who wishes to be prequalified for a construction project?
4. Give an example of an irregular or nonconforming bid.
5. Under what conditions might the low bidder be able to withdraw the bid, even though the bids have already been opened?
6. What unique risks are taken by a contractor who is not licensed?

10

CONSTRUCTION CONTRACT DOCUMENTS

CONSTRUCTION CONTRACT DOCUMENTS play an important role in the development of a project. They provide the bridge between the owner's conceptual image of a project and the actual construction of the physical facility. This vital link is provided by project designers: architects, engineers, or both. On many projects the roles of owner, designer, and constructor are played by different firms or individuals, often parties who have never worked together before. The common bond for these parties is provided by the construction contract documents, which may include the construction agreement (figure 10.1), drawings, general conditions, supplementary provisions, technical specifications, and addenda. These documents are prepared by the designer and become the vehicle through which the owner and the contractor communicate.

The designer is generally the first party selected by the owner, and the contractor is often selected after the designer has completed the contract documents. It is common for designers to be paid on a fee basis, which historically has been determined as a percentage of the cost of construction; however, a variety of payment methods may be devised.

Although this is not generally a serious concern, the question of design ownership may arise. On public works projects, the design invariably belongs to the owner, as dictated by law. On private projects, design ownership is established by the contract between the owner and the designer. Usually architects retain such ownership rights if the owner-designer contract does not address this issue.

On private works projects, the arrangement between the owner and the architect does not affect third parties. Consequently, third parties can still end up using the design that legally belongs to the owner or the designer. This problem can be avoided if the owner obtains a copyright, which is good for 50 years. The owner can also retain ownership of the design if the design was made for hire. This generally means that the designer must be an employee.

CONSTRUCTION AGREEMENT

This agreement and contract, made and entered into at Denver, Colorado, this *21st* day of *March, 20*11, by and between: *Little People Play and Care Providers,* a private corporation hereinafter designated as the "Owner," and *Rocky Mountain Constructors, Inc.,* hereinafter designated as the "Contractor."

WITNESSETH:

That whereas the Owner has heretofore caused to be prepared Call for Bids, Definitions, General Instructions to Bidders, Special Instruction to Bidders, Affidavit of Pre-Qualified Bidder, Contractor's Proposal, Specifications for Construction, Construction Drawings, Addenda, and Performance Bond Form, hereinafter referred to as "Contract Documents" for the construction of the *Administrative Offices for Little People Play and Care Providers* Project and the Contractor did on the *7th* day of *March, 20*11, file with the Owner a proposal to construct said Project and agreed to accept as payment therefore the sum of

Nine Hundred Sixty-Five Thousand and no/100 Dollars, AND

WHEREAS, the said Contract Documents fully and accurately describe the terms and conditions upon which the Contractor proposed to furnish said equipment, labor, material, and appurtenances and perform said work, together with the manner and time of furnishing same, AND

WHEREAS, the duration of construction will be as set forth in the Specifications for Construction, with construction work to commence on the date set forth in the Notice to Proceed.

IT IS THEREFORE AGREED, first, that a copy of said Contract Documents be attached hereto and do in all particulars become a part of the Agreement and Contract by and between the parties hereto in all matters and things therein set forth and described; and further, that the Owner and the Contractor hereby accept and agree to the terms and conditions of said Contract Documents as filed as completely as if said terms and conditions and plans were herein set out in full.

IN FAITH WHEREOF, witness the hands and seals of both parties hereto on the day and year in the Agreement first above written.

By: _____   Title: _____
    (Owner's Authorized Representative)

By: _____   Title: _____
    (Contractor's Authorized Representative)

Witnessed By: _____

FIGURE 10.1
Example of a construction contract.

DRAWINGS

Drawings are an important component of the construction contract documents. The drawings, also known as the *plans* or *blueprints*, are the primary vehicle by which the physical, quantitative, or visual description of the project is conveyed. The drawings are organized in a fashion that follows to some extent the physical sequence in which the construction work will be performed. The general categories on building projects are as follows:

- General information and site work
- Structural
- Architectural
- Plumbing
- Heating, air-conditioning, and ventilation
- Electrical

Each of these topic areas usually has a separate numbering system. The pages in each section are usually numbered consecutively, with the page number preceded by a letter designating the section. For example, the architectural pages may be numbered A1, A2, A3, and so forth. Different types of information are presented in the different sections. Thus, a clear understanding of the organization of the drawings will enable a quicker assessment of the requirements for a particular component of the project. The type of project will dictate the nature of and extent to which detailed information must be provided. Although the building materials to be used in a structure may be identical to those used in other structures, the actual configuration of the structure will be unique. Thus, the details for some portions of the project may be similar to those used on other projects, but the drawings will constitute a unique project.

Most drawings that are presented represent orthographic projections of walls, wall sections, and so on. Orthographic projections are advantageous since they can be scaled to obtain information directly. In scaling a drawing, one always refers to the scale being used and checks that the drawing size was not altered, such as through photocopying at 50 percent, after it was drawn. This can generally be checked easily if the dimensions of some portions of the drawing are provided. Under some conditions, such as for piping systems, isometric drawings more clearly present the desired information.

The general or site section of the drawings gives overall information about the project, including property lines, roadways, access routes, and the location of the structure as it relates to the site. The survey control monuments or location points for the project are identified in this section. All relevant elevations, grade lines, slopes, and boring log information are shown. Details concerning landscaping and site utilities that are not part of the primary structure are also shown.

Each section of the drawings typically begins with a list of the standard symbols and standard abbreviations used in that section. Each drawing page is typically accompanied by information concerning the scale and the date drawn, along with approval signatures.

The architectural section consists of drawings that show the finishing treatment for the various components of the project. For buildings, this often includes a floor plan for each level of the project, a unique designation for each area, various wall sections to clarify architectural treatments, room finish schedules, door schedules, window schedules, building elevation drawings, and reflected ceiling plans. Standard notations are generally used to cross-reference the details, wall sections, and floor plans.

The structural drawings show notes and the typical structural details—whether steel, concrete, or timber—to be used for the project. All the major structural components from the foundation to the roof are shown, along with major connections. As with the architectural drawings, the structural drawings provide numerous details of sections cut through major members, such as concrete columns and beams, to show reinforcing steel placement, anchoring details for steel columns, connections between steel and concrete, and so forth.

The mechanical drawings show the various locations of piping runs and the details related to elbows, valves, meters, controls, and the like. Similarly, the electrical segment shows all drawings to convey information about the proper installation of the electrical components of the project. Care must be taken in the preparation of these drawings to make sure the information is complete. The electrical contractors who will bid on the project will focus primarily, if not exclusively, on these drawings. Any omissions will probably not be noticed or caught by someone cross-referencing between the different sections of the drawings.

The "roll of plans" is often a very bulky document. This is beginning to change, however, as some designers and owners are making drawings available via compact discs and websites. Just as the traditional "blueprints" gave way to "blue line" drawings, so too, the blue line drawings gave way to "half size" drawings, and these will eventually give way to electronic images. For some owners, the drawings and bid packages are virtually paperless.

PROJECT MANUAL

The *project manual* consists of the bidding documents, general conditions, supplementary provisions, and the technical specifications. These documents are often contained within a single binder or "book." This simplifies the handling of the documents and provides greater assurance that some items are not misplaced. As with the drawings, some owners and designers are now making these available via compact discs and website addresses.

GENERAL CONDITIONS

The *general conditions,* often referred to as the *boilerplate,* augment the construction contract and outline the rules under which the project will be built. They establish the rights, authority, and obligations of the contracting parties: the owner, the owner's representative, and the contractor.

Various standard conditions have been developed by different groups. Some common ones are as follows:

- General conditions developed by the American Institute of Architects (AIA).
- Owner/Contractor Standard Agreement & General Conditions prepared by 21 construction industry associations (ConsensusDOCS).
- General conditions developed by the Associated General Contractors (AGC) and the American Society of Civil Engineers (ASCE).
- General conditions developed by the U.S. government (federal acquisition regulations).
- General conditions developed by the Engineers Joint Contract Documents Committee (National Society of Professional Engineers, American Consulting Engineers Council, ASCE, Construction Specifications Institute).

Most parties involved in the construction process prefer the use of standard general conditions. Like standard specifications, standard general conditions have the advantage of being familiar to all parties, and the wording is clearly understood. This saves time and effort in redrafting new general conditions for each job. Furthermore, these general conditions have often been court-tested so that the legal interpretation is known. On most building construction projects and on projects designed primarily by architects, the general conditions are often those developed by the AIA, referred to as AIA Document A201 (refer to Appendix).

SUPPLEMENTARY CONDITIONS

Also known as special provisions or special conditions, *supplementary conditions* are more specific for the job being constructed. They serve the function of amending and augmenting the general conditions and thus tend to be more specific. Topics addressed in the supplementary conditions include the following:

- The number of copies of the contract documents to be received by the contractor.
- The type of surveying information to be provided by the owner.
- Which materials the owner will provide.
- Specific information about material substitutions.
- Changes in insurance requirements.
- Requirements concerning the phasing of construction.
- Examination of the site.
- Start date for construction.
- Requirements for project security.
- Requirements for temporary facilities.
- Specific procedures for submitting shop drawings.
- Cost-reporting requirements.
- Job schedule requirements.
- Special cleaning requirements.

- Traffic control requirements.
- Discovery of artifacts of cultural or historical value.

Some supplementary provisions may eventually become part of the "boilerplate." The owner may take a supplementary provision and subsequently decide that the provision should apply on all contracts. For example, suppose an owner has a construction site that is suspected of containing artifacts from an encampment of one of the early Spanish explorers. The owner might want to develop a supplementary provision for this subject, and once it is deemed time-tested, it may become part of the general conditions. Such a provision might state the following:

> If the Contractor or any workers discover evidence of possible scientific, prehistoric, historical, or archaeological importance, the Contractor will stop all activity in the vicinity of the discovery and promptly notify the Owner's representative by telephone giving the nature and location of the findings. Written confirmation shall be forwarded within 24 hours. Until the finding is examined, the Contractor shall exercise care so as not to damage artifacts or fossils uncovered during excavation operations. The Owner will ensure that the discovery is studied immediately by archaeological or paleontological authorities, within a timeframe not to exceed 72 hours. The Contractor agrees that there will be no claim for additional payment or for an extension of time because of any delays in the construction progress due to the discontinuance of work or removal of any remains or artifacts for the first three days. Thereafter, the contractor shall be compensated in terms of contract amount and project duration as adjusted through a formal change. After the site of the discovery has been examined, the Contractor will be notified in writing to resume construction operations, or that adjustments in the construction sequence may be required. The Owner reserves the right to terminate the contract if the find is determined to be significant.

This provision might become a general conditions provision if the owner has several projects for which these terms might apply. As soon as a supplementary provision is no longer unique to one project, consideration might be given to adding it to the general conditions.

SPECIFICATIONS

The term *specifications* is often used very broadly to include all the contract documents, with the exception of the drawings. This would include the following:

- Invitation to bid
- Instructions to bidders
- General conditions
- Supplementary conditions
- Bid proposal form
- Bid bond form
- Contract bond form
- List of prevailing wages (may be part of the supplementary conditions)
- Noncollusion affidavit
- Technical specifications

Technical Specifications

The *technical specifications* are needed to cover the qualitative items of a project. This is information that is not easily shown on the drawings, which are more quantitative in nature. The technical specs are written descriptions (as opposed to being drawn) of the quality of the various aspects of the construction project. Technical specs will generally follow a standard format of providing the following information:

• General: stipulates ground rules for the work to be performed and defines the scope of work to be performed within the specification section.
• Product or products: describes the product or products (materials, equipment, accessories, components, fixtures) and the development and manufacturing process to be used in producing them.
• Execution: describes the preparation, workmanship, installation, erection, and application procedures to be employed along with quality requirements and performance criteria that must be satisfied.

Although the written information could be included directly on the plans, this would detract from the information shown on the plans as they will quickly become cluttered with this verbiage.

A drawing may adequately show how a basement is to be constructed, but the qualitative aspects have to be described further in the specs. Information to be shown in the specs for such a basement might include the following:

• Quality of concrete
• Quality of aggregate
• Quality of workmanship (mixing, placing, curing, forming)
• Quality of material used for damp proofing
• Description of material for pipe drains
• Preparation of soil foundation
• Type of backfill
• Compaction requirements

The specs are used to modify or clarify what is shown on the drawings. Occasionally there is a conflict between the plans and the specifications. This can easily happen because the specification writer and the draftsperson are rarely the same individual. Thus, a change made on one document may not be communicated to the person in charge of the other document. The plans or specs may even contain a conflict within themselves if there is a large project with several spec writers and draftspersons. The contract should stipulate how such conflicts are to be interpreted. There are two general means of resolving the conflicts.

1. In case of a conflict between the plans and specifications, the specifications will govern. (It is often assumed that the specifications should govern over the plans, but the conflicts may be contained within different parts of the technical specifications. The special provisions may also address how conflicts between the plans and specifications are to be resolved. AIA Document A201-1997 does

not specifically address this other than to state that the contractor is to promptly bring conflicts to the architect's attention.)

2. In case of a conflict between the plans and specifications, the conflict will be resolved by the architect.

Instances may occur in which an item is mentioned in one document (plans or specs) but not in the other. The contract should clarify how this discrepancy should be resolved. Often the contract will state that if an item is included in one of the documents, it is to be assumed that it is covered in both. For example, a particular valve may be noted on the plans but not mentioned in the specifications. Under the above interpretation, the omission of the information from the specification gives no relief to the contractor; that is, if it is in the plans, it is treated as if it were also included in the specifications.

An example of a provision in which it might be particularly difficult for a contractor to interpret the inherent risks is as follows: "Should inconsistencies exist such as the Drawings disagreeing within themselves or with the Specifications, the better quality and/or greater quantity of work or materials shall be estimated upon, performed, and furnished unless otherwise ordered by the Architect in writing during the bidding period." If inconsistencies are not detected before bidding, such oversights can be costly.

If conflicting, ambiguous, or vague information is noted by the contractor, the contractor will want to obtain a quick clarification of the information. This is generally accomplished through a *request for information,* commonly known as an RFI, that outlines the question to be clarified by the architect (refer to Article 3.2.1 of AIA Document A201-1997).

Organization of the Technical Specifications

With the exception of the drawings, most of the contract documents will be included in one binder, which is often referred to as the *specifications.* In this binder, the technical specifications will be in the later portion. That is, the technical specifications are the last items included, generally consisting of more than half the book.

The organization of the technical specifications, like that of the drawings, follows the general order of the construction process. The technical specs usually begin with the site-type items and conclude with the finish items. As with the drawings, the technical specs are separated into sections for quick reference. In building construction, most technical specifications are organized into divisions. These divisions usually segregate the technical specs by craft jurisdiction and into segments that conveniently package similar types of work for subcontracting. Thus, the subcontractors on a project need be concerned only with the divisions of the technical specs that affect them.

A format for organizing the specifications for building construction was developed by the Construction Specifications Institute (CSI) and is used very widely. When the CSI format is used, the user can quickly become oriented to the document. A new format obviously will be more confusing to the user. The CSI format is broken down into the following broad categories:

- Division of bidding and contract requirements

> Prebid information
> Instructions to bidders
> Information available to bidders
> Bid forms
> Supplements to bid forms
> Agreement forms
> Bonds and certificates
> General conditions of the contract
> Supplementary conditions
> Drawings index
> Addenda and modifications

- Division 1: General requirements
- Division 2: Sitework
- Division 3: Concrete
- Division 4: Masonry
- Division 5: Metals
- Division 6: Wood and plastics
- Division 7: Thermal and moisture protection
- Division 8: Doors and windows
- Division 9: Finishes
- Division 10: Specialties
- Division 11: Equipment
- Division 12: Furnishings
- Division 13: Special construction
- Division 14: Conveying systems
- Division 15: Mechanical
- Division 16: Electrical

General Information about Specifications

For a specification to serve its purpose, it must satisfy some basic criteria, including the following:

- Technical accuracy and adequacy.
- Definite and clear stipulations.
- Fair and equitable requirements.
- A format that is easy to use during bidding and construction.
- Legal enforceability.

Whether these criteria are satisfied depends on the type of specification that is written. There are various types of specifications, some of which are used more frequently than others.

Design Specifications

Design specifications are also called material and workmanship specifications, method and materials specifications, and prescriptive specifications. In this type of specification, a particular kind or type of material is to be used, a particular dimension is required, the installation instructions are given, and so forth. If the spec concerns a method, it will state in detail exactly what the contractor is to do to satisfy the requirement. With this spec, the desired result may not occur, even though the contractor fully complied with the spec.

By using this type of spec, the owner warrants by implication that the specs will produce the desired results if they are followed by the contractor. Thus, the contractor is not liable if the desired end result is not obtained. The Spearin Doctrine states that the contractor is not liable for performance when the specifications have been followed. Essentially, the Spearin Doctrine, based on an often cited 1918 case (*United States v. Spearin,* 39 S. Ct. 59), states that there is an implied owner's warranty of the accuracy and adequacy of the drawings and specifications. The contractor cannot be held responsible for defects in the drawings and specifications.

Examples:

The wall shall be constructed of 2×4 studs spaced at 16 inches on center.
Two-inch batt insulation shall be used in the wall.
A Gentrix Model 373H air conditioner shall be provided and installed.

The adequacy of the design specifications was at the root of *Pittman Construction Company v. Housing Authority of New Orleans* (169 So.2d 122). In this case the Housing Authority of New Orleans (HANO) was a public corporation formed to construct low-income housing. The plans and specifications for one of those projects took 20 months to prepare. The specifications described in minute detail exactly what work was to be done, including the quality and quantity of all materials and the precise manner and sequence of performance. At some point during construction, the contractor, Pittman, reported subsidence in the soil in varying degrees. Eventually the subsidence reached about 30 inches in some locations. Pittman repeatedly advised HANO of the problems and requested written instructions about what should be done; the response was consistently that Pittman was to continue "in accordance with the plans and specifications." HANO's representative had certified in 23 monthly work estimates that the work had been done in accordance with the contract provisions. When the next payment request was submitted, it showed that 98.1 percent of the work had been done and approved by HANO's representative. Pittman was then told that payment was being withheld until Pittman corrected a number of "deficiencies." Pittman then walked off the job and sued HANO for materials furnished, work performed, and retainage withheld. HANO contended that Pittman had abandoned the project and owed HANO for project completion. The court ruled in favor of Pittman, stating that failure to pay Pittman constituted a breach of contract for which Pittman was justified in its refusal to continue to work. Furthermore, HANO had warranted implicitly that the plans and specifications would be adequate and that the damage that occurred was due to an inadequate design. In short, the contractor cannot be held liable for the end result of a design specification if the contractor adheres to the specification.

Design specifications do not require that performance be as expected. This was demonstrated in *Fanning and Doorley Construction Co., Inc. v. Geizy Chemical Corporation* (305 F. Supp. 650). Geizy hired Metcalf & Eddy Engineers (M&E) to design and supervise improvements on one of its plants. In the contract documents M&E included a design specification stating that the joints of the pipe system should be made with asbestos-rope caulking and "Causplit Mortar by Pennsalt Inc." The contract for the work was awarded to Fanning and Doorley (F&D). When a portion of the pipe work was completed, it was subjected to a test, which it failed. The joint work had been done as specified, but the pipes had not passed the leakage test. The manufacturer eventually came up with a successful joint that utilized a caulking that differed from the original specifications. F&D finished the project with the new method. Then Geizy asked F&D to correct the work in places where the joints had failed the earlier test. F&D refused to do the work without additional compensation. Geizy refused to pay F&D the balance due on the contract, and F&D filed suit. Geizy stated that F&D had to correct the faulty workmanship, but F&D argued that the failure existed in the specifications, not in the workmanship. The court ruled in favor of F&D. Essentially, the owner is responsible for the outcome of design specifications when the contractor has complied with those specifications.

Performance Specifications

With a performance specification, the results or the performance of the finished product, rather than the specific methods and materials used to construct the product, are specified. The product satisfies the spec as long as it does the job. Since this spec focuses on the end product rather than the means of getting the product, this form of spec is growing in popularity.

This specification does not stipulate the method to be used to obtain the desired results. However, the spec may offer suggestions that may be employed to obtain those results. Of course, the contractor is not obligated to accept the suggestions. Note that if the architect on a project gives specific verbal directions on how a task is to be done, the specification then becomes a design specification.

Examples:

The wall shall be constructed to support a vertical load of 300 pounds per lineal foot (plf).
The system shall develop a total output of 20,000 BTU.
The 28-day compressive strength of the concrete shall be 4000 psi.
The wall shall have an insulation value of R-19 or greater.

In these specifications the responsibility for design rests with the contractor. The end results may be stated in various ways. For example, the performance of the product may be described in terms of quality, actual in-place operation conditions, finish, color, appearance, tolerance, clearance, noise level, and the like.

Since only the performance criteria must be met, the contractor is responsible for selecting the methods and materials. If this selection proves to be inadequate, the contractor is liable and the work must be redone at the contractor's expense. This type is strongly preferred by owners because it tries to tap the ingenuity and creative talents of contractors to the greatest extent. An innovative contractor may be able to satisfy the spec more inexpensively than other contractors can. This lower cost will be reflected in the contractor's bid. Of course, an innovation developed after bidding only increases the profit of the contractor.

Under this type of spec, the contractor is limited to the performance as specified and not beyond. If the owner requires the contractor to provide more than is specified, an extra will probably be claimed by the contractor.

Performance and Design Specifications

This type of spec is one in which the contractor is instructed how to do a task, and then told to warrant that the results will be satisfactory.

Examples:

The wall shall consist of 2×4s spaced at 24 inches on center, and it shall support a vertical load of 500 plf.

The concrete shall consist of ___, and the 28-day compressive strength shall be no less than 4000 psi.

If the contractor follows the procedures as specified, the contractor will not be bound by the performance portion of the specification. That is, both portions of the specification cannot be enforced by the owner. Obviously, the use of this type of spec is to be avoided.

Closed Specifications

A *closed specification* requires a specific item or system. The purpose of this type of spec is to ensure that only products of a particular type are used. As a rule, this type of spec is more frequently found in the private sector because it is in principle not legal on public works projects. The reason for this is that a closed spec eliminates the chance for competition; that is, the express inclusion of one item implies the exclusion of all others.

Closed specs give strong advantages to manufacturers who have their products specified. Since there is no competition, the result of this type of spec is to drive up the cost of construction.

A closed spec can be either a design or a performance specification. For example, the specification can stipulate that a particular model of a particular manufacturer be used on the project. It is still considered a closed spec if two models from two manufacturers are named. In public works projects, at least three manufacturers' models must be named to avoid the designation of being closed. Another type of specification could be a performance specification that describes the end performance so succinctly and precisely that only one model of one manufacturer can be used to satisfy it. This is a more devious and indirect method, but it is still considered a closed spec. On public works projects, this type of closed specification is found more often since its closed nature is veiled to a greater extent. Thus, the legal question of the enforceability of this type of specification is not as clear.

Proprietary Specifications

A *proprietary specification* is a type of closed specification. It specifically states what is to be provided without any allowance for alternatives. This is a unique type of a design specification. It is common for a proprietary specification to prescribe the use of a particular model of a particular manufacturer.

Example: The panel shall be a type E085-P31 Kemply by Kemlite.

Multiple Proprietary Specifications

This may be an open or a closed specification. It is a design specification as well. In a *multiple proprietary specification* the models of more than one manufacturer are specified.

Example: The wall shall be constructed of Milcor "C" studs by Inland Ryerson or Cee studs by Wheeling.

As was mentioned earlier, if at least three manufacturers are named, it is common to consider this an open spec. However, it should be noted that it is still a restrictive specification.

Open Specifications

Open specifications are nonrestrictive in that they permit a wide variety of choices. Public projects should be bid under this type of spec, naming at least three manufacturers. The products of various manufacturers should be acceptable whether or not they are actually mentioned by name. This type of spec is desired by owners as it gives contractors the widest opportunity to get the lowest prices for delivering the project.

Or Equal Specifications

This is essentially a modification of the proprietary specification in that it is a proprietary spec followed by the words *or equal.* Or equal specs should be avoided by spec writers. The contractor could make a substitution and claim that the substitute is an equal. Resolving the conflict will be difficult.

Example: Provide a Hentin Electric Model #370 thermostat or equal.

There are ways of avoiding the conflicts that this type of specification frequently generates, including the following:

- Name the specific acceptable brands and the model numbers and delete the words *or equal.* (This may be a closed spec.)
- Name many acceptable brands and models. This will require that the spec writer be familiar with all the different models listed.
- Let the contractor name an alternate. The use of the substitute may be accompanied by an addition or deduction from the cost of the project. The contractor will

have to submit a full description and technical data on the new item. Of course, the base bid must include only what is specified. A bidder should not include any pricing or cost adjustments for unapproved substitutions in the bid, as a base bid that is contingent on the acceptance of a substitution will probably be rejected. If the substitute is used, however, the substitution can be incorporated in the contract if it is accepted before the contract is signed. If it is accepted after the contract is signed, a change order can be issued.

- Request substitutions up to a given time before the bid date. To be effective, the designer must allow sufficient time for the submittal of relevant information concerning the substitution, time to review the submittal information, and time to issue an addendum to the different bidders.

The case of *Camp et al. v. Neufelder et al.* (95 P. 640) illustrates the problems that can be presented by an "or equal" clause. This case resulted from the renovation work on a Seattle building in 1905. E. C. Neufelder awarded a contract to F. McClellen and Co. to perform the renovation work. Neufelder then sublet a portion of the work to the partnership of E. H. Camp and H. Teroller. The dispute arose over the installation of recessed lights in the sidewalk. The specification read as follows:

> All sidewalk lights to be 3″ × 3″ reflecting prism lens set in cement; all frames to what is known as bar-lock construction. All joints must be made and guaranteed water tight. These lights shall be of the W. B. Jackson make, or equal, and shall be constructed to carry a safe load of 350 pounds per square foot.

The general conditions also stated that the decision of the architect was to be final. Work was progressing smoothly on the project. When Camp and Teroller were nearing the point where the recessed lights could be installed, they notified the architect that they intended to submit for approval an alternative brand of prism light that they considered equal to the type specified. The architect refused to entertain any substitute requests since he felt the Jackson lights had no equal. Under protest, the Jackson lights were installed as specified. Camp and Teroller then filed suit because they had not been permitted to use the less expensive substitute. They contended that the lights they wanted to use were equal to the Jackson lights. The owner defended the architect by claiming that substitutes would be considered only if the Jackson lights were not available. The owner further stated that the contract provisions were clear in that the decisions of the architect were final. Note that the equivalence of the two types of lights was not at issue. The court decided in favor of Camp and Teroller, stating that the substitution could not be based solely on the availability of the Jackson lights. In addition, while the court agreed that the architect had the power to make final decisions, the architect had acted in a capricious manner by refusing to consider requests for substitutions. Camp and Teroller should have been given the opportunity to demonstrate the equivalence of the lights they wanted to use.

Or Approved Equal Specifications

These are open specifications in that they give all acceptable products an opportunity to be considered. The ideal form of this type of spec lists the brands and model numbers of various manufacturers followed by the words *or approved equal.*

This places the determination of the acceptability of a substitution directly with the architect or engineer. It provides the potential for cost savings in that free competition is fostered between the various manufacturers of the products being specified.

This type of spec is not without problems. Suppose a contractor has identified a product that is felt to satisfy all the criteria implied in the specification, but does not have time to acquire the necessary technical data from the manufacturer and obtain the approval of the architect prior to the time of bidding. The contractor must then decide if the lower cost of the potential substitute is to be used in the base bid. (What price should be used?) The contractor will be taking a gamble if the lower price is used without the prerequisite approval. (Will the architect approve the substitute?)

It must also be borne in mind that the architect is liable if a substitute is accepted and later proves to be inadequate. Consequently, architects are reluctant to accept substitutes. The approvals should be screened very carefully.

All-Inclusive Specifications

Occasionally a contractor will read a specification that leaves many questions unanswered. Various items may or may not be required, depending on the whim of the owner's representative. Some example phrases that cause difficulties for contractors include "as directed by the engineer," "to the approval of the architect," "the architect's decision will be final," and "to the architect's satisfaction."

From the perspective of the spec writer, such phrases save tremendous amounts of time in the spec writing process, but they are a real problem for contractors trying to interpret them. Such phraseology gives the owner's representative a free hand to use this type of specification as a club whenever an interpretation is to be made. Obviously, this may not have been the primary motivation for the specification, but if an architect develops a reputation for using the specification in this way, it will be reflected in the base bid in higher prices. A potentially harsh specification is as follows:

Example: "The intent of the specs is to provide ___ (a given product)."

This is an example of an "architectural intent" clause. Some architects have taken such phrases to mean that their "intent" is to be construed from the existing drawings and specifications. Thus, the constructor may be asked to include items that were omitted from the contract documents, but which were intended to be included by the architect. In this case, the contractor is essentially asked to bid on a project without being fully informed of all the tasks that will be required.

Reference Specifications

Reference specifications are found in the technical specifications and make items, established tests, or formal procedures a part of the contract documents by reference. It is common to have a specification that will establish the performance of a product as measured by a standard or accepted test procedure. This can be part of a design or performance specification. This type of spec is generally used to ensure that a product conforms to industry-accepted test criteria. Such test procedures are established by groups such as the following:

- American Association of State Highway and Transportation Officials (AASHTO).
- American Water Works Association (AWWA).
- American Institute of Timber Construction (AITC).
- American Concrete Institute (ACI).
- American Society for Testing Materials (ASTM).
- American National Standards Institute (ANSI).
- National Fire Protection Association (NFPA).

Reference specs not only specify quality, they also set up a standard procedure by which the acceptability of the finished product can be determined.

Standard Specifications

The term *standard specifications* refers to an entire set of technical specifications that have been developed by an owner. Once developed, such specs can be used for many similar types of projects. It is common to have standard specifications in highway, bridge, and utility construction. The standard specs will apply to the entire industry. They are often adopted by state agencies and are modified only to satisfy unique conditions. This saves time in spec writing and requires contractors with several projects with the same owner to familiarize themselves with only one set of specifications.

Specification Problems for Contractors

Regardless of the type of specification used, there are instances when the contractor will be perplexed.

The specs may specify a model that is no longer being made, and so the contractor will not know what to include in the bid for this item. This problem occurs when the spec writer does not keep up with the products being specified. This occurs particularly when the spec writer is using the cut and paste method of putting together the specs.

Some words are difficult to interpret, such as the following:

- *Any:* as in "eliminate any leaks." (Will there be a broad interpretation?)
- *Either:* as in "the Contractor shall paint either side." (Can the contractor choose?)
- *And/or:* as in "the Contractor shall remove debris and/or lumber." (Is there a choice?)
- *Etc.:* as in "the Contractor is responsible for site clearing, etc." (What else is intended?)
- *Use:* as in "use a model 37 anchor." (It is better to use the words *provide and install.*)
- *As shown:* as in "install fittings as shown." (Is this really shown? Where is this shown?)
- *Reasonable time:* as in "give notice within a reasonable time." (It is better to give the specific time allotted.)

The following are known as "murder," "weasel," or "escape" clauses:

- "to the satisfaction of the architect"
- "where directed by the architect"
- "unless otherwise directed"
- "from an approved source"
- "at the owner's discretion"

Example: "Install shoring in all trenches if so directed by the architect."

This type of spec is full of guesswork for the contractor. Will the architect require shoring? If so, how much and where? Will the other bidders exclude shoring in their bids? The end result may be that the owner pays for shoring whether or not it is actually required, or that the contractor loses money on a project where the bid excluded shoring, but shoring was required.

This type of problem can be avoided. The shoring portion of the contract could be let by a unit price. This unit price would apply only if the shoring was actually required. Another procedure would be to require shoring in the specification and later ask for a deduction by change order if shoring is not required. Another method would be to have an alternate in the bid that would provide for an addition or a deduction from the base bid, depending on the use of shoring.

CONTRACT OBLIGATIONS DURING CONSTRUCTION

After the contractor has formally entered into a construction contract, a variety of obligations that require careful attention are imposed on the contractor. While many of these obligations are not directly related to the actual construction process, they are a means of communicating to the owner that the project is being delivered in accordance with the plans and specifications.

Submittals

The specifications will outline the quality standards to be attained on the project. The quality of the work performed will be evaluated primarily through on-site inspections. The quality of the materials incorporated in the project can be verified through a number of means. Independent tests can be made in the factories before shipment or after installation. In some cases the materials can be readily identified by their trade names, with adequate information being provided in the standard manufacturer's literature, commonly known as cut sheets. Regardless of the means by which the quality of materials and equipment is defined, the contractor must convey this information to the owner. This is typically done through the submission of the relevant information for the owner's approval. The nature of such submittals will vary depending on the type of material or equipment. The information must be sufficiently detailed so that the owner can make an informed decision about the adequacy of the item in question. Once the owner has approved the submittal of an item, the contractor can requisition the item and incorporate it in the project. A submittal provision may read as follows:

> If the information shows any deviation from the contract requirements, the Contractor shall, by a statement in writing accompanying the information, advise the Owner of the deviation and state the reason therefor. It shall be the Contractor's responsibility to ensure there is no conflict with other submittals and to notify the Owner in any case where the submittal may concern work by another contractor or the Owner. The Contractor shall ensure coordination of submittals among all related subcontractors and materials suppliers. Information shall be submitted in time to allow one month to review and return to the Contractor without interfering with the accepted construction schedule.

Submittals may include cut sheets, working drawings, shop drawings, descriptive data, certificates, methods, calculations, materials samples, test data, schedules, progress photographs, procedural descriptions, and manufacturer's instructions. The process of making submittals is streamlined considerably when the designer itemizes the specific items for which submittals are required. Unfortunately, this is not a common practice among all designers. These submittal items will typically be identified by name, along with the applicable specification section. Submittals are not generally required for materials that are specifically identified in the specifications. However, if one of several materials can satisfy a specification, or if an equal material is proposed by the contractor, submittals will be required. The owner will evaluate the submitted information and communicate his or her decision to the contractor within a stipulated time period. The actions of the owner on submitted items are generally to approve them as submitted, to approve portions of the submittal, to give a conditional approval, or to reject the submittal and require a resubmittal. Eventually, approval must be received on all items. The contract may require the contractor to provide the entire file of approved submittals to the owner at project completion.

The procedure by which submittals are generally submitted, reviewed, and approved is shown in figure 10.2. Note that considerable time may be involved for a single submittal to pass through each of the steps, especially if a submittal is initially rejected. Thus, it is important for the general contractor to identify the items that must be submitted early in the process. Delays due to late submittals can easily be averted if submittals are given prompt attention. During the early stages of construction on a project, it is common for one individual to be given the task of keeping track of the submittals to ensure that the procedure runs smoothly.

The contractor must be careful that all submitted items satisfy the specifications. If the contractor makes a submittal that does not conform to the specifications and does not identify the material as being in nonconformance, the owner's approval of the submittal does not release the contractor from liability for using that material. This may be addressed in the contract with provisions such as the following:

> Acceptance by the Owner of any submitted information regarding materials and equipment that the Contractor proposes to furnish shall not relieve the Contractor for any responsibility for any errors therein and shall not be regarded as an assumption of risks or liability by the Owner, and the Contractor shall have no claim under the

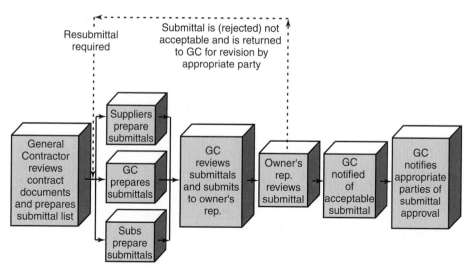

FIGURE 10.2
Typical procedure to process submittals.

contract on account of the failure or partial failure or inefficiency or insufficiency of any plan or method of work or material and equipment so accepted. Such acceptance shall be considered to mean merely that the Owner has not made an objection to the Contractor using, while assuming full responsibility therefor, the plan or method or work proposed, or furnishing the materials and equipment proposed.

As in this provision, the owner may wish to be absolved of any liability associated with the acceptance of submittals. Such provisions are not always effective in shifting all liability to the contractor, as some courts are reluctant to enforce them.

The submittal record on one construction project may give some idea of the extent of effort required to process submittals. The project was a four-story cast-in-place concrete building that was about 25 percent complete. At this stage, a total of 691 submittals had been processed. Of these, 38 percent were related to concrete, 14 percent were related to mechanical, 13 percent were related to electrical, 7 percent were related to metals, 6 percent related to sitework, 5 percent were related to general requirements, and the remaining 17 percent were in the areas of waterproofing, doors/windows, masonry, wood, finishes, equipment, elevators, and special construction. The disposition of these submittals may also be of interest. Of these 691 submittals, 8 percent were rejected without notations and another 20 percent were rejected with markings and notations on them. Thus, 28 percent of the 691 submittals required at least a second submittal. Of the 691 submittals, 72 percent were approved, with 11 percent requiring some type of confirmation of field conditions, 32 percent contained markings and notes that must be addressed (contingent approval), and 29 percent were approved as submitted. It was noted that of the submittals that were approved as submitted, approximately one-third were not reviewed by the owner's representative, as the contractor simply confirmed that the materials as specified were being provided. While the nature of the distribution of the types of submittals will vary from project to project, and the disposition of the submittals will vary by contractor and designer, the experience of this project gives an illustration of the results of the submittal process.

Shop drawings are a unique type of submittal and require considerable effort on the part of the supplier or installation contractor. Shop drawings include drawings, diagrams, layouts, schematics, illustrations, graphic presentations, schedules, and other materials that describe specific portions of the work to be performed. The shop drawings are the connecting link between the designer's drawings (construction drawings) and the construction of the specific facility component. The shop drawings become an extension of the design, but the responsibility for their preparation lies with the supplier or contractor. Shop drawings are needed because the designer cannot economically prepare the construction documents with the level of detail that is ultimately required and still allow some flexibility for the contractor.

The use of shop drawings in the construction industry is extensive; that is, virtually all projects require at least a few shop drawings. Shop drawings may be needed for such manufactured items as guardrails, structural steel, sign posts, built-in furnishings, precast concrete, concrete forms, falsework, shoring, cofferdams, dewatering, and so forth. Field examinations are often needed to finalize or verify measurements. Such verification is especially important on rehabilitation or restoration projects.

Submittals, including shop drawings, are considered necessary to control the final construction of facility components, and these must be processed through an approval process as depicted in figure 10.2. Whether prepared by a supplier or subcontractor, the general contractor is expected, and often contractually required, to review and approve submittal materials prior to submitting the information to the owner's representative. In truth, contractors vary in their practices as to the level of detail with which they actually review submittal materials from subcontractors and suppliers. Some contractors readily admit to simply "processing" submittal materials from suppliers and subcontractors, placing heavy reliance on others for preparation accuracy.

Who bears ultimate responsibility for the approval of submittals? For the Hyatt Regency skywalk collapse in Kansas City in 1981, the ultimate responsibility for the failure was placed on the design engineer for failing to perform detailed calculations and analysis on the structural connection that was changed by the contractor and submitted for approval. Other courts have supported this opinion, but in some cases the courts have held that the contractor is responsible for construction safety, and that the designer/engineer does not assume any of the contractor's responsibilities. To avoid being held responsible for submittal approvals, some contracts include provisions that make the contractor responsible for "satisfactory performance of work despite an advance approval of materials or method of work." It is also important that if a submittal constitutes a deviation from the construction drawings, the contractor must clearly note this deviation when providing the submittal for approval.

In some cases, the contract language may have little bearing on the court's interpretation of affixing responsibility for submittal approval. In *Jaeger v. Henningson, Durham & Richardson* (714 F.2d 773), the architect approved a submittal for a 14-gage landing pan when the contract documents required a stronger 10-gage pan.

The court held the architect responsible for failing to exercise reasonable care in conducting the review when the change was an obvious deviation from the contract.

Most contract documents do not dictate when submittals are to be submitted for review by the owner's representative. Surprisingly, many contracts do not stipulate the time period in which submittals are to be reviewed by the owner's representative, although industry practice places the time of review at about two or three weeks. Nonetheless, the process is much clearer if the review time is specified. The actual contract conditions might dictate the amount of time that is considered reasonable. In *John Grace Co. v. State University* (472 NYS 2d 757), a seven-day review period was considered too long. Projects of short duration may require very short periods for review, even if the time of review is not stated.

The timely review and approval of submittals is obviously important for construction project progress. In some cases, construction actually stops due to the lack of approval of specific submittals. Consequently, it may come as no surprise that many contractors proceed with construction activities even though some relevant submittals have not been approved. This is a practice that puts the contractor at risk, but many contractors contend that this practice is common for them, especially when they have some confidence that the submittal will be approved.

In recent years, some contractors have indicated that they have noticed a shift in practices related to submittal approvals. This is especially true of submittals on materials that are considered "equal" to those specified in the contract. Some contractors contend that designers have become more rigid in what they will approve, to the point of essentially mandating that materials that exceed the specifications are required. They contend that this is a ploy by which changes in the project are introduced at no cost to the owner. A contractor takes a risk in such an environment by bidding on materials that are not clearly specified.

Operating and Maintenance Manuals

Projects that include mechanical equipment usually include a contractual requirement for the contractor to provide operating and maintenance manuals for the equipment. These materials are for the owner's benefit during the actual occupancy and use of the facility. Equipment for which such manuals are required should be clearly noted in the contract documents. The specific information to be included in the manuals should also be stated. Typically, these manuals include information such as: the name and location of the manufacturer, the manufacturer's local representative, the nearest supplier, and the spare parts warehouse; applicable accepted submittal information; recommendations for installation, adjustment, start-up, calibration, and troubleshooting procedures; recommendations for lubrication; recommendations for step-by-step procedures for all modes of operation; complete internal and connection wiring diagrams; preventive maintenance procedures; complete parts lists by generic title and identification number, with exploded views of each assembly; suggested spare parts; and disassembly, overhaul, and reassembly instructions. Because of the importance of this information, the owner may link payment to the receipt of the manuals. An example provision states.

Before receiving payment for more than 50 percent of the purchase value of any equipment, the Contractor shall deliver to the Owner three sets of acceptable manufacturer's operating and maintenance manuals covering that item of equipment or equipment assembly provided under this contract.

Contractors are similarly inclined to contractually state that they can withhold payment from subcontractors and suppliers for failure to provide the necessary materials for applicable operating and maintenance manuals.

As-Built Drawings

The drawings in the bid documents generally represent a reasonable approximation of the final appearance of a project. However, not all information included in the original drawings will be accurate in describing all aspects of the project. The differences between the original drawings and what is finally constructed can be attributed primarily to changes authorized by the owner and errors in defining the existing conditions. While these differences may not appear significant, they may later become crucial. For example, utility lines may not be located exactly as shown in the drawings for a variety of reasons, for example, to avoid taking out a tree. Even though the lines may be placed only a few feet from the originally planned location, subsequent underground work may be based on the location of those lines. Thus, some accuracy is desired in showing their location. If the location is not accurately identified, a subsequent excavation effort may sever the lines, resulting in property damage and possible injury. To avoid such problems, owners often require the contractor to maintain an updated set of drawings that show the exact locations of all in-place items. These drawings are referred to as the *as-built drawings*. Constant updating of even small changes is recommended, because details are often lost due to failure of memory, or as memory becomes clouded by other, more important matters.

As-built drawings are most helpful when alterations or modifications are made to a structure. An example provision requiring as-built drawings is as follows:

> The Contractor shall maintain a neatly and accurately marked set of record drawings showing the final locations and layout of all mechanical and electrical equipment, piping and conduit, structures, and other facilities. Drawings shall be kept current weekly, reflecting the impact of all applicable change orders, adjustments to accommodate equipment, and construction adjustments. Drawings shall be subject to the inspection of the Owner's representative at all times, and progress payments or portions thereof may be withheld if drawings are not current. Prior to acceptance of the work, the Contractor shall deliver to the Owner two sets of these record drawings accurately showing the information required above.

Until the project is complete, the contractor will essentially be working with two different sets of drawings. One is the as-built drawings, which show what has actually been put in place; the other is the record set, which is essentially the set of plans on which the bid was made. The record set will be useful if litigation occurs. The record set will then be used to compare what was actually constructed with what the contractor originally anticipated.

REVIEW QUESTIONS

1. What is the underlying purpose of the general conditions, or boilerplate, of a construction contract?
2. What is the general purpose of supplementary provisions or special conditions in a construction contract?
3. What is the significance or value of using the CSI format in preparing the technical specifications for a building project?
4. Give an example of a design specification.
5. Give an example of a performance specification.
6. Can the performance aspects of a specification result in that specification being closed? Explain.
7. How can a proprietary specification be made into an open specification?
8. Describe how the general conditions might reflect the bias that might be introduced by the party or parties drafting the documents. Give an example of such possible bias in the documents.

11

UNIT PRICE, COST-PLUS, AND LUMP SUM CONTRACTS

ALL CONSTRUCTION CONTRACTS invariably contain certain basic ingredients. The methods by which the contractor will be paid may vary, but the other aspects of the contracts are generally very similar. Since the working relationship between the owner and the contractor is essentially independent of the project, standardized forms of agreements are common.

TERMS OF AGREEMENT

Contracts should be drafted to ensure that specific topics are addressed. A construction contract should identify the contractor, the owner, and the designer. The scope of work for which the contract is being drawn should also be defined. The project may be defined in a descriptive fashion, along with specific reference to documents such as the construction drawings and specifications.

The contract may stipulate that the construction project must be completed by a given date or within a given number of calendar or working days of the notice to proceed. Failure to complete the project within the stipulated time may result in liquidated damages (a reasonable estimate of costs for late project delivery) charged against the contractor. Conversely, the contract may also state that an incentive payment will be made to the contractor for each day the project is completed ahead of schedule. If the contract contains an award for early completion, it may also stipulate a penalty (not just liquidated damages) for late completion.

The contract will state how payments are to be made to the contractor. Generally, payments are made on a monthly basis. However, on small projects, the payment for the entire project may occur after the project is completed. On other projects, the contractor may be given an agreed sum (for example, 30 percent of the contract amount) before construction begins, with the understanding that the final payment will be made after project completion. A variety of arrangements can be devised.

The basis on which payments will be made is the essential aspect that differentiates the types of construction contracts. The three types of construction contracts are the unit price contract, the cost-plus contract, and the lump sum contract. There is no best form of contract. The nature of the project and the specific needs of the owner will determine the form that is most suited for the project.

Finally, the contract must contain the signatures of the contracting parties.

UNIT PRICE CONTRACTS

The primary feature of unit price contracts is that the pricing for the various units of work is determined before the start of construction. For such contracts, the owner will estimate the number of units included for each element of work. The following will serve as an example:

Item	Unit	Quantity	Unit Price, $	Amount Bid, $
Excavation	Cubic yard	2,500	18.00	45,000
Aggregate	Ton	1,200	14.00	16,800
Piling	Linear feet	500	30.00	15,000
Reinforcing steel	Pound	8,000	0.45	3,600
Mobilization	Each	1	10,000.00	10,000
Concrete	Cubic yard	2,000	200.00	400,000
Total				490,400

Note that the contractor need only determine the unit price bid for the various items in the contract. When the extensions are made (quantity times unit bid price), the sum of these extensions (the total) is used to determine the low bidder. The contractor must be careful when calculating the amounts to be bid so that all anticipated costs are included in the bid items. For example, the overhead and profit on the project must be included within the unit price bids. Perhaps extensive forming will be required for the concrete, yet there is no unit price item for forming. Thus, the contractor must include the forming costs in the other bid items, most likely the unit price for concrete.

When are unit price contracts appropriate? These contracts are used when the project is fairly well defined, but the actual quantities may be difficult or impossible to estimate with accuracy until after construction has started. Civil projects are typical examples where a unit price contract may be appropriate. The quantities that are most often difficult to estimate relate to earthwork. This may include uncertainty about the amount of excavation required, the amount of fill required, or even the amount of concrete required in a foundation or footing. The unit bid price is utilized as a means to establish the payment to be made to the contractor. With this arrangement, it is obvious that the total construction cost of a project will be at least slightly different from the total price that was originally used to establish which contractor was in fact the low bidder.

A *balanced bid* is one in which the anticipated costs for the various bid items are accurately reflected in the unit prices that are submitted. *Unbalancing* a bid is a method used by some contractors in which the unit prices of the various bid items are altered so that they do not reflect the true costs of those items. Contractors may use unbalancing to give them an advantage on some aspect of the project. A couple of examples will illustrate the use of unbalancing. Suppose a balanced bid by a contractor is as follows:

Item	Unit	Quantity	Unit Price, $	Amount Bid, $
Excavate sand	Cubic yard	8,000	4	32,000
Excavate rock	Cubic yard	2,000	20	40,000
Fill material	Ton	4,000	12	48,000
Total				120,000

Assume that this is a fairly simple contract and that the three bid items above are the only items for which payment will be made. Obviously, the contractor must distribute all project-related costs to these three bid items, along with an appropriate allowance for profit and general overhead. Note that no direct payment is permitted for mobilization. Assume that the contractor has distributed the estimated mobilization cost of $6,000 equally among the three bid items. However, the contractor may wish to unbalance the bid, rationalizing that most of the mobilization costs will occur at the beginning of the project, so that they should really be allocated to the work items performed early in the project. The unbalanced bid might appear as follows:

Item	Unit	Quantity	Unit Price, $	Amount Bid, $
Excavate sand	Cubic yard	8,000	4.50	36,000
Excavate rock	Cubic yard	2,000	19.00	38,000
Fill material	Ton	4,000	11.50	46,000
Total				120,000

Note that the unbalancing does not alter the total amount bid. In this example it is assumed that the sand will be excavated before excavation of the rock. The unbalancing would be very similar if the contractor simply wanted to distribute costs so that the owner's money would be used to finance the project. Obviously, the contractor must use company resources until the first payment is received. With the unbalancing, the contractor will soon be in a positive cash flow position. To some extent, this form of unbalancing can be observed in almost every unit price contract. Most owners will accept unbalancing of this type to some extent, as many contractors do not have a strong cash position, and a contractor who is hurting for money often does not deliver the final product in an ideal fashion.

Another reason some contractors use unbalancing is to take advantage of an error in the owner's estimate. For example, suppose the bidder in the above balanced bid has determined or suspects that the owner has miscalculated the quantity of sand and rock. Specifically, the contractor feels confident that more rock will be encountered than the owner has estimated. The contractor may decide to take advantage of this error by unbalancing the bid as follows:

Item	Unit	Quantity	Unit Price, $	Amount Bid, $
Excavate sand	Cubic yard	8,000	1	8,000
Excavate rock	Cubic yard	2,000	32	64,000
Fill material	Ton	4,000	12	48,000
Total				120,000

In this example, the owner is not adversely affected by the unbalancing if the quantity estimates are accurate. However, suppose the actual quantities are different from the estimated quantities. The cost of the project to the owner could be as follows:

Actual Cost on an Unbalanced Bid Project

Item	Unit	Quantity	Unit Price, $	Amount Paid, $
Excavate sand	Cubic yard	5,000	1	5,000
Excavate rock	Cubic yard	5,000	32	160,000
Fill material	Ton	4,000	12	48,000
Total				213,000

Actual Cost on a Balanced Bid Project

Item	Unit	Quantity	Unit Price, $	Amount Paid, $
Excavate sand	Cubic yard	5,000	4	20,000
Excavate rock	Cubic yard	5,000	20	100,000
Fill material	Ton	4,000	12	48,000
Total				168,000

It is obvious that the owner's error in estimating quantities will be more costly to the owner than was the original bid total. However, the cost impact is much greater when the bids have been unbalanced. In this example, the difference between the balanced contract and the unbalanced one is $45,000.

Does the owner have any recourse when a bid is unbalanced? In most cases the contractor is merely redistributing funds so that a disproportionate amount of the payment occurs earlier in the project. Most owners will permit this type of front-loading without considering the possibility of using another bid. By contrast, unbalancing that is designed to exploit an estimating error can be very costly if the owner is not prepared. However, owners are not helpless. Most unit price contracts state that the unit price for a given item may be renegotiated if the actual quantity varies from the estimated quantity by more than a stated percentage, typically 20 to 25 percent. For example, a contract provision might state, "Should the total as-built quantity of any major pay item required under the contract exceed the estimate contained in the proposal therefor by more than 25 percent, the work in excess of 125 percent of such estimate will be paid for by adjusting the unit price." An owner can also interpret an obviously unbalanced bid as being irregular and, on those grounds, reject the bid. Another method that may be used by the owner is to resort to a change order that will delete a grossly unbalanced bid item. For example, the owner might decide to delete the rock excavation from the construction contract and award that portion of the work to another contractor. This may constitute a breach of contract if the deleted work is a large portion of the contract.

As noted, unbalancing of bids may be employed by a contractor for several reasons. It was also noted, particularly when the unbalancing is extreme, that the contractor does assume some risks. Owners may be reluctant to reject a bid as irregular. However, when the potential cost of a project is considerable as a consequence of unbalancing, this is a risk (bid rejection) that is always taken by the contractor. The instructions to bidders may include a provision that the disqualification of a bidder and the rejection of the submitted bid is an option the owner can exercise "if any bid prices are obviously unbalanced." On the other hand, a change order which eliminates the unbalanced work items poses an even greater risk to the contractor. In addition to the dangers or risks associated with unbalancing, the issue of ethics may arise. While most contractors might consider it acceptable to unbalance in order to conceal a specific pricing strategy from competing contractors, many consider extreme unbalancing unethical. Most owners will not take exception to unbalancing that will help the contractor maintain a healthy cash flow on the project.

How are payments to the contractor determined? Payment for work items is a relatively simple matter. The unit price for each item of work is established by contract. The only matter that must be resolved is the quantity of work that is actually performed. To determine this quantity, the contractor will measure the in-place quantities and request payments on the basis of that measurement. The owner, however, must ensure that the quantities reported by the contractor are accurate. This means that an independent measurement must be made by representatives of the owner. For unit price contracts compared with other forms of payment, the owner may have to increase the field staff to attend to the additional administrative tasks involved in double-checking or verifying the contractor's stated in-place quantities. Determining the number of in-place quantities includes measuring in-place materials, tabulating the delivery tickets of bulk deliveries, counting truckloads of materials delivered or removed, and a variety of other methods. Since the cost of construction is directly linked to these determinations, considerable effort is often devoted to the task to ensure that the appropriate payment amounts are being made.

If the owner makes any changes in the project design, these modifications can be handled by change order. The procedure for handling change orders should be clearly described in the contract documents.

What are the disadvantages of a unit price contract? One shortcoming is that the owner is not certain of the actual cost of the project until the project is completed. The extent of this shortcoming is directly related to the accuracy of the estimated unit quantities. Additional staff requirements are posed by this form of contract so that the owner can certify in-place quantities. In general, the plans must be reasonably complete in order for the bidders to develop unit prices for all the bid items.

Gjehlefald v. Drainage District Number 42 et al. (212 N.W. 691) was an interesting case involving a unit price contract. This case concerned a contract to make improvements in a drainage system. Gjehlefald was deemed the low bidder with a total price of $143,000. This total was based on the sum of all the unit price extensions applied to the estimated quantities. For overdepth excavation work, Gjehlefald's bid stated that this work would be priced at "50 percent above the attached schedule." When the contract was sent to Gjehlefald for his signature, he did not notice that the wording for the overdepth pricing had been changed to "50 percent of the prices specified in Exhibit A [the attached schedule]." Other changes had also been made to Gjehlefald's bid submittal. These changes remained unnoticed when Gjehlefald signed the contract. When overdepth work was encountered, the unit price paid by the drainage district was considerably lower than Gjehlefald had expected. He sued to obtain the unit price payments as bid. On the issue of the discrepancy between the bid and the contracted unit price values, the court ruled that since Iowa had no laws to the contrary, a contract was constituted when the submitted bid was accepted by the drainage district. Since the contract existed at the time of acceptance, the unit price for the overdepth work had to be paid at 50 percent above the unit price listed in the schedule.

If the owner elects to change the number of units of work on a unit price contract, it would appear easy to determine the price to be paid. This may not be so simple if the effort associated with the work can vary in difficulty. On a contract involving debris removal from a riverbed, a contractor was to be paid a single unit price on the volume of debris removed. In determining the unit price to bid on the project, the contractor determined that there were three levels of difficulty. The easy removal would be in the rural areas where dozers could be used. In the residential and commercial areas, some hand work would be required. In the most difficult area, the river, the debris would have to be handled one piece at a time. The contractor estimated the cost of doing the work in all three types of areas and then spread these costs evenly over the volume of debris that was to be removed. The contractor, Tompkins & Company, was awarded the contract, but the owner soon made changes. The contractor was asked to remove additional debris along a greater portion of the river, and the owner elected to delete some of the debris removal in the easier rural areas. The contractor asked for more compensation since the ratio of easy and difficult debris removal had been altered. When the owner did not agree, the case was presented to the Corps of Engineers Board of Contract Appeals (ENGBCA No. 4484, 85-1 BCA). The Board agreed that the contractor was entitled to additional compensation. The contractor had made a careful assessment of the conditions when the bid was prepared. Changing these conditions entitled the contractor to a change in the unit price.

The case of *Depot Construction Corp. v. State of New York* (224 N.E.2d 866) concerned a lump sum contract in which unit prices were set by the owner. Depot was awarded a contract to construct a building for the Manhattan State Hospital for a sum exceeding $6 million. The owner's engineer determined that this project included 500 cubic yards of rock excavation, and for any variations from that quantity, the contract would be adjusted at a rate of $10 per cubic yard. The contract sum would also be adjusted at a rate of $22 per cubic yard for rock excavation in piers and trenches that varied from 600 cubic yards. The total rock excavation was

2982 cubic yards, which exceeded the anticipated amount. The contractor was paid for the additional rock that was excavated, but the contracted rate of pay was not considered sufficient to cover the actual costs of excavation. Depot sued to get the unit prices adjusted. Depot argued that excess rock was beyond the reasonable contemplation of the contracting parties, and therefore was not bound by the unit prices which should be adjusted. The state of New York stated that the contract was clear concerning the unit prices and that no provisions called for unit price adjustments for large variances in quantity. Depot argued that the scope of the work was misrepresented, but the state denied this. The state had taken 17 borings for the excavation area and had used those findings to compute the quantity of rock to be excavated. Furthermore, bidders were advised that the information on the borings was "presented in good faith, but it is not intended as a substitute for personal investigation, interpretation, or judgement of the Contractor." The court ruled in favor of the state. No additional payment was made to Depot as the risk involving the rock quantity clearly rested with the contractor. The state had not misrepresented the rock excavation as Depot claimed.

COST-PLUS CONTRACTS

A cost-plus contract is one in which the contractor is reimbursed for most of the direct expenditures associated with a particular project plus an allowance for overhead and profit. It is common for the allowance for overhead and profit to be based on a percentage of the costs. If the allowance for overhead and profit is reasonable, the contractor is almost assured of not losing money.

Many contracts reimburse the contractor for the direct project costs plus a percentage of the costs of overhead and profit. Other payment methods have also been used. For example, the contract may specify "cost plus a fixed fee." This type of arrangement removes the incentive for the contractor to increase costs in an attempt to increase the overhead and profit allowance.

Another approach may involve "cost plus a percentage of costs with a guaranteed maximum." With this form of contract, the owner is assured that the total cost of a project will not exceed a stated amount. If the cost exceeds the stated amount, the contractor will bear those costs. If the cost is lower than the stated amount, some contracts will provide that the savings will be shared between the owner and the contractor in a predetermined manner.

Payment for work done is a simple matter. The contractor and the owner need simply agree on the validity of the various cost reimbursements that are requested. It is imperative that the contract clarify which costs will be reimbursed.

When is a cost-plus contract appropriate? In general, these contracts are used when the actual costs of a project or portions of a project are difficult to estimate with accuracy. This may occur when the plans are not complete, or when the project cannot be accurately portrayed. It also may occur when a project is to be completed within a fairly short time period and the plans and specifications cannot be completed before construction starts.

Another type of project for which the cost-plus arrangement is well suited occurs when the true nature of the project cannot be accurately described before construction begins. This may occur in remodeling or renovation projects that contain many unknowns.

A cost-plus contract does not lend itself well to competitive bidding. This form is used almost exclusively in the private sector.

A cost-plus contract, unlike the other forms of contract, does not place the owner and the contractor in an adversarial relationship. However, this form mandates that the contractor be trustworthy.

If many changes are anticipated in the design of a project as construction is under way, a cost-plus contract is appropriate. Changes can be easily incorporated into the scope of work under contract. In fact, the procedure for the reimbursement of costs for original contract items and subsequent changes is the same.

What are the disadvantages of cost-plus contracts? The most serious disadvantage, compared with the other forms, is that the owner has little idea of what the actual cost of the project will be. Also, the owner must maintain additional staff to monitor the progress of the contractor. The emphasis of the staff will be primarily on costs. All reimbursable costs must be carefully documented, and the owner's staff must ensure that the costs being reimbursed were actually incurred.

It might appear that a cost-plus contract would be quite simple, and in concept it is not complex. The general intent is for the contractor to furnish, and the owner to compensate the contractor for, all labor, equipment, material, and supervision expenditures incurred to complete the project according to the construction documents. Simply stated, the contractor is to be reimbursed for the construction expenditures plus a fee as determined on the basis of a percentage of these costs, when the contract is a cost-plus-a-percentage-fee agreement. The contractor's reimbursable costs usually fall into the following categories:

- Labor (wages, employee benefits and contractor contributions to taxes, unemployment compensation and Social Security).
- Materials and Equipment (the cost of items physically incorporated into the project, and the cost to transport them to the site).
- Subcontractor Costs (based on payments made directly to specialty contractors).

This seems to be a straightforward type of arrangement, but there are many areas in which there may be a disagreement of whether the expenditure is a direct cost item, or if it is included in the fee. It is generally assumed that the reimbursable costs include such expenditures as sales and use taxes on items included in the project, site cleanup, rental costs of heavy equipment, temporary utilities (electrical, water, and telephone), permit fees, and land-use fees. Although there will be exceptions, the costs that are not reimbursable include hand tools, insurance premiums, home office expenditures (unless the costs can be directly allocated to the project), interest on money borrowed by the contractor, contractor profit and overhead (salaries of home office personnel), and the cost of rework necessitated due to defective workmanship. The best way to avoid any confusion is to clearly spell out or delineate those costs which are reimbursable and which costs are included in the fee. This includes the costs of dumpsters, drinking water, ice, cups, temporary toilets, and so forth.

LUMP SUM CONTRACTS

The lump sum contract is probably the most common type of contract used in the construction industry, particularly in building construction. Of the various forms of construction contracts, the lump sum, or fixed-price, contract is the simplest. The contract essentially states that the contractor will produce the project as designed for a stated specific sum.

On most large lump sum contracts, the contractor will be paid on a monthly basis. The contract is simply for the work put in place. The value of the various work items is ideally established before the start of construction. The contractor is generally asked to break down the project into a variety of work items and to allocate the appropriate payment to be made for each item. This payment schedule, or schedule of values, will become the basis for all payments throughout the project. Naturally, the sum of all the values must equal the amount stated in the contract as a lump sum.

In the discussion of unit price contracts, the issue of unbalancing was discussed at length. With lump sum contracts, a form of unbalancing can also occur. This arises out of the preparation of a payment schedule, or schedule of values, in which the contractor manipulates the cost distribution so that the owner pays more for early work items and less for work items that occur later in the project. Note that this unbalancing is done almost exclusively to get the owner to finance more of the construction effort. Since the sum of the items on the payment schedule must equal the contract amount, this unbalancing places less risk on the owner. However, if the payment schedule is severely unbalanced and the contractor defaults after receiving several periodic payments, the owner is at considerable risk. Thus, it is important that the owner carefully evaluate the schedule of payments submitted by the contractor.

If changes are made to the contract, negotiations between the owner and the contractor will establish the payment to be made to the contractor for such work. Thus, unlike cost-plus contracts, the cost of each change order must be negotiated (similar to a separate contract) between the owner and the contractor.

When is a lump sum contract appropriate? The project and design will generally dictate whether this approach is viable. First of all, the plans must be fully completed so that the contractor can estimate quantities accurately.

Owners with a limited budget prefer this form of contract because it is the only one that yields a fairly accurate indication of the final cost of a project. Unless changes are made in the project, the amount stated in the contract will be the amount actually paid by the owner.

What are the disadvantages of lump sum contracts? With these contracts, the construction of the project is delayed while the plans are being completed. Also, errors in the plans will be costly because they will result in extras. Thus, while there is an incentive to complete the plans, there is also an incentive to get the construction effort under way as soon as possible. A compromise is needed as an early construction start may mean costly errors and omissions in the plans and specifications.

JOB ORDER CONTRACTING

Job order contracting (JOC) was initially developed for government agencies in the 1980s for renovation, repair, and construction projects. When a facility owner has extensive facilities that require routine and unscheduled maintenance and repairs, JOC offers some strong advantages, as it essentially ensures an owner that a contractor will be "on call" for any task to be done. Some have referred to JOC as "blanket" contracts, as they are designed to cover any tasks that must be performed. JOC is generally done through multi-year contracts. Since a contractor is on board for virtually any task that comes along, there is a huge reduction in the time consumed in design work and procurement. Under JOC it is common for the contractor to have an ongoing presence in some of the owner's facilities.

JOC emphasizes teamwork between the contractor and the owner and is ideally suited for projects of varying sizes that have poorly defined delivery dates and unknown quantities. When entering into a JOC the owner will want to be very careful in contractor selection to ensure that performance can be assured; that is, JOC contractors are not generally selected on the basis of price alone, unless governmental statutes establish this as the sole criteria. The owner will want to enter into a JOC with contractors who are reliable, efficient, safe, technically competent, and can guarantee good quality work, while delivering products on a timely basis and within budget.

The employment of JOC started in the Defense Department, but it has been adopted by various public agencies, including those at the federal, state, and municipal levels. JOC is also widely used in the private sector, especially by owners with substantial facilities.

A key element of JOC has to do with pricing. At the outset, the contracting parties will agree on the prices to be charged for specific items or the hourly rates for certain personnel. The list of prices can be quite extensive; this is preferred, as this will ensure that the contracting parties need not worry about negotiations on most prices. Many unit prices will be established in the contract, with quantities remaining unknown until the work is actually defined. The JOC agreement will be based to a large extent on the established agreed-upon prices, along with a factor that will allow for contractor overhead and profit. At the time of the agreement, there may not be a specific project envisioned by either party. The contractor may be guaranteed a minimum amount of work, and there may also be a maximum stipulation for the expenditures to be made in any one year. The contractor is not assured that every item of work will be done by the company, as the owner can always award individual contracts to other firms. This option that is available to the owner provides a strong incentive for the JOC contractor to perform efficient and high-quality work that fully satisfies the owner.

For the owner, JOC ensures a quick contractor response to needs that arise without notice. Since the terms of the arrangement are written to cover a wide variety of circumstances and types of work, the owner can issue a work order on short notice and have strong assurance that the JOC contractor will be able to devote immediate attention to the required task. This eliminates the need to advertise for bids and to negotiate a contract each time that a new construction need is identified, and this essentially eliminates the issuance of change orders. Because of these characteristics of JOC, it is ideally suited for small construction projects, repair jobs, and maintenance work on facilities. JOC is especially advantageous when emergencies arise. For the owner, there is little risk when implementing JOC, as the owner is not bound to authorizing any specific task to the JOC contractor. An unscrupulous JOC contractor might simply find very little work being assigned under the JOC agreement.

An owner might be tempted to award two JOC contracts at the same facility. This implies that the owner is unsure of the level of performance that one or both of the contractors can and will deliver. It is probably unwise to enter into a JOC contract with a firm that is not trusted, or that is regarded as an unknown entity in terms of the performance to be expected.

CONCLUDING COMMENTS

There are three basic differences in the types of contracts that a contractor can enter. The fundamental differences lie in the ways that payment for services is established. Each form of contract has its own advantages, and consequently, certain types are particularly well suited for given projects. The lump sum contract can be considered only if the scope definition is clear as demonstrated in a complete design. Unit price contracts are better suited than lump sum contracts when the actual quantities (earthwork) cannot be accurately estimated. The cost-plus contracts can be utilized on any type of project, unless public policy forbids this type of contract. Another element that is introduced when cost-plus contracts are considered is the amount of trust or confidence that can be placed in the contractor.

Contract Type	Basis for Payment	Basis of Profit	How Changes Are Addressed
Unit price	Measured quantities of work performed	Included in unit prices	Negotiated separately
Cost-plus	Receipts, documentation of expenses incurred	Portion of fee beyond incurred costs	Automatically addressed as part of the contract
Lump sum	Work performed as defined in the schedule of value	Included in pay items in schedule of values	Negotiated separately

REVIEW QUESTIONS

1. Describe conditions in which unbalancing a bid is done to maximize profits. Describe conditions in which unbalancing is done to minimize risk to the contractor.
2. Contrast and compare the extra work provisions of lump sum contracts, unit price contracts, and cost-plus contracts.
3. What unique aspects of a project are most appropriately accommodated by a unit price contract?
4. What unique aspects of a project are most appropriately accommodated by a lump sum (fixed-price) contract?
5. What unique aspects of a project are most appropriately accommodated by a cost-plus contract?
6. Discuss the implications of the type of contract for the staffing requirements of the owner.
7. Discuss the implications of the type of contract for the degree of completeness needed in the design.
8. Describe a situation in which job order contracting might be a desired approach of contracting.

12

CHANGES

In spite of the fact that architect/engineers spend many months of effort on the design of a project, modifications are invariably sought later. Some of these changes may be necessitated by items that were inadvertently left out of the original design. Others arise as a result of changes in the design that are sought by the owner. This usually occurs after the owner has had a greater opportunity to evaluate the original design. The total elimination of changes to a design, while desirable, is not realistic, in that the design phase would have to be extended considerably. Some changes are made after the design is ostensibly complete, but prior to the receipt of bids. These changes are made in the estimating phase of a project in the form of addenda. Changes made through addenda become part of the bids that are received and are therefore automatically included in the construction contract. Changes made after contract award must be negotiated apart from the contract. Although most projects seem to be plagued by changes made by owners, it is important to note that an owner does *not* have an inherent or implied right to make unilateral changes to the contract. That right must be obtained contractually.

When an owner makes a change to the contract, the contract time and amount may or may not be affected. In most instances, changes tend to increase the amount of compensation but do not generally include an extension in the time. Changes are often referred to or interpreted as extras.

Since changes are expected on most construction projects, the parties to the contract must be fully knowledgeable about the specific contract terms concerning changes. The specific wording of contract provisions for changes is very important.

CHANGES CLAUSES

It is common for changes clauses to include the following elements:

- The owner has the right to make changes within the general scope of the contract.
- The contractor is obligated to perform the work necessitated by the change.
- The change must be in written form and must be signed.
- An adjustment to the contract price and/or contract duration will be assessed by some means, or can be predetermined.

The changes made by the owner will become a modification of the original contract. The terms that are required to enter a legitimate contract must also be in existence when a change order is put into effect. Each change must be carefully described with the necessary drawings and relevant technical specifications. The negotiation of the cost of changes to the owner is quite different from the terms that were in existence when the original contract was formulated. If the terms of the original contract were determined through a competitive bidding process, the owner had the advantage of having several construction firms competing for the work. When the costs of changes are negotiated, however, there are no competing firms. Thus, the owner should have very knowledgeable advisors or very trustworthy contractors to avoid undue charges for changes.

The ConsensusDOCS acknowledge that the contractor or the owner might initiate a change in the contract. It is clearly stated that any change that will impact the contract price or the construction duration must be formalized in a change order. The provisions state the following:

§8.1.1 The Contractor may request or the Owner may order changes in the timing or sequencing of the Work that impacts the Contract Price or the Contract Time. All such changes in the Work that affect Contract Time or Contract Price shall be formalized in a Change Order.

§8.1.2 The Owner and the Contractor shall negotiate in good faith an appropriate adjustment to the Contract Price or the Contract Time and shall conclude these negotiations as expeditiously as possible. Acceptance of the Change Order and any adjustment in the Contract Price or Contract Time shall not be unreasonably withheld.

CHANGE ORDERS

Whenever a change is issued by the owner, it is usually referred to as a *change order*. It is common to use the terms *change* and *change order* interchangeably on many projects. This is unfortunate in that a contemplated change is quite different from a change that has been specifically ordered by the owner. A change order is really a change that carries with it a specific directive for the contractor to perform that work. In a sense, a change order is a minicontract to perform a specific item of work. In reality, a change order is an adjustment made to the original contract, and as such it must satisfy all the prerequisites for a contract. The only thing that sets a change order apart from a typical contract is that the parties have already contractually agreed that the owner can make the changes, and that the contractor must perform the work called for in the change order.

FIELD CHANGE

Field Change No. *6*

Project: Date issued: *Nov. 15, 2011*
>   Clay County Civic Center
>   3499 Stonewall Jackson Way
>   Augusta, GA

Contractor:
>   Edward J. Billings, President
>   RST Constructors
>   Charleston, SC

Description of the Change:

>   *The designer and contractor mutually agreed that the stair nosing detail should be changed as noted on the attached drawing.*

Section of Specifications or Drawings Affected:

>   *Division 6*

Rationale or Reason for the Change:

>   *Greater safety is provided with this modification and ease of construction is also facilitated.*

It is mutually agreed that this change will not result in a change in contract price or project duration.

APPROVED: ACCEPTED:

By: *Adam B. Designer*   By: *George D. Builder*
  (Owner/Owner's Representative)   Contractor (Authorized Signature)

Date: *Nov. 17, 2011*   Date: *Nov. 17, 2011*

FIGURE 12.1
Example of a field change.

One means of avoiding confusion with contemplated changes and those which have not been fully negotiated is to refer to the unresolved changes as change proposals or modification proposals. A change order is a directive from the owner to execute the terms of a change proposal. If the change has no impact on the contract's duration or amount, it is common to refer to it as a *field change* (figure 12.1). Field changes can often be authorized by personnel in the field without direct owner approval. Field changes are typically minor, but are required to facilitate the construction effort.

Initially, the general contractor will be notified of the owner's contemplation of a particular change. This potential modification is generally described in the same level of detail that was used in the original contract documents. This level of detail is necessary in order for the general contractor and affected subcontractors to determine the cost of the change. Change orders are similar to contracts and must therefore meet the same criteria as the original contract. The primary issue to

CHANGE ORDER

Project: Date issued: _Nov. 15_, 20_11_
    Clay County Civic Center
    3499 Stonewall Jackson Way
    Augusta, GA

Contractor:
    Edward J. Billings, President
    RST Constructors
    Charleston, SC

Original Contract Amount .... _$970,460_
Number of Previous Change Orders: #__1__ to __4__
Cumulative Value of Previous Change Orders .... _$ 13,330_
Adjusted Contract Price Prior to this Change Order .... _$ 983,790_
Number of this Change Order: #__5__
Net Increase (decrease) of this Change Order: .... _$5,150_
Adjusted Value of the Contract after this Change Order..... _$988,940_

Original Contract Duration .... __360__ calendar days
Net Change in Contract Duration
for Previous Change Orders.. __14__ days
Adjusted Contract Duration Prior to This Change Order........ __374__ calendar days
Net Increase (decrease) of this Change Order:...................... __0__ days
Adjusted Duration of the Contract after this Change Order.... __374__ days

APPROVED: ACCEPTED:

By: ___John D. Owner___   By: ___George D. Builder___
   Owner (Authorized Signature)   Contractor (Authorized Signature)

Date: ____Nov. 22, 2011____   Date: ____Nov. 17, 2011____

FIGURE 12.2
Example of a change order.

be resolved with changes is the amount that the contractor is to be paid for the changed work, and the amount that the duration of the contract will be altered by the change. Preparation of the cost of a change is similar to the price determination of any construction work. As such, the contractor must include in the proposed price: the cost of labor; fringe benefits; payroll taxes, material (including applicable sales tax); subcontractor expenses and markup; rental equipment (including applicable sales tax); operating and maintenance costs for owned equipment; field overhead (stipulated percentage); liability and compensation insurance; home office overhead (stipulated percentage); and social security and unemployment insurance. This price is then quoted to the owner for review. Should the owner find the price and time adjustment acceptable, a formal change order will be issued (figure 12.2). The contractor will then be required to perform the changed work. The change order provides the owner with the ability to adapt to actual conditions in order to achieve the desired end result.

In this example, the contract duration is unaltered, but the contract amount is increased by $5,150 to a new contract total of $988,940. Note that previous change orders have altered the original contract price and the duration.

While most changes increase the cost of construction, others may actually result in cost reductions. This will depend on the nature of the change being proposed. The owner may very well decide to reduce the scope of a project in order to realize some savings, especially if other changes have resulted in significant cost increases. These decisions will obviously be tempered by the owner's budget.

When the modification proposal is received by the general contractor, a careful evaluation will be made of the potential impact of the change on the cost and duration of the project. The general contractor also will require subcontractors impacted by the change to submit quotations. Once the information is obtained from subcontractors, suppliers, and in-house personnel, the general contractor will submit a price quotation to the owner. If the quotation is acceptable, the owner will issue a change order. If the price is deemed unacceptable, the owner may ask the general contractor to reconsider the price, possibly by providing details to show a justification for the amount of the impact being projected. If this negotiation is successful, the owner will then issue the change order. If the price and/or schedule impact cannot be agreed upon, the owner may decide to cancel the prospects of the change altogether, or the owner may decide to issue a change order and have the price established, on a cost-reimbursable (cost-plus) basis. These stages of change order evolvement are shown in figure 12.3.

Establishing the agreed amount that the contractor is to be paid for changed work can be resolved in several ways. The options may be spelled out quite clearly in the contract. For example, the contract may state, "The cost for extra work performed by the General Contractor will be determined by either (1) an agreed lump sum, (2) an agreed unit price, or (3) an actual field cost plus the outlined percentage for lump sum work." Ideally, the owner would like to have an agreed lump sum price established for the changed work. Thus, the owner will often require the contractor to submit a breakdown of the labor and materials. This will help the owner to make an informed judgment about the reasonableness of a price. If the owner and contractor cannot agree on a price, the owner will often invoke another changes provision that states that the contractor will be reimbursed for "the actual cost of the work which will include the cost of materials, labor, payroll taxes and insurance, builder's risk insurance, bond, subcontracted work, and other itemized direct jobsite expenses as previously approved by the owner." In addition to the actual costs, the contractor will generally be entitled to a fixed percentage, or possibly a step-wise allowance of "20 percent for the first $10,000, 15 percent on the next $10,000, and 10 percent on the balance over $20,000." The markup permitted on the subcontracted work is less than for work performed by the general contractor's own forces. As a rule, the markup that the general contractor can claim on the subcontracted work is roughly half of that performed by the general contractor's own employees.

Contractors generally have a reasonable degree of comfort in developing the cost of the work to perform a change. Even when this is not possible, the owner and the contractor can proceed with the change with the understanding that the

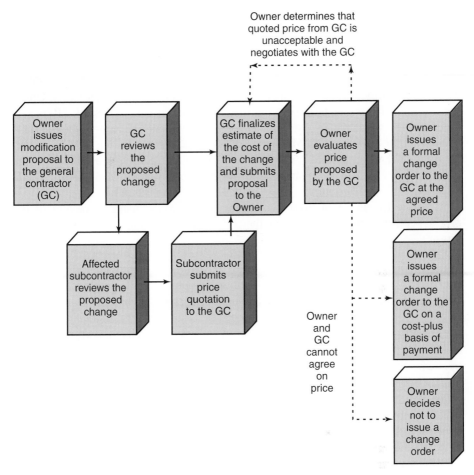

FIGURE 12.3
Process of evolvement for change orders.

contractor will be entitled to the costs reasonably incurred in performing the changed work, as well as a pre-agreed percentage to cover the costs of overhead and profit. The other impact of changes is that of time. Obviously, if critical activities are impacted by a change, the project duration might very well be increased. Contractors will generally know about the potential time impact of a change, but determining this cost is wrought with uncertainty. Contractors may assume that the time impact of a change will be negligible, but they will prefer a contract provision whereby they reserve the right to subsequently seek an equitable price adjustment if the work directly impacts portions of the project not being changed. The impact of a change on unchanged work will be difficult to determine at the time a change order is being negotiated. Without a contract provision granting the contractor the right to pursue subsequent impact costs, an allowance must be made when the cost of the change is initially negotiated.

In some instances, changes might be made without formal change order being issued. While this is not a suggested practice, circumstances might arise (such in emergency situations) whereby the contracting parties agreed to a change with an understanding that this would be formalized at a later date. The contractor might also request a change necessitated by an interpretation made by the designer or the owner that impacted the cost and/or time of construction. Under these circumstances, it is imperative that the contractor act promptly. Requests for a change very late in the construction process may have no success. This is noted in the Construction Management Association of America (CMAA) Document A-3 that states that "No request by the Contractor for a change in the Contract price or time . . . shall be allowed if such request is made after final payment."

Despite the fact that prior agreements concerning changes are part of the original contract, much litigation in the construction industry is directly related to this area.

CARDINAL CHANGES

Although the construction contract usually grants the owner the right to make changes to the contract, this is by no means an unlimited right. Changes made by the owner must be within the general scope of the original contract. Changes that are not within the general scope are known as *cardinal changes,* and these are not covered in typical changes clauses. That is, cardinal changes are beyond the scope of the contract. When is a change not within the general scope of the original contract? The definition is not clear. It occasionally becomes the central issue to be decided in the judicial system.

Since a cardinal change is not within the scope of the original contract, it is considered a breach of the contract to force a contractor to perform such a change. The rationale for defining this as a breach of the contract is that the terms of the contract are altered to an extent where the original contract is essentially supplanted by another contract, rather than simply being modified.

On private projects, cardinal changes are not a problem if the contractor and the owner both agree to the terms of those changes. On public works projects, the issue is more complex. Even when a cardinal change is made to a public works contract and the contractor agrees to perform the work for an agreed price, problems can still arise. The primary problem is that the contractor may not be able to enforce payment for performance on a cardinal change. The rationale is that since a cardinal change is not within the scope of the original contract, the change constitutes a new contract and should be awarded by competitive bidding. Thus, any individual (usually a competing contractor) can bring suit against the public agency to bar payment for work done on a cardinal change.

Minor changes are not considered to be cardinal in nature. As was mentioned earlier, the actual definition is not clear, and courts must occasionally determine if the definition applies to a particular change on a contract. In general, the definition of a cardinal change hinges on two issues. First, a cardinal change is assumed if the essential identity of the project is altered. For example, changing a 2-story building to a 10-story structure would be considered a significant alteration, and the definition of a cardinal change should apply. A change that calls for a different size window in one room of a building would obviously not be considered cardinal. Courts have regularly upheld changes in quantities up to 25 percent as being within the scope of the contract. Of course, the cumulative effect of many smaller changes, if significant, could be interpreted as constituting a cardinal change. Second, a cardinal change is one in which the method or manner of the anticipated performance is so drastically changed that essentially a new agreement is made. On public works projects, this type of change in the work method would probably not be contested by others, and the contractor would still be paid for performing the work. Usually the contractor would be the party claiming a cardinal change when the method of performance is changed appreciably and to the contractor's detriment.

Cost overruns were examined in *U.S. Fidelity and Guaranty Co. v. Braspetro Oil Services Co.* (97 Civ. 6124) to determine if they were outside of the scope of the original contract for the construction of an oil production facility for Braspetro in Brazil. U.S. Fidelity provided the performance bond on the project. The consortium to construct the facility apparently underbid the project, as cost overruns were realized early in the project. Braspetro declared the contract in default and sought financial recovery from the surety. U.S. Fidelity claimed that the project was materially changed by the owner, and this voided the performance bond. The court noted that the contract allowed for changes of up to 25 percent of the contract, but the changes actually amounted to only 5.5 percent of the contract. The surety had to reimburse Braspetro for the cost overrun.

Why would a contractor not want to perform the work called for in a change, even a cardinal change? The contractor may not have the expertise to perform the requested work. The change may cause the contractor to use up or exceed the firm's bonding limits. The contractor may not have the financial or other resources to perform the work, or the contractor may have other projects that will require the resources that would otherwise be utilized to perform the work associated with a cardinal change.

If a change on a public project is considered cardinal in nature, the contractor should immediately inform the owner in writing. If the owner insists that the contractor perform the change under the provisions of the changes clause in the contract, the contractor should proceed under protest. If this is not done, the contractor will be limited in the options for recovery as stated in the changes clause. If the contractor protests and subsequently is successful in having the change declared cardinal, damages may be claimed for the owner's breach of the contract. If the contractor performs the work requested in a change order without protesting, a later claim by the contractor that the change was cardinal will probably be denied.

THE WRITTEN CHANGE ORDER

Changes clauses may state that payment will be denied for work performed for changes in the plans or specifications unless the change is specifically ordered in a signed change order from the owner. Whether it is specifically stated or not, most owners consider that payment will be made only when changes are ordered by the owner in writing.

Since the wording is usually clear in the written requirements for changes, can a contractor ever force payment for a change performed in response to a verbal directive? Suppose a contract is let for the construction of a multistory building. The contract states that "changes in the plans and specifications must be in writing." On this project the owner needs to finish some areas by certain dates, but this was not conveyed in the contract documents. As the finish work is being done on the structure, beginning at the bottom floor and working upward, the owner verbally directs the contractor to stop all work on the third floor and immediately go to the eighth floor and finish it first. Then the contractor is directed verbally to finish the sixth floor. After that the contractor is free to finish the building in any sequence. Should the contractor be paid for this disruption in the work sequence? The plans and specifications have not been altered. This was a change only in the method and manner of performance, and the contractor should be compensated. Owners may try to avoid this problem by placing a provision in the changes clause that includes changes in "the method and manner of performance."

It is a good practice for the contractor to inform the owner in writing whenever compensation will be sought for directives that are considered to be changes in the contract.

Does a changes provision apply in a strict manner? A changes clause may state that no payment shall be made "for extra work unless done pursuant to a written change order." Can the contractor ever collect for performance that was initiated by a verbal directive? Suppose a construction contract was let on a steel structure with a large amount of mechanical equipment. The contract contained a changes clause that specifically stated that payment for extra work would be made only "pursuant to a written change order." While performing work on the project, the contractor caught an error in the design. The ductwork was to penetrate a wall at a particular location that also was to be occupied by structural steel. Clearly, both could not simultaneously occupy the same space. The owner's representative was notified of the problem and the additional costs to be incurred in correcting it. The owner's representative orally directed the contractor to change the work. The work was performed so that the ductwork was routed around the steel structure. Subsequently, the contractor requested payment, but the owner denied the request. Should the contractor collect?

In similar cases the courts have tried to distinguish between extra work and additional work. *Extra work* consists of work that is outside and entirely independent of the contract. Essentially, it is work that need not be performed in order to satisfy the terms of the original drawings and specifications. *Additional work* consists of work that must be undertaken to meet the contract requirements and without which the work requested in the original contract could not be completed. For work to be considered additional and for the contractor to be entitled to added compensation, the following criteria should be met. The change

- should not have been anticipated,
- was not open to observation, and
- could not be readily discovered until work under the contract was undertaken.

There are other conditions under which compensation for verbally directed changes is granted, even though the changes clause requires that they be in writing. In emergency conditions, courts regularly state that the written requirements can be waived.

What constitutes compliance with a changes clause mandating the use of a signed written order? It must be remembered that changes are a common occurrence on virtually all construction projects. To ease the paperwork, most architects and other owners' representatives use specific change order forms when communicating with contractors, but rarely make specific reference to these forms. Since specific references are not generally made, courts have upheld various other changes. These include changes communicated in (1) letters, (2) transmittal notices, (3) revised drawings, (4) revised specifications, (5) notations on shop drawings, (6) job minutes, (7) field records, and (8) daily reports. These documents may not all be available to the contractor in the normal course of conducting business with the owner. Thus, the contractor is not sure if the owner's files will support a claim for a change. This may not be known until the pretrial discovery procedures have taken place.

There is another court interpretation under which a contractor may successfully obtain payment for verbally directed changes, in spite of the existence of a clause requiring a written change order. This interpretation considers the words *change order* to be separate words rather than a phrase. By this interpretation, a change may be written in a sketch of some sort and the order to perform can be given verbally. This is not a common means of seeking recovery for changes work.

The changes clause may stipulate a signed change order. The courts have occasionally been asked to determine the definition of *signed*. Is an order signed if it contains a person's initials, professional registration stamp, or typed name? The courts have addressed these issues, but their findings have varied. On such issues, the courts generally do not try to isolate one term and define it. The owner's actions may be taken into consideration to determine if a change exists. Thus, the principle of estoppel may be utilized to determine that the actions of the parties have indeed satisfied the conditions of creating a change for which the contractor is to be reimbursed. When a doubt exists about the definition of a change, the contractor should write to the owner and make it clear that a verbal directive or another form of communication has been or will be construed as a change order. If the owner does not take issue with this interpretation, the courts will probably construe it as a change order.

THE AUTHORITY TO ISSUE A CHANGE ORDER

Since most change orders result in added costs to the owner, most contracts are fairly specific about the particular parties who are empowered to issue change orders. Often this power is removed from the architect, or the changes may have to be jointly issued by the owner and the architect. For example, the contract may state, as does Article 7.3.1 of AIA Document A201-1997, that the power to issue a change order must include specific authorization by the owner. The specifications may include a section that defines the owner and may broaden the term with a phrase such as "duly authorized representatives."

On most building projects, the architect acts as the owner's representative, granting payment requests, rejecting inferior work by the contractor, interpreting drawings, and in general enjoying a great deal of power. Contractors often assume erroneously that since the architect has all these powers, the architect has the power to issue changes as well. Such logic would be a mistake on most projects. Note that in the AIA general conditions, the changes clause does not permit the architect to independently issue a change order. That clause requires that the owner's signature also appear on change orders.

If a contractor does not understand the changes clause, work may be performed for which no compensation will be made. However, some court decisions have upheld the rule of equitable consideration. This is essentially an interpretation by the courts that the words or actions of the owner concerning a change order amount to a waiver of the exact contract requirements. This is frequently the first avenue under which contractors seek recovery.

When recovery is sought, contractors claim that the owner waived the requirement of a written change order. Some major questions that the courts must then answer are, Was this requirement waived by someone in authority? and Can this be waived on a public works project? It is a much clearer case if the owner knows about a directive to change work, is aware that the change work is being performed, and does not object. If these criteria are met, a waiver of compliance with the specific terms of the changes clause is assumed.

CHANGE VERSUS ANOTHER NEW CONTRACT

Most contracts for construction will stipulate that changes be initiated by written change order. If a change is directed orally, another interpretation may give the contractor some relief. Suppose the owner approaches the contractor and verbally orders that certain work be performed. The contractor performs the work and then asks for payment. The owner refuses to pay since the changes clause clearly states that all change orders must be in writing. The contractor then sues the owner, but does not base the claim on the technicalities of a change order. Instead, the contractor claims that the verbal directive was simply an offer and that performance by the contractor consummated the unilateral contract. Thus, it is considered to be a new and separate contract, with the criteria of offer and acceptance having been satisfied.

CONCLUSION

Numerous tactics have been presented by which contractors may seek recovery for changed work, even when the specific requirements of the changes clauses have not been satisfied. It should not be inferred that the contractor need simply decide which tactic to follow in order to win a particular case. Each case has unique circumstances, and another seemingly similar case may serve as a precedent and deny recovery to the contractor.

It should also be remembered that cases are won by lawyers, and that the owners and contractors are merely the winners or losers of record. Whenever possible, litigation should be avoided. Thus, whenever a contractor interprets a directive as a change order, it should be made clear to the owner that compensation is expected. Directives that do not alter the contract amount rarely end in litigation.

Whenever a directive for a change is given to the contractor, the contractor should ask the following questions:

- Should I perform?
- Is this proper authorization?
- Is this within the scope of the contract?
- Is this directive in the proper form?
- If I perform, will I get paid?

If the contractor feels that all these questions can be answered satisfactorily, prompt written notice should be given to the owner concerning the cost impact of the change. It is extremely unwise simply to perform the work and later seek compensation.

On major change orders, the surety should be kept informed. The contractor should carefully review the provisions of the contract with the surety. Usually the surety need not be notified of minor changes in the work, but changes that amount to more than 10 percent of the original contract amount may require that prompt notification be given.

Example Cases

A brief description of a few cases dealing with changes will illustrate the different circumstances under which changes can result in litigation.

Majestic Builders Corp. v. Mount Airy Baptist Church Housing Corp., Inc. (430 F. Supp. 1376) involved the construction of low-income housing in Washington,

DC. Majestic was the general contractor for the contract with Mt. Airy Church, which had obtained a mortgage for the project. Shortly after construction began, it became apparent that part of the foundation would have to be supported by caissons not included in the original contract documents. A formal change order was issued. However, Mt. Airy Church was not able to increase the mortgage to include the cost of the added work. This made Mt. Airy Church directly responsible for the added construction costs, but the church refused to pay for this work. Majestic then filed suit. Mt. Airy Church claimed that inadequate consideration existed for the change order to be binding. The court did not agree and stated that the contractor was being asked to perform work that it had no obligation to perform under the original contract.

The promise that a change order would be forthcoming became the grounds for a suit in *Darrell J. Didericksen & Sons, Inc. v. Magna Water and Sewer Improvement District* (613 P.2d. 1116). Didericksen contracted with the Magna Water and Sewer Improvement District to relocate a sewer line to circumvent the routing of a planned highway to be constructed by the Utah Department of Transportation (DOT). However, when Didericksen started work, it was asked by Magna to begin in an area different from the one shown in the contract documents. This was done to assist in the coordination efforts with the state DOT. Magna had not issued a change order, but promised to do so. Relying on Magna's promise, Didericksen began the construction work and continued until it came into direct conflict with the highway construction. Since no change order had been issued, Didericksen stopped its efforts. Magna then sued the contractor. The court ruled in favor of the contractor, stating that the contractor had a right to begin work without a formal change order, but that stoppage of its work was also justified, since the work undertaken without a change order might have reduced its right to payment for performing the changed work. In other words, working without payment might otherwise have been interpreted as a contractor waiver of the payment requirement.

The contract provisions for following specified changes procedures were not enforced in *City of Baytown v. Bayshore Constructors, Inc.* (615 S.W.2d 792). Bayshore entered into an agreement with the city of Baytown to install specified sewer lines within 150 days. The agreement stated that certain procedures were to be followed for all changes authorized by the owner. The specified procedure was used on one change, but ceased to be used after that. Numerous changes were then authorized which did not follow the specified changes procedures. When the contractor submitted a bill for $312,411 for the extra work, the city refused to pay. The contractor then filed suit. The court noted the extent of the changes made and ruled that the plans and specifications were sufficiently inadequate to constitute a breach of contract. Since the city had breached the contract, it was in no position to have the court enforce the changes procedures. The added costs were awarded to the contractor.

Excessive changes were the primary issue in *C. Norman Peterson Co. v. Container Corp.* (173 Cal.3d 348B). Container Corporation of America (CCA) contracted with Peterson to modernize one of its mills. The project duration was 18 months, 14 of which would take place while the mill was operating. During construction, CCA redesigned portions of the work. In addition, the 4-month shutdown period was reduced. During the shutdown, some of CCA's drawings were determined to be inadequate, prompting various verbal changes directives to be made. Hundreds of these changes were made during construction, without a paper trail to document the alterations. When Peterson requested payment for the extra work, CCA refused to pay. Peterson then sued for compensation. The court ruled that the extent of the changes could be regarded as an abandonment of the original contract, and stated that Peterson was entitled to reasonable payment for the work it performed.

Understanding the contract terms is one of the fundamental requirements of successful contracting. The failure to understand the contract can be costly, as was shown in the case of *James A. Cummings v. Young* (589 So.2d 950). Cummings, the general contractor on the Florida Power & Light Company's Miami district office building, subcontracted with Bob Young, Inc., for some site work. Young was asked to relocate several large boulders and place them in a pile toward the back of the site. Young then learned from the architect that boulders were to be placed at specific locations on the site as part of the landscaping plan. Young refused to move the boulders again without additional compensation. Cummings insisted that Young comply with the contract and move the boulders to the designated areas. A request by Young to receive compensation for moving the boulders a second time was denied. Young stated that he never knew that he would be expected to move the boulders two times. Young stated that he felt he would be asked to remove and relocate the boulders, which was done. Cummings stated that the architectural drawings clearly showed the locations of the boulders, and that Young was contractually bound to place the boulders in their final location. Since Young refused to do the work, Cummings hired another firm to do the work, and then Cummings backcharged Young for the work done on its behalf. The court agreed that Young was bound by the contract to place the boulders in their final locations. Cummings could successfully withhold the backcharged amount from Young.

Changes may be the necessary result of incomplete plans and specifications. This was clearly exemplified in *Standard Construction Co. v. National Tea Co.* (62 N.W.2d 201). On March 23, 1946, Standard entered into a contract to build a large, multipurpose warehouse, bakery, and office building for National. The contract award had been accelerated to avoid a building ban that was to become effective two days later, on March 25. The ban did not apply to projects on which some construction work had already started. To satisfy this deadline, the design phase had been drastically shortened, resulting in numerous errors and omissions in the contract documents. The construction contract was for a "total cost, not to exceed $1,245,000." This upper limit of the total price was soon abandoned as change orders were issued. Among the 145 change orders issued, four became the source of dispute. One dealt with compaction costs of $71,815.66, which Standard claimed were not anticipated. The other three contested change orders related to costs incurred as a result of acceleration. Standard had paid overtime to workers to hasten the project completion; it had also rented equipment for longer periods than originally estimated and had incurred higher costs for concrete placement as a result of winter conditions. Standard claimed that it had advised against the overtime and the winter concrete work, but National insisted that the work had to be done in a hurry. National argued that some of these costs should have been anticipated as being typical in the industry. The court decided in favor of Standard, ruling that the plans were grossly in error; for example, the building size was increased by about 25 percent, and the project was constructed of reinforced concrete rather than steel as in the original plans. In addition, change orders were occasionally revised by telephone, thus waiving the procedures established by the contract. The problems for National began with the hastily prepared, incomplete plans and were further aggravated by the 24-hour limitation imposed on Standard to bid the project. This culminated in drastic alterations that were subsequently ordered by the owner.

REVIEW QUESTIONS

1. Discuss the premise that the owner has an inherent or implied right to make reasonable changes in a project.
2. What is a cardinal change? What are the risks to an owner's representative who authorizes a cardinal change? What are the risks to a contractor who performs the work required for a cardinal change?
3. It is common for contracts to stipulate that changes in the work be authorized in writing by someone in authority. Under what conditions might a verbal change be considered valid?
4. How does a field change differ from a change order?
5. What is the difference between additional work and extra work?
6. Examine the AIA documents and the ConsensusDOCS that are included in the Appendix and determine the similarities and differences in the provisions related to change orders. Specifically examine AIA provisions §7.1.1, §7.1.2, and §7.2.1, and ConsensusDOCS §8.1.1 and §8.1.2.

13

CHANGED CONDITIONS

IDEALLY, WHEN A construction contractor enters into an agreement with the owner, all the conditions of the contract are clearly outlined and understood. These conditions are generally stated in the various contract documents, including the plans, specifications, addenda, general conditions, and supplementary provisions. With the information that is provided, the contractor generally assumes that the project can be constructed in the manner described. This is based in a large part on the assumption that the conditions at the project location are essentially as they have been represented. Unfortunately, the actual conditions are not always as they have been described. When this occurs, the costs of construction can be expected to increase, and the contractor should be entitled to an equitable adjustment in the contract amount.

The differences in site conditions that are most commonly encountered involve subsurface conditions. These differing conditions may include the presence of a high water table, unstable foundation materials, rock where softer materials were expected, and undisclosed utility lines. Not surprisingly, differing site conditions are a common source of litigation on civil works projects and large military construction sites. Differing site conditions may also be of the aboveground variety. For example, the survey markers or the stated locations of existing buildings or trees may be in error. When the construction contract is for a remodeling project, the differing site conditions may relate directly to conditions that could not be established without destructive testing or demolition work to expose items such as hidden structural members and piping systems.

Most owners prefer that the contract amount be fixed when the contract is signed. When this is done, the owner is fully aware of the final cost of the project. However, this is not realistic on most projects, as many changes may be initiated by the owner, or changes may be necessitated by circumstances that are revealed during construction. These costs may be diminished to some degree if the incidence of differing site conditions can be minimized, or if the costs associated with differing site conditions can be borne by the contractor.

The primary means of minimizing the incidence of differing site conditions is for the designer to be diligent in investigating the project site and to represent the site conditions fully and clearly in the contract documents. An additional measure by which the owner can reduce such costs is to shift the responsibility for identifying differing site conditions to the contractor. To some extent these risks should be borne by the owner, as there are some conditions under which it would be unfair for the contractor to be forced to bear the full burden of responsibility. This is a topic that frequently results in litigation.

PREBID SITE INVESTIGATIONS

The information to bidders invariably contains information about the existing site of the construction project. This generally includes information in the form of boring logs and results of tests on the indigenous soils. Although this information may be prepared with care and may be an accurate representation of the actual site conditions, it is common practice to stipulate that the contractor make an independent site visit and investigation. This will shift some of the responsibility to the contractor. For example, the plans may show a project site as a vacant lot, while a visit may reveal that the lot has been used as a dump site by others. A provision in the specifications may stipulate that the contractor is responsible for clearing the site. If the contractor did not make a site visit before bidding on the project, it is not likely that an adequate amount was included in the estimate for clearing the site, or that the contractor will have grounds for a claim against the owner for these added unanticipated costs. A site visit would have revealed that additional clearing costs would be incurred. Therefore, most contract documents contain provisions, usually in the general conditions, regarding independent site investigations made by the contractor. The following provision is typical of provisions used to stress the contractual importance of site visits:

> Before submitting a bid proposal, the Contractor is to carefully examine the site of the proposed work. The submission of a bid will be considered proof that the bidder has examined the site and understands the conditions to be encountered in performing the work.

Some contracts include harsh language regarding differing site conditions. The contract documents may include a clause by which the owner denies any liability and responsibility for the actual conditions. Such onerous provisions do not have automatic force and effect, as the courts do not favor such clauses and often will not enforce them to the fullest intent.

DIFFERING SITE CONDITIONS

As previously mentioned, the term *differing site conditions* is typically applied to subsurface conditions. On renovation projects, it may apply to any existing conditions that are uncovered only after construction work commences. These are conditions which cannot be readily evaluated through a cursory visit. However, this does not mean that the contractor is automatically entitled to additional funds if the conditions are not exactly as they are represented. To evaluate subsurface conditions, a contractor might go to a site with a backhoe and dig test pits or conduct additional borings on the site. Of course, these test pits and borings represent a random sampling of the site, and the results may not be conclusive. In addition, a claim against the owner may be considered valid if insufficient time was provided during the bidding period for extensive site investigations to be made, or if access to the site was limited to the extent that site conditions could not be fully evaluated.

What constitutes a differing site condition? There are two broad general categories under which differing site conditions can exist: (1) The actual conditions differ materially or are at variance with the conditions indicated in the contract documents, and (2) the actual conditions are of an unusual nature and differ materially or are at variance with what the contractor should have reasonably anticipated. Note that this does not mean that the condition is one the contractor did not expect, but rather that it is one the contractor should not have reasonably expected. Thus, the condition need not be a geological freak, but must simply be one that was not foreseeable based on examination of the contract documents and the site investigation; that is, it involves conditions a prudent contractor would not anticipate. An additional criterion is that the cost of construction will be adversely affected by the changed conditions. The contractor will invariably request added compensation as a result of encountering a differing site condition. The following provision contains typical elements of a differing site conditions provision:

> Should the Contractor encounter subsurface or latent conditions at the site materially differing from those shown on the plans or indicated in the specifications or if conditions at the site differ materially from those ordinarily encountered and generally recognized as occurring in work of the character provided for in the contract, the Contractor shall immediately give written notice to the Owner's representative of such conditions before they are disturbed and in no event later than 14 days after such conditions are observed. The Owner's representative will promptly investigate the conditions, and if such conditions are determined to be materially so different as to cause an increase or decrease in the Contractor's cost thereof, or the time required for performance of any part of the work under the contract, an equitable adjustment will be made and the contract modified in writing accordingly.

Many of the differing site provisions used by public agencies on engineering projects tend to be fairly standardized. The following provision is typical:

> During the progress of work, if subsurface or latent physical conditions are encountered at the site differing materially from those indicated in the Contract or if unknown physical conditions of an unusual nature, differing materially from those ordinarily encountered and generally recognized as inherent in the work provided for in the Contract, are encountered at the site, the party discovering such conditions shall promptly notify the other party in writing of the specific differing conditions before they are disturbed and before the affected work is performed.
>
> Upon written notification, the Engineer will investigate the conditions, and if he or she determines that the conditions materially differ and cause an increase or decrease in the cost or time required for the performance of any work under the contract, an adjustment, excluding loss of anticipated profits, will be made and the contract

modified in writing accordingly. The Engineer will notify the Contractor of his/her determination whether or not an adjustment of the contract is warranted.

No contract adjustment which results in a benefit to the Contractor will be allowed unless the Contractor has provided the required written notice.

The ConsensusDOCS provisions related to the discovery of unforeseen conditions, specifically mention that changed conditions are related to conditions that differ materially from what is shown in the contract documents, or from what would be reasonably anticipated. The provision states the following:

§3.16.2 "If the conditions at the Worksite are

a. subsurface or other Physical conditions which are materially different from those indicated in the Contract Documents or
b. unusual or unknown Physical conditions which are materially different from conditions ordinarily encountered and generally recognized as inherent in Work provided for in the Contract Documents,

the Contractor shall stop Work and give immediate written notice of the condition to the Owner and the Architect/Engineer."

The ability of a contractor to recover expenses incurred as a result of differing site conditions may depend on the specific criteria being satisfied. The contract sum may be adjusted only if "the Contractor used reasonable diligence to fully inspect the work site" and/or "the concealed items can be considered extra work to the extent that additional new construction beyond the scope of the contract documents is required. Otherwise, any cost adjustment associated with the concealed items will not be considered." With such wording, it is necessary for the contractor to demonstrate that the concealed conditions required extra work beyond the scope of the contract in order to receive additional compensation. Simply encountering conditions that require the contractor to change work methods or equipment will not guarantee recovery of costs; additional new construction must be required. For example, soft ground discovered on a scraper earth-moving project that can be commenced only by changing to a dragline operation for the excavation may not result in an adjustment to the contract. This may not be interpreted as being new construction. If the soft ground condition necessitates the driving of piling for a foundation project for which no piling was originally specified, the criterion of new construction is satisfied and compensation should be received.

As stated in the provisions cited above, specific procedures must be followed for the contractor to be assured of receiving compensation after encountering differing site conditions. The primary requirements are that the owner's representative be given a prompt notification of the discovery of the differing site conditions, and that the conditions not be disturbed before the owner's representative has had an opportunity to observe them. The architect or owner's representative must be permitted to verify the conditions, observe the changes if necessary, and begin to keep accurate records on this portion of the contract. When these requirements are met, the contractor is in a much better position to receive an equitable adjustment for added costs incurred as a result of the changed conditions.

The contractor need not demonstrate or even infer that there was any fraudulent intent on the part of the owner. Nor does the contractor need to claim that the contract has been breached, or that there has been a failure to achieve a meeting of the minds. When a differing site conditions provision exists, the primary point that the contractor must make is that the site conditions were not foreseeable and could not be reasonably known at the time of bidding.

Contractors assume considerable risk when they are barred from making claims for differing site conditions. An example of a provision in which such risk is borne by the contractor is as follows:

> The Contractor is cautioned that details shown on the subsurface conditions are preliminary only. The Owner does not warrant or guarantee the sufficiency or accuracy of the information shown or the interpretations made as to the type of materials and conditions to be encountered. The Contractor is cautioned to make such independent subsurface investigations as deemed necessary to become familiar with the site. The Contractor shall have no claim for additional compensation or for an extension of time for any reason resulting from the actual conditions encountered at the site differing from those indicated in the subsurface information.

It should be obvious that without a provision for differing site conditions in the contract, the contractor may be assumed to be at risk if changed conditions are encountered. This added risk will be recognized by most contractors and will be reflected in their bids. Thus, the owner may be paying for differing site conditions because of the absence of a provision for differing site conditions, even though none may be encountered. By including such provisions, however, the contractor can bid lower, since there is a greatly reduced need to allow for contingencies. Without such provisions, an unwary contractor may default on the contract if insufficient funds have been allowed in the bid to cover such contingencies. This will be ruinous to the contractor and will necessitate the costly letting of a separate contract by the owner.

The absence of a provision for differing site conditions does not automatically preclude compensation for expenses incurred as a consequence of encountering such conditions. Some contracts are written so that differing site conditions may be addressed as part of the extra work clause. The following is an example of an extra work provision which is written to potentially provide for the recovery of costs for unforeseen work, such as differing site conditions:

> The Contractor shall perform unforeseen work, for which there is no price included in the contract, whenever it is necessary or desirable in order to complete fully the work as contemplated. Such work shall be performed in accordance with the specifications and as directed and will be paid for as provided.

The inclusion and wording of differing site provisions are important in determining the allocation of risk. For example, the following provision is used in the construction contracts of a large metropolitan government:

> The Contractor agrees that he [or she] has included in his or her bid prices for the various items of the contract any additional costs for delays, inefficiencies, or interferences affecting the performance or scheduling of contract work caused by, or attributable to, unforeseen or unanticipated surface and subsurface conditions.

Note that with this provision, the contractor is not permitted by contract to subsequently recover any of the costs associated with delays or inefficiencies on the job as a result of differing site conditions. Since even the costs associated with equipment that is idle while the contractor awaits further direction in dealing with a changed condition can be substantial, this provision places considerable risk on the contractor. This risk will undoubtedly be reflected in the bids of contractors who prudently include this potential construction cost in their estimates.

Example Cases

Cases involving changed conditions cover a multitude of circumstances. The changed conditions might consist of unexpected rock encountered during excavation, excessive amounts of water encountered, or even incorrectly located utilities discovered during excavation. The actual nature of the differing site conditions plays less of a role in litigation than does the contract wording. No matter how different the conditions may be, the contract must be carefully examined to determine whether the contractor may be entitled to added compensation. The following cases will illustrate this point.

In *Donald B. Murphy Contractors, Inc. v. State of Washington* (696 P.2d 1270), an excessive amount of rainfall was claimed as the changed condition. Donald B. Murphy Contractors entered into an agreement with the state of Washington for the construction of two adjacent sections of Interstate 90, east of Issaquah, Washington. The project scope included the laying of new traffic lanes, the demolition of two bridges, the construction of two new bridges, and the construction of several detours. During construction, a record amount of rainfall occurred, causing floodwaters to reach an all-time high level. The floodwaters severely damaged the excavation work Murphy was doing at the time. The worst damage was incurred as a result of the destruction of a diversion culvert that had been designed by the state. The damage had to be repaired before construction could resume. The state paid Murphy for the replacement cost of installing a new culvert. Murphy brought suit for additional funds for the added costs incurred as a result of the delay. Murphy contended that the state had designed the culverts and had thereby warranted their performance. The state was thus to blame for the damage caused by water when the culvert failed. The record amount of rainfall constituted a changed condition as well. The state claimed that the culvert was for temporary use only, and that for typical rainfall it would have been adequate. The state also claimed that the heavy rainfall constituted an act of God. The court ruled that the state had not breached its implied warranty of the culvert design and was not liable for delays due to adverse weather. No damages were awarded to Murphy. Thus, in this 1985 case, unusual weather was not construed to be a changed condition. Murphy testified that a state employee had admitted that the state was at fault; however, the court ruled that this was inadmissible.

The 1995 case of *Millgard Corp. v. McKee/Mays* (49 F.3d 1070) was a dispute that related to two conflicting provisions. McKee/Mays had the contract to construct a jail and courthouse. It subcontracted the pier drilling for the caisson foundation to Millgard. Prior to bidding, McKee/Mays provided Millgard with a copy of a log of soil borings, which was for "information only" and was not to be regarded as "part of the Contract Documents." The owner and the architect specifically disclaimed any responsibility for the accuracy of the soil information. In the subcontract agreement, there was a "concealed conditions" provision that stated, "Should concealed conditions encountered in the performance of the work below the surface . . . be at variance with the conditions indicated by the contract documents, or should unknown physical conditions . . . be encountered, the contract sum shall be adjusted." Millgard encountered a considerable amount of quicksandlike material at a depth of 5 to 15 feet in most of the holes that were dug. Millgard expected to be compensated for the poor soil conditions through the concealed conditions provision. McKee/Mays refused to pay and the court battle ensued. Millgard felt the disclaimer clause should be thrown out. McKee/Mays argued that the disclaimer was relevant, especially with the statement that the soil report was not part of the contract documents. The court ruled that the disclaimer was effective. Furthermore, Millgard had assumed a considerable risk by not independently testing the soil. Finally, the court stated that when two provisions clash, the more specific one should be considered to take precedence. The disclaimer clause was upheld.

A demolition project in Dade County was embroiled in the differing site conditions dispute of *Hendry Corp. v. Metropolitan Dade County* (648 So.2d 140). The Florida project was to demolish the old Rickenbacker Causeway bascule span. In addition to providing them with a set of demolition plans and specifications, bidders were given access to the 1941 plans of the original bascule span. Hendry Corporation, one of the bidders, based its bid on the conclusion that the pilings supporting the bridge were concrete, relying on its own visual observations and past experience. The original 1941 plans did not provide this information. Hendry was the low bidder and was awarded the contract. During construction, Hendry discovered that the pilings were actually made of wood, and that subsurface debris from the original construction project was making demolition more difficult. Hendry requested compensation for the differing site conditions, but Dade County refused. Hendry filed suit contending that Dade County had a duty to disclose all relevant project information to the bidders. Even if it made no actual misrepresentations of fact, Hendry contended Dade County was obligated to disclose facts it knew through its superior knowledge. Dade County responded that it had not made any inaccurate representation, but simply that it had made "no representation" about the piling material. In this case, the court ruled that Dade County did not misrepresent any information and did not have a duty to provide the information. Hendry should have conducted a more detailed on-site investigation and could have accurately made the determination of the site conditions.

The case of *Condon-Johnson v. Sacramento Municipal Utility District* (57 Cal. Rptr. 3d 849) tested the validity of trying to shift all responsibility for changed conditions onto the contractor. On a Sacramento project for the construction of concrete foundations for piers, Condon-Johnson and Associates, the contractor, relied heavily on the information about soil conditions as described in the contract documents, and the analysis report indicating that rock strength was expected to be in the range of 3600 to 7300 psi. The contract contained a standard differing site conditions clause that gave the contractor relief if conditions differed materially from what was shown. The contract also contained a provision that stated it was "the sole responsibility of the Contractor to evaluate the jobsite and make his own technical assessment of subsurface soil conditions." Despite the provision granting relief, the contract also stated that "no additional compensation or payments" would be made if "soil conditions are different from that assumed by the contractor." When it began to work on the project, Condon-Johnson encountered rock that had strength of about 13,000 psi. Condon-Johnson filed suit when the District refused to pay for the added costs encountered due to the change in rock conditions. The court determined that by using the standard differing site condition provision, the risk was to be borne by the public agency, and that Condon-Johnson was entitled to added compensation. The court rejected the argument that the additional provisions vacated the relief provided in the differing site conditions provision.

Morrison & Lamping v. State of Oregon (357 P.2d 389) involved the construction of slightly over five miles of highway. A portion of the highway to be constructed by Morrison & Lamping crossed irrigation ditches. Morrison & Lamping was familiar with such situations and had allowed for delays and shutdowns during the irrigation operations. Shortly after construction began, irrigation operations began causing problems for the construction equipment. Although an allowance had been made in the bid, the extent of the delays caused by the irrigation was not anticipated, and the volume of water was much greater than expected. Morrison & Lamping requested $39,000 for the additional costs, but the state refused to pay. When the suit was filed, Morrison & Lamping claimed that it had expected some extra water from the irrigation, but had expected the water to be confined to the irrigation ditches and not to flood the area between the ditches. The court ruled that the conditions were not unanticipated, and that Morrison & Lamping had simply made an error in judgment for which no added compensation was due. It stated that the irrigation operation was anticipated, and that the only issue was the amount of the water that had been misjudged by the contractor.

Differing site conditions provisions commonly state that if changed conditions are discovered, the owner is to be notified promptly in writing. This was also stated in a contract that Brinderson Corp. entered for the construction of a wastewater treatment plant for the Hampton Roads Sanitation District of Virginia. Brinderson encountered unusually wet soil and severe weather that seriously delayed project completion. Although the owner's resident engineer was aware of the differing site conditions and had ample time to investigate the conditions, the contractor had failed to give prompt notice in writing. The owner did not want to pay for the costs encountered due to the differing site conditions, because the contractor had not given notice in writing, in accordance with the contract. The court concluded that the notification did not have to be in writing, because the conditions were clearly communicated to the owner and Brinderson had allowed adequate time to investigate. Essentially, the notification requirement had been satisfied since the owner was made aware. Thus, complying with the letter of the contract was not deemed to be an essential requirement in *Brinderson Corp. v. Hampton Roads Sanitation District* (825 F.2d 41).

Provisions that require contractors to make their own independent assessments of site conditions that will offer no compensation for differing site conditions contain considerable risk for contractors. Such provisions commonly state that the owner does not assume responsibility for the accuracy of the site conditions that are described in the contract documents. The extent to which such provisions are enforced was tested in *P.T. & L. Construction Co. v. State of New Jersey* (531 A.2d 1330). P.T. & L. was awarded the contract for the construction of 1.4 miles of interstate highway. In the bid documents, the Department of Transportation (DOT) disclaimed any responsibility for the accuracy or completeness of the data provided to describe the site conditions. There was also a statement indicating that there would be no entitlement to added compensation due to differing site conditions. While the plans were being prepared, the DOT received a letter from a consultant warning of the possibility of encountering saturated soil. This letter was not made available to the bidders. When P.T. & L. encountered saturated clay, the construction duration and the costs of construction rose considerably. When the contractor asked for additional compensation, the DOT refused and P.T. & L. filed suit. The court ruled that the disclaimer could not be enforced if the owner provides inaccurate information, or if it withholds relevant information from the bidders. Once the DOT was made aware of the potential problems associated with soggy soil, there was an obligation to disclose this to all the bidders. The court refused to enforce the provision that would deny compensation to P.T. & L. The contractor could recover its increased costs.

Differing site conditions do not materialize only from below ground conditions. This was evident in *Robert W. Carlstrom v. German Evangelical* (662 N.W.2d 168) that pertained to a roofing project on a church. The German Evangelical congregation of Jordan, Minnesota, sought bids from roofing contractors. They invited interested contractors to inspect the roof, but the church refused to remove any attic insulation, for fear that this would damage the insulation and increase costs. When bids were opened, Robert W. Carlstrom Company, Inc., was awarded the contract for $213,910. Carlstrom started to work and, as the old roof was being removed, observed that the structural integrity of the roof was compromised. Carlstrom verbally notified the church that additional work (change order) was needed to correct the condition. Carlstrom invoiced the congregation for $51,680 to cover the cost of the additional work. When the congregation refused to pay for the change (since the notification was verbal and not in writing, as the congregation alleged was contractually required), Carlstrom filed suit. The church argued essentially that Carlstrom, as an experienced roofing company, should have known about the unusual condition. The court concluded that the contract did not require written notification of concealed conditions, but that some type of notification must be given within 21 days of the discovery of the condition. Carlstrom was found to be entitled to the additional compensation necessitated by the changed conditions.

The circumstances surrounding *United States for Davies & Sons v. Blauner Construction Co. et al.* (37 F. Supp. 968) are typical of many claims involving differing site conditions. Blauner, the general contractor on a post office building, subcontracted the heating and plumbing to Davies. Blauner provided Davies with a copy of all bid documents before execution of the subcontract. The plans disclosed

the location of rock and rock ledges at three different locations on the site. The locations of four test pits, which had revealed the presence of the rock and rock
ledges, were also noted. When Davies began excavation to install piping, the crew
immediately encountered rock. After rock was encountered at several locations,
Davies notified Blauner's superintendent that added compensation would be expected. Davies rented a compressor and rock-breaking equipment and also purchased explosives. The president of Blauner then notified Davies that a change
order would be forthcoming, but the change order was never issued. Davies completed the work and requested additional compensation of $5,425. Blauner refused
to pay, and Davies filed suit. Davies's claim was based on the promises to issue
change orders and the fact that rock was never discussed when the subcontract
agreement was signed. Blauner stated that the contract was a lump sum contract,
and that the plans clearly showed that rock existed at several locations. The court
ruled in favor of Blauner, stating that Davies had failed to recognize the obvious
warning signs that rock would be encountered. The subcontract stated that the work
was to include "all digging in connection with the plumbing, also the catch basins."

In *Cruz Construction Co., Inc. v. Lancaster Area Sewer Authority*
(439 F. Supp. 1202), rock also caused problems for the contractor. Cruz entered a
contract for the construction of a sanitary sewer system. The owner had estimated
that 8050 cubic yards of rock would be encountered, but the actual amount of rock
was 27,124 cubic yards. Cruz requested additional compensation, since more
expensive equipment and methods had to be utilized. Cruz filed suit when payment was denied, contending that it had been delayed by representatives of the
sewer authority and did not have sufficient time during bidding to investigate the
site adequately. The court ruled in favor of the sewer authority, stating that its estimates of the rock quantities had not been fraudulently determined. An additional
point was that Cruz did not request the added compensation when the rock was
encountered. Cruz had based its claim on the precedent of *Pennsylvania Turnpike
Commission v. Smith* (39 A.2d 139), which found for the contractor when the
commission knew about the subsurface conditions and knew that the estimates
were in error. Cruz did not prove fraud in its case.

In *Wunderlich et al. v. State of California-Department of Public Works* (423
P.2d 545), subsurface conditions were alleged to be different than were shown in
the contract documents. Wunderlich had a contract for the construction of
14.4 miles of highway. Wunderlich had been furnished information from test borings in recommended borrow pits. Wunderlich also made a brief inspection of the
materials at the borrow pits. The borrow pit, known as the Wilder pit, was selected
by Wunderlich. Samples had been taken from this pit by Wunderlich, and the
material appeared to be adequate. However, this pit failed to produce the quantity
of desired material. The contract documents stated that the Wilder pit contained
sand and gravel, but the proportions of those materials were not given. Additional
materials were obtained from other sites. The material excavated consisted almost
entirely of sand and was not suitable for use. Wunderlich then filed a claim for
additional compensation for breach of warranty, since the Wilder pit did not provide sufficient acceptable material. The court ruled in favor of the state, indicating
that the information provided was not a misrepresentation of facts known by the

state. The test borings were merely indications from which deductions might be drawn. The contract essentially required the contractor to make sure there was enough acceptable material.

An interesting comparison can be made between the Wunderlich case and *E. H. Morrill Co. v. State of California* (423 P.2d 551). Morrill, a general contractor, entered into a contract to construct the Mono-Inyo Conservative Facility for the state of California. The contract contained a special conditions section which described the subsurface conditions of the site. The conditions were stated as consisting of boulders varying from 1 to 4 feet in diameter, and dispersed 6 to 12 feet in all directions. The documents also stipulated that the contractor was to make an independent investigation of the site. When excavation was begun, considerably larger boulders were encountered, and their spacing was much closer than was shown in the bid documents. Morrill filed suit for added compensation, claiming that the conditions were different from those shown on the plans. The state contended that an independent investigation by the contractor was required to confirm the conditions. Morrill stated that the site investigation made did not reveal any information that was counter to that which had been described. The court ruled in favor of Morrill, stating that the preponderance of added boulders of larger size necessitated the use of larger equipment and a change in the method of excavation. Compared with the Wunderlich case, the following factors are noteworthy: (1) The materials were described generally in *Wunderlich* and specifically in *Morrill;* (2) sand and gravel, as in *Wunderlich,* are common materials, while the boulders encountered in *Morrill* are uncommon; (3) the conditions of *Wunderlich* did not alter construction methods, while *Morrill* involved a change in construction methods; (4) the soil test representation and the disclaimer in *Wunderlich* appeared together, while they appeared in different locations in the documents of *Morrill;* and (5) Wunderlich made no independent soil tests (other bidders did), while Morrill did make such tests.

REVIEW QUESTIONS

1. One of two possible situations must generally exist for a differing site condition or changed condition to be valid. What are they?
2. By contract, what must generally be done by the contractor to ensure that payment will be received for performing work required by a changed condition?
3. Under what conditions might a contractor successfully claim a differing site condition, even though the information to bidders stated that the contractor had to perform an independent site investigation?
4. What are the implications for a contractor when the contract documents state that, by submitting a bid, the contractor certifies that a site visit was made, and full responsibility is assumed for all subsurface conditions?
5. Examine the AIA documents and the ConsensusDOCS that are included in the Appendix and determine the similarities and differences in the provisions related to differing site conditions. Specifically examine AIA provisions §3.7.4 and ConsensusDOCS §3.16.2.

14

MATTERS OF TIME

TIME OF COMPLETION is usually a major aspect of construction contracts. The owner generally has a specific need for the project and may have developed specific arrangements for the use of the completed facility on a certain date. An office building may be leased to future tenants with an established move-in date. Schools usually must be completed in the summer, before the beginning of the fall term. Retail stores are often completed so that the new businesses are ready to compete during peak shopping seasons. To help owners obtain their projects when needed, most contracts specifically include provisions about the time schedule.

CONSTRUCTION DURATION

A contract may stipulate the actual number of working days that will be available for the contractor to complete the work. This is not without problems, as *working day* is not universally defined. Such contracts should carefully define this term. Some contracts use calendar days to measure the amount of time available to the contractor to complete the work. Other contracts may stipulate the date on which construction is to be complete. This is a fixed completion date contract which has the shortcoming of an undefined start date.

The definition used for *contract time* or *contract duration* may be determined by several factors. A working day schedule may be most appropriate when the site conditions are subject to delays caused by weather, differing site conditions, and the like. A fixed completion date may be specified when the owner needs a completed project by a specific date. The general preference of the owner for this type of contract may also be a factor. Federal agencies tend to use calendar day schedules; state highway agencies often, though not exclusively, prefer to use working day schedules; and many private owners prefer either calendar day or fixed completion date contracts. The time of year a construction project is to begin, the geographic location of a project, and the complexity of the work may also influence the choice of time definition in the contract, particularly where weather is a strong factor.

It is obvious that weather can adversely affect a construction project, especially during the early stages of construction. This impact is of an ongoing nature on earth-moving and highway construction projects. Weather constraints can be addressed to some extent by scheduling construction projects during the drier summer months. However, projects of longer duration may be unable to avoid some unseasonable weather conditions. Cold weather conditions are addressed in some contracts, often through the incorporation of a winter exclusion period or winter exception period. A winter exclusion period consists of a block of time in the winter months during which no contract time is consumed. Winter exclusion periods are most commonly used by public agencies in the northern and eastern states, where severe winter weather can halt a construction project for extended periods. Owners not including winter exclusion periods in their contracts tend to use working day schedules. Working day schedules typically extend construction durations for severe weather delays. Federal agencies do not typically use winter exclusion periods in their contracts. Winter exclusion periods provide a reduced risk to contractors who anticipate little construction progress during the winter months. Conflict may still occur if the contractor actually performs work during the winter exclusion period, as some owners charge the number of days worked during this period against the contract duration. Some owners justify this practice by contending that they must have their representatives present on the site whenever work is performed. Whether days worked during the exception period will count against the contract duration should be clearly stated in the contract documents to avoid subsequent disputes.

PROJECT SCHEDULE AND PRECONSTRUCTION CONFERENCE

When a construction contract is awarded, owners generally want some assurance that the contractor can fulfill the terms of the contract in the allotted time. To demonstrate this, the contractor may be required to submit a schedule showing the sequence of activities for performing the work. Although requirements for construction schedules vary considerably, the following is typical of the requirements of some public agencies:

> After award of the contract and prior to starting work, the Contractor shall submit to the Owner a satisfactory progress schedule or critical path schedule which shall show the proposed sequence of work, and how the Contractor proposes to complete the various items of work within the number of working days set up in the contract or on or before the completion date specified in the contract. This schedule shall be used as a basis for establishing the controlling item of construction operations and for checking the progress of the work. The controlling item shall be defined as the item which must be completed either partially or completely to permit continuation of progress. The Contractor shall confer with the Owner at regular intervals in regard to the prosecution of the work in accordance with the progress schedule or critical path schedule.

The ConsensusDOCS address the preparation of a project schedule. Note that the acceptance by the owner of the contractor's schedule places an obligation on the contractor to comply with the schedule, and for the owner to also abide by the schedule, or the construction duration might be altered:

> §6.2 Before submitting the first application for payment the Contractor shall submit to the Owner, and if directed, its Architect/Engineer, a Schedule of the Work that shall show the dates on which the Contractor plans to commence and complete various parts of the Work, including dates on which information and approvals are required from the Owner. On the Owner's written approval of the Schedule of the work, the contractor shall comply with it unless directed by the Owner to do otherwise or the Contractor is otherwise entitled to an adjustment in the Contract Time. The Contractor shall update the Schedule on a monthly basis at appropriate intervals as required by the conditions of the Work and the Project.

The owner may take the initiative in developing a schedule and place the contractor in the position of taking issue with it. This might be stipulated in a provision as follows:

> The Owner will furnish the successful bidder a progress schedule developed in the determination of contract time. Such progress schedule may be used as the approved contract progress schedule or the successful bidder may submit another progress schedule for approval.

With such a provision, the owner's schedule will become the approved project schedule if the contractor fails to provide one.

Often the owner may require that this schedule be submitted before the preconstruction conference at which contract administration, planning, and interaction details are fine-tuned. Some owners require that the schedule be submitted within a designated number of days after the contract award. Some require that the schedule be submitted prior to the start of construction work. As a practical matter, the schedule should be available for the owner's review before the preconstruction conference, as this is the appropriate time to discuss the means by which the contractor will meet the contractual requirements. If the schedule will be used as a monitoring tool, it is advisable that the contractor and owner subject it to close scrutiny, and that this dialogue take place prior to construction.

Various nuances of the project, particularly those pertaining to the interrelationship of the contractor and the owner, must be addressed in the early stages of contracting. The forum commonly used for this is the preconstruction conference. Many contract documents require the contractor to attend such a conference. An example of a clause specifying this conference is as follows:

> After receipt of the notice of award and prior to the beginning of construction, the Owner and the Contractor shall establish a mutually agreeable date on which a preconstruction conference will be held. The Contractor shall have present at the preconstruction conference the project superintendent and other representatives or responsible officials who will be involved during the construction of the project, including representatives of any subcontractors. Officials of local, county, and municipal governments, representatives of affected utility companies, and other affected agencies will be requested by the Owner to attend in order that a working understanding can be established, thus providing for the

coordination of the work among the various parties and allowing the work to proceed with minimum delay.

The topics discussed at the preconstruction conference vary from project to project. The initial intent is to identify the key individuals who will be involved in the project. Discussions invariably focus on questions arising from the plans and specifications, payment procedures, matters related to right-of-way, compliance with permits that have been issued, unusual conditions, erosion control requirements, pollution controls, traffic control, unusual hazards, and so forth. Clearly, these are items which should be addressed before construction begins. The risk of misunderstandings will be greatly reduced if the contracting parties resolve identified problems before construction begins.

SCHEDULING BASICS

It is often said that construction is schedule-driven because time is so crucial to the successful undertaking of a project. While entire textbooks have been written on the subject of scheduling, a brief introduction will be provided here. This will be explained via a simple project that consists of a fence repair job at a city park. The project consists of five distinct activities, including: purchasing paint and flowering plants, repairing the board fence, painting the board fence, planting the flowers, and cleaning up the site. This schedule is shown in the following figure. The five activities are shown along with the estimated time (stated in days) to perform each activity. The schedule is read from left to right and the activity links show which activities depend on other activities. Note that the schedule shows that, as soon as the board fence is repaired, the painting of the board fence and the planting of the flowers can begin. By observation, the project can be completed in 11 days.

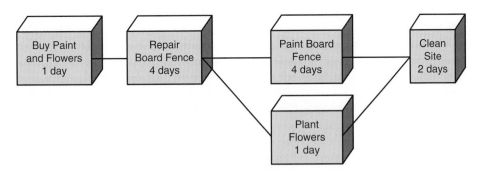

Although this schedule is much simpler than a construction schedule consisting of hundreds of activities, this shows the basic principles involved in a critical path method (CPM) schedule. In the five-activity schedule shown, there are four critical activities. A critical activity is one that must start as soon as its preceding activity has been completed, and it must be completed in the allotted time or the project will not be completed in the scheduled 11 days. The only noncritical activity is Plant Flowers, as this activity takes place concurrently with Paint Board Fence. The activity of Plant Flowers is said to have a "total float" of three days, as planting flowers can start three days later than scheduled and the project will still be completed in the period of 11 days. This total float exists because painting the board fence will take longer than planting flowers, and the delay in planting flowers (up to three days) merely uses up the float. Note that once the planting of flowers has been delayed by three days, there is no more total float, and this activity will also become critical.

WHO OWNS THE FLOAT?

The question of float ownership is an issue that has been debated extensively. To begin to answer the question, it is important to first examine how this is addressed in the contract. If this is not addressed in the contract, the contractor might well assume that all float belongs to the contractor. This is based on the premise that the schedule is a rough portrayal of how a project will be constructed, and that any float that exists will be needed by the contractor to make day-to-day adjustments in the work. The owner, on the other hand, might regard the float to be completely at the disposal of the owner, or that the contractor and owner have joint ownership of the float. When joint ownership is assumed, the float essentially belongs to the first party who utilizes the float.

Assume that the owner issues a change order on a project. The only activities impacted by the change are noncritical activities and, according to the schedule, the project duration will not be altered. The owner will regard the impact of the change to be minor, as the change consumed only float time, meaning that the contractor is not entitled to a time extension on the project. The contractor might have a completely different view, as the contractor's options in terms of methods, means, techniques, sequences, and procedures have been greatly limited by the change. In the absence of clear contract verbiage about the ownership of float, a claim could ensue. The contract might state that float "is not considered to be owned by either the Owner or the Contractor. The amount of positive float and its relative position within the overall progress schedule will be factors to be considered during negotiations to resolve time extensions." This type of provision does not guarantee float ownership to any party. Instead, the context of the position of the impacted activities must be considered to determine the actual ownership of the float.

THE CONSTRUCTION SCHEDULE

It is common for contracts to stipulate that the contractor must submit to the owner's representative a progress schedule prior to commencing with construction work. The information in this schedule is a significant means by which the contractor can show how the project is to be executed. Consider a project where the contract duration is given as 360 days, but the contractor's schedule shows the project duration as 300 days. Suppose the owner makes a change in the design of the project; the contractor's schedule now shows that the project duration has been extended 10 days by the change to 310 days. Can the contractor make a successful claim for the extra 10 days? According to the AIA A201 Document, it is not likely that any added compensation will be granted for the time extension, as the contract duration is fixed and the schedule is merely informational. This is where specific agencies might make different determinations. For example, on federal projects it is more likely that added compensation might be granted, as the contractor's construction schedule forms the new or revised contract duration. The contract provisions will vary regarding the terms under which the construction schedule is submitted. Some contracts require the schedule to be submitted for the owner's approval, while other contracts regard the schedule as informational only. The specific treatment of the schedule in the contract must be carefully examined to determine the impact of changes and delays.

CALENDAR DAYS

A calendar day represents every day that takes place, including weekends and holidays. There is no confusion with this definition. The definition of *calendar days* is more generally understood, but disagreements can still occur. For example, when does the counting of the calendar days begin? This potential problem exists on all contracts and can be avoided by clearly stating in the contract documents when the "clock starts running" to define project duration. Building projects, as opposed to civil projects, are more likely to have the project duration defined in calendar days. This is because weather problems are less likely to adversely impact construction progress. That is, calendar days are preferred when time extensions, such as those resulting from adverse weather or owner-caused delays, are not generally anticipated.

COMPLETION DATE

Project duration can be established with a fixed completion date. The nature of the project will often dictate that the owner would like to occupy and utilize a completed facility at a certain point in time. For a school building, it is understandable for the completion date to be set at a date that is prior to the beginning of the fall school term. A retail store owner may stipulate that a facility be completed prior to the Thanksgiving holidays when the Christmas shopping begins. State highway departments have been known to specify a completion date on roadways to ensure that the roads can be put to use prior to major events, such as a world's fair.

The establishment of a completion date does not, in and of itself, establish a clear value of the project duration. It sets the completion date for the project, but it does not establish the starting date. Thus, the duration is dependent on the start date, which is generally not known prior to the bid date. This can present a dilemma to a construction firm that is trying to estimate the cost of a project. The cost is generally directly impacted by the project duration. Suppose the bid date is set. The Owner may then have 30 to 60 days to actually award the contract. Once the contract is awarded, the start date is generally defined in the notice to proceed, either, for example, within a prescribed number of days, or on a stipulated date. In this instance, the duration for the project can vary considerably, ranging from a start date that is within a few weeks of the bid date, to a time that is more than two months after the bid date. The additional two-month delay may add more than two months to the project duration. Consider the impact of this two-month delay if the rainy season begins within this time frame. Other costs may also be incurred, due to resources being held in abeyance as the contractor waits for the notice to proceed. Clearly, there is some risk involved in the completion date contract in that the duration is not clearly defined.

WORKING DAYS

The definition of working days on a working day contract is of considerable interest to the contracting parties. The owner generally will monitor and make determinations of the amount of time to be charged against a contract. The contract provisions should clearly define working days and indicate how time will be charged. It is typical to define a working day as any day except Saturdays, Sundays, and holidays on which the contractor performs work or could have performed work.

Since holidays are not universally recognized, it is appropriate that they be specifically identified, particularly on public works contracts. Federal holidays such as New Year's Day, Martin Luther King Day, Presidents' Day, Memorial Day, Independence Day, Labor Day, Columbus Day, Thanksgiving Day, and Christmas Day are widely accepted in the public sector. While there is often consistency in the definition of some holidays, other holidays, especially state holidays, may also be defined in the contract, and these vary from state to state. For example, Good Friday and the day after Thanksgiving may be contract holidays. Unique state holidays include Benningston Battle Day (Vermont), Texas Independence Day, West Virginia Day, and Seward's Day (Alaska). Several southern states observe Robert E. Lee's birthday, Confederate Memorial Day, and Jefferson Davis's birthday. As many as 14 observed holidays have been noted in some documents. Although these holidays should be specifically noted in the contract documents, it is common for many public agencies not to list the observed holidays. This omission can result in problems during the construction phase. For example, inspectors may not be available on a holiday on which the contractor elects to work. The contractor may also be prohibited from working on observed holidays.

Although the provisions may vary considerably, some owners state in their provisions that Saturdays, Sundays, and holidays will be counted as working days if they are worked. In most contracts, however, these days, particularly Saturdays, belong to the contractor.

As the construction effort progresses on a project, the owner typically will monitor the number of days worked and provide periodic reports to the contractor, indicating the number of cumulative working days that have been charged. These reports are generally issued on a weekly basis, although some owners provide the reports on a biweekly or monthly basis. Upon receipt of the periodic report, the contractor is given a specified time period, generally from 7 to 15 days, in which to take exception to the working days that have been charged. Failure to take exception within this period will generally constitute an acceptance of the information in the report and a waiver against a subsequent challenge of the report.

On working day contracts, days on which weather conditions delay construction progress are typically excluded from time charges. As with most contract provisions, a few exceptions may be encountered.

Several of these issues are illustrated in the following provision of a state highway agency:

> When the contract provides a specified number of working days or a completion date with a guaranteed number of working days, the charging of working days shall start when the Contractor begins actual construction work, and in no case later than 10 days after the execution and approval of the contract, unless otherwise provided in the contract.
>
> A working day shall be defined as any calendar day between May 1 and December 15 inclusive except Saturdays, Sundays, or holidays observed by the Contractor's entire workforce. The length of a working day will be determined by the Engineer from the number of working hours established by actual job practice by the Contractor for the current controlling item, except that not less than eight hours will be considered in the determination. A full working day will be charged for any day described in the foregoing on which conditions are such that the Contractor could be expected to do a full day's work on the controlling item. A full working day will be charged on days when the Contractor could be working on a controlling item, but elects not to work, or elects to work elsewhere.
>
> No allowance will be made for delays or suspension of the work due to the fault of the Contractor.
>
> The Engineer will determine which days are workable.
>
> (a) A partial working day of one-quarter, one-half, or three-quarters shall be charged under the following conditions:
>   (1) When weather conditions do not permit the completion of a full day's work on the controlling item.
>   (2) When job conditions due to recent weather do not permit full efficiency of the workers or equipment assigned to the controlling item.
>   (3) A shortage of help which is beyond the Contractor's control prevents reasonable progress on a controlling item.
>   (4) When any condition over which the Contractor has no control prevents completing a full day's production on the controlling item.
> (b) No working day shall be charged under the following conditions:
>   (1) When adverse weather prevents work on the controlling item.
>   (2) When job conditions due to recent weather prevent work on the controlling item.

> (3) When work has been suspended by an act or omission of the Owner.
> (4) When strikes, lockouts, extraordinary delays caused by utility and rail-road work, extraordinary delays in transportation, or inability to procure critical materials suspend work on the controlling item, as long as these delays are not due to any fault of the Contractor.
> (5) When any condition over which the Contractor has no control causes suspension of work on the controlling item.
>
> One copy of the Weekly Report of the Resident Engineer will be mailed to the Contractor's office weekly. Any disagreement with the working day charges shown must be expressed in writing to the Engineer within seven days of receipt of the Report giving detailed reasons for the disagreement. The final resolution of such disagreement will be made by the Engineer.

This provision provides for a weekly report to be submitted to the contractor with an allowance of one week for the contractor to take exception to the report. Note that the contractor will be charged for any day on which work is performed, including Saturdays, Sundays, and holidays. The following provision is similar, with a few variations:

> A working day is defined as a calendar day exclusive of Saturdays, Sundays, or state legal holidays when the contractor can prosecute the work. Working days will not be charged if the contractor elects to work on Saturdays, but will be charged if he or she is permitted to work on Sundays or state legal holidays. Time charges in the form of working days will start when the contractor begins actual construction work and in no case later than 10 days after the written notice to proceed with the work. Working days will not be charged under the following conditions:
>
> (1) If the contractor elects not to work, when the condition of the ground, weather conditions, or other conditions beyond the control of the contractor make it impossible in the opinion of the Engineer to carry on any work in accordance with the work schedule.
> (2) When operations are suspended due to an act or omission on the part of the Owner.
> (3) On working day and calendar day contracts no time will be charged for work performed during December 16 to March 15 inclusive. When the contractor performs work during the period from December 16 to March 15 inclusive, authorization must be secured from the Engineer sufficiently in advance of the proposed work to provide for proper inspection.

In this provision the contractor is given the option of working during the winter exception period without being charged contract time. In addition, the contractor is not charged for working on Saturdays, only on Sundays and holidays.

In some contracts no mention is made of charging time for work performed on Sundays, on holidays, or during a winter exclusion period. Failure to address this issue clearly introduces ambiguity into the contract. It should be recognized that allowing the contractor to work on otherwise nonworking days without charging contract time reduces the risk to the contractor, and greatly increases the flexibility of the contractor in organizing and executing the construction work. There is generally some cost to the owner when the contractor elects to work on nonworking days. For example, the owner's representative and staff may be required to monitor construction activities performed on the nonworking days. These staff costs will generally accrue at a premium rate and should be included in the owner's budget if such a practice is to be permitted.

LIMITATION OF OPERATIONS

An owner may have reason to include a contract provision that specifically prohibits work on certain days, such as Sundays or holidays. When such a provision is present in the contract documents, the contractor simply does not have these days available for work performance. Such provisions may place a hardship on a contractor who wants to make up lost time on a project by working on holidays or Sundays. These provisions are reasonably common, particularly among public owners. Saturdays are generally permitted as makeup days. A few owners also require that no work be performed the day before or the day after a holiday. Most clauses indicate that the contractor may request permission to work on these days, but that such permission is not guaranteed. Saturdays are generally not included as days on which no work can be performed. The following public agency provision is a good example of these clauses:

> Unless otherwise specified, the Contractor shall not carry on construction operations on Sundays or holidays, unless of an emergency nature. The Contractor may work on Martin Luther King Day, Washington's Birthday, Patriots Day, and Columbus Day without first obtaining permission of the Owner. Permission to work on Sundays and other holidays must be obtained prior to the Sunday or holiday. In addition, if so directed, the Contractor shall suspend all work, other than maintaining the roadway for traffic, on all portions of the project open to traffic, and such other times as the Engineer deems necessary. Working days or calendar days will not be charged against the Contractor during any such suspension period.

The inclusion of these work restrictions clauses provides an additional risk for contractors, because the contractor cannot work on the specified days without the owner's permission. If that permission is not granted, the work schedule can be shortened only by working longer workdays or employing more workers. These may be undesirable alternatives in some instances. At any rate, the contractor's flexibility is reduced by such a provision.

The contractor cannot begin construction work before being directed to do so. This directive from the owner is called the *notice to proceed* or the *letter of intent*. Without this formal communication from the owner, a contractor who begins the work effort may not be paid if the owner subsequently decides to cancel the project. As has been mentioned, the contract start date is usually stated in the notice to proceed or is related to the date of the notice.

LIQUIDATED DAMAGES

The ConsesusDOCS (§6.1.3) state it very succinctly that "Time Limits stated [in the Contract to perform the Work] are of the essence of this Agreement." If time is contractually of the essence, and if the contractor fails to perform within the contract time, the owner is entitled to damages. If no other provisions exist in the contract, the contractor is assumed to have breached the contract. This can lead to long and expensive court battles. This can be avoided through the inclusion of a liquidated damages provision in the contract. This is essentially an agreement, at the time of drawing up the contract, about the cost to the contractor for each day that the project extends beyond the contract time. If the contract time is stated as being 180 days, the contract may state that the contractor must pay $1,000 per day for each day the project delivery extends beyond 180 days. This amount is called *liquidated damages* and avoids legal entanglements later. Liquidated damages provide compensation to the owner for financial and other losses resulting from delayed completion. Since it is difficult to determine with accuracy the actual value of the losses, a predetermined sum is used in lieu of actual damages.

Liquidated damages are included in the contracts of most owners. Although the damages apply to each day of late completion, a day may be either a calendar day or working day. Owners who employ different means of defining contract duration often use different values of liquidated damages for calendar day and completion date contracts, versus working day contracts. The amount of liquidated damages assessed will be based largely on the amount of the total contract. The following provision of a public agency illustrates this point:

> Should the Contractor fail to complete the work within the time agreed upon in the contract or within such extra time as may have been allowed by extensions, there shall be deducted from any monies due or that may become due the Contractor, for each and every calendar day for completion date contracts, or working day for working day contracts, that the work shall remain uncompleted, a sum specified as follows:

Original Contract Amount		Daily Charge	
From More Than	To and Including	Calendar Day	Working Day
$0	$50,000	$125	$275
50,000	100,000	175	300
100,000	300,000	225	475
300,000	500,000	375	750
500,000	1,000,000	475	1,200
1,000,000	————	750	1,750

> This sum shall be considered and treated not as a penalty but as fixed, agreed, and liquidated damages due the owner from the Contractor by reason of inconvenience to the public, added cost of engineering and supervision, maintenance of detours and other items that have caused an expenditure of public funds resulting from the contractor's failure to complete the work within the time specified in the contract.
>
> Permitting the Contractor to continue and finish the work or any part of it after the time fixed for its completion, or after the date to which the time for completion may have been extended, shall in no way operate as a waiver on the part of the Owner of any of its rights under the contract.

The amounts charged per day vary between owners, even for contract values that are the same but tend to be comparable in general magnitude. The contract may also stipulate that the liquidated damages will be reduced to 50 percent of the stated amount per day if the project can be used by the public (for example, a highway or street), even though the project is not otherwise complete. Such provisions, although not common, recognize the legal argument that liquidated damages should not apply to projects that can be used by the public or owner even when the definition of substantial completion is not satisfied.

Note that the amount is called liquidated damages and is usually specifically noted as not constituting a penalty. Penalty provisions generally are not enforceable unless a bonus clause also exists. Thus, it is important that the amount stipulated as liquidated damages bear some resemblance to the actual anticipated costs of late completion; that is, the amount stated in the contract need not be close to the actual incurred costs of late completion, but must instead be a reasonable approximation of anticipated costs, given the information at the time of the contract award.

The case of *Bethlehem Steel Corp. v. Chicago* (350 F.2d 649) is a good example of how liquidated damages are interpreted by the courts. Bethlehem was awarded a contract to supply, erect, and paint the structural steel for a portion of the South Route Superhighway in Chicago. There were a series of contracts involved in this large undertaking. The Bethlehem contract provided for a liquidated damages amount of $1,000 per day for late completion. Although Bethlehem was granted some time extensions on the contract, it still completed its portion of the work 52 days beyond the extended duration. While Bethlehem was late in its completion of the contract, there were other contractors who completed their portions earlier than contractually required. This resulted in the roadway being opened at the originally projected time. When Bethlehem was charged the $52,000 for liquidated damages, it felt that the owner had not incurred any actual damages, and that the liquidated damage provision should not be enforced. The court did not agree with Bethlehem. Since Bethlehem signed the contract with the liquidated damages provision, the court said it would not rewrite the contract. It is perhaps easy to understand the court's view. Had the owner incurred greater actual damages than the $1,000 per day, Bethlehem would surely have insisted on enforcing the contract provision.

In the eyes of the law, a clear distinction exists between a sum that represents liquidated damages and one that constitutes a penalty. Liquidated damages amounts represent a reasonable reflection of the anticipated costs incurred by the owner for late project completion. A penalty is assumed to exist when the sum has been set at an amount that is sufficiently above the anticipated costs of late completion to constitute a punitive measure. Such punitive sums may be desirable from the owner's perspective, in that they provide an added incentive for the contractor to deliver the project within the contracted project duration.

As has been stated, penalty provisions may be unenforceable without a bonus clause. Suppose an owner determines that the anticipated costs for late completion will accrue at a rate of $1,500 per day. The contract could be enforced if the liquidated damages were set at $1,500 per day. The contract could also be enforced if the contractor were assessed a penalty of $2,500 per day for late completion, provided that the contract also contained a provision by which the owner would pay the contractor $1,000 per day for early completion.

Liquidated damages are used when the calculation of actual damages is complex and difficult to determine. This is also an efficient and effective means of avoiding legal entanglements when project completion is delayed. Most construction contracts have liquidated damages provisions, and as long as the sum charged per day is not unreasonable, the courts tend to uphold the legality of these clauses.

A liquidated damages provision essentially limits the amount of recovery by the owner for late project delivery. Additional claims by the owner, if the amount of liquidated damages is less than the actual costs incurred, may not be honored. This was shown in *J. R. Stevenson Corp. et al. v. County of Westchester* (493 N.Y.S.2d 819), involving the construction of a county courthouse. The contract contained a no-damage-for-delay clause and a liquidated damage provision setting a daily rate of $300 for late project completion. The county placed a claim against the contractor for the actual damages and the liquidated damages. The contractor refused to pay the actual damages. The court decision hinged on the intent of liquidated damage provisions. The ruling was that reasonable liquidated damage provisions preclude the recovery of actual damages, even though the actual damages may be in excess of the liquidated amount. The county was not in a good position to claim that the liquidated damage amount was unrealistic, since it had drafted the contract documents and was therefore in control of the amounts stipulated in the contract.

Mattingly Bridge Co., Inc. v. Holloway & Son Const. Co. (694 S.W.2d 702) involved a general contractor and a subcontractor. Holloway was a prime contractor on a road project with concrete construction being subcontracted to Mattingly. The subcontract completion date was stipulated as being November 15, 1971, and Holloway's completion date was set for December 1, 1971. The general contract and the subcontract set the liquidated damages at $750 per day. The project was accepted on July 19, 1972. Holloway was assessed liquidated damages for 17 2/3 days by the owner. However, Holloway claimed that it was entitled to liquidated damages from Mattingly for the period from November 15, 1971, to July 19, 1972. Mattingly filed suit. The court stated that liquidated damages for a breach should be included in the contract when they are difficult to ascertain. The court ruled in favor of Mattingly, stating that it would be unreasonable to enforce the letter of the subcontract agreement. Instead, the court found that Holloway was entitled to liquidated damages from Mattingly for the period from November 15 to December 1, in addition to the 17 2/3 days. With this decision, Holloway received the full benefit of the liquidated damages provision. Any excess damages granted would have constituted a windfall.

TYPES OF DELAYS

Construction projects are delayed by numerous causes, including strikes, adverse weather, late decisions by the owner, delays caused by other contractors, unforeseen changes that affect completion time, unavoidable casualties, restraint by a government or government agency, unsuspected subsurface conditions, discovery of Indian artifacts during excavation, and discovery of an endangered species (beetle, flower, minnow, etc.) on the construction site.

Delays have a detrimental impact on all the contracting parties. Because of delays, owners receive their projects later than desired and as a result lose some of the revenue the project would have generated. Understandably, owners would like to be compensated (liquidated damages) whenever a contractor makes a late delivery of a project. Contractors are also adversely affected by delays. The primary result of delayed completion is increased construction costs. This is incurred through increased overhead costs, loss in productivity, and demoralization of the workforce (loss of esprit de corps).

Causes of delays can be categorized into three unique groups:

- Delays caused by the contractor or the contractor's agents.
- Delays caused by the owner or the owner's agents.
- Delays caused by force majeure or acts of God.

Delays caused by the contractor may be of such magnitude as to give the owner just cause to terminate the contract. However, this is a drastic measure. Generally, contract termination is the end result of a series of efforts on the part of the owner to get the contractor to perform. If a contractor is not progressing as required, the owner should first notify the contractor that if work does not proceed satisfactorily, the owner will exercise the contractual option of taking over the project. This option, if exercised, is not always the same as termination, since it can be applied only to a portion of the job. Although the owner may have the contractually granted option of completing a portion of the project, this is generally not practical. In most instances, the owner will have to decide on the merits of terminating the entire contract. Regardless of how the project is eventually completed, the owner will claim damages against the contractor for contractor-caused delays.

All too often contractors claim that construction delays were a direct result of actions of the owner, or the failure of the owner to act promptly on critical construction matters. Although the impact of an owner-caused delay can be as severe as that of any other type of delay, contractors do not have the same rights of termination that are enjoyed by most owners. Basically, the only time a contractor can terminate is when the owner delays in issuing a certificate of payment, or when there is a delay in making payments.

Most standard construction contracts contain provisions that provide for an extension of the contract time for owner-caused delays. In addition, these contracts may stipulate that the contract amount can also be changed for owner-caused delays. However, in practice, added monies for such delays are often received through a formal claims procedure. Whenever a contractor anticipates making a claim for added compensation for owner-caused delays, the following steps should be followed: (1) Keep an up-to-date progress schedule that is approved by the owner; (2) maintain an accurate job diary outlining the relevant facts about the delay; (3) give written notice to the owner indicating that a delay has been incurred; and (4) request a written notification from the owner that a time extension has been granted.

It is obvious that added costs are generally incurred when the project completion is delayed. What is the consequence of delays that occur but do not prevent the project from completing on time? This question was examined in *D'Angelo v. State of New York* (362 N.Y.S.2d 283). On a road-building contract, Triple Cities Construction Company (family business owned by D'Angelo) was delayed by the owner. Despite the owner-caused delays, the project was still completed ahead of schedule. D'Angelo sought compensation for the delays. The court determined that D'Angelo had incurred real costs as a result of the owner-caused delays and "could rightfully expect to operate free from needless interference by the State, and, therefore, they are entitled to compensation where, as here, they could have completed their work ahead of schedule and thereby saved substantial sums of money, absent the delays caused by the State." A similar decision occurred in *Grow Construction Co. Inc. v. State of New York* (391 N.Y.S.2d 726) in which the court ruled that the owner's interference resulted in an increase in the "performance of the contract."

The third type of delays are those which are commonly referred to as having been caused by acts of God. As a general rule, neither of the contracting parties can successfully claim damages for such delays, but there are exceptions. Suppose the owner delays in approving submittal materials, such as shop drawings, to the extent that a project in Boise is delayed by a full two months. Suppose also that all concrete foundation work was scheduled to take place during September and October. With the delay, the concreting operations are theoretically shifted to November and December. However, with the stipulation that no concrete is to be placed in weather below 40 degrees, the actual concreting is spread over November, December, January, February, and March. The owner would acknowledge that two months of the delay were the direct cause of the late approvals, while the remaining delays were acts of God (cold weather). Is the contractor entitled to damages for acts of God? In this case the answer is yes, because the contractor would not have incurred any delay if it had not been for the owner's actions.

Whenever a contractor is about to undertake a new construction project, the possibility of adverse weather conditions should be considered. If the weather conditions encountered during the construction phase are typical of the weather for that region during that time of year, the courts may rule that the weather should have been anticipated. A contractor on the Gulf Coast may include a contingency for a hurricane in the schedule. A contractor in North Dakota would be well advised to anticipate severe cold weather during the winter months. A contractor could hardly be successful in claiming for damages or time extensions resulting from hot weather in the Mojave Desert. The weather must generally be of an extraordinary or unanticipated nature, for example, a 12-inch rainfall within a 24-hour period.

Whenever a delay occurs, a contractor may make a claim for a time extension, a monetary settlement, or both. Delays for strikes and bad weather usually result in time extensions only. A delay caused by a subcontractor will probably result in compensation only (the contractor can place a claim against the sub). Change orders issued by the owner are the most common means by which compensation is coupled with a time extension.

NO-DAMAGE-FOR-DELAY CLAUSES

Owners do not like to see their projects affected by delays. They also do not like to pay damages when they have contributed to or have been solely responsible for delays. To accomplish the latter wish, owners may include a contract provision which is intended to bar a contractor from claiming for delays. These clauses, although very harsh on contractors, are generally enforced if the contract is specific in its wording. The general interpretation by the courts is that if the delay or obstruction was within the contemplation of the parties at the time the contract was entered, the no-damage clause is valid. An example of the relevant portion of a no-damage-for-delay clause reads as follows:

> No payment or compensation of any kind shall be made to the contractor for damages because of any hindrance or delay from any cause in the progress of work, whether such hindrance or delay be avoidable or unavoidable. Any finding by any administrative officer, arbitrator, and/or judge that a delay was caused either wholly or in part by actions of someone other than the Contractor shall only entitle the Contractor to equivalent extensions of time.

A similar provision reads as follows:

> Neither the Owner nor the Contractor shall be entitled to damages for any delay caused by the Owner in the performance of the work under the contract. In such event, however, the owner shall grant the Contractor an extension of time.

Contractors should be aware that no-damage-for-delay clauses can have a severe monetary impact on contractors who could otherwise seek recovery. Such a provision is called exculpatory and is typical of an adhesion agreement. That is, it excuses one of the parties to the contract from liabilities which that party would otherwise incur. As a rule, such clauses are upheld by the courts. However, the court interpretations are very narrow, in that the courts tend to limit these clauses to their literal terms. In spite of this tendency, these clauses protect many owners from liability.

Exceptions to No-Damage-for-Delay Clauses (No Owner Protection)

With no-damage-for-delay clauses, owners try to avoid liability for losses caused by delays, interruptions, and even interferences that occur during the construction phase. It must be borne in mind that if the provisions use terms such as *reasonable delays* and *ordinary delays,* the courts will interpret them in a very narrow manner against the party seeking release from liability. There have been a few instances when no-damage-for-delay provisions have not been enforced by the courts. These were instances in which the wording of the provisions was interpreted so strictly that the clause did not apply to the delays in question.

Delays Not Contemplated

Delays that are not contemplated by the contracting parties at the time of contract negotiation do not exempt the owner from liability. The most common delay of this type is the denial of access to the job site to the contractor.

Example. Suppose a contract was let for remodeling the exterior of a consular building. The contractor was about to begin work when protestors demonstrated in front of the consulate. The contractor was forbidden to begin work as long as the unrest continued. In spite of a no-damage-for-delay provision, the contractor recovered, because this delay was not contemplated by either contracting party.

Active Interference

Acts or omissions of the owner which actively interfere with work progress are often beyond the protection of a no-damage-for-delay clause. The courts have ruled that the existence of such a provision does not give the owner license to cause delays willfully. The term *delay* refers to situations in which the contract completion date is extended. The term *interference* relates to performance problems that result in increased costs, whether or not the completion date is changed.

Example. Suppose a negotiated contract was let to construct a small office building. The contractor was issued a notice to proceed, but the plans for the foundation were not complete. The contractor was forced to wait for the plans, which were being drafted by the owner's personnel. The court ruled that the owner had interfered.

Delays of Unreasonable Duration

Unreasonable delays may not be covered by no-damage-for-delay provisions. What constitutes an unreasonable delay? This question must be answered by the courts. The courts take the perspective that lengthy delays are not contemplated, and as a result, exculpatory clauses do not afford the owner complete protection. If the delay is excessive, the contractor may have legal grounds for termination, but the contractor may prefer to continue and simply be paid for costs incurred as a result of the delay.

Example. Suppose a contract was let to install an attractive entry gate to a country club. The contractor was to begin work on this $30,000 job as soon as a gas line was relocated. The contractor had to wait nine months before the relocation occurred. The court ruled that the delay was unreasonable and was comparable to abandonment by the owner. Note that a delay of nine months on a large job may not be considered unreasonable.

Fraud or Bad Faith

A party cannot escape liability under a no-damage-for-delay clause when the delay is caused by that party's intentionally false statements or acts. Of course, on public works projects this is against public policy.

Example. Suppose an owner negotiated a contract with a contractor. During the negotiations, the owner made various assurances about the conditions at the site, the completeness of the plans, the existence of a building permit, and so on. When the contractor began work, it was discovered that the plans were not finished and that no permits had been obtained. The contractor had relied on the owner's statements when the contract was negotiated and was therefore granted recovery.

No-damage-for-delay clauses can be extremely harsh on the contractor. Fortunately, many owners do not incorporate such provisions in the contract. When such provisions are included, contractors should be fully aware of the implications.

The case of *Goss v. Northern Pacific Hospital Association of Tacoma* (96 P. 1078) illustrates the strength of such a provision. Goss verbally agreed to construct several buildings for the hospital. It was also agreed that a separate contract would be let for the plumbing and heating facilities for all the buildings. Goss started the construction work and subsequently entered into a written agreement with the hospital. A no-damage-for-delay provision was included in the contract. When the buildings under construction were ready for the installation of the plumbing and heating systems, the specialty contractor defaulted, causing considerable delays for Goss. Goss sued the hospital for the added costs incurred as a result of the defaulting specialty contractor, rather than simply accepting the contractually agreed time extensions. Goss questioned the validity of the contract, since some of the construction work had already begun at the time the agreement was signed. The hospital stood behind the no-damage-for-delay provision, even though the delay was caused by an independent contractor. The hospital argued that since Goss had signed the contract, Goss had ratified the contract, which did not allow for damages. The court ruled in favor of the hospital, stating that validity of the contract was not affected by the signing of the contract after work had begun.

Nelse Mortensen & Co., Inc. et al. v. Group Health Cooperative of Puget Sound (586 P.2d 469) also involved a contract that included a no-damage-for-delay provision. This occurred on a project to construct and remodel medical facilities. The work was to take place in phases and was to be closely coordinated so that hospital operations would not be excessively disrupted. The new structure adjoined the operating hospital. The contract documents consisted of the standard AIA forms, with the provisions regarding damages for delays being struck, and a provision added by which no damages were due for causes beyond the control of the contractor. During this project, Mortensen was allegedly delayed on 146 occasions and sought damages for these delays. Seventy-eight of the delays resulted from change orders, and 37 were caused by interpretations that were not made by the owner within 15 days, as required by contract. Mortensen claimed that the hospital had caused the hindrances and that the delays were unreasonable, entitling the firm to damages. The hospital based its defense on the grounds of the contract provisions. The court ruled in favor of the hospital, in part because the provision allowing the contractor to claim for damages had been deleted. Mortensen's case was weakened by the fact that it demanded compensation five months after substantial completion and had not allowed for the added costs of delays when agreeing to the change orders. The delays, totaling three months, were not excessive and should have been contemplated.

In *City of Seattle v. Dyad Construction Inc.* (565 P.2d 423), a dispute occurred involving a sewer line project. The contract included a provision that stated that delays were to be compensated "for a period equivalent to the work time lost." The sewer line was being installed along Seola Beach, with the trench being dug through sand and gravel in the tide flats as staked by the city's survey crew. The path of the trench ran between the base of a bluff on one side and the tide flats on the other. The work could be performed only during a favorable low tide. During the trenching operation, a landslide occurred behind the backhoe performing the excavation work. Dyad stopped work and asked the city to redesign the alignment of the pipe location so that it was farther out on the beach. The Washington State Labor and Industries safety inspector also considered the project unsafe. Eight months later the city approved a new plan for the sewer location. The project was completed four months after originally scheduled. Dyad sued for damages caused by the delay, which occurred through no fault of its own. The city claimed that the contract was clear and that the contractor was entitled only to a time extension. Dyad claimed that the provision dealt only with foreseeable delays at the time the contract was signed. Not only were the delays that were encountered unforeseeable, they were also of unreasonable length. The court ruled in favor of Dyad, stating that the contractor was entitled to compensation. Although a no-damage-for-delay provision is usually upheld, there are extenuating circumstances that place limitations on the application of the provision. The court stated that it is implied that the owner will not hinder or delay the contract. Dyad was entitled to damages as well as a time extension.

An unusual situation arose in the case of *Atlanta Economic Development Corp. v. Ruby-Collins* (425 S.E.2d 673). In this case, Ruby-Collins, Inc., was a general contractor that entered a lump sum contract with Atlanta Economic Development Corp. (AEDC) to widen a street and extend a culvert. The bid package contained a standard form AIA agreement that was executed, and a bid package prepared by the designer. The contract form in the bid package contained a no-damages-for-delay provision. This contract form was never signed; it also did not have any entry of the sum to be paid for the work. After project completion, Ruby-Collins sought delay damages from AEDC. The request was denied by AEDC and a suit was filed. The court ruled that the contract was ambiguous, since the contract form was not filled in and since it was not signed. The court ruled that the ambiguity of the contract could be ruled in favor of AEDC, so it determined that Ruby-Collins was entitled to delay damages.

EXTENSIONS OF TIME

Construction contracts try to control the contract time, which is usually closely regulated. This is evidenced by the common employment of liquidated damage provisions and penalty-bonus provisions and the practice of retaining a percentage of the money due the contractor. These provisions are not harsh compared with no-damage-for-delay provisions, in that they essentially dictate that the contractor is entitled to claim for extensions of the contract time for any delay beyond the contractor's control.

When the contractor is delayed for a reason that is not his or her own fault, the contractor should consider requesting an extension to the contract time. Should requests be made for delays caused by a subcontractor? The answer is generally no, since the subcontractor is under the direct control of the contractor. If a delay is caused by late delivery of materials, an extension of the contract time may be requested. This too may be unsuccessful, because the contractor has an agreement directly with the materials supplier.

Time extensions may be granted for a number of conditions. Most construction contracts allow for time extensions. A typical provision will indicate that additional contract time will be granted when the contractor is delayed because of "unforeseeable causes beyond the control and without the fault or negligence of the contractor" and which the contractor is "unable to prevent." Such causes are often referred to as *force majeure,* which include acts of God, expropriation of facilities, changes in applicable law, war, earthquake, change of government, rebellion, civil disturbances, sabotage, riots, floods, unusually severe weather, fires, explosions, strikes, or other similar occurrences. Although most owners grant additional time in the contract under different provisions, many limit the ability of the contractor to recover monetary damages for such delays. One contract stated, "No delay or failure in performance by either party hereto shall constitute default hereunder or give rise to any claim for damages if, and to the extent, such delay or failure is caused by force majeure. Unless such force majeure substantially frustrates performance of the Contract, force majeure shall not operate to excuse, but only to delay performance."

The ConsensusDOCS stipulate that the contractor shall be entitled to time extensions for several specific causes. These are included in the provisions as follows:

§6.3.1 If the Contractor is delayed at any time in the commencement or progress of the Work by any cause beyond the control of the Contractor, the Contractor shall be entitled to an equitable extension of the Contract Time. Example of causes beyond the control of the contractor include, but are not limited to, the following: acts or omissions of the Owner, the Architect/Engineer or Other; changes in the work or the sequencing of the Work ordered by the Owner, or arising from the Owner that impact the time of performance of the Work; transportation delays not reasonably foreseeable; labor disputes not involving the Contractor; general labor disputes impacting the project but not specifically related to the Worksite; fire; terrorism, epidemics, adverse governmental actions, unavoidable accidents or circumstances, adverse weather conditions not reasonably anticipated; encountering Hazardous Materials; concealed or unknown conditions; delay authorized by the Owner pending dispute resolution; and suspension by the Owner under Paragraph 11.1.

EXCUSABLE DELAYS

Failure to perform by a specified date or within a reasonable time is often excused because the defaulting party may be contractually excused for the delays, and as a result, an extension of time is granted. A general rule is that the construction time is extended one day for each excused day of delay.

Excusable delays include acts of God, labor strikes, flooding, embargoes, epidemics, national emergencies, acts of third parties, riots, changes, and unusually severe weather. As was noted earlier, the definition of *severe* as it pertains to weather is subjective. Extensions of contract duration are not automatic; extensions of contract time must be requested in the manner prescribed in the contract. Usually this request must be in writing and must be submitted within a given time after the delay has occurred.

Extensions of time are important to the contractor. A time extension in a contract may save a contractor money by avoiding the costly need to accelerate work, or may eliminate the assessment of liquidated damages. It is prudent to seek legitimate time extensions, even if it appears that the project will be completed by the originally contracted completion date. These extensions of time, even if not considered necessary, will provide a valuable cushion that will be appreciated if other, inexcusable delays are subsequently encountered. Regardless, it is imperative that the contractor carefully document each incident or event for which a time extension is to be potentially requested (refer also to Article 8.3 of AIA Document A201-1997).

Delays caused by weather are perhaps the most commonly occurring delays and are of the greatest importance to contractors. Many construction contracts state that the contract time will be extended for weather delays, but the nature of these provisions must be fully understood. For example, the granting of time extensions may hinge solely on whether the contract time is measured in working days, calendar days, or completion dates. The following provision illustrates this point:

> If the contract time is on a calendar day basis or has a fixed calendar date for completion, no extension of time will be considered for unsuitable weather or conditions resulting therefrom.

It is typical for time extensions for weather delays to be granted in working day contracts. Some owners include provisions that grant additional contract time for weather delays only if the weather is unusually severe. An example of such a provision is as follows:

> Rains or other inclement weather conditions and related adverse soil conditions will be considered as the basis for granting of a time extension only when such conditions are unseasonable, provided that the project records indicate that they did in fact delay one or more controlling items of work.

Under these provisions, adverse weather will not automatically constitute grounds for receiving time extensions. The delays must be caused by weather that is unusually severe or unseasonable.

Some contracts state that time extensions will be granted for adverse weather, but no definition of this term is given. This is a common deficiency in many contracts. Without a definition, the determination of unseasonable weather may be made by referencing locally recorded weather conditions over an extended period. Less risk is imposed on a contractor, however, if the definition is related to the impact of the weather conditions on the construction effort. Such definitions may state that an excusable weather delay has occurred if a stated percentage (such as 50 or 60 percent) of the workforce was unable to work, a stated percentage (such as 50 or 60 percent) of the work day could not be worked, or significant progress could not be made on a critical or controlling activity. The following is an example of a provision that provides an objective means of defining delays caused by weather:

> When delay occurs due to reasonable causes beyond the control and without fault of the Contractor, including but not restricted to "acts of God," . . . the time of completion of work shall be extended in whatever amount is determined by the Engineer to be

equitable. An "act of God" as used in this article is construed to mean an earthquake, flood, cyclone, or other cataclysmic phenomenon of a nature beyond the power of the Contractor to foresee or make preparation in defense of. A rain, windstorm, or other natural phenomenon of normal intensity, based on United States Weather Bureau reports, for the particular locality and for the particular season of the year in which the work is being prosecuted, shall not be construed as an "act of God" and no extension of time will be granted for the delays resulting therefrom.

The following provision grants time extensions for weather delays, but only when certain criteria are satisfied. This provision is fairly general and can result in disputes unless more objective criteria are added.

> Delays caused by weather or seasonal conditions shall be anticipated and will be considered as the basis for an extension of time only when the actual work days lost exceeds the number of days that would normally be lost due to weather conditions for that time of year.

In this provision, adverse weather is defined as conditions which are abnormal for the period of time and could not have been reasonably anticipated. Adverse weather is more clearly defined and is not left to be decided solely by the owner's representative or architect. Since this provision varies considerably between different agency contracts, particular attention should be given to this matter before beginning the construction effort. If a provision states that "requests for extension of time shall be filed in writing by the Contractor with Owner not more than 30 days following the termination of the delay," the contractor should evaluate all delays on a project at least on a monthly basis. If the request period is seven days, a weekly review of delays must be made.

A weather delay may not result in a contract extension if all aspects of the time extension provision are not satisfied. It is common for this type of provision to state that a request for a time extension must be made within a specified period. The time in which to request a contract time extension may range from a period as short as seven days after the delay to any time before completion. With such varying provisions, contractors must be fully aware of the provisions governing each project.

ACCELERATION

Acceleration can occur in two ways. The first is called *actual acceleration* and consists of a direct order by the owner to hire additional workers, work overtime, or work extra shifts on the project. The second is *constructive acceleration;* it does not result from a direct order, but is construed as acceleration because of the owner's refusal to permit or grant time extensions for an excusable delay. An assertion of acceleration is clear if one party directs another party to accelerate. The contract may provide for acceleration pay. If no provision is made for payment for acceleration, the contractor must protest when acceleration occurs.

Acceleration is not always clear-cut. As in most legal cases, there are two differing viewpoints. Suppose a contractor has fallen behind schedule on a project. The contractor has applied for time extensions stemming from excusable delays.

Requests for time extensions were made but were denied by the owner, and the owner asked the contractor to complete the project at the originally contracted date. The contractor must proceed with the work under protest. This could be viewed by the contractor as constructive acceleration, while the owner is simply enforcing the original completion date. If the contractor obtains a court decision finding that the delays were excusable, the owner will be charged with acceleration costs.

Another example helps illustrate this point. Suppose the owner issues a change order on a project but does not permit the contract time to be extended. This, too, may be regarded as acceleration. However, if acceleration claims are going to be made by the contractor, this must be formally communicated to the owner.

In *Continental Heller Corporation v. U.S. Government* (GSBCA No. 6812), constructive acceleration was claimed in spite of the fact that a time extension was granted. This case involved a contract that Heller had for the construction of a federal office building and courthouse in San Jose, California. As part of the work procedures planned by Heller, the material excavated from the proposed basements would be used as fill material for the elevated parking lot. However, because of unusually heavy rainfall, when the excavation work began, it was apparent that the moisture content (saturated condition) of the soil made the excavated material unsuitable for fill material. The contractor requested a time extension to delay the construction activities and allow the soil to dry out. The government refused to consider the request until it was substantiated by sources such as the U.S. Weather Bureau and by a critical path method (CPM) diagram that showed the delay. When no response was forthcoming from the owner, the contractor continued working through the bad weather. To stay on schedule, the contractor removed the excavated material and imported the fill material for the parking lot. The contractor then placed a claim for the additional costs of excavation. Sixteen months after the initial time extension request was submitted, the government granted the added time desired. However, the government refused the claim for the added costs incurred by the contractor. The contractor appealed to the board of contract appeals, claiming that the delay in issuing a response to the request constituted constructive acceleration. The board agreed that the refusal to grant an earlier time extension forced the contractor to deviate from the originally planned excavation method. Essentially, the government's action in not responding amounted to an insistence that the contractor stay on schedule. The contractor had carefully documented the fact that additional costs were incurred and that the original method would have cost less.

It is important that the contractor be fully aware of the potential impact of project activities. Notices of delays, changes, and work suspensions must be made promptly. The contractor cannot claim for these items after final payment has been received. Once final payment has been accepted, claims are generally no longer permitted. Prompt notice must be given so that the other party (owner or government) has an adequate opportunity to develop its side of the case. With proper notice, corrective action may be taken. Also, this will avoid surprise claims after the job is completed.

SUSPENSION OF WORK CLAUSES

The suspension of work provision is the contractual equivalent of a breach of contract action for delays. Suspension clauses are seldom included in private contracts. It might be a good idea for contractors to consider their inclusion or to negotiate for them. Such clauses generally state that if prompt notice of the suspension is not given, the suspension claim will not be valid.

The general conditions provisions in most owners' contracts contain clauses that give them the ability to suspend work. Suspensions may result from an action of the contractor or the owner. The most important components of these provisions concern the general nature of suspensions and the ability of the contractor to recover costs associated with owner suspensions. If the suspension provision does not permit the contractor to recover damages caused by an owner's suspension or project delay, contract contingencies will increase, as demonstrated by increased bids. When addressed in the contract, contractor recovery of damages is typically restricted to instances where the contractor is not at fault.

Most contracts contain provisions granting the owner power to suspend work when the contractor is not in compliance with the contract. Another cause for suspending construction work relates to the discovery of historical or archaeological artifacts. Other causes for suspending a contract may also be enumerated. The following provision is typical of the suspensions clauses used by many state highway agencies:

> If the performance of all or any portion of the work is suspended or delayed by the Engineer in writing for an unreasonable period of time (not originally anticipated, customary, or inherent to the construction industry) and the Contractor believes the additional compensation and/or contract time is due as a result of such suspension or delay, the Contractor shall submit to the Engineer in writing a request for adjustment within 7 calendar days of receipt of the notice to resume work. The request shall set forth the reasons and support for such adjustment.
>
> Upon receipt, the Engineer will evaluate the Contractor's request. If the Engineer agrees that the cost and/or time required for the performance of the contract has been increased as a result of such suspension and the suspension was caused by conditions beyond the control of and not the fault of the Contractor, its suppliers, or Subcontractors at any approved tier, and not caused by weather, the Engineer will make an adjustment (excluding profit) and modify the contract in writing accordingly. The Engineer will notify the Contractor of his or her determination of whether an adjustment of the contract is warranted. No contract adjustment will be allowed unless the Contractor has submitted the request for adjustment within the time prescribed. No contract adjustment will be allowed under this clause to the extent that performance would have been suspended or delayed by any other cause, or for which an adjustment is provided for or excluded under any other term or condition of this contract.

This provision allows the contractor to recover costs associated with suspensions, but does not permit the contractor to profit from any suspensions. For the contractor to recover costs, an appropriate request must be made within the stipulated time period. Note that this above clause does not enumerate the causes for suspensions that are deemed to be for an "unreasonable period of time."

The specific wording of a suspensions clause should be read with care. Recovery of costs may be barred in the provision. The following example is a portion of such a provision:

> The Contractor agrees to make no claim for extra or additional costs attributable to any delays, inefficiencies, or interference in the performance of this contract occasioned by any act or omission to act by the Owner except as provided in the agreement. The Contractor also agrees that any such delay, inefficiency, or interference shall be compensated for solely by an extension of time to complete the performance of the work in accordance with the provision in the Standard Specification. In the event the Contractor completes the work prior to the contract completion date set forth in the proposal, the Contractor hereby agrees to make no claim for extra costs due to delays, interference, or inefficiencies in the performance of the work. The Contractor further agrees that he or she has included in the bid prices for the various items of the contract any additional costs for delays, inefficiencies, or interference affecting the performance or scheduling of contract work caused by or attributed to . . .

Included among the items for which no additional costs can be recovered are: the failure of a public body to issue a permit; labor strikes; shortages of supplies of materials; various climatic conditions, including hurricanes, earthquakes, and floods; an increase in contract quantities; failure of the owner to provide right-of-way parcels; unforeseen subsurface conditions; and stop orders issued by the owner. When specific criteria can be met, compensation is limited to documented, additional, direct field costs, and escalation of the costs for labor, materials, and rental equipment. Even when compensation is permitted, additional costs for home office overhead, idle equipment, and profit are specifically excluded. This is a very restrictive provision which severely limits the ability of the contractor to recover for delays and suspensions caused by the owner.

Although suspensions clauses are generally written to contractually give the owner authority to suspend the construction effort, a few contracts grant similar powers to the contractor. Although these contracts are rare, one public agency provided for the following type of contractor-initiated suspension:

> The Contractor will be allowed to suspend operations for a period not to exceed 14 days annually in order to provide vacation time for his or her employees. These 14 days may be divided into no more than two separate periods of vacation time.

The contract which included this provision required the contractor to request the owner's approval of any suspension at least 30 days prior to the planned suspension. The owner also reserved the right to deny such suspension requests when it was deemed to be in the best interest of the public. If the permission was granted, no time would be assessed against the contractor for the duration of the suspension.

In some provisions, the contractor is given the right to suspend work activities if the owner fails to make prompt payments to the contractor. Such provisions are not common in public works documents.

TERMINATION

Provisions for termination of construction, which are included in most contracts, typically occur in two varieties: termination for default of the contractor and termination for convenience of the owner. In a termination for convenience provision, the owner may reserve the right to terminate the contract at any time, regardless of the percentage of completion of the project, if this is determined to be in the best interest of the owner. Some owners may limit the termination to instances where court injunctions or national emergencies prevent the owner from completing the contract. In recent years, termination for convenience has been used by the U.S. government on several military base closings, causing the stoppage of some major projects. A typical termination for convenience clause will allow the contractor to recover costs incurred for work completed up to the point of termination. It is also advisable for such a clause to define costs for which the contractor can seek recovery. The relatively thorough provision of one public agency follows:

> The Owner may, by written order, terminate the contract or any portion thereof after determining that for any reasons beyond either Owner or Contractor control he or she is prevented from proceeding with or completing work as originally contracted, and that termination would therefore be in the public interest. Such reasons for termination may include, but need not necessarily be limited to, executive orders of the President relating to prosecution of war or national defense, national emergency which creates a serious shortage of materials, insufficient funds by the Owner due to extenuating circumstances, orders from duly constituted authorities relating to energy conservation, and restraining orders or injunctions obtained by third-party citizen action resulting from national or local environmental protection laws or where issuance of such order or injunction is primarily caused by acts or omissions of persons or agencies other than the Contractor.
>
> When the Owner orders termination of a contract effective on a certain date, all completed items of work as of that day will be paid for at the contract bid price. Payment for partially completed work will be made either at agreed prices or by force account methods described elsewhere. Items which are eliminated in their entirety by such termination shall be paid for as provided for elsewhere in this specification.
>
> Acceptable materials, obtained by the Contractor for the work but which have not been incorporated therein, may, at the option of the Owner, be purchased from the Contractor at actual cost delivered to a prescribed location, or otherwise disposed of as mutually agreed.
>
> After the receipt of Notice of Termination from the Owner, the Contractor shall submit, within 60 days of the effective termination date, his [or her] claim for additional damages or costs not covered above or elsewhere in these specifications. Such claim may include cost items such as reasonable idle equipment time, mobilization efforts, bidding and project investigative costs, overhead expenses attributable to the project terminated, legal and accounting charges involved in claim preparation, Subcontractor costs not otherwise paid for, actual idle labor costs if work is stopped in advance of termination date, guaranteed payments for private land usage as part of original contract, and any other cost or damage item for which the Contractor feels reimbursement should be made. The intent of negotiating this claim would be

that an equitable settlement figure be reached with the Contractor. In no event, however, will loss of anticipated profits be considered as part of any settlement.

The Contractor agrees to make his [or her] cost records available to the extent necessary to determine the validity and amount of each item claimed.

Termination of a contract or portion thereof shall not relieve the Contractor of any contractual responsibilities for the work completed, nor shall it relieve the Surety of its obligation for and concerning any just claims arising out of the work performed.

Note that this provision provides for a fair settlement with the contractor for the value of the terminated contract.

In a provision for termination for contractor default, the owner lists the conditions under which the owner will terminate the contract and make a claim against the contractor for damages. These provisions generally provide for serious consequences for the contractor. The construction effort can be stopped, and payments to the contractor may be suspended. The contractor is liable for damages incurred by the owner through the contractor's defaults. In addition, the contractor's performance bond is at risk. Most owners include a termination for default provision in their contracts.

In addition to stating the conditions under which the contract may be terminated for default, most provisions stipulate that notification will be given to the contractor prior to termination. Some provisions may also indicate that the surety will receive an additional notice after termination but before the owner assumes the work. The following public agency provision is typical of a termination for default provision:

If the Contractor: (1) fails to work under the contract within the time specified, or (2) fails to perform the work with sufficient workers and equipment or with sufficient materials to ensure the completion of said work within a specified time, or (3) performs the work unsuitably or neglects or refuses to remove materials or to perform anew such work as shall be rejected as unacceptable and unsuitable, or (4) discontinues the prosecution of work, or (5) fails to resume work which has been discontinued within a reasonable amount of time after notice to do so, or (6) becomes insolvent or is declared bankrupt, or commits any act of bankruptcy or insolvency, or (7) allows any final judgment to stand against him [or her] unsatisfied for a period of 48 hours, or (8) makes an assignment for the benefit of creditors, or (9) is determined to be in violation of the provisions of the contract relative to hours of labor, wages, equal opportunity, character and classification of workers employed, or (10) for any other cause whatsoever fails to carry on the work in an acceptable manner, the Owner may give notice in writing to the Contractor and to his [or her] surety of such delay, neglect or default, specifying the same.

If the Contractor, within a period of 10 calendar days after the date of such a notice, shall not proceed in accordance therewith, then the Owner shall, upon written certification by the Owner's Representative of the fact of such delay, neglect, or default and the Contractor's failure to comply with such notice, have full power and authority to forfeit the rights of the Contractor and at its option to call upon the surety to complete the work in accordance with the terms of the contract. In lieu thereof, the Owner may take over the work, including any and all materials and equipment on the ground as may be suitable and acceptable, and may complete the work by or on its own force account, or may enter a new agreement for the completion of the said contract in an acceptable manner. All costs and charges incurred by the Owner, together with the cost of completing the work under contract, shall be deducted from any monies due or which may become due on such contract. In case the expense so incurred by the Owner shall be less than the sum which would have been payable under the contract if it had been completed by the Contractor, then the said Contractor shall be entitled to receive the difference subject to any claims for liens thereon which may be filed with the Owner, or any prior assignment filed with it. In case such expenses shall exceed the sum which would have been payable under the contract, the Contractor and the surety shall be liable and shall pay to the Owner the amount of such excess.

Once the contractor has been notified of the termination for default, there is still an opportunity to remedy the default. The corrective action must take place in the prescribed time period. Although 10 days is the most common response period stated in such provisions, the time generally ranges from 5 to 15 days. A common deficiency in many contracts is to exclude the response time. Another aspect of this provision that is often omitted is the amount of time the surety will be given to take over the work before having the owner assume responsibility for having the work performed. For the few owners who include a response time for sureties, the times range from 5 to 30 days, with the most typical period being 10 days.

When an owner terminates a contract for default, it is imperative that legitimate grounds for the termination are demonstrated. If the termination is ruled to be unsubstantiated, the termination will generally be considered to be for convenience, stipulating that the contractor is entitled to payment for work performed. This issue was central to the case of *Bison Trucking & Equipment Company v. U.S. Army Corps of Engineers.*

Bison Trucking was hired to make repairs at Buckhorn Lake at Fort Rucker, Alabama, that were necessitated by erosion. A significant amount of work pertained to the preparation and compaction of a new pipe bed. As the work was being done, problems arose when the desired compaction could not be achieved due to wet soil. There were disagreements among the contracting parties as to which party was responsible for the costs associated with addressing the water problem. After considerable work had been done on the pipe bed, Bison informed the Corps that the compaction of the bed was suitable for pipe installation. The Corps disagreed, stating that excess rainfall had collected on the bed after the compaction tests were performed. Bison did not agree, but on April 4 asked the Corps to inform Bison of the specific locations where additional density tests were to be conducted. The Corps did not respond to this request, but insisted that Bison perform the required compaction tests and reiterated the need for Bison to take responsibility for the dewatering. On April 23 Bison informed the Corps that the additional work had been done and that the pipe bed met the compaction requirements. On May 1 the Corps insisted that the compaction tests done by Bison were not representative of the site conditions and gave formal notice that Bison had 10 days to perform. On May 3 Bison asked the Corps to identify where compaction tests were to be taken to satisfy the Corps, and it stated that the failure of the Corps to provide proper direction was preventing Bison from performing in accord with the contract.

The Corps did not respond to the written request, and on May 18 issued a letter terminating the contract for default by Bison. The Armed Services Board of Contract Appeals (ASBCA) stated that termination for default must be made only for good cause and with solid evidence. Thus, the Corps had the burden of proving default. The ASBCA ruled that there was no evidence of default or abandonment. Furthermore, it placed considerable weight on the fact that the Corps never responded to Bison's written requests on April 4 and May 3, which was interpreted as the Corps actually impeding performance on the project. Bison had not defaulted on the project, as it was never shown that Bison's work would not be completed by the contractual completion date of June 12.

A termination case developed out of a 100-mile Texas gas pipeline installation project in which the general contractor, Driver Pipeline Company, Inc., was delayed by heavy rainfall. The owner, Mustang Pipeline Company, refused to grant a time extension for the adverse weather. Subsequently, the owner's consulting engineer signed a certificate that enumerated the grounds of contract termination by default, including insufficient workers on the project and inadequate equipment to complete the project in a timely manner. Driver challenged the termination. The court learned that Mustang had given additional time (30 days) to another contractor due to the heavy rain, and that the consulting engineer had no idea of the number of workers required on the project or the type of equipment needed to finish the project. In fact, the consulting engineer had never been to the project site. The court determined that Mustang had not shown that Driver was unable to complete the project on time and that the termination was unjustified.

While most construction contracts give the owner the authority to terminate a contract, few give equal authority to the contractor. The ability of the contractor to terminate the contract is a powerful tool that provides some assurance that the owner will not interfere in the contractor's work effort.

PROJECT COMPLETION

The time for the completion of construction projects is usually stated in the contract. There is still some latitude, however, in determining what constitutes completion. Two terms are frequently used in defining completion. *Substantial completion,* as one might infer, implies something less than absolute or final completion. Substantial completion is typically determined by the owner's representative. A project is substantially complete when the owner can occupy it (refer to Article 9.8 of AIA Document A201-1997). The Construction Management Association of America (CMAA) Document A-3 defines substantial completion as the date when the project "has progressed to the point that it is sufficiently complete in accordance with the Contract Documents so that the Owner may fully occupy and use the Project or designated portion thereof for the use for which it is intended, with all of the Project's parts and systems operable as required by the Contract Documents." Owner occupancy is a milestone in a project and is generally established

---

CERTIFICATE OF OCCUPANCY

County of Marion
Department of Building Inspection

This Certificate issued pursuant to the requirements of Section 109 of the Uniform Building Code certifying that at the time of issuance this structure was in compliance with various ordinances of the County regulating building construction or use. For the following:

Use of Classification: _Office Building_   Building permit No.: _003278990_

Type Construction: _Wood Frame, Brick Veneer_

Use Zone: _C-24_

Owner of Building: _RTC-Consultants_   Address: _1228 Grant St., La Grange_

Building Address: _1964 Bristol Drive, La Grange_

Date: _Dec. 14, 2011_   By: _P. B. Monson_

Building Inspector

FIGURE 14.1
Example of a certificate of occupancy.

by a certificate of occupancy that states that the facility is suitable for occupancy (figure 14.1).

What is the purpose of defining a project as being substantially complete? From a practical point of view, it means that the project can be occupied and used by the owner despite the need for contractor corrections of minor deficiencies. Unfortunately for the contractor, this also means that work operations can be obstructed and delayed by premature occupation. However, liquidated damages will not be assessed against the contractor if the project is substantially complete within the stated contract time. Legally, substantial completion defines the date from which the registration of liens is counted by lien statutes.

Essentially, substantial completion is the same as practical completion. At the time of substantial completion, the last periodic payment is made to the contractor. After substantial completion, the only funds remaining to be returned to the contractor are the amounts withheld by the owner as retainage. The retainage amount is commonly 5 to 10 percent. The retainage is released upon final completion. In terms of the physical project, the primary distinction between substantial completion and final completion is that a few minor noted items must be corrected before the release of any of the retainage. These minor work items are typically documented by the owner's representative and distributed to the general contractor and the subcontractors. This documentation of minor deficiencies of the project is commonly called the *punch list*. The certificate of substantial completion will be similar to the certificate of occupancy, but will generally make specific mention of the punch list items to address. Correcting the punch list items will generally result in the release of all the retainage. Some owners release part of the retainage even though the contractor has not taken care of all the punch list items. Such owners simply assess the amount of money required to complete the project and withhold only the amount that represents the cost of taking care of the punch list items.

The items listed in the punch list are not necessarily the limit of the contractor's obligation to complete the contract. For example, the contract provisions might state, "The failure to include any items on such punch list does not alter the responsibility of the Contractor to complete all work in accordance with the Contract Documents." As a practical matter, the punch list is the primary document that indicates the remaining work items to address after substantial completion.

IMPACT OF DELAYS

Construction delays generally adversely affect construction progress. Most disputes arise out of delays that are at least partially the fault of the owner. As has already been shown, such delays can be due to suspension of work, slow owner responses to the contractor's questions, slow processing of shop drawings and other submittals, failure to provide timely access to the construction site, differing site conditions, change orders, and other actions of the owner. Most contracts provide additional contract time when owner-caused delays occur. If the contract does not contain a no-damage-for-delay provision, the contractor will also have a good chance to receive monetary compensation for owner-caused delays.

The amount of monetary compensation that is justified for an owner-caused delay is difficult to assess, because the contractor may sustain added costs on work items that were not directly affected by the delay. These added costs are on unchanged work items or items not directly associated with the delay, but are nonetheless a consequence of the delay. These costs are the result of the "ripple effect," which is essentially the principle that the cost impact of a change or delay on one work item is not limited to that item, but has an effect on various portions of the project.

For example, suppose one major activity on a project is delayed so that the project duration is extended for two months. The owner may readily concede to extending the project duration two months, but what monetary compensation would be appropriate? The contractor will point out that the field administrative staff salaries and other job site overhead expenses have continued for an additional two months. If the time extension prolongs the project into the winter season or into a rainy season, productivity losses may be considerable. This may mean that some workers will be asked to work overtime in order to finish the project on time. If additional workers are hired, losses in efficiency can be anticipated because of the unfamiliarity of the newer workers with the site, and because of worker crowding. If labor agreements expire or enter into a new wage era, the contractor will have to pay higher wages than were originally budgeted. The sequence of the work may also have to be changed. Equipment may remain idle during a delay, resulting in added costs to the contractor. Prices for some materials might have risen during the extended time period. Some materials may deteriorate during a long delay and require replacement. The contractor may be forced to obtain additional sources of financing if periodic payments are delayed, or if the release of the retainage is delayed. Many of these costs will be difficult for the contractor to document with accuracy.

The ripple effect may cause the contractor to incur costs that were not anticipated. During the 1940s and 1950s these impact costs were largely treated as nonreimbursable. This practice was a result of the 1942 landmark Supreme Court case *United States v. Rice* (317 U.S. 61), in which Rice, the contractor, was denied impact costs for the increased wages that resulted when a project was delayed. Rock was encountered that was in excess of the amount noted in the plans, delaying the project and extending the construction effort into the winter season. For years the so-called Rice Doctrine prevented contractors from successfully claiming for impact costs. However, this doctrine has been eroded over the years through a variety of court decisions and the modification of contract provisions that provide compensation for ripple effects or impact costs. The best chance of receiving compensation for justifiable ripple effects is through the maintenance of accurate and detailed records. At the same time, the contractor must abide by the contract terms concerning notification procedures and time constraints.

HOME OFFICE OVERHEAD

Home office overhead includes the costs of operating and maintaining the home office. These costs include the salaries of company officers, estimators, accounting personnel, and secretarial personnel. Other expenses include: the rent, lease, or mortgage payments on the home office premises; the utility expenses; the costs of supplies; and the costs of the company vehicles of the home office personnel. While some expenses, such as the salary of an accountant, may be charged on an hourly basis to a particular project, most home office expenditures are charged against the project on the basis of a predetermined percentage of the project costs. There may be little consistency between seemingly similar firms in terms of the manner in which home office overhead is allocated to a project.

This issue is particularly perplexing when an owner-caused delay is encountered. Again, careful documentation may give an indication of how to allocate some costs, but not all home office expenditures can be traced easily. For example, how should a project be charged with the salary of an estimator who prepares estimates on various projects during a delay, but is not successful in acquiring an additional project? In other words, how should home office expenditures be handled when they cannot be directly allocated to any one project?

Home office overhead is often a problem area in regard to compensation. One method for determining the appropriate home office overhead for a project is the Eichleay formula. This is a controversial formula that is accepted by some owners, but staunchly rejected by others. The Eichleay formula is relatively straightforward and consists of the following:

$$OH_P = (BILLINGS_P/BILLING_{ALL}) * OH_T$$

where OH_P = overhead allocable to a project P

    $BILLINGS_P$ = all contract billings on project P

    $BILLINGS_{ALL}$ = billings on all company projects during the construction of project P

    OH_T = total overhead incurred during project P

$$OH_{DELAY} = (OH_P/DURATION_P) * (DAYS\ OF\ DELAY)$$

where OH_{DELAY} = home office overhead for the delay

    $DURATION_P$ = duration of project P, including the delay

The Eichleay formula may not be accepted by all owners, but it provides one approach by which unabsorbed home office overhead can be quantified. When one is using the Eichleay formula, it should be recognized that the duration of the project is the total duration, which includes the delay period. The billings for the project in question include the total billings, not simply the original contract sum. An examination of the formula may indicate that the results will be biased under certain conditions. It may be more appropriate in some circumstances than in others. In some instances the contractor may consider the results of the Eichleay formula to understate the unabsorbed home office overhead costs, while in others the owner may regard the calculation as overstating these costs. Modifications of the Eichleay formula, or different approaches to establishing these costs, may be worth considering.

The basic or unmodified Eichleay formula has its foundation on the premise that the home office overhead is evenly distributed over all the company revenues. This means that the overhead, when stated as a percentage of the revenues, is the same for a small project and for a large project, regardless of their duration. A simple example will demonstrate the use of the Eichleay formula. Assume that a company has total annual billings on all its projects of $53,000,000 and that the total overhead in a given year is $1,800,000. One of its projects was contracted for $7,000,000 and it was delayed 55 days beyond its original 240-day duration to 295 days. The Eichleay computations are as follows:

    $OH_P = (BILLINGS_P/BILLINGS_{ALL}) * OH_T$

    $OH_P = (\$7,000,000/\$53,000,000) * \$1,800,000 = \$237,735.72$

    DAILY CONTRACT OVERHEAD $RATE_P$ = $237,735.72/295 days = $805.88/day

    OH_{DELAY} = $805.88/day * 55 days = $44,323.40

According to these computations, the contractor would be entitled to $44,323.40 to cover the delay costs of home office overhead.

While the overhead computations may be relatively simple, the overhead costs will not be realized if certain criteria are not met. This was demonstrated in *Singleton Contracting Corp. v. Harvey*. Singleton was a contractor that contracted with the Department of the Army to perform work at two army reserve centers. The contract stipulated that Singleton was to obtain insurance and provide proof of insurance for the entire contract duration. Proof of insurance was to be obtained prior to commencing construction work; a certificate of insurance was to be provided at the preconstruction conference. This was not done. It was also apparent that the plans were flawed and that considerable work on the drawings was required. Although actual construction work was not being done as the drawings were corrected, Singleton submitted invoices to cover the costs of materials and bonds purchased for the contract. The government refused to pay the invoices, since no certificate of insurance had been provided. Eventually, the government terminated the contract for convenience. By this time, revised drawings had not been provided to Singleton, and Singleton had not presented the government with a certificate of insurance. The court realized that Singleton was aware of its failure to provide a certificate of insurance and never showed any attempt to satisfy this obligation. As a result, Singleton was jointly accountable for the delay, and was therefore not entitled to be compensated for its unabsorbed overhead during the contract period.

REVIEW QUESTIONS

1. In defining project duration, at what point does the time of construction begin? When does it end?
2. Why are provisions included in contracts that state that time is of the essence?
3. Contrast actual acceleration and constructive acceleration.
4. Describe the implications of compensation (in terms of time extensions and money) for delays caused by the owner. The general contractor. Acts of God.
5. What basic exceptions in no-damage-for-delay clauses permit the contractor to be awarded damages?
6. What are the implications of not having a liquidated damages provision in a contract when the contractor does not finish in the allotted time?
7. What is the significance of substantial completion? That is, what contractual issues are frequently linked to substantial completion?
8. What is the significance to the contractor of final completion?
9. What distinguishes substantial completion from final completion as far as construction progress is concerned?
10. Examine the AIA documents and the ConsensusDOCS that are included in the Appendix and determine the similarities and differences in the provisions related to construction schedule. Specifically examine AIA provisions §3.10.1 and ConsensusDOCS §6.2.
11. Examine the AIA documents and the ConsensusDOCS that are included in the Appendix and determine the similarities and differences in the provisions related to time extensions. Specifically examine AIA provisions §8.3.1 and ConsensusDOCS §6.3.1.
12. Examine the AIA documents and the ConsensusDOCS that are included in the Appendix and determine the similarities and differences in the provisions related to contract termination by the contractor. Specifically examine AIA provisions §14.1.1, §14.1.2, §14.1.3, and §14.1.4, and ConsensusDOCS §11.5.1, §11.5.2, and §11.5.3.

15

PAYMENTS

THE CONSTRUCTION INDUSTRY is unique in regard to payments. While most payments for purchases in other industries are made at the time of delivery, in the construction industry periodic payments are typical. These payments are generally made on a monthly basis.

The practice of making periodic payments is based on one factor: The cost of construction is high compared with most other types of purchases. If a contractor were forced to finance the entire cost of a project through to final completion, the costs would go up considerably, partly because there would be a dearth of competition. Most construction contractors would not survive if the cash flow from the owner was suddenly halted or delayed until project completion. It is a considerable burden for some contractors simply to finance the construction effort from one month to the next. Although periodic payments are made on almost all major construction projects, it is not an inherent or implied right of the contractor to receive periodic progress payments from the owner; this is a right that is granted contractually.

The following provision is a typical example:

> Partial payments will be made to the Contractor once each month as work progresses. Payment shall be based on estimates of the value of work performed. Monthly partial payment periods will end at the close of the 25th day of the month. Five (5) percent of the amount of the contract completed shall be retained on each estimate until payment of the Final Payment. However, when the Contractor has completed at least 95 percent of the work, the Owner may prepare an estimate of the cost to complete the work remaining. The Contractor may be required to furnish consent of the Surety before the retained amount is reduced to less than 2 1/2 percent.

In practice, the application for payment is usually submitted at the end of the month or by the tenth of the month after the month when work was performed. This application is reviewed by the owner's representative to determine whether the payment request bears a fair resemblance to the work actually performed.

UNIT PRICE CONTRACTS

The timing of payments on unit price contracts is similar to that of payments on lump sum contracts. However, certain aspects of unit price contracts make the payment procedure different.

On unit price contracts, payment is based on the unit prices as bid and the precise measurements of in-place field quantities. These measurements form the very basis for making payments to the contractor, and the final or eventual cost of the project to the owner is linked directly to them. After the architect/engineer reviews the application and verifies the measurement of the quantities, a certificate of payment is issued to the owner. The owner is then expected to pay for those in-place items by the stipulated date. On some contracts, late payments to the contractor are subject to interest charges.

COST-PLUS CONTRACTS

On cost-plus contracts the payments made to the contractor are based on the actual expenditures made on a project by the contractor. As a general rule, the contractor is reimbursed for direct expenditures, plus an allowance for profit and overhead. Note that the reimbursement of the contractor is based on fully documented expenditures. This administrative task could be quite burdensome, depending on the level of detail required by the owner.

Although the payment procedure may seem straightforward, the contract must state the specific nature of all expenditures for which the contractor will be reimbursed. Some common expenditures for which reimbursements are made include the following:

- Materials costs (both temporary and permanent).
- Subcontractor costs.
- Field labor costs.
- Owning and operating costs of plant and equipment.
- Field overhead (project superintendent, field supervisors, clerks, inspectors).
- Transportation costs for workers, materials, and equipment.
- Small tools, fuel, and utilities.
- Consultants.
- Miscellaneous expenditures (surety bonds, insurance premiums, taxes, permits, vacation and sick leave allowances, travel to project for home office personnel, pension and retirement allowance).

While most on-site expenditures are reimbursed, reimbursement for home office personnel is typically excluded. In addition to these constraints on reimbursements, the percentage fee that is applied to the expenditures for profit and overhead may vary depending on the nature of the expense. For example, the percentage fee allowed for expenditures for direct labor and materials may be 20 percent; for subcontractor expenditures, it may be 10 percent; and for field overhead expenditures, no fee allowance may be permitted.

For payments to be made to the contractor, payment requests submitted to the owner must be accompanied by documents that support the various expenditures for which reimbursement is sought. Thus, the owner's representative must be familiar with the work being performed and the various contractual issues concerning reimbursement.

LUMP SUM CONTRACTS

On lump sum (or fixed-price) contracts, the payment procedure is somewhat different. Since the contract already stipulates the amount, it is not necessary for the owner or architect to measure all in-place quantities exactly. However, this does not mean that the contractor's payment requests do not need to be verified to some extent. To avoid the monthly negotiations that could take place with each progress payment request, it is common for the contract to stipulate that before performing the work, the contractor submit a schedule of values or cost breakdown of all work items for which payments will be requested. Naturally, the sum total of all these work items will be equal to the contract amount. Once the breakdown is received by the owner or architect, the schedule of values is evaluated for reasonableness. If the contractor excessively front-end-loaded the schedule of values, the owner may request that the contractor resubmit it. It is not safe for the owner to assume that in the end everything will work out. An unscrupulous contractor who has been overpaid may be inclined to abandon the project. Therefore, the owner must look for vested interests in the project. The schedule of values will also be examined for reasonableness, and the degree of front-end-loading will be noted. Once the negotiations have been completed, the schedule of values will become the basis on which payments are made to the contractor. Essentially, the only verification needed on each payment request is proof that the work has been performed.

The ConsensusDOCS include these points in the provisions as follows:

§9.1 Within twenty-one (21) Days from the date of execution of this Agreement, the Contractor shall prepare and submit to the Owner, and if directed, the Architect/Engineer, a schedule of values apportioned to the various divisions or phases of the Work. Each line contained in the schedule of values shall be assigned a value such that the total of all items shall equal the Contract Price.

Clearly, a lump sum contract gives the contractor some latitude to front-end-load the billings. Not all the costs incurred will be paid directly by the owner. Ideally, the owner would like to pay the contractor on the basis of the *value* of construction in place. The contractor, by contrast, would like to be paid on the basis of the cost of construction. These are rarely the same. Consequently, the contractor must "bury" these costs in other job items. Some of the costs will occur early in the project, and the contractor will want to front-end-load on the billings to cover those expenses. Expenses of this sort include insurance premiums that are due at the beginning of the project, permits that must be obtained prior to construction, bond premiums that are paid at the project start, miscellaneous early fees, and mobilization (if not allowed as a pay item). Some front-end-loading is often permitted by owners who recognize that the contractor will be short of funds early in the project. Early in the construction phase, the contractor would be happy to be paid for the actual cost of the project. This will change later in the project when profits begin to be realized. The owner, by contrast, would like to pay for the actual value of the project as each payment request is made. The contract will generally stipulate exactly when the payment requests are to be submitted to the owner's representative. The contract will also provide information on when the payments are to be made to the contractor. Thus, it is crucial that the payment request be submitted on a timely basis (figure 15.1).

Assuming that the contract specifies periodic payments, the contractor will be entitled to be paid for the value of the work done. Generally, the contractor establishes the percentage of work performed on each item listed in the schedule of values and submits it to the architect/engineer for approval. The architect/engineer may decide to visit the site to verify the work performance. The payment request may also include materials that have been delivered but not installed. Some contracts specifically exclude payment for materials not installed, with provisions such as, "payment shall not include any allowance for materials and equipment not incorporated in the work."

Payment was a key issue in *Argeros & Co., Inc. v. Commonwealth of Pennsylvania* (447 A.2d 1065). Argeros had contracted with the Pennsylvania DOT to paint five bridges. The bid documents indicated that one bridge consisted of "approximately 180 tons" of steel. Argeros discovered that the weight was closer to 260 tons. When the error was discovered, Argeros notified the DOT, which instructed Argeros to perform the work. Argeros complied and subsequently requested an additional payment of $6,900. The DOT refused to pay the requested amount, and a suit was filed by the contractor. The DOT contended that the specifications cautioned bidders that the quantity estimates were approximate, and that the DOT assumed no responsibility for them. The court ruled in favor of the DOT. It noted that Argeros had stated in its contract that it had prepared its bid on the basis of an independent examination of the bridges and had not relied on the information provided by the state.

Lump sum contracts may contain substantial incentives if the facility owner wants to encourage early completion. This is becoming increasingly common on highway and bridge projects. Every day of early completion results in less driver aggravation, less wasted fuel, and less driving time. Thus, every day of early completion is an added benefit to the traveling public. Departments of transportation and cities are increasingly offering incentives to have their projects completed early. For example, a recent contract for renovating the Hood Canal Bridge in Washington resulted in the bridge being closed for over a month. The detour route for the traffic resulted in a substantial added distance for the travelers. To encourage early completion, the contractor was offered $75,000 per day for early completion. There was also a liquidated damages provision in the event the bridge did not open in the contract time.

A "No Excuse Bonus" may also become a part of the contract in order to stress the importance of completing on time. For example, on a public project in Tennessee, a contract provision stated,

> For the completion and acceptance by the Department of this project, in substantial conformity with the plans as determined by the Department, on or before the Project Completion Date/NO EXCUSE BONUS Completion Date of December 31, 2010 the Department will award the Contractor a NO EXCUSE BONUS of two-million dollars ($2,000,000).

MONTHLY PAYMENT REQUEST

Project: _Warehouse Big Place_

To: _____Warehouser Marketers_____
From: _High Country Builders_
Address: _Coyote Junction, CO_   Date: _____3/10/11_____

Current Contract Summary:

Original Contract Price:	_$380,000_	Contract Number:	_53-01_
Adjustment for Change Orders:	_±$11,300_	Account Number:	_E55-141-30_
Adjusted Contract Price:	_$391,300_	Estimate Number:	_3_
		Period:	_02/01/11 to 02/28/11_

Performance Summary on Schedule of Values:

Payment Item Description	Adjusted Value	% Complete	Prev. Earned	This Period	Total Earned	Retained	Work Remaining
Mobilization	$12,000	100%	$12,000	0	$12,000	$1,200	0
Site Clearing and Prep.	34,000	100%	34,000	0	34,000	3,400	0
Foundation	66,000	100%	66,000	0	66,000	6,600	0
Electrical Rough-in	13,000	100%	3,000	$10,000	13,000	1,300	0
Block Walls	87,000	100%	24,000	63,000	87,000	8,700	0
Steel Cols and Joists	48,000	60%	9,000	19,800	28,800	2,880	$19,200
Windows and Ext. Trim	18,000	80%	2,500	11,900	14,400	1,440	3,600
Roofing	26,100	80%	0	20,880	20,880	2,088	5,220
Doors	14,300	0%	0	0	0	0	14,300
Security System	22,800	40%	3,900	5,220	9,120	912	13,680
Painting	23,000	10%	0	2,300	2,300	230	20,700
Fencing	13,000	0%	0	0	0	0	13,000
Clean-up	2,100	0%	0	0	0	0	2,100
Demobilization	12,000	0%	0	0	0	0	12,000
Total	**$391,300**		**$154,400**	**$133,100**	**$287,500**	**$28,750**	**$103,800**

Payment Summary

Total Amount of Work Done	$287,500
Less 10% Retained	28,750
Amount Due to Date	258,750
Less Previous Payments	138,960
Amount Due This Estimate	$119,790

Prepared by: _____John Pool_____

Approved by: _____Ray Westall_____   _10 March '11_
                 (contractor)   (date)

Approved by: _____
      (owner's authorized representative)   (date)

FIGURE 15.1
Example of a request for payment.

The contract emphasized that the completion date for which the no excuse bonus applied was fixed, and that the completion date would not be adjusted for any "reason, cause or circumstance whatsoever, regardless of fault, save and except in the instance of a catastrophic event (e.g., tornado, earthquake or declared state of emergency), directly or substantially affecting the project as determined by the Department." The contract further stipulated that the contracting parties understood that the construction duration might be impacted by work deletions, change orders, work disruptions, differing site conditions, utility conflicts, design changes/revisions or defects, extra work, right-of-way issues, permitting issues, environmental issues, acts of suppliers, acts of subcontractors, actions of third parties, holidays, suspensions of operations, adverse weather, and so on, but that these were to be contemplated and would not extend the completion date. It was further stipulated that the contractor was solely responsible for incurring any costs of acceleration necessitated by these anticipated delays to deliver the project on the prescribed completion date.

It should be noted that the contract stipulated that failure to complete the project on the set date would result in the assessment of liquidated damages of $5,000 per day for every day that the project was completed after the contractual completion date. This constitutes the disincentive portion of the "no excuse bonus and disincentive" payment contract. On a similar Florida project, the no excuse bonus was set at $3,000,000 with liquidated damages accruing at a rate of $50,000 per day, with the maximum penalty set at 60 days or $3,000,000.

PROJECT CLOSEOUT

For any successful project, the contractor must show that the completed project meets the terms of the agreement between the owner and the contractor. This is demonstrated to the owner as a project nears completion and is commonly called project closeout. It is at the project closeout phase that the typical work momentum ceases. If this is drawn out excessively, the contractor may very well experience the loss of considerable profit. Thus, it is important for the contractor to be adequately prepared to turn the project over to the owner. Project closeout is primarily focused on ensuring the quality of the completed project. To some, project closeout simply means that the architect will walk through the project and ensure that all aspects of the project satisfy the contractual requirements; but project closeout is more than that.

In actuality, preparations for project closeout should begin at the start of construction and continue until the project is finally accepted by the owner. It is advisable for the contractor to perform self-inspections of the work, and that this be done throughout the construction phase. It is not wise for the contractor simply to assume that the architect will find the deficiencies and point them out. Suppose the architect is asked to inspect the work on a multistory building. The architect inspects one room and finds 30 deficiencies. It is very likely that the architect will stop the inspection process at that point and tell the contractor, "You are obviously not ready; call me when you are ready for a serious inspection."

When the contractor has essentially completed the construction work, a formal request is made to the architect for an inspection. Conducting such an inspection is referred to as punch listing a project. A *punch list* is a list of items that must be corrected before the project is acceptable to the owner. The final punch list is normally developed on a joint job visit conducted by the contractor and the architect/engineer. Punch list items are often of a minor nature (figure 15.2), including smudges in paint, caulking on window panes, missing cover plates on wall outlets, and mud in carpets, but they can cause considerable delay in the final payment if they are not corrected promptly.

In addition to the preparation of the punch list, the contractor may be required to demonstrate that the facility functions in a satisfactory manner. For example, the contractor may be required to test various systems, including heating, ventilation and air conditioning (HVAC), plumbing, electrical, elevator, escalator, Americans with Disabilities Act (ADA) requirements, communications systems, and other similar systems. Tests performed on such systems generally are witnessed by the owner's representative. The owner will also expect to receive warranties on appliances and facility components, operating and maintenance manuals on all major pieces of equipment that include manufacturer's information, trouble-shooting tips, parts lists, maintenance requirements, and so forth.

The closeout procedure will also include such items as: providing the owner with the permanent keys for the locks; the warranty for the project; affidavits that workers, subcontractors, and suppliers have been paid; lien releases; prevailing wage certificates; as-built drawings; and a complete submittal file for the project.

FINAL PAYMENT

A contractor's final payment is directly linked to final completion or final acceptance of the project by the owner. This is a crucial step, because the contractor has little incentive to perform additional work if all payments have been made. Since final acceptance is so important, the owner must verify the adequacy of the work performed. This is accomplished by verifying the punch list items have been addressed. Once all the items on the final punch list have been corrected, the owner is free to make final payment to the contractor and release the retainage. Figure 15.3 shows the steps in this process.

Substantial completion is important in that many aspects of the project are somehow linked to it. One of the most important aspects pertains to the final retainage withheld by the owner. This is also included in the ConsensusDOCS as follows:

> §9.6.4 Upon acceptance by the Owner of the Certificate of Substantial Completion, the Owner shall pay to the Contractor the remaining retainage held by the Owner for the Work described in the Certificate of Substantial Completion less a sum equal to two hundred percent (200%) of the estimated cost of completing or correcting remaining items on that part of Work, as agreed to by the Owner and Contractor as necessary to achieve final completion. Uncompleted items shall be completed by the Contractor in a mutually agreed upon timeframe. The Owner shall pay the Contractor monthly the amount retained for unfinished items as each item is completed.

PUNCH LIST

Tri-City Office Towers
217 South Avocet Road

ITEM NO.	DESCRIPTION	SUB RESPONSIBLE	DATE FOR COMPLETION	COMPLETED	REMARKS
1	Room #104				
	Door #154 is sticking at top	Walnut Door	3/16/11		
	Adjust ceiling tile around sprinkler head at SE corner	The Right Tile	3/16/11		
	Straighten cover plate for the light switch	Main Electric	3/16/11		
	Touch up paint on north wall	4A Painters	3/16/11		
	Remove paint spots from cabinet	4A Painters	3/16/11		
	Patch drywall at cabinet	Sheetrockers, Inc.	3/16/11		
2	Room #105				
	Remove paint stain from carpet	4A Painters	3/16/11		
	Replace cracked cover plate	Main Electric	3/16/11		
3	Room #106				
	Wallpaper is marred near door	Jack's WP	3/16/11		
4	West Entry Walkway				
	Smooth out truck ruts	Green Plantings	4/06/11		
	Laurel needs to be replaced	Green Plantings	4/06/11		
	Clean up debris in area	Green Plantings	4/06/11		

FIGURE 15.2
Example of a punch list.

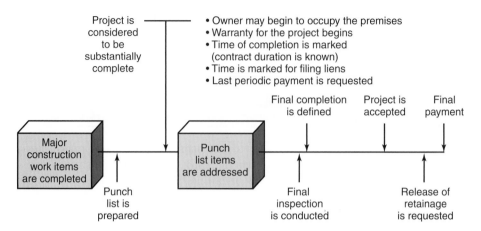

FIGURE 15.3
Typical project completion phase.

On federal projects, the Prompt Pay Act requires that the government make payment to contractors within 30 days of the date of submission of a properly prepared invoice. When payment is not made within the prescribed time period, the government is obligated to pay interest on that amount at a rate established by the Secretary of the Treasury. Similarly, the contractor is to pay suppliers and subcontractors within 15 days of receiving payment from the government, less an allowable retainage amount. The Act also obligates the contractor to pay the subcontractors who have satisfactory performance not later than seven days after receipt of payment from the government. When subcontractors are not provided with prompt payment, an interest penalty (as high as 2 percent per month) is assessed against the contractor for late payment. Similar provisions are to apply to the agreements that subcontractors have with sub-subcontractors. Over 40 states have enacted prompt pay acts that have the same intent—contracting parties are to receive payment within a reasonable time of performing the services. At least 30 states have enacted prompt pay legislation that applies to private projects. It is common for contracts to include provisions that require owners to make prompt payments to the contractors. The prompt payment is embodied in the ConsensusDOCS as follows:

> §9.2.1 The Owner shall pay the amount otherwise due on any payment application, as certified by the Architect/Engineer, no later than twenty (20) Days after the Contractor has submitted a complete and accurate pay application, or such shorter time period as required by applicable state statute.

RETAINAGE

The owner typically holds back or retains a portion of the money earned by the contractor as an incentive for the contractor to complete the project properly and promptly. Since the billings can be in error, the retainage represents a necessary cushion for the owner if the billings exceed the value of the project. However, the retainage is generally considered to be a substantial incentive for the contractor to bring the project to a rapid completion. The retainage on a project may range from 5 to 20 percent (5 percent or 10 percent is typical), while the contractor's net profit on the job is generally about 2 percent. Thus, the retainage is needed to realize a profit on most jobs. The incentive to complete the project and finish the punch list items is usually strong, as the value of uncompleted items tends to be small.

The following is a typical retainage provision:

> In making progress payments, there shall be retained 10 percent of the estimated amount until final completion and acceptance of the work.

There is a trend to use provisions such as the following:

> If the owner's representative (architect/engineer) at any time after 50 percent of the work has been completed finds that satisfactory progress is being made, he or she may authorize any of the remaining progress payments to be made in full.

When full payment is authorized following 50 percent completion, the retainage withheld may be considerably reduced. If retainage is normally 10 percent, the retainage withheld at final completion will be only 5 percent, if retainage is no longer withheld after the project is 50 percent complete.

As was mentioned earlier, the retainage is released after final acceptance (see figure 15.3). Could the retainage be released at substantial completion? Substantial completion implies that there are still some small items of work that require the contractor's attention. If the contractor fails to complete those items, the owner can use the money held in retainage to pay for that work. The punch list essentially separates the last periodic payment from the release of the retainage. Thus, retainage should never be released prior to final completion. Note that the final payment (the last periodic payment), excluding the retainage, may be paid after substantial completion.

Should the full retained amount be withheld until final acceptance? The retainage is regarded as the owner's assurance that sufficient funds are available to complete the punch list items in the event that the contractor fails to do so. When the punch list is prepared, it is appropriate to estimate the cost of correcting the deficiencies. With retainage amounts of 5 to 10 percent, there are generally ample funds to address the punch list items. Retainage amounts of two to three times the estimated cost of addressing the punch list items are considered adequate. Any additional amounts held as retainage can be released to the contractor.

Final payment is predicated on satisfying all the conditions as spelled out in the contract. In one Texas case, a contractor, Solar Applications Engineering, Inc., was ruled as having substantially completed the work on a truck stop. A punch list was prepared by TA Operating Corporation, but Solar did not agree with all the items on the list. Solar was notified that final payment would be withheld until the items on the list had been completed. These issues were eventually resolved, and Solar felt that it had addressed the listed items. The contract was eventually terminated because Solar had not submitted an all-bills-paid affidavit that was a contractual requirement of final payment. Solar then sued for breach of contract. The court determined that Solar had not complied with the contractual requirement to submit an all-bills-paid affidavit and was therefore not entitled to the final payment.

There are many nuances to final payments that can be introduced from project to project. For example, in the case of *Western Concrete Structures Co. v. James I. Barnes Construction Co.* (23 Cal. Rptr. 506), the owner tried to resolve the issue of a claim and the final payment with a single check. The final payment was to be $28,636.20, but there was an acknowledged outstanding settlement on a claim for $6,400. Barnes sent a check to Western Concrete for $28,636.20 with a letter stating that the check constituted "payment in full for services." Western responded with a letter stating that they did not accept the check as satisfying the full indebtedness of Barnes, but that the check would be accepted as partial payment. Western also asked Barnes to stop payment on the check if they did not fully agree with the conditions under which Western was accepting the check. Barnes did not stop payment on the check, assuming that by depositing the check, Western would essentially agree that the payment was being made "in full." The court ruled that by failing to stop payment on the check, Barnes still had a legitimate debt of $6,400 to pay to Western.

The Uniform Commercial Code (UCC) now permits some parties to accept checks marked "payment in full" and still preserve their rights to additional compensation. For example, in New York, a supplier receiving a check marked "payment in full" may include the words "notwithstanding the foregoing, supplier accepts this payment without prejudice and with full reservation of its right to assert a claim for . . ." Although this was determined in a case involving a supplier in *Braun v. C.E.P.C. Distributors, Inc.* (77 A.2d 358), some jurisdictions extend the same interpretation to contractors. Thus, the proper endorsement on a check marked "payment in full" may nullify the intent of the restrictive wording accompanying the check.

Suppose the contractor defaulted and the surety completed the project. Is the surety entitled to the retainage? The courts have generally ruled yes. Just as the owner would have used the retainage funds to finish the job, the surety can also claim the retainage for this purpose. This is practical only if there are no outstanding debts. The impact on the owner should be such that the cost of the project does not exceed the contract amount.

A contractor may obtain the retainage without paying for some labor or materials. Of course, the workers and suppliers can sue the contractor, but this may not be their best option. The surety can be sued for the amount owed.

Suppose a bank extended a loan to a contractor. The loan was conditioned on the fact that progress payments were assigned to the bank. The contractor defaulted on the contract, and the bank contended that it had an equitable lien on the money held as retainage. The surety wanted to have the same funds. Who is entitled to the retainage? The courts generally hold that surety rights are superior to those of banks. The reasoning is that if the owner had completed the job, the construction costs would have been paid for with the retainage.

Suppose a contract stipulated that 10 percent retainage be withheld on a project. However, the owner failed to hold back any retainage as stated in the contract. Subsequently, the contractor defaulted on the project. The surety could win a release from its obligation (performance bond) to complete the project.

On a government project, the contractor defaulted and the surety paid out $275,000. The surety then sued the owner for the retainage. The court awarded the retainage to the surety, ruling that the owner could have paid off the debts with the retainage. Therefore, if the surety pays off the debts with its own resources, it can then claim the retainage.

On another project the contractor defaulted, and the face value of the bond was insufficient to pay all the laborers and suppliers. The retainage was claimed by laborers, suppliers, and the surety. The court ruled that the laborers and suppliers had a superior claim to the retainage. The purpose of the surety is to pay such debts, and it should not compete with laborers and suppliers for the excess funds.

On another project the general contractor was failing to make payments to the subcontractors. The surety notified the owner that the subs were not being paid. Later, the contractor defaulted. The surety was responsible for completing the project, but it successfully sued the owner for the amount paid to the contractor after the notification.

SUBCONTRACTOR PROGRESS PAYMENTS

Subcontractors have their contracts with the general contractor. Thus, progress payments are made to subcontractors if the subcontract stipulates them. In most contracts with subcontractors, the subcontractor receives payment from the general contractor when the general contractor is paid by the owner. The courts have ruled, however, that this "pay when paid" provision may not be absolute. If a reasonable time elapses without the general contractor being paid, the general contractor may be liable for making payments to the subcontractors for work properly completed, even if the owner has not issued payment to the general contractor. This has been held by courts in Arizona, California, Massachusetts, North Carolina, Ohio, Oklahoma, Oregon, and Texas.

Typical subcontract agreements have provisions that state subcontractors will be paid after (1) completion of the work, (2) acceptance by the architect/engineer, and (3) full payment for the work by the owner. Some courts have ruled that this provision obviously intends that the general contractor not finance the cost of construction in the event of an owner default. This has been held by courts in Louisiana, Georgia, Virginia, Kentucky, and Connecticut. On this issue, New York is unique in that the courts have ruled both ways.

The pay-when-paid provisions in subcontracts are intended to avoid having the general contractors finance the costs of the construction project. If owners pay promptly, the provision has little negative impact on a project. This is where construction firms (general contractors and subcontractors) should assess the historical payment practices of owners. If an owner has a history of delaying payments, these risks should be reflected in their bids on the project. It should be clearly communicated to owners that the practice of making delayed payments is an added risk for contractors and that there is a cost associated with it. This cost should rightfully be placed on the owner.

Another payment provision incorporated in some subcontracts is the pay-if-paid provision that stipulates the general contractor has no obligation to pay the subcontractors if the owner becomes insolvent and fails to pay the general contractor. While the pay-when-paid and pay-if-paid provisions may appear to be similar, there are some differences. The pay-when-paid provision is intended to delay payments until the owner has paid the general contractor, while the pay-if-paid provision is a means by which the general contractor can avoid making any payments to the subcontractor in an owner default.

Saxon Construction v. Masterclean of North Carolina (641 A.2d 1056) was an interesting case in which a subcontractor, Masterclean, failed to complete the work as required by the subcontract. The work entailed the removal of asbestos from four buildings. After the asbestos was removed from two buildings, Masterclean refused to do any further work. Saxon Construction, the general contractor, then hired another subcontractor to remove the asbestos from the remaining two buildings. Masterclean then asked to be paid for the work it had performed. Generally, the defaulting subcontractor will be charged with the excess cost of the subcontracted work. In this particular instance, Saxon managed to obtain the services of another subcontractor at a price below that of Masterclean. Saxon sought damages from Masterclean for their breach of the contract and wanted to avoid payment to Masterclean. The court refused to award damages to Saxon since they had not incurred any added costs. On the other hand, the court determined that Saxon had to pay Masterclean only for the work it had done, and that Masterclean was not entitled to any of the savings that Saxon had realized.

The lack of payment to a subcontractor may not constitute a breach of contract if withholding payment is justified. This was addressed in *K & G Construction v. Harris* (164 A.2d 451). K & G was the owner and general contractor on a housing project; Harris was the subcontractor, hired to perform excavation and earthmoving. It was a standard procedure for the subcontractor to submit requests for payment on the 25th of the month and to receive payment on the 10th of the following month. On August 9, 1958, a bulldozer operator employed by Harris drove too close to one of the houses and caused a wall to collapse. K & G had received a payment request from Harris on July 25, but refused to make payment to Harris on August 10. Harris continued to work until September 12 and then stopped work on the project due to nonpayment. K & G withheld payment to cover the costs of repairing the house that was damaged. This resulted in a lawsuit. Harris claimed that nonpayment was a breach of contract. The court found otherwise. The court determined that Harris had breached the contract twice, once when the house was seriously damaged (failing to perform work in a workmanlike manner), and again when Harris stopped working on the project. In essence, the nonpayment of Harris was justified, and Harris could not then terminate the contract due to nonpayment.

REVIEW QUESTIONS

1. How are payments for manufactured goods different from payments that are typically made on construction projects?
2. What is the purpose of retainage?

3. Discuss the premise that retainage is not needed if the contractor provides the owner with a payment bond.
4. What payments are typically linked to substantial completion? Final completion?
5. What is the most common means by which an owner can breach a contract?
6. How are the amounts due for periodic payments on unit price contracts determined?
7. How are the amounts due for periodic payments on lump sum contracts determined?
8. Examine the AIA documents and the ConsensusDOCS that are included in the Appendix and determine the similarities and differences in the provisions related to prompt payment of the contractor. Specifically examine AIA provisions §9.6.1 and ConsensusDOCS §9.2.1.
9. Examine the AIA documents and the ConsensusDOCS that are included in the Appendix and determine the similarities and differences in the provisions related to retainage once a project is deemed to be substantially complete. Specifically examine AIA provisions §9.8.5 and ConsensusDOCS §9.6.4.
10. Examine the AIA documents and the ConsensusDOCS that are included in the Appendix and determine the similarities and differences in the provisions related to the schedule of values. Specifically examine AIA provisions §9.2 and ConsensusDOCS §9.1.

16

WARRANTY

ONCE THE OWNER has accepted a completed construction project, the owner wants assurance that everything in the project will function as intended. That is, the owner would like to have the contractor maintain some responsibility for the project even after it has been completed. This is usually provided to the owner in the form of a warranty, or guaranty.

A *warranty* constitutes certification that an aspect of a contract is in fact as it was promised to be. Usually this certification stipulates the time period for which the warranty is valid. Such warranties are often required of manufacturers. On construction projects, the job specifications often require the warranty to be in writing. If there is a breach of the warranty agreement, the warrantor becomes liable for damages. Warranties are similar to guarantees, but guarantees generally also stipulate the means by which restitution will be made in the event of a defect. For defects related to warranties, the remedy is not automatic.

Owners prefer contractual warranties so that they do not have to prove contractor negligence. However, the scope of these warranties is not always clear. A typical construction contract may contain several distinct warranties, each establishing different obligations in terms of scope and effect. This causes confusion, and the owner may correct faulty items that the contractor was obligated to correct. Conversely, the contractor may correct items for which no obligation existed.

On completed construction projects, a maintenance bond (warranty) may be required to ensure compliance with the contract. A maintenance bond guarantees that the contractor will rectify defects in workmanship or materials for a specified time after project completion. A 1-year maintenance bond is normally included in the performance bond without any additional charge. It is during this first year of occupancy that the contractor is obligated to correct any work that is defective or not in compliance with the contract documents.

PATENT AND LATENT DEFECTS

When a construction project has been completed, the owner officially accepts it as having met the contract conditions. As already described in Chapter 14, the initial acceptance is at the point of substantial completion, the time at which the project is sufficiently completed that it can be occupied and used by the owner for its intended purpose. At the point of substantial completion, a punch list is developed that describes the minor items of work that remain to be done. After the punch list items have been satisfactorily addressed, the project is officially completed and the final payment is then made to the contractor. After final completion, the contractor has no additional obligations under the contract, except for addressing any defects that are identified after this point in time.

When a project is accepted, it is imperative that the owner carefully inspect the project for any defects. It is far simpler to resolve defects before the final payment is made, than to ask the contractor to return at a later date to correct deficiencies. In fact, the contractor may not have an obligation to correct all deficiencies once the project has been accepted. This is especially true of patent defects.

It is important to understand the distinction between patent and latent defects. Patent defects are deficiencies that are clearly able to be observed or can be detected with a reasonably thorough investigation. These are deficiencies that can generally be identified without doing any destructive work on a facility or its components. Examples of patent defects include such deficiencies as an excessively large gap between the bottom of a door and the threshold, a wall paint that is the wrong shade, poorly matched edges of wallpaper, paint smudges on carpeting, a concrete slab with an improper slope, a sidewalk that is two inches narrower than specified, poorly abutted base trim, and so on. Note that some patent defects can be observed with a quick scrutiny while others might be detected only after specific types of measurements are made. As noted, once a project has been accepted by the owner, it is generally assumed that the contractor has no further obligations for correcting patent defects. It is the patent defects that should be clearly noted on the punch list for the contractor to correct. Patent defects are often of a cosmetic nature, but not exclusively. That is, the principle of caveat emptor or "let the buyer beware" applies specifically to patent defects. In essence, there is generally no recourse once patent defects have been accepted.

Latent defects are the more problematic concerns for contractors. These are defects that cannot be readily observed, even through intensive inspection procedures. It is often only through destructive testing that such defects might be discovered. Since these are difficult to identify at the time of project acceptance, latent defects continue to be the responsibility of contractors after the owner has begun occupancy. Examples of latent defects include improperly installed flashing that will eventually result in water intrusion, improper design of mechanical systems, soil subsidence, improperly installed roofing, excessive voids in concrete components, mold growth resulting from the installation of wet or moist materials, and so on. There are many types of latent defects, but most tend to be related to water permeability of in-place materials and the structural integrity of components.

There are clear differences in the law between latent and patent defects. For example, in Florida, any lawsuit related to patent defects must be filed within the statutory period of four years of project completion, contract termination, or project abandonment, whichever is latest. That is, if a patent defect is noted prior to the issuance of the certificate of occupancy, the owner will have four years in which to file the lawsuit.

For latent defects, there are two timeframes that must be understood. One pertains to the statute of repose and the other pertains to the statute of limitation. The statute of repose is generally the longer period of time, and it stipulates the time that the owner has to discover any latent defect after the completion of the project. These statutes are state-specific and can vary considerably. For example, the statute of repose for the observation of latent defects ranges from 4 to 15 years in the 50 states, with most states setting the timeframe from 7 to 10 years. The statute of limitation is the time allotted to actually file a lawsuit after a latent defect has been discovered. The time to file such lawsuits for latent defects is generally from one to three years from the point of discovery of the damage. There will be differences between the various states, but their intent is basically the same. The particular nuances of any timeline differences and specifics on procedures to follow should be clearly understood from one state to the next.

EXPRESS WARRANTY

A typical contract requires the general contractor to warranty workmanship and materials. For example, a warranty provision may state, "Contractor hereby warrants to the Owner that all materials and equipment furnished and the workmanship used to incorporate them in the project will conform fully to the standards stipulated in the Contract Documents. Any materials, equipment or workmanship that fails to conform to the quality standards outlined in the Contract Documents shall be considered defective. All deficiencies will be promptly corrected or replaced by the Contracter at no additional cost."

The contractor will have breached the warranty even if the faulty materials and workmanship cannot be detected by the contractor's or owner's diligent inspection. The general contractor also is responsible for the performance of the subcontractors and the materials and equipment provided by suppliers.

If the owner specifies equipment and the contract states that the general contractor warrants its performance, the courts generally will not enforce the warranty obligation placed on the contractor. However, if the contractor assisted the owner in the selection of the equipment, the responsibility may be shared with the owner.

It is common for the contractor to be required to guarantee the sufficiency of the work for a stated time period. The following wording includes the typical components of such provisions:

> The Contractor warrants that all materials, equipment, or supplies furnished and all work performed under this contract will be new, of specified quality, free from faults and defects, free from faulty design, and of sufficient size and capacity and of proper

materials to meet in all respects the requirements of the contract. The Contractor shall obtain for the benefit of the Owner all standard warranties of subcontractors, suppliers, and manufacturers of all materials, equipment, or supplies manufactured, furnished, or installed. Original copies of these warranties shall be furnished to the Owner. The Owner shall promptly notify the Contractor in writing of the discovery of any failure, defect, or damage. The Contractor shall, at its own expense, promptly repair or replace all such materials, equipment, or supplies which fail to conform to the warranties in any respect if such failure is discovered within one year of acceptance. Any item on which warranty work is performed shall have its warranty reinstated for the period of one year from the date of said repair or replacement.

Another type of warranty provision is as follows:

For the period of 365 days commencing on the date of acceptance of the work, or that portion of the work described as substantially complete, upon the receipt of notice in writing from the Owner, the Contractor shall promptly make repairs arising out of defective materials, workmanship, or equipment. The Owner is hereby authorized to make such repairs if, 10 days after the giving of such notice to the Contractor, the Contractor has failed to make or undertake the repairs with due diligence. All expenses in connection with such repairs made by the Owner shall be charged to the Contractor.

The ConsensusDOCS are brief in the stipulation that the contractor warrants that the work quality, materials, and equipment will be delivered in conformance with the contract documents. The provision states the following:

§3.8.1 The Contractor warrants that all materials and equipment shall be new unless otherwise specified, of good quality, in conformance with the Contract Documents, and free from defective workmanship and materials.

The Contractor further warrants that the Work shall be free from material defects not intrinsic in the design or materials required in the Contract Documents.

It is important that the contractor and the owner be aware of the timing of the warranty period, which is not always clear-cut. The warranty period typically begins at substantial completion, but it could begin after final acceptance by the owner. If it begins at final completion, the contractor has an obligation for an extended period compared with substantial completion. The time should be clearly stated in the contract documents. One must know the contract wording. In the state of Texas, the statute provides that for construction defects, subcontractors are liable for up to 10 years from the time of completion of the work. The measurement of this time period was tested in the case of *Gordon v. Western Steel Co.* (950 S.W.2d 743) involving subcontracted work on a condominium in Corpus Christi. The work of two subcontractors, Western Steel Co. and Braselton Construction Co., was finished in August 1984 and the project was substantially completed on June 1, 1985. In later years, defects were noted and the developers filed suit on October 21, 1994. The suit was filed in excess of 10 years after the work had been completed by the subcontractors, but it was filed less than 10 years after project completion. The court ruled that the subcontractors could not be sued if 10 years had passed since their work was completed. The suit was dismissed.

The contractor should also be aware of the warranty period of products from subcontractors and suppliers. The subs and suppliers may give a 1-year warranty of their product or their work that begins on the date of performance or manufacture. The lower-tier warranty may expire before the warranty of the general contractor expires. It is the general contractor's responsibility to have protection from liability by coordinating the warranty periods of the subs and suppliers with those required by the owner.

The general contractor's warranty requires the contractor to bring defective work up to the mandated quality requirements as stipulated in the contract. In order for the contractor to be liable for the work, notice of the defect must be received by the contractor during the warranty period. The warranty, however, is not a blanket guaranty, covering all types of deficiencies. Generally, it excludes responsibility for design flaws, natural catastrophes, and mistakes of other prime contractors.

IMPLIED WARRANTY

If the construction contract contains no express warranty provision of compliance with the drawings and specifications, such a warranty is automatically inferred or implied. An *implied warranty* does not mean that the contractor guarantees that the completed project will be suitable for its intended purpose. Of course, owners would like that interpretation and often make it the basis for a legal claim against contractors, but the courts have rejected this position.

According to the Uniform Commercial Code, a party who sells goods to another implicitly (without expressly saying so in writing) warrants that the particular product is fit for the particular purpose intended.

Many cases involving implied warranties concern homeowners. A few examples will illustrate some of the major issues that have been addressed in the courts. *Columbia Western Corp. v. Vela* (592 P.2d 1294) was an Arizona case in which two families purchased homes from Columbia Western, the builder and seller, in 1974. Both houses that were purchased developed similar problems—cracking walls—shortly after the owners began occupancy. The homeowners hired a soils engineer who said the problem resulted from dry soil that existed during the time of construction and started to swell beneath the footings as surface water infiltrated the area. The court ruled that it would not enforce the doctrine of caveat emptor (let the buyer beware) and that an implied warranty had been breached by the builder-seller.

In the Maine case of *Banville v. Huckins* (407 A.2d 294), the court determined that there is no difference where implied warranties are concerned between a contractor who contracts with a landowner and a contractor who builds on his or her own property for the purpose of a subsequent sale. Paul Huckins and Thomas Buchnam acquired a parcel of land and built a house on it. In the latter stages of construction they signed a sales agreement with Ronald Banville. The Banvilles took possession after construction was completed. However, it soon became apparent that the water supply was unsatisfactory for drinking, bathing, or laundry. In addition, the basement flooded with as much as 10 inches of water after rainstorms. The Banvilles sued for breach of implied warranty of habitability. The court ruled in favor of the Banvilles, stating that a reasonable person would not live in a house that was subject to flooding after every rainstorm.

The Alabama case of *Capra v. Smith* (372 So.2d 317) was similar to the Banville case in that it involved a builder-vendor. Ruth Capra, a real estate broker, built a house on speculation. It was Capra's first attempt at constructing a house. Upon completion, Capra sold the house to the Smiths. Two years later the basement wall collapsed. The Smiths sued Capra for breach of implied warranty of workmanship and fitness for habitability. Capra argued that the implied warranty did not apply since she was a builder-vendor and this was her first house construction. The court did not agree and stated that a builder-vendor is "just as liable for his first house as his one hundredth." The number of houses built by a firm is immaterial.

A similar Oklahoma case was decided on the basis that the caveat emptor doctrine did not apply. In *Jeanguneat v. Jackie Hames Construction Co.* (576 P.2d 761), the Jeanguneats purchased a new home, but found that the water was unsuitable for use. They had to have a new well dug, and incurred damage to their yard, clothes, personal effects, and water heater. Hames, the contractor, contended that there is not an implied warranty of fitness when the sale involves a house that is already completed. The court did not agree.

In *Petersen v. Hubschman Construction Co., Inc.* (389 N.E.2d 1044), the Petersens entered into a contract with Hubschman for the purchase of a parcel of land and the construction of a new home. The Petersens provided a deposit of $10,000 and also provided some of the labor and materials for the house. When Hubschman deemed the house to be complete, the Petersens refused to accept the house because of various defects, none of which rendered it dangerous or uninhabitable. The Petersens sought the return of their investment through a suit. The court ruled in favor of the Petersens. It was inferred by the court that since the house was not complete when the contract was signed, the issue of habitability was not the deciding doctrine. Since the house was not constructed to the Petersens' desires, they did not have to accept it.

In the Florida case of *Hesson v. Walmsley Construction Co.* (422 So.2d 943), a house and lot were purchased by the Hessons. After one year, cracks began to develop in the house, presumably from settlement of the house. The Hessons sued the construction company for breach of implied warranty of habitability. The company argued that the construction of the house was not at issue; the cause of the cracking was the lot, which had nothing to do with the quality of construction. The company contended that the subsurface conditions causing the settlement were not covered by the implied warranty. The court ruled that the implied warranty did apply, since the new house and lot were sold as a package.

In the case of *Burger v. Hector* (278 So.2d 636), a house was constructed on a parcel of land for the landowner. The defects that occurred after the owner moved into the house were not covered by the warranty because the land was deemed to be contributory to the problem.

There are rigid time constraints about the length of time that a warranty is valid. In the case of *Liptak v. Diane Apartments* (167 Cal. Rptr. 440), the statute of limitations was defined in the dispute. Warren and Toups, a subcontractor, agreed to do the grading and filling of land on two adjacent lots. They completed the site work in 1967. The substantial completion of the houses occurred in 1972. In March of 1978, heavy rains caused some earth movement that resulted in significant damages to the home owned by the Liptak family. The Liptaks brought suit, claiming that Warren and Toups had been negligent in the performance of their work. The court had to decide on a more fundamental issue, namely, whether the statute of limitations had passed. The law is clear that in California one cannot be sued if one "develops real property or performs or furnishes the design, specifications, testing, or construction of an improvement to real property more than ten years after the substantial completion of such development or improvement." Naturally, the court had to determine if the work of Warren and Toups was separate and apart from the work of building the homes. If their work was separate, the warranty period would run until 1977, and if it was tied to the house construction, it would run until 1982. The court ruled that the site work performed by a subcontractor was separate from the house construction, and that the warranty period of 10 years had expired. The court implied that Warran and Toups were not the developer, so their work was completed when the site work was finished. The implication was that the warranty of only the developer was valid until 1982.

The Illinois case of *Minton v. Richards Group of Chicago through Mach* (452 N.E.2d 835) provides considerable insight into the protection offered to new homeowners. The Mintons entered into a contract with the Richards Group for the construction of a new house. Several subcontractors worked on the home under the supervision of Richards. One of the subcontractors was International Decorating, Inc., which had responsibility for painting. The paint began to peel less than 90 days after the Mintons assumed possession of the house. They sued the Richards Group, which had dissolved, and also named International in the suit. International claimed that it had no contract with the Mintons and was not liable for any implied warranty. The court disagreed. The court stated that the purpose of an implied warranty is to protect innocent purchasers, and it would therefore extend the implied warranty to subcontractors who perform faulty work when there is no possible recourse against the builder-vendor.

Similar protection was offered to the second owners of a house in *Terlinde v. Neely* (271 S.E.2d 768). In 1973 the Johnsons purchased a house from Neely, the developer. Substantial settlement in the house occurred in 1976. Neely paid the Johnsons $230 in exchange for a release from further liability. That same year the Johnsons sold their house to the Terlindes. Additional settlement occurred, causing wall cracks, floor settlement, brick veneer cracks, and other problems. Repair estimates ranged from $6,000 to $23,000. The Terlindes sued Neely for breach of implied warranty. The court ruled in favor of the Terlindes. The release of liability did not apply to subsequent purchasers as this was counter to the principle on which implied warranties are based.

In *San Luis Trails Association v. E. M. Harris Building Co., Inc.* (706 S.W.2d 65), the extent of the rule of habitability was addressed. E.M. Harris Building Co., Inc., had developed a subdivision of Jefferson County, Missouri. An association of the homeowners was established to hold title to the common areas, including the streets. When defects in the streets were noted, the association filed suit against Harris for breach of implied warranty. The court decided that implied warranties extend to purchasers of new homes in cases where ordinarily prudent purchasers fail to detect flaws in construction. The court noted that the association had not claimed that there was any faulty construction in a house or to any portion of or attachment to a house. Therefore, the implied warranty did not apply.

Contractual relationships or the lack of them were the principal issues in *San Francisco Real Estate Investors v. J. A. Jones Construction Co. et al.* (524 F. Supp. 768). Harold A. Berry Associates was hired by Hill Crest Square, Ltd., to design an office building and parking garage in Cincinnati. Berry in turn hired the firm of Hixon, Tarter & Merkel to be the principal architect. A contract was entered with Jones for the construction of the building. Hill Crest sold the building to San Francisco Real Estate Investors in 1971, shortly after the construction was completed. Leaks developed in the parking structure in 1978 and were repaired at a cost of $159,000 to San Francisco Real Estate. A suit was then filed against Jones and both design firms. The court ruled that the new owner had no contractual relationship with the contractor and the two design firms, and was barred from recovering any amounts from those firms.

With an implied warranty, the contractor is obligated only to provide the owner with a facility that complies with the plans and specifications. If the facility does not perform properly after having been constructed according to the plans and specifications, the owner's grievance is with the designer. Thus, a contractor who is involved in design work will warrant, pro rata, the sufficiency of the functioning of the final product. Thus, exposure to liability is increased when a contractor performs design work.

The quality of the construction documents is also the subject of dispute in numerous cases. One such case was *Bradford Builders, Inc., v. Sears, Roebuck & Co.* (270 F.2d 649). Bradford Builders had entered a contract with Dade County Schools to build a school building that was enclosed by a fence. Bradford subcontracted the fence work to Sears, and Sears sub-subcontracted the work to Jack Wilson. The subcontract with Sears stated that there was to be no sub-subcontracting of any of the work; however, Bradford was aware of the arrangement between Sears and Jack Wilson and did not protest. Jack Wilson was provided with a full set of drawings and specifications. One of the drawings erroneously showed the fence being located near the conduit, but not actually crossing the conduit. With the error in conduit location going unnoticed, Wilson damaged the conduit and electric cable while doing the fence work. Bradford sued Sears, contending that they erred in sub-subletting the work and in not discovering the mistake in the drawings. Wilson stated that he had done the work in accordance with the design drawings. The court decided against Bradford, stating that a contractor cannot be held responsible for problems arising from defects in the plans and specifications. The issue of the sub-subcontracted work was dismissed, as Bradford was aware of the arrangement and failed to express any opposition to it.

If a subcontractor fails to comply with the drawings and specifications, the express and implied warranty may be breached. In the case of *Biscayne*

Roofing Co. v. Palmetto Fairway Condominium Assoc. (418 So. 2d 1109), the roofing subcontractor on a condominium project substituted, without authorization, a substandard roofing material. The court ruled that the damages caused by the faulty roof must be paid for by the owner-developer. It was then the owner-developer's prerogative to seek compensation from the general contractor and, similarly, for the general contractor to seek redress from the subcontractor.

The knowledge of the contractor can also be considered by the courts. In *Don Siebarth Pontiac v. Asphalt Road Building* (407 So.2d 42), an automobile dealer contracted with Asphalt to lay asphalt roads and other surfaces at the dealership. No work was done to the subgrade or the subsurface. Before the work was completed, cracks began to appear on the asphalt surface. The dealer had the entire work redone at a cost of $25,000 and sued Asphalt. The road builder contended that since no work had been done to the subgrade, the implied warranty did not apply. The court said that a contractor is liable when it should have known through the exercise of good judgment that the latent defects in the subgrade should be corrected before performing the work.

Changes in the contract are an area where the responsibility is not clearly vested in one party. As a result, if modifications to the design are made by the contractor, such changes should be approved in writing by the owner and the designer. If the contractor makes a design change, any positive assurances of the effect of the change may be interpreted as a warranty of performance.

Whenever the contractor performs corrective work, it is important that accurate records be kept of all related costs. Some or all of the work may be beyond the scope of the warranty. The contractor may successfully seek full or partial reimbursement if the work or part of it is not within the scope of the warranty clause.

The rules of implied warranty do not apply to professionals in the same manner as they do to those who perform construction work. This was addressed in *State of New Mexico v. Gathman-Matotan Architects and Planners, Inc.* (653 P.2d 166). The New Mexico Department of Finance hired W. C. Kruger and Associates to draft a plan for changes to improve the state penitentiary. One recommendation addressed the remodeling of a bay window in the central control area. The steel gridwork in the window was to be replaced by large panes of bulletproof glass. The drawings for the suggested improvements were to be prepared by another firm, Gathman-Matotan Architects and Planners, Inc. In 1980 inmates at the penitentiary rioted and gained control of the central control area by breaking the glass of the new bay window. During the riot, 35 people were killed and severe damage was inflicted on the building. The state filed suit against Gathman-Matotan Architects for breach of contract, claiming that there was an implied warranty that the glass that was specified would protect against an inmate takeover. The court ruled in favor of the design firm, stating that an implied warranty is meant to protect buyers of goods and cannot be extended to contracts for professional services. Warranties of professional services essentially stipulate that the work performed by the professional is consistent with the standards of the profession. Failure of the work product will not generally result in a liability for the professional if the work is consistent with the standards of the profession at that time.

COMPLIANCE WITH THE CONTRACT DOCUMENTS

The quality of the work to be provided on a construction project is to be detailed in the contract documents. Once the contractor has complied with the contract documents, the owner is then obligated to pay for the work. This is ostensibly a simple arrangement, but complications can develop when questions about the degree or level of compliance arise.

When there is strict compliance with the contract requirements, there is seldom an issue that must be resolved. The complications arise when there is compliance that is close to the contract requirements, but not full compliance. In some cases, owners have refused to accept work that did not fully satisfy the contract requirements. The courts have agreed with such decisions, especially when safety was the primary concern. That is, if failure to fully comply results in a product that is less safe, courts have tended to err on the side of caution and have agreed with owners who have insisted that the work be redone without added compensation.

The courts have been less supportive of owners when contractors have substantially complied with the contract requirements. This is especially true in cases in which the cost of removing the in-place work and replacing it far outweighs the need for correcting the defect, as this would be particularly burdensome on the contractor. This might be deemed to be "economic waste" in which full conformance with the contract should not be mandated.

For some items of work, the quality of the completed or in-place items is readily expected to deviate somewhat from the contract requirements. For example, the specifications might stipulate that the 28-day compressive strength of concrete be a minimum of 3500 psi, or that the soil compaction be at least at 95 percent compaction based on the Modified Proctor Test. It is recognized that the delivered performance will not be exactly as specified. For example, if the minimum concrete strength is to be 3500 psi, the contractor will probably deliver concrete that is above 3500 psi, as any concrete that is below 3500 psi would be in noncompliance. Thus, in order to avoid delivering any concrete that does not meet the minimum requirement, the contractor will be inclined to deliver higher strength concrete.

Some owners recognize that minimum requirements (as in the concrete specification of a minimum of 3500 psi) on some materials will result in higher costs to them. Since the contractor does not want any of the in-place concrete to fail to meet the specifications, a higher quality concrete will be sought. This will have a higher cost than if the concrete must achieve an average of 3500 psi. In addition, the owners also recognize that contractors may increase their concrete bids to account for the possibility of having to replace some of the concrete that does not satisfy the specification. This realization about the cost impacts of these specifications has resulted in some owners using incentive/disincentive specifications.

An incentive/disincentive specification is one in which an owner will accept material of differing quality, but this is done by paying less for work of lower quality and paying more for material of higher quality. One such specification that has been used is associated with concrete pavement smoothness that is measured by a lane profile index (LPI). The technical specifications might stipulate that there will be no adjustment in the price paid to the contractor if the LPI is in the range of 14 to 16. Since lower LPI values indicate better quality surface conditions, an additional payment is made to the contractor if the LPI values for the pavement are less than 14. For example, the added payment for completing pavement with an LPI of 10 might be determined with the formula as follows:

$$\text{Incentive Payment (per square yard)} = \$(196 - (14 * \text{LPI}))/60 \text{ sq. yd.} = \$0.93/\text{sq. yd.}$$

Thus, the LPI value would result in an incentive or additional payment of $0.93 per square yard of pavement. The specification might stipulate that the maximum additional payment to be made would be $1.40 per square yard, meaning that no additional compensation will be earned (beyond the $1.40 per square yard) when the LPI is below 8. The disincentive portion of the specification may stipulate that the cost per square yard will be reduced when the LPI value exceeds 16. The reduction amount for pavement with an LPI value of 18 might be determined with a formula such as follows:

$$\text{Disincentive Payment (per sq. yd.)} = (224 - (14 * \text{LPI}))/60 = -\$0.47/\text{sq. yd.}$$

Note that the formula results in a negative value which means that the contractor's payment is reduced by this amount. The specifications might also stipulate that an LPI value greater than 22 will be unacceptable, meaning that the work must be redone.

Similar incentive/disincentive specifications can be created for concrete strength, soil compaction, asphalt thickness, and a host of other materials that can vary in quality. The concept of using incentive/disincentive specifications is that work need not be redone and this will save time. In addition, if quality is better than specified, the product life is often longer, so the owner actually acquires a product for which additional payment has been earned.

OWNER'S ACCEPTANCE OF WORK

Most federal contract documents state that the government's inspection and acceptance of the finished work constitute a waiver of claims against the contractor, except those resulting from latent defects or fraud. For example, on a concrete runway project, the contractor failed to saw expansion joints to a full depth of six inches. The government contended that the error was not discovered prior to final inspection, because the joints were covered with joint sealer shortly after the cutting had been done. The government sued the contractor to recover the costs of correcting the error, claiming that this was a latent defect. The contract stated that acceptance was final "except as regards latent defects." It was clear that the contractor had erred, but it was determined that the defect was patent and not latent. Since the defect was patent, the board determined that the government's acceptance of the work nullified the subsequent claim [Federal Construction Co., ASBCA No. 17599, 73-1 BCA 10,003 (1973)].

Private works contracts often state that the owner's inspection and acceptance of the work does not waive any claims that result from the contractor's defective performance. The courts, however, are reluctant to enforce such provisions rigidly. The general interpretation, regardless of implied or express contract provisions, is that the owner cannot accept work without taking exception and later demand that the contractor correct defects that should have been detected during a reasonable inspection. This interpretation specifically does not apply to latent defects, which remain the responsibility of the contractor even without any contract provisions.

On private works projects, the owner assumes a considerable risk if the facility is accepted, the contractor is paid for the services rendered, but lien releases are not obtained from those who have contributed to the effort. Without the lien releases, it is possible that an unpaid subcontractor or supplier might still place a lien on the property.

Acceptance of the contractor's shop drawings is a different matter entirely in that it does not relieve the contractor of liability. In spite of the information shown on the shop drawings, acceptance does not relieve the contractor of responsibility for complying with the plans and specifications. The only exception here occurs when the contractor can clearly demonstrate that the owner was fully aware that the shop drawings altered the original plans.

To what extent is the contractor absolved of responsibility for faulty work if it is not identified by an owner's inspectors? This question was raised in *J. A. Tobin Construction Co. v. Kemp* (718 P.2d 302). Tobin had a highway contract with the Kansas Department of Transportation (KDOT). The work was inspected by KDOT inspectors. The work in question consisted of the installation of reinforcing steel. The inspectors failed to notice that the steel reinforcing had been improperly installed. A year later, prior to final acceptance, cracks formed on the concrete surface. An investigation of the cause of the cracks revealed the faulty placement of the reinforcing steel. The contractor corrected the work at a cost in excess of $20,000 and then requested a change order for the corrective work. KDOT denied the request, and Tobin filed suit. The contractor contended that the failure of the inspectors to note the error in steel placement absolved it of blame. The contractor's argument was that the state's inspection services were for the benefit of the contractor, while the state thought otherwise. The court found for the state, ruling that the contractor was not in a position to put full reliance on the inspectors. In essence, the court was stating that the initial approval of the work by the inspectors did not constitute final acceptance.

REVIEW QUESTIONS

1. What is caveat emptor, and how do courts currently apply it to construction warranties?
2. What are latent defects, and are they covered by warranties?
3. Are patent defects covered by warranties?

4. How are the principles of warranties applied to the work of design professionals?
5. Discuss the impact of the acceptance of shop drawings, submittals, or work put in place on the validity of warranties.
6. Examine the AIA documents and the ConsensusDOCS that are included in the Appendix and determine the similarities and differences in the provisions related to the contractor's warranty. Specifically examine AIA provisions §3.5 and ConsensusDOCS §3.8.1 and §3.8.2.

17

CONSTRUCTION INSURANCE

CONSTRUCTION INSURANCE IS a complex subject that entails protection for or coverage of various parties and types of injury or damage. The cost of insurance for a typical contractor ranges from 1.5 percent to about 15 percent of the total costs of construction. The nature of the tasks undertaken and the risks involved directly affect the cost of insurance to the contractor.

There are various types of coverage to protect workers, the public, the property of others, the project, the vehicles, and so forth. There can be much confusion if the contractor is not fully familiar with the nature of the coverage, the nature of the risk involved, or the various types of insurance available. The contractor must be wise in acquiring insurance coverage. This goes beyond buying at the right price. With the myriad forms of coverage available, the contractor must be careful to avoid wasteful double coverage and also to avoid any gaps in coverage that leave the contractor unprotected.

To help assess the risk on a project, a contractor is well advised to get an insurance agent to review the contract documents in detail. This review should disclose the risks inherent in the job and the risks the contractor is being forced to assume by contract. Thus, it may be wise for the contractor to obtain as much coverage from one insurance carrier as possible. This will minimize the chance of gaps and duplication in the coverage and may result in a better price for the coverage.

INSURANCE TERMS

The insurance industry uses a number of terms which are not frequently used in other industries. An understanding of these terms is important to anyone who needs to be conversant with construction insurance.

Subrogation

Normally it is assumed that after a loss on an insured item, the insurance company pays the claim and the matter is settled, with perhaps a slight impact on future insurance premiums. This is not entirely the case if the loss was caused by a third party. Under the rights of *subrogation,* the insurance company can seek recovery from the third party that was responsible for causing the loss. This does occur, and it can present problems if a subcontractor is sued through subrogation. Literally, *subrogation* means the substitution of one person for another in claiming a lawful right or debt. To the insurance company, this means that if it pays on a claim, it gains the insured's right to sue.

Premiums

The payment or consideration for an insurance contract is called a *premium.* The premium guarantees that the insurance policy will be effective for a stated period of time.

Dividends

Insurance companies measure their success by comparing their premium receipts with their actual expenditures for administering insurance policies and settling claims. If an insurance company has had a good year, it should have a large surplus of cash. To keep their clients, such companies frequently return portions of the surplus to clients with relatively low losses. These returns are called *dividends.* Dividends are usually paid by insurance companies on an annual basis, but these companies are under no obligation to pay dividends. If an insurance company has had a bad year, it may not pay dividends. When purchasing insurance, contractors should determine if the insurance company pays dividends when profits permit this practice.

As was mentioned earlier, the dividend must reflect the basic administrative expenses incurred by the insurance company. These expenses vary with the volume of premiums paid and do not reflect actual losses resulting from injury cases. A typical cost of administration may be as follows:

45 percent of premiums paid on a total premium of $25,000
15 percent of premiums paid on a total premium of $250,000

These administrative costs are charged against the insured's account, along with the actual costs associated with claims. An administrative fee of 17 percent of the amount paid for losses may cover the costs of handling the cases. When these costs have been totaled, calculations can be made to see if the insured is entitled to a dividend. Note that the insurance company cannot guarantee a dividend; the dividend is dependent on the overall profits of the company. One insurance company derived the following schedule:

If premiums are $25,000, a dividend will be paid if the losses on claims are below $11,752 (47 percent).

If premiums are $250,000, a dividend will be paid if the losses on claims are below $181,624 (73 percent).

From this schedule it is clear that a large firm is in a better position to receive a dividend, even though the loss prevention program in the smaller firm may be better.

The receipt of a dividend sends a message to management that its loss prevention efforts are paying off. However, the receipt of a dividend in one year does not mean that the firm's total insurance premium will be decreased the following year (see the section on experience modification rating).

Loss Ratio

The *loss ratio* is a quotient which represents the level of success that the insured has had in minimizing losses. This ratio is simply the insurance company's costs for handling a company's claims divided by the total insurance premium paid by that company. Naturally, an insurance company will be reluctant to provide insurance coverage to a client whose loss ratio is 1.00 or higher.

Direct and Indirect Funds

Insurance is a topic that tends to be ignored by many people. Perhaps these people believe that it is too complicated to understand, or that the coverage is all that must be understood. At any rate, there are serious misunderstandings about insurance.

Indirect expenditures and routine expenditures of cash are often invisible to management. Insurance premiums fall into this category. The payment of insurance premiums becomes a routine and is not regarded as an area where profits can be improved. It is too often assumed that insurance premiums are a necessary evil. Eventually, routine insurance premiums are regarded as part of overhead and management expense. The end result is that top managers never consider the total cost of insurance. This means that management does not seriously assess the expenditures made on insurance.

Direct funds are more *visible or readily identified* as being related to insurance. For example, the outflow or inflow of cash, particularly large sums, is very visible to contractors. Dividend checks from insurance carriers fall into this category. The dividend check is often a single payment that is made toward the end of the year. In many cases the check is quite large; a contractor may receive a dividend check for $80,000 from an insurance company. This will impress most managers, who may then infer that their loss prevention program is operating well, ignoring the fact that the weekly insurance premiums of $8,000 over the past year amounted to more than $400,000. Thus, the seemingly large dividend checks are readily noticed, while the routine premium payments are paid without further thought.

Self-Insurance

Some contractors recognize the high cost of insurance and decide that they should play the role of an insurance company. When a firm acts as its own insurance company, it is practicing self-insurance. Not all companies can be self-insured. Laws stipulate that to qualify for self-insurance, a firm must meet certain minimum standards, particularly in the area of financial stability. A careful investigation must be conducted to demonstrate the financial stability of the firm. Reasons for becoming self-insured include the following:

> It provides a direct and immediate incentive for a contractor to reduce the cost of claims.
>
> The company is in a better position to conduct a detailed follow-up study on all claims.
>
> Interest is earned on company reserves set aside to meet possible claims in the future.
>
> The company can process claims at a lower cost than can the insurance carrier.

Although few contractors can qualify for self-insurance, another option is available: contractors may band together as a group and become self-insured. When this is done, it is paramount that all the contractors be carefully screened. This should be done before a contractor is added to a self-insurance group. The group should also be empowered to expel members whose losses are excessive. Otherwise, contractors with the better safety programs are forced to carry contractors with poor programs.

Wrap-Up Insurance

Wrap-up insurance is insurance coverage for a project that is provided by one insurance company. With wrap-up insurance, the owner traditionally would obtain the insurance coverage otherwise provided by the general contractor, subcontractors, architect, engineers, and owner. In recent years, this coverage has been referred to as an owner-controlled insurance program (OCIP). With this type of insurance, all insurance is provided by a single insurance company. Since the owner provides the insurance, the contractors are essentially asked to bid or quote prices for the construction effort without insurance. The possible benefits of a wrap-up insurance policy from the perspective of the owner are as follows:

> Lower total premiums through volume purchases.
> No gaps in or duplication of coverage.
> Simplicity of administration.
> Easy settlements (no subrogation suits).
> No disputes between insurance companies.
> Unified project-specific loss control and safety effort.

Wrap-up insurance historically has been obtained for large projects, but in recent years the use of OCIPs has extended to smaller projects. In general, this type of insurance is still a rarity. It has been used on projects such as the construction of the Bay Area Rapid Transit (BART), the United Nations Building, the New York World's Fair, the Seattle World's Fair, and the Washington Metropolitan Area Transit Authority (WAMATA).

Although there have been successes in using wrap-up insurance, there are also disadvantages. Assume that an owner will provide wrap-up insurance on a project. Bids will be let on a competitive basis. Contractors with good safety records cannot use their experience as a bidding advantage. The safety records of the competing contractors will play no role in determining the low bidder.

Although it is inferred that wrap-up insurance means that all the insurance coverage is provided by the owner, this may not be true. Contractors are well advised to conduct an independent review of the coverage being offered under a wrap-up insurance policy. This review may reveal deficiencies in the coverage. If this is the case, the contractor will want to obtain the necessary added coverage independently.

WORKERS' COMPENSATION

Until 1900 most industrial injuries were paid for by the party at fault. In many cases this party was the injured worker. For the injured party to be covered by the employer, it was necessary to file suit and prove that the injury was due to the employer's negligence. At that time employers were immune from guilt if a common-law defense applied. These defenses consisted of the following: (1) contributory negligence of the worker—under this defense, the worker could not recover from the employer if the worker was negligent to any degree, regardless of employer neglect; (2) assumption of risk by the worker—if it could be shown that the worker knew about the inherent danger of the task, the employer was not liable; and (3) negligent acts of third parties—employers were relieved of liability if the injury was caused by a fellow worker. Even if the employer was held liable, there was generally a $25,000 limit to this obligation.

Workers' compensation laws eliminated many of the ills of common-law defenses. Today workers' compensation laws have been enacted in all 50 states and in all the provinces of Canada. Under workers' compensation insurance, compensation is granted for disability and medical treatment for injuries resulting from accidents occurring as a result of employment, regardless of fault. Workers' compensation insurance falls under the jurisdiction of the individual state (not the federal government) and as a result varies from state to state. Some state laws do not cover agricultural workers, domestic workers, or firms with only a few employees. Under workers' compensation, in some states the employer may be immune from suit; that is, workers' compensation insurance may be the sole remedy for the injured worker. In some states the employer is subject to a suit only if the worker waives any workers' compensation benefits. In some states the injured worker or the workers' compensation carrier, after workers' compensation benefits are provided, can file suit against third parties whose negligence caused the injury. Some states give immunity from suit to fellow workers. If a subcontractor does not provide workers' compensation for the workers, the general contractor is often held responsible. Thus, the general contractor should require the subcontractor to furnish proof of purchase (a certificate of insurance) of workers' compensation coverage.

Under workers' compensation laws, workers are invariably covered while at work. This has been interpreted quite broadly, and some courts have ruled that employees are covered when they are attending company picnics or parking their cars at a place of employment. However, workers' compensation does not provide blanket coverage. Compensation benefits may be denied if a worker is guilty of willful misconduct, or if intoxication contributed to the injury.

Employers have few options when providing workers' compensation coverage for their workers. In most states workers' compensation insurance is underwritten by private companies. Five state governments support the workers' compensation program through a monopolistic fund. The monopolistic states include North Dakota, Ohio, Washington, West Virginia, and Wyoming. The only other option available to companies in these states, if they qualify, is to become self-insured.

Benefits

Workers' compensation coverage can vary. Several endorsements are available, including the following:

1. *All states endorsements.* This provides coverage to employees working out of the home state. This does not obviate the need for compliance with the state laws. This type of endorsement is generally available through monopolistic state funds.
2. *Special maritime endorsement.* This provides coverage on navigable waters for employees aboard vessels.
3. *Dockworkers and harbor workers.* This provides coverage for workers on boats and docks.
4. *Extralegal or additional medical.* This endorsement is desirable when the employer feels that the state provisions for medical coverage are not sufficiently broad.
5. *Voluntary compensation.* This provides coverage to employees when not in the course of normal work activities, for example, during attendance at company-sponsored athletics.

The benefits under workers' compensation include coverage for medical expenses incurred, hospitalization costs, and disability payments. Injured workers are generally eligible for disability benefits after a given number of work days (usually five) have been missed as a direct result of the injury. These benefits are usually based on a percentage (typically 75 percent) of the worker's average wages. For "schedule" injuries, a predetermined specific monetary reward is given to the worker. For example, a fixed sum may be paid to workers who suffer a permanent loss, such as the sight of one eye, a finger, or a leg; every body part is assigned a specific value.

If the benefits received are not considered adequate, the injured worker can file a claim. These claims are handled by state administrative agencies. The amount of a workers' compensation award can be appealed, although few award cases are actually decided by the courts.

Premiums

The premiums for workers' compensation coverage are based on payroll. The manual, or basic, rate is determined at the state level and reflects the past losses encountered in each craft category. The manual rate is stated as a percentage of the payroll, or as a dollar amount for every $100 of payroll. Since workers' compensation insurance is governed by state legislation, there is often considerable difference between the costs and benefits related to workers' compensation insurance. Within each state, a manual rate will be established for each craft, based in large part on the loss history of that craft in the state. For example, the 1999 manual rate for masonry workers was 5.79 percent in Indiana and 25.38 percent in Rhode Island. The range in the 1999 manual rates was broad, being as low as 2.89 percent for interior electrical wiring in Indiana to 191.36 percent for structural steel erection ("dwelling 2 stories") in Minnesota. Within a state, the rates will vary by craft. For example, in Oregon the 1999 manual rate was 17.51 percent for general carpentry, 48.65 percent for structural steel erection, 10.94 percent for masonry, 6.96 percent for plumbing, and 4.86 percent for electrical wiring. For 1999, the national average for the manual rate of carpenters was computed to be 16.76 percent and for structural ironworkers it was 35.60 percent. While some of the manual rates appear to be high, there has been a declining trend in the magnitude of the manual rates. This decline is attributed to several factors including: the strength of the economy, workers' compensation reform efforts, effective safety programs, and others.

Experience Modification Rating

The *experience modification rating* (EMR) is relevant to certain types of insurance coverage, especially workers' compensation. This is a rating that directly affects the insurance premiums paid by a contractor. As the name implies, it considers the past performance or experience of the insured. Since cost reports for many claims will not be complete (many cases remain open) for the previous year, they are not included in the determination of the experience modification rate. EMRs are determined by a bureau that sets the manual rates for workers' compensation. The rates reflect losses incurred over a period of three years beginning prior to the previous year. The information from the immediately preceding year is not included in this calculation. Thus, the mod rate for 2001 will reflect the loss record of an employer for 1997, 1998, and 1999. Thus, heavy losses by a company will affect the ratings for three years. The rating reflects actual amounts spent on claims and also funds held in reserve. In essence, the mod rating, as it is often called, adjusts the basic premium to reflect the success of the contractor or insured in minimizing losses.

If the EMR for a particular coverage is 0.68, the insured's premium will be only 68 percent of the regular premium (based on the manual rates). However, if the mod rating is 1.60, the contractor will be assessed a 60 percent penalty above the regular premium to compensate for some of the past losses on insurance claims.

Not only do the premium rates vary between states and between crafts, they also vary between employers. Companies must meet specific criteria in order to be experience-rated. In most states firms must pay a given minimum in premiums in order to qualify for an experience rating. In effect, small firms and those that are too young (less than four years old) to have a claims history are given modification rates of 1.00. The mod rate is applied to the employer's total insurance premium. The calculations for a company using the manual rate may appear as follows:

Work Type	Manual Rate	Total Paid in Wages, $	Total, $
Masonry	3.05	260,000	7,930.00
Roofing	6.89	370,000	25,493.00
		Sum	33,423.00
		Mod rate	0.78
		Premium paid	26,070.00

In this example, the EMR was 0.78, from which it can be inferred that the employer implemented an effective loss prevention program. Note that the premium actually paid was $26,070, an amount that was more than $7,000 below what the premium would have been if the modification rating had not been applied. It must be recognized that the modification rate can also result in higher total premiums. The values for EMRs generally average about 0.90, but the range of these values may be from as low as 0.20 to over 2.00.

The determination of EMRs is a complex task. The rates reflect not only financial losses from injuries, but also the frequency of injury cases. The cases are weighted so that a single catastrophic injury will not dramatically affect the mod rate beyond a predetermined amount. As a rule, many low-cost injuries are weighted higher in terms of dollar-for-dollar impact than are the less frequently occurring high-cost injuries. For example, in considering the frequency and severity of losses, the first $750 of a claim is considered a component of the frequency, while the remainder is considered part of the severity.

Experience modification rates provide employers with an incentive to reduce accident costs. With the lag of three years that is represented by the modification rates, a company may be paying very high premiums based on its past history, with no regard to its current successes in loss prevention.

Because EMRs are based on losses and claims history, there is a tendency for many owners and general contractors to utilize the EMR as a measure of the safety effectiveness of a firm. While there is some rationale behind this practice, it is important that the limits of this practice be understood. The computation of the EMR is highly sensitive to the hourly wages paid to workers in a firm and the total amount of payroll. For example, assume the EMR of company A is 0.77 and that the EMR of company B is 0.64. The intuitive claim that company B would make is that it has the better claims history and that it is the "safer" company. While this may be the case, it may also be the case that company B is a larger company, and that it pays higher wages than company A. It may even be possible that company A is sufficiently small that the EMR computation for this size of firm is at its lowest possible value with 0.77. Thus, if firms are being compared on the basis of their EMRs, care should be exercised to ensure that the companies are similar in terms of types of trades employed, the state for which the EMR is computed, hourly wages paid, and the total payroll amount.

Reserves

When a loss occurs, the insurance company will investigate the claim (if the loss is large or potentially large) and pay for the loss. In most property damage claims, the case is closed after the insurance company pays for the loss. If the loss involves an injury, the case may be more complex, particularly if the injury will take a long time to heal. For example, a back injury may heal within a week or may lead to permanent disability. When a case or claim may extend into the future with unknown cost impacts, insurance companies set aside an estimated amount of money from which the claimant will be paid. The account set aside is called a *reserve*. The amount in the reserve account is considered when the insurance company assesses the losses for each contractor or client.

Workers' Compensation Fraud

Workers' compensation is established to provide financial protection to workers who are injured or who become ill as a consequence of their employment. Because the financial benefits can be substantial, especially in permanent disability cases, and because the costs of treatment are often high, it is perhaps not surprising that fraudulent activities have been noted with workers' compensation. To counter these practices, some states have enacted harsher laws in conjunction with their workers' compensation reform movement. Fraudulent acts may be initiated by workers who fake injuries, by employers who misclassify the crafts of workers in order to reduce their premiums, or by physicians who deliberately falsify their findings about an injury or illness. There is another practice of some contractors to pay workers on a piece-rate basis, and then claim that the workers are actually independent contractors. Such workers are then left with no insurance coverage. Workers' compensation reform is focused on each of these fraudulent practices.

Retrospective Rating Plans

Another form of loss-sensitive insurance is available to some large firms. This is the *retrospective rating plan,* which takes into account an insured's loss record and adjusts the premiums to reflect the results. Unlike experience modification ratings, which reflect the record over a period of three years, retrospective rating plans are usually based on a period of one year. This is often referred to as cost-plus insurance.

Under a retrospective rating plan, the premium is calculated and prepaid at the beginning of each year, and is adjusted upon expiration of the policy to correspond with the actual loss experience for the policy year. How is the prepayment determined? Generally, an estimate is made of the employer's labor hours for the period of one year for each work classification. An adjusted manual rate is then applied to the project labor hours.

At the end of the policy year, adjustments are made to reflect the incurred costs for the policy year. The calculations of the insurance company's costs may be as follows.

The firm will charge a basic fee to cover administrative costs and ensure against excess losses above a specific maximum amount. This could amount to 18.5 to 20 percent of the prepaid amount. Losses for the year are charged against the account. These charges amount to 100 percent of the losses, up to a maximum limit per occurrence. A charge is assessed against the account for the handling of claims, which may be about 14 percent of the costs of the claims. Additional costs such as taxes, which vary by state, are then added, amounting to perhaps 3 percent of the losses.

These charges are totaled and compared with the prepaid premium amount. If the costs exceed the premiums paid by the contractor, the difference is paid to the insurance company. If the costs are lower than the prepaid premium amount, the insurance company pays the difference to the contractor. In a sense, this is similar to self-insurance with a guarantee against catastrophic losses. The minimum possible premium costs are about 22.5 percent of typical or standard premiums, while the maximum premium cost is 128 percent of the standard premium. This form of insurance is not attractive to most small contractors.

Reporting Losses

When an injury occurs on the job, it is imperative that the injured party be given immediate treatment. Following this effort, it is important to complete the necessary reporting forms promptly. These accident forms should be available at the job site, and even minor injuries should be reported. The insurance company may try to deny coverage if reporting is not prompt. Small cases that are not reported may later result in large claims. Thus, the insurance company wants to be informed about all injuries and potential claims. If no initial report has been made, the workers' compensation board may question the veracity of a written or oral report submitted months later. If the claim is large, the insurance company may try to establish that the injury occurred at a location not associated with employment. Cases such as this are often resolved in court.

EMPLOYER'S LIABILITY POLICY

Employer's liability coverage is associated with workers' compensation insurance. It has the same format as a typical liability policy. This type of coverage protects the employer against any obligations the employer may have under the workers' compensation law. If an injured worker decides to bring suit under the law of negligence rather than applying for workers' compensation benefits, the policy will provide protection for the employer. However, as a practical matter, an injured worker can rarely file a successful suit against the employer; this type of suit can be more successfully pursued by the survivors of the worker.

KEY-MAN INSURANCE

Key-man insurance is insurance obtained on company principals. This type of policy insures a company against the heavy losses that can result from the untimely death of a principal. This is simply a large life insurance policy that is owned by the company (the benefits are tax-free). The benefits from such a policy can be used in several ways, including the following: (1) buying the deceased's interest in the firm, (2) providing a means for the company to continue paying a salary to the family, (3) ensuring the financial ease of getting a successor. The essential purpose of key-man insurance is to insure against the loss of profits if a business is suddenly deprived of the managerial skills and experience of an important party. The policy may include coverage for disability as well as death. Coverage benefits should lighten the financial shock of the loss of a principal's services and make it easier to obtain the services of a successor who is qualified. Key-man insurance increases the security of partnerships and corporations.

COMPREHENSIVE GENERAL LIABILITY

Liability insurance provides protection from third-party lawsuits, including property damage and bodily injury. Liability is an obligation imposed by law. It can be incurred from several types of incidents. The most typical sources are as follows:

Injury to nonworkers on the site (caused by commissions or omissions of the prime contractor).
Contingent or indirect liability (acts of parties such as subs for whom the prime is responsible).
Damage caused after a project is completed or accepted.
Damage caused by the contractor's mobile equipment.
Injury to employees (liability may result if benefits under workers' compensation are not granted).

Liability of the contractor for personal injury or property damage of third parties exists when it is shown that the contractor caused or contributed to the conditions resulting in the accident, and that the contractor had a duty to address the conditions. The comprehensive general liability insurance is specifically for bodily injury and property damage sustained by third parties.

The general contractor should be careful in choosing the liability insurance coverage to be purchased by the company. Liability policies do not provide protection for damage to the contractor's own property. Liability coverage generally includes claims caused by operations performed for the insured by an independent contractor or a subtradesperson. Since the sources of liability are varied and since gaps and overlap are possible, it is generally advisable to have the same insurance carrier for all liability coverage. Since the owner is at greater risk if the contractor has insufficient or no liability coverage, it is common to include contract requirements for such coverage. The following provision is typical:

> The Contractor shall carry such public liability and property damage insurance that will protect the Contractor and the Owner from claims for damages for bodily injury, including accidental death, as well as for claims for property damages, which may arise from operations under the contract whether such operations be by the Contractor or by any subcontractor or anyone directly or indirectly employed by either party. The limits of coverage shall be as stated.

Most contracts necessitate that liability coverage be obtained because the owner often shifts additional risk to the contractor. This shift in risk is done through indemnification clauses and less common provisions such as the following:

> In the event that any suits, actions, or claims are brought against the Owner, money equal to the "claim" amount may be withheld from payments due the Contractor under and by virtue of this contract as may be considered necessary by the Owner for such purpose. Money due the Contractor will not be withheld when the Contractor produces satisfactory evidence that adequate public liability and property damage insurance has been obtained.

The effect of this provision is that in the absence of insurance coverage, the owner regards the retainage and any other money due the contractor as a substitute for insurance. These funds may be utilized if the contractor does not have insurance coverage.

The contract will typically specify the type of insurance that the contractor must provide, the limits of the coverage, and the duration over which the insurance coverage is to be in effect. The ConsensusDOCS regarding insurance coverage (in part) are as follows:

> §10.2.1 Prior to the start of the Work, the Contractor shall procure and maintain in force Workers' Compensation Insurance, Employers' Liability Insurance, Business Automobile Liability Insurance, and Commercial General Liability Insurance (CGL). The CGL policy shall include coverage for liability arising from . . . operations . . . products-completed operations . . .
>
> §10.2.4 The Contractor shall maintain completed operation liability insurance for one year after acceptance of the Work, Substantial Completion of the Project or to the time required by the Contract Documents, whichever is longer. Prior to commencement of the Work, the contractor shall furnish the Owner with certificates evidencing the required coverage.

There are two basic types of liability coverage, namely, *occurrence-based* and *claims-made*. *Occurrence-based* provides coverage for claims that arise during the policy period, regardless of when the claim is actually filed. Under occurrence-based coverage, a party might be exposed to a health hazard during the policy period, suffer as a result of that exposure after the policy period has ended, and successfully obtain compensation from the insurance policy. Thus, the coverage exists even if the claim is filed years after the policy period has ended, as long as the hazardous exposure occurred during the active policy period. *Claims-made* coverage provides protection only for claims that are filed during the period when the policy is in effect; that is, claims-made policies provide less extensive coverage. Under a claims-made policy, a hazardous exposure and subsequent ailment will not be covered if the claim itself is filed after the policy period has ended. During the liability crisis of the mid-1980s, many policies were written as claims-made. Fortunately, insurance companies have generally prospered in the past decade and occurrence-based coverage is now again available.

The premiums for liability coverage are paid for in various ways. Liability coverage is experience-rated, and if all coverage is with one carrier, a good record on one policy may offset a bad record on another. Liability coverage may include exposures, perils, and hazards. However, liability policies are subject to exclusions, and these should be known and understood.

The Commercial General Liability policy states limits in terms of a "combined single limit" for bodily injury and property damage combined. The policy indicates multiple limits that apply and are shown on the policy as follows:

Per Occurrence Limit: $1,000,000
General Aggregate Limit: $2,000,000
Products/Completed Operations Aggregate Limit: $2,000,000
Personal Injury: $1,000,000

The per occurrence limit is the most the policy will pay for any one loss, with the commonly accepted limit being $1,000,000. The general aggregate limit, which is typically twice the per occurrence limit, is the most the policy will pay, regardless of the number of occurrences, during the policy period. The general aggregate applies only to claims not falling within the products/completed operations coverage. As noted, there is a separate aggregate limit for products/completed operations claims, and this is also often stated as being twice the per occurrence limit.

Some contracts state that liability coverage is required, but do not indicate the limits of coverage. Good judgment must be used by the contractor to determine the limits of coverage needed. When the limits of coverage are stated in the contract, the following provision is typical:

The public liability insurance shall be in the amount of at least $1 million for each occurrence, and at least $2 million in the aggregate.

Just as there are limits of liability coverage, there are also provisions for deductibles. Higher deductible amounts reduce insurance premiums and also place greater responsibility on the contractor. There are several basic types of coverage included in the standard commercial general liability policy.

Premises/Operations Liability

Most liability claims in the construction industry fall under the category of *premises/operations*. The premiums are levied on the basis of the amount spent on payroll. Liability under this coverage can come from two sources: operations and premises.

Operations coverage provides liability protection from injuries or property damage arising from business operations in progress.

Example. Suppose a carpenter dropped a piece of wood from a high-rise building and it fell on a visitor to the site. The operations coverage would respond.

Example. Suppose the overspray from a painting operation coated a car parked on the street. The operations coverage would respond.

The premises portion of the policy covers personal injury or property damage resulting from buildings or premises owned by or under the control of the insured.

Example. Suppose children are attracted to a contractor's project on a weekend or after working hours. If one of the "visitors" steps on a nail or is injured in some other way, the premises coverage would respond to the injury.

Contractor's Protective Liability and Owner's Protective Liability

The contractor's protective liability policy covers incidents caused by operations performed for the insured by an independent contractor, usually a subcontractor. This coverage is essentially designed to provide coverage for contingent liability. The premiums are based on sublet or subcontracted amounts. It provides the general contractor with automatic insurance for contingent or secondary liability resulting from sublet operations. The policy will respond when the subcontractor's primary limits are inadequate or void because of nonpayment of a premium. The owner's protective liability policy is obtained by the general contractor for the benefit of the owner. This provides coverage for the owner if the owner is named in a suit.

Example. Suppose a subcontractor's worker was working above the ground level and accidentally dropped a tool on a visitor who was walking below. The subcontractor would be covered by his or her general liability policy. The general contractor would be covered by the contractor's protective coverage (that portion above the limits in the sub's policy). The owner would be covered by the owner's protective liability coverage (that which is not covered by the contractor's protective coverage).

Example. Suppose a visitor on a construction site was injured when she tripped on a subcontractor's welding lead. The claim was for $33,000. If the subcontractor's limit of liability was only $25,000, the remaining $8,000 would be sought from the contractor's liability policy. This policy would function as if the general contractor had obtained an umbrella policy on the subcontractor's liability policy.

Completed Operations and Product Liability

The completed operations portion of this two-part coverage covers incidents arising from completed or abandoned operations caused by an occurrence away from premises owned or rented by the insured. The product liability portion covers incidents arising from insured products caused by an occurrence away from the insured's premises and after physical possession of the product had been relinquished to others.

This coverage takes effect when the *premises/operations* insurance coverage ends. When does this occur? In general, it occurs when all operations to be performed by or on behalf of the insured contractor under the contract have been completed. It can also occur when the portion of work in question has been put to its intended use. Completion, not acceptance, is the usual guideline.

Example. Suppose materials are delivered to a job site. The materials are then installed in the project. After six months the materials are found to be defective, and the defect was the direct cause of damage to other materials in the structure. The manufacturer will be responsible for replacing the defective materials. The damage caused to the other materials is covered by the manufacturer's product liability insurance.

Example. Suppose the construction of a retail store was recently completed. Shortly after the store opened for business, a ceiling tile fell on a customer. The completed operations coverage would respond to the injury.

Example. After a building was completed and occupied by the owner, it became apparent that the plumbing installation in one area of the structure was defective. The piping system failed and caused damage to some of the drywall surfaces. The mechanical contractor was forced to replace the faulty piping system and repair the wall damage. The mechanical contractor's completed operations coverage paid for the damage to the drywall. Note that the piping system was *not covered.*

Contractual Liability

As the term implies, *contractual liability* is liability that is assumed by contract. Contractual liability is often liability that would be someone else's liability in the absence of the contract wording. In order to assess the risk, it is important to examine the contract carefully. The "hold harmless" agreement in the contract is where this liability is conveyed to the contractor. There are three basic types of hold harmless agreements.

Limited Form Indemnification (Limited Hold Harmless Agreement)

Under this form of contractor's negligence agreement, the owner is held harmless for claims caused by operations or negligence of the contractor or subcontractors. This coverage may be for damage to property due or claimed to be due to negligence of the contractor or the subcontractors, employees, or agents of the contractor. This wording is not unduly harsh, because the contractor is generally considered liable for such causes in the absence of this type of contract provision.

In other words, this form of indemnification does not give the owner any additional protection. The following is an example of such a provision:

> The Contractor shall indemnify and save harmless the Owner from all suits, actions, or claims of any character brought because of any injuries or damages received or sustained by any person, persons, or property due to the operations of the Contractor; or because of or in consequence of any neglect in safeguarding the work; or through use of unacceptable materials in constructing the work; or because of any act or omission, neglect, or misconduct of the Contractor.

The ConsensusDOCS are balanced, in that the contractor indemnifies the owner for the negligent acts or omissions of the contractor, and the owner indemnifies the contractor for the negligent acts or omissions of the owner. These provisions state the following:

> §10.1.1 To the fullest extent permitted by law, the Contractor shall indemnify and hold harmless the Owner, the Owner's officers, directors, members, consultants, agents and employees, the Architect/Engineer and Others (the Indemnities) from all claims for bodily injury and property insured under Subparagraph 10.3.1, including reasonable attorneys' fees, costs and expenses, that may arise from the performance of the work, but only to the extent caused by the negligent acts or omissions of the Contractor, Subcontractors or anyone employed directly or indirectly by any of them or by anyone for whose acts any of them may be liable. The Contractor shall be entitled to reimbursement of any defense costs paid above Contractor's percentage of liability for the underlying claim to the extent provided for under Subparagraph 10.1.2.
>
> §10.1.2 To the fullest extent permitted by law, the Owner shall indemnify and hold harmless the Contractor, its officers, directors, members, consultants, agents and employees, Subcontractors or anyone employed directly or indirectly by any of them or anyone for whose acts any of them may be liable from all claims for bodily injury and property damage, other than property insured under Subparagraph 10.3.1, including reasonable attorneys' fees, costs and expenses, that may arise from the performance of the work by Owner, Architect/Engineer or Others, the Owner shall be entitled to reimbursement of any defense costs paid above Owner's percentage of liability for the underlying claim to the extent provided for under Subparagraph 10.1.1.

Intermediate Form Indemnification
(Intermediate Hold Harmless Agreement)

Under this form of joint negligence, the owner is held harmless when both parties are negligent. With this form of indemnification, the contributory negligence of the owner is waived and the entire burden for claims rests with the contractor. This does have an advantage to the owner, in that joint liability for a loss does not result in a dispute about the proportionate share of the responsibility to be borne by each party; that is, the contractor is fully responsible when both the owner and the contractor are at fault. The contract wording may be as follows:

> The Contractor shall indemnify and save harmless the Owner from all suits, actions, or claims of any character, resulting from injuries or damages received or sustained by any persons or to any property as a result of, in connection with, and pursuant to the execution and performance of the contract, whether such injuries to persons or damage

to property are due or claimed to be due to the negligence of the contractor, subcontractors, the owner, architect, engineer, or their agents. The Contractor shall be responsible for all such claims except such claims in which the injury or damage shall have been occasioned by the sole negligence of the Owner.

Broad Form Indemnification (Sole Negligence of Indemnitee)

Under this form, the owner is held harmless against all losses caused by or contributed to by the owner, architect, or others. Essentially, the contractor is asked to pick up the tab for anything that might happen. This passes the owner's liability to the contractor. This is potentially very harsh on the contractor and may be unenforceable in some instances. The wording of such a provision might state: "The Contractor agrees to indemnify, hold harmless, and defend the Owner from and against any and all liability. . . The provisions of the paragraph shall apply regardless of the fault, negligence or strict liability of the Owner." This form is rarely used since, in most cases, it goes against public policy and is found unacceptable and unenforceable by most courts.

Exclusions and Limitations

Although general liability policies give broad coverage, there are exclusions and limitations. These exclusions and limitations tend to narrow the scope of coverage, but are common when specified hazards are better insured elsewhere, or when certain risks are prohibitively expensive, or when some hazards are considered to be completely uninsurable. Examples of exclusions are as follows:

Losses not caused by an occurrence; a specific incident must exist or a loss must be traceable to a definite time, place, and unexpected cause.

Losses for which there is no tangible property damage.

Drivers on an air track drill; they will not be covered by automobile liability.

Watercraft away from the premises; this is best handled in a marine policy.

Aircraft, usually; this is best handled in an aircraft liability policy.

Employee injuries; coverage is already provided in the workers' compensation policy.

Nuclear hazards; this is a special risk that may be covered through a pool of carriers.

War risks, a standard exclusion.

Contractual liability; this must be defined in the policy for coverage.

Faulty design, maps, drawings, or specifications.

Damage to property out of which the occurrence arose (coverage exists for the repair of damage to other property but not to the property causing the damage).

Expenses of withdrawal or recall of the insured's product or work (product defects are not covered).

Failure to perform as intended (the coverage does not guarantee performance; it covers only losses caused by the malfunction).

Professional liability.

When obtaining liability coverage, it is advisable for contractors to obtain all policies from one insurance company. This will reduce the chance of overlaps or gaps in coverage. The exclusions will be more clearly understood if only one insurance company is involved.

A composite rating can be obtained for liability coverage. Instead of premiums being based on the various factors (payroll, contract amount, subcontracted amount), a composite rate is determined. The premium is basically the same, but it is based on only one factor. This simplifies bookkeeping and the estimation of insurance costs for projects that are being bid.

The specific coverage that is required on a project should be clearly stated in the contract. If the owner is remiss in requiring a particular type of coverage, the contractor may make a similar oversight. A provision such as the following will clarify this for the contractor:

> The liability policy shall include coverage for bodily injury, broad form property damage (including completed operations), personal injury (including coverage for contractual and employee acts), blanket contractual, independent contractors, products, and completed operations. Further, the policy shall include coverage for the hazards commonly referred to as XCU (explosion, collapse, and underground). The products and completed operations coverage shall extend for one year past acceptance, cancellation, or termination of the work.

The coverage requirements may also include special coverage, such as asbestos or railroad coverage. These coverages will be unique to the project. It is important that the contractor be fully aware of such special coverage requirements. For example, the cost of coverage on projects where asbestos problems must be abated may exceed 10 percent of the contract amount. An oversight in including such costs in a bid can be devastating to a contractor.

Umbrella Excess Liability

Umbrella excess liability coverage extends the limits of liability coverage. Umbrella coverage is one of the better insurance buys. This is essentially a separate insurance policy that uses the limits of the basic liability policy as the deductible amount. Sufficiently high umbrella limits can eliminate, for all practical purposes, the question of the adequacy of the amount of insurance carried.

The primary type is a policy that provides coverage in excess of the existing primary insurance. The existing policies are not eliminated, but become the underlying layer above which the umbrella provides excess limits for the same hazards insured under the primary policy.

The other type of umbrella coverage is for excess coverage on self-insured hazards. This is for self-insured firms. Under this plan, the deductible amount is generally much lower than the limits of liability coverage mentioned earlier. The self-insured contractor maintains responsibility for a deductible amount on the umbrella policy of about $10,000 to $25,000 per occurrence. This form of umbrella coverage is used for property damage liability, completed-products liability, personal injury liability, automobile liability, and so on. The umbrella policy goes into effect as soon as the loss per occurrence exceeds the deductible amount. The basic liability policy may have limits of $500,000 for bodily injury and $250,000 for property damage, and the umbrella coverage will be stated at some limit beyond that. With the umbrella policy, this coverage can be extended, usually in increments of $500,000. The umbrella policy will kick in if the losses exceed the limits of the basic liability policy.

With liability coverage in hand, the contractor is afforded protection from third-party suits. However, the contractor will not have protection for self-owned equipment. Other types of insurance must be obtained.

BUILDER'S RISK OR COURSE OF CONSTRUCTION INSURANCE

The contractor will want to have insurance on a project as it is being built, because a project is susceptible to a variety of losses while it is under construction. The foremost threat to most projects is fire. *Builder's risk insurance* is the type of coverage the contractor will want to acquire. This will cover the structure during construction. The general contractor is responsible for obtaining this insurance coverage, because the prime contractor is responsible for the entire project until it has been accepted by the owner. Thus, builder's risk or basic fire insurance is obtained. Builder's risk coverage may be required by provisions such as the following:

> The insurance required of the Contractor shall include an All-Risk Builder's Risk policy which shall provide fire and extended coverage, vandalism, and malicious mischief coverage for an amount equal to 100 percent of the completed value of the entire project and shall be written in the Owner's and Contractor's name. Such coverage shall be kept in full force and effect until all work is fully completed and accepted by the Owner.

Several types of coverage are available. The policy can be written in the name of the contractor, the owner, or a subcontractor. The specifications generally spell out the contractor's specific obligations. The contractor must make sure the insurance protection is in force for the full period of time during which the contractor is exposed to risk of losses. For materials and equipment, the policy should provide coverage from the time the material or equipment leaves the manufacturer. This loss potential continues to exist after the materials have been installed in the project. The policy should continue in force until the contractor is no longer responsible. This will occur when (1) the contractor no longer has an interest in the project, (2) the project has been accepted by the owner, (3) the owner occupies the structure, or (4) the policy expires.

The general contractor generally provides the risk insurance. However, the owner may provide this coverage if there are several contractors on the project. If the owner provides the coverage, the contractor should still review the policy to make sure everything is covered. If some gaps in coverage are noted, the contractor may want to augment the coverage obtained by the owner.

Premiums on Builder's Risk

There are two basic ways in which builder's risk premiums are assessed. The first is the *reporting form.* This is an open-ended policy that covers all the contractor's work (all projects under construction). New projects are added automatically as they are obtained. The insurable value of the work in place is determined for all jobs covered, and this information is reported monthly. The reports must be accurate or the contractor may become liable for losses, because the insurance company may deny full coverage if information has been falsified in the reports. This method of assessing builder's risk premiums is used very infrequently. It is particularly desirable or advantageous when the project value is very low during the initial stages of construction.

The second way of assessing and making premium payments is the *completed value* form. Under this type of format, the rates are adjusted on individual projects to suit the degree of coverage provided. The basic premium is based on an assumed constant increase in the value of the project. This policy is taken out at the project start. There is a separate policy for each project, and a lump sum is paid before construction begins. The completed value form is very common. The policy is based on the total project value, while the premium is based on a reduced average value of the project for the duration of construction. As a result, the overall premium provides coverage for the project duration for one-half of the project value. There is no reporting required, as a straight-line increase in value is assumed. This method of assessing premiums is not advantageous to the contractor if the construction work starts slowly, or if some of the more expensive items are added very late in the project.

Regardless of the method used to assess premiums, these policies will typically have a provision for a deductible amount of $500 to $1,000. Contractors who are good negotiators find that the premium rates can vary without reducing the coverage.

With careful planning, the premiums (based on the contract value) may be reduced. This can occur if the following steps are taken:

Exclude excavation and sitework and the value of the foundation.
Exclude items such as removal of existing structures.
Raise the deductible amount.
Use lower limits of coverage, but be certain the coverage is adequate.

In a similar manner, the premiums may be increased over what is typical. This can occur if the following steps are taken:

Include owner-furnished material.
Anticipate inflation costs in rebuilding after a loss.
Include the value of existing buildings on alteration jobs.

The premiums may also be increased due to bad builder's risk experience of the contractor. Depending on this experience, the actual premium may be increased by as much as 40 percent.

The objective is not primarily to pay a low premium, but to obtain all the coverage needed to protect the contractor from losses. To achieve this, the contractor must understand the potential losses on the project. For example, steel is susceptible to wind damage, wood is susceptible to fire, and excavation work is susceptible to liability.

Standard Builder's Risk

The standard builder's risk policy is a basic plan that protects projects only against direct losses caused by fire and lightning. This includes damages to the facilities and materials adjacent to or connected to the building and covers temporary structures, materials, machinery, and supplies. The policy also usually covers vandalism and malicious mischief such as pilferage, burglary, larceny, and theft.

At an added cost, the policy may be extended to include damage caused by windstorms, hail, riot, aircraft, nonowned vehicles, smoke, and so on. The policy usually excludes damage from floods, earthquakes, landslides, subsidence, and boiler explosions. The inclusions are specifically noted; risks that are not listed are generally not covered. The risk coverage may be void if the project is abandoned for 60 days.

All-Risk Builder's Risk Insurance

The title of this type of policy is a misnomer. It is not truly an "all-risk" policy; instead, the policy specifically notes the exclusions. This is a policy with broader coverage than the standard builder's risk policy. Although not all perils are covered, those that are not covered are noted in the policy. This type of insurance covers materials, apparatus, and supplies pertaining to the project before delivery, while in transit, after delivery at the job site, and after installation. The policy may also cover the contractor's tools and equipment. Exclusions from an all-risk policy may consist of freezing, explosion of steamboilers, glass breakage, rain, snow, earth movement, floods, and nuclear radiation.

The contractor generally should arrange for the builder's risk coverage, as the owner may exclude items for which coverage is desired, such as flood and earthquake. In addition, the owner may make the contractor responsible by contract, but may not specify or require insurance.

Installation Floater

The builder's risk policies discussed in this chapter are based on fire as the primary exposure. However, not all projects have high fire hazards. This is when an *all-risk installation floater policy* is most appropriate. It is much less expensive than the other risk policies.

After project completion, the builder's risk or installation floater policy is canceled. The owner should then have a permanent insurance policy in effect. Generally, the risk insurance is terminated when the certificate of substantial completion is issued. It is vital, however, that the policy not be terminated early.

EQUIPMENT FLOATER INSURANCE

An *equipment floater insurance policy* covers equipment that is not licensed. This includes equipment (pumps, air compressors, small tools, etc.) and off-road vehicles such as dozers, loaders, scrapers, graders, compactors, excavators, cranes, and air track drills. The equipment that is covered is that which tends to "float" from project to project. The major sources of loss are theft and vandalism. Other coverage that is generally included is damage from fire, landslide, collision, tornado, flood, explosion, windstorm, and overturning. Note that this type of coverage does not include a liability component. The equipment floater typically excludes registered vehicles (highway use), waterborne equipment, and damage incurred from the overloading of equipment (no coverage for human error). The policy can include owned, borrowed, leased, or rented equipment. Large equipment is usually insured at its depreciated value. There is usually coverage for wherever the equipment floats. When the total value of equipment in a fleet is above a maximum amount, the policy can be designed to include all equipment without an equipment schedule. That is, for large equipment "spreads," each piece of equipment is automatically covered when purchased.

The premiums are based on the equipment value, with adjustments made for the judged exposure to risk (work in a floodplain, rough terrain, etc.), the type of work done, the contractor's past loss experience, the contractor's reputation, the dispersion of work, specific exclusions of coverage, and specific inclusions of coverage; it can also be of an all-risk form. As a rule, a deductible amount will apply.

AUTOMOBILE INSURANCE

The contractor also needs to obtain conventional automobile insurance for all licensed vehicles operated on the public road and highway system. The coverage is similar to that obtained for privately owned cars. Particular coverages to consider include collision, comprehensive liability, and uninsured motorists. Coverage is often also available for emergency road service or towing, car rental, disability, loss of earnings, and so forth. Limits of $1,000,000 for bodily injury and property damage are indicative of the requirements imposed in many public works contracts. State law may also dictate the minimum coverage that must be obtained. Thus, in the absence of contract requirements, a prudent contractor will investigate mandatory insurance requirements for the states in which work is to be undertaken. Note that the insurance coverage may stipulate coverage for owned, rented, or nonowned automobiles and trucks. As with general liability coverage, liability limits are written on a combined single limit basis for both bodily injury and property damage.

CONTRACTOR-CONTROLLED INSURANCE PROGRAMS (CCIPs)

The increasing costs of insurance premiums have led some contractors to seek other, more innovative, approaches to control construction risks. One such method is a contractor-controlled insurance program (CCIP) in which the general liability, umbrella liability, workers' compensation, and builder's risk exposures of all subcontractors are covered under a single policy. This pooling of the risks results in a single insurance carrier and a policy that is managed by the general contractor. This "one-stop shopping" offers a number of advantages that account for its increasing popularity. Although owners have historically been the party obtaining wrap-up insurance, a recent trend is for some large contractors to obtain wrap-up policies. These CCIPs occur on more, smaller projects than has been common in the past with owner-provided wrap-up policies. OCIPs were generally employed on only very large projects. Projects valued as small as $50 million have been covered with CCIPs. The prudent contractors who provide wrap-ups are very proactive in loss control and can often realize significant savings or profits as a result of these ventures.

Since there is a single policy, with limited gaps in coverage and virtually no double coverage, the overall costs of CCIPs are competitive. The economies of scale result in an overall better buy than if the general contractor and all the subcontractors independently obtained their own insurance coverages. With a single policy, the general contractor has to be concerned with only one policy and the insurance carrier similarly has significantly reduced administrative costs; a win-win scenario is realized by the general contractor and the insurance carrier. If the contractor has a low loss experience on a project with CCIP coverage, the insurance underwriter will be quite willing to provide coverage for a subsequent project. In fact, the limits of coverage might even be increased, as well as the total coverage. Note that with a single insurance carrier, the contractor will be wise to insist on strong support from the company safety representative to assist in the goal of no project losses.

With a single insurance carrier, the attention of the insurance representative will be on potential losses. Thus, the attention might very well be focused on the work of a small subcontractor who might be considered at risk. This subcontractor might receive little attention if separate insurance policies were written on the project.

The contractor will be able to control some of the risk that is assumed versus the risk that is covered by insurance. For example, the contractor can make changes in the deductible amounts on the policy and might also elect to participate in a retrospective rating plan. Claims handling is also generally more efficient, as there is no bickering between different insurance carriers. The claims handling is focused on the merits of the claim, and, if valid, the benefits can be made available in a short time. Note that with CCIPs, insurance coverage can be made available to firms (small subcontractors) that would have difficulty in obtaining the coverage on their own.

CCIPs are not without their disadvantages, or potential disadvantages. It has been alleged that a subcontractor might not be inclined to work safely if covered under the insurance umbrella of the general contractor. This is a serious concern, but it is also one that can be addressed. In the selection process, the general contractor will be wise to include past and projected safety performance as a selection criterion for subcontractors. Of course, the subcontract agreement should similarly address the need for the subcontractor to implement an aggressive safety program. Subcontractors with historically strong safety records will be less inclined to prefer CCIPs, as they cannot fully take advantage of their safety records when pursuing future work, unless this safety record is used as a selection criteria.

CCIPs also are said to require a great deal of administration effort on the part of the general contractor. The general contractor supposedly also assumes a stronger role in the safety arena with the use of CCIPs. This increases the contractor's liability exposure. The general contractor also incurs a decidedly larger workload as a result of CCIPs (broader coverage of the policy), but this effort must be considered in light of the benefits. If the general contractor determines that the effort is minimal, or at least less than the benefits realized, CCIPs will be favored. On the bright side, a general contractor with a strong focus on project safety will often be able to take the credit for good safety performance, a plus for the workforce.

CERTIFICATE OF INSURANCE

A *certificate of insurance* is a means by which a contractor can demonstrate to the owner that specific forms of insurance have been obtained. It is customary to require the contractor to show that there has been full compliance with the insurance requirements. The following is a typical provision:

> Before any work embodied in the contract will be permitted to be performed, the Contractor shall furnish two copies of a certificate of insurance as evidence that the required liability insurance has been obtained.

Since this certificate is the only assurance the owner has that the contractor has obtained the required insurance, it is advisable to require contractors to furnish a certificate of insurance. Some owners require the certification to be filed on a form furnished by the owner. While this may be inconvenient to the insurance carrier and the contractor, it does eliminate the need to interpret the different types of certificates that might otherwise be prepared. Similar certificates of insurance should be required of the various subcontractors by the general contractor. The general contractor's insurance carrier may require that the certificates of insurance be provided by each of the subcontractors, or the cost of such coverage may become the burden of the general contractor. The certificate of insurance will typically include information such as: the name and address of the insurance company; name and address of the insured; type of policy (workers' compensation, general liability, property damage, excess liability, automobile liability); policy number; policy period; limits of liability; date of issue; signature of authorized insurance representative; and a statement about the terms of cancellation or major changes in the policy, especially the time frame in which advance written notice must be provided to the owner for cancellation.

PROVISIONS ON INSURANCE CANCELLATION

Insurance coverage is valuable only as long as the insurance policy is in effect. The cancellation of an insurance policy can be devastating if a loss occurs during a period for which no coverage exists. Although the owner may not be in a position to stipulate that insurance policies cannot be canceled, added protection is assured if the owner requires prior notification of such a cancellation. Provisions such as the following should be considered for inclusion in all contracts:

> All insurance coverages required in this contract must include an endorsement whereby the insurer agrees to notify the Owner at least 30 days prior to nonrenewal, reduction, or cancellation.

Essentially, it is advisable to require that prior notice be given of any major change in coverage. Failure of the contractor to provide ongoing insurance coverage may also be addressed in the contract. The following provision addresses the consequences that may arise if an insurance policy is allowed to lapse:

> In the event the Contractor fails or refuses to renew any insurance policy, or if any policy is canceled, terminated, or modified so that the contractual insurance requirements are no longer being maintained, the Owner may refuse to make payment of any further monies due under this contract or refuse to make payment of monies due or coming due under other contracts between the Contractor and the Owner.

This provision should send a clear message to the contractor that all insurance coverage must remain intact for the full duration of the contract.

SUMMARY OF INSURANCE COVERAGES

Insurance coverage, to address every potential loss, involves the purchase of a number of types of insurance policies. The information in figure 17.1 helps to summarize the basic insurance needs of most construction projects. With this information it should be easy to formulate the type of coverage required for the following scenario. Suppose a worker is operating a front-end loader with bad brakes. The loader is being operated on a slope and it begins to roll down the incline. In the loader's path is a vehicle belonging to a job visitor. The loader strikes the car and continues to roll toward the plate glass window that has just been installed in the new storefront. The operator jumps from the loader when his efforts to stop the loader are futile. The operator suffers a sprained ankle as a result of his jump. The loader crashes through the plate glass window and comes to a stop when it strikes a block wall. Four potential insurance policies may be activated in this scenario. The injured operator would be covered by workers' compensation insurance. The damaged vehicle would be covered by the *premises/operations* portion of the general liability policy. The damage to the storefront, the project being constructed, would be covered by the builder's risk policy. If the loader sustained any damage, it would be covered by the equipment floater policy. Note that these are the primary coverages, but subrogation rights might result in subsequent lawsuits between the insurance companies.

Type of Policy	Type of Loss Covered	Party Generally Acquiring Policy
Workers' compensation	Injuries to employees	Employers (GC, subs, owner, et al.)
Employer's liability	Employer liability for employee injuries	Employers (GC, subs, owner, et al.)
Key-man insurance	Death of company principal	Company
Builder's risk	Damage to project being constructed	General contractor or owner

Commercial general liability
    Premises/operations
    Owners and
    contractor's protective
    Completed
    operations and product
    Contractual
	Injury and property damage to third parties	General contractor, subcontractors

Umbrella excess	Injury and property damage to third parties	General contractor, subs where applicable
Equipment floater	Loss or damage to equipment	Equipment owner
Automobile	Loss or damage to an automobile	Automobile owner

FIGURE 17.1
Summary of construction insurance coverages.

REVIEW QUESTIONS

1. What is subrogation, and why is it important to insurance companies?
2. Under what conditions might a construction firm receive dividends on its insurance policy?
3. What are reserves? How do reserves influence the experience modification rating of a firm?
4. Describe the determination of and importance of an experience modification rating to a construction company.
5. Discuss the merits of comparing the loss ratio with the experience modification rating of a firm to measure the effectiveness of the firm's loss-control program.
6. Under what circumstances might an owner decide to use wrap-up insurance? What are some disadvantages of having wrap-up insurance on a project?
7. What is the basis of workers' compensation insurance?
8. In what type of situations might a firm be best advised to purchase key-man insurance?
9. An accident on a construction project was described as follows: A general contractor's worker was drilling holes on the third floor of the project. The vibration of the saw caused a board to work its way to the edge of the building and fall off. The board struck a mason who was working on the next floor down. The injured worker fell against a stack of loose bricks. These bricks then fell from the scaffolding, breaking a plate glass window on the first floor

and striking a car that was parked on the public street below. Discuss the different types of insurance that might be involved in this incident.

10. Describe the essential differences between intermediate form and broad form indemnification clauses.

11. Discuss the differences in the payment of premiums in the reporting form and the completed value form of builder's risk insurance.

12. Examine the AIA documents and the ConsensusDOCS that are included in the Appendix and determine the similarities and differences in the provisions related to indemnification. Specifically examine AIA provisions §3.18.1 and ConsensusDOCS §10.1 and §10.1.2.

13. Examine the AIA documents and the ConsensusDOCS that are included in the Appendix and determine the similarities and differences in the provisions related to the contractual insurance requirements. Specifically examine AIA provisions §11.1.1, §11.1.2 and §11.1.3, and ConsensusDOCS §10.2.1 and §10.2.4.

18

SUBCONTRACTORS AND SUBCONTRACTS

SUBCONTRACTORS ARE VERY important to the successful completion of most construction projects. Even on fairly simple building projects, it is common for as many as 20 to 30 subcontractors to be employed; on larger and more complex projects more than 100 subcontracts may be awarded. Subcontractors bring to the project unique skills and talents that the general contractor typically does not possess. On a building project the general contractor often has the necessary in-house capabilities to perform concrete work and major structural work, but lacks the ability to undertake the specialized interior work. On a project such as a water treatment plant, the general contractor may have the in-house capability to perform the major mechanical work, but lack the expertise to perform the structural work. In some cases the general contractor may be able to perform both functions. Regardless of the general contractor's skills, portions of virtually every project will be subcontracted to firms that possess specialized skills. Because of the use of specialty firms or subcontractors, the role of general contractors is broadened considerably in the area of management. The various subcontractors must be carefully coordinated so that projects are constructed in an efficient manner.

GENERAL CONTRACTOR-SUBCONTRACTOR RELATIONSHIP

Most construction projects are undertaken through contracts that owners enter into with general contractors. While the general contractor on a project presents one contracting entity with which the owner deals on a day-to-day basis, much of the actual construction work is performed by specialty contractors or subcontractors. Despite the obvious importance of subcontractors on most construction projects, subcontractors have their agreements with the general contractor. This contractual arrangement is preferred by owners to assure that one party assumes the overall responsibility for

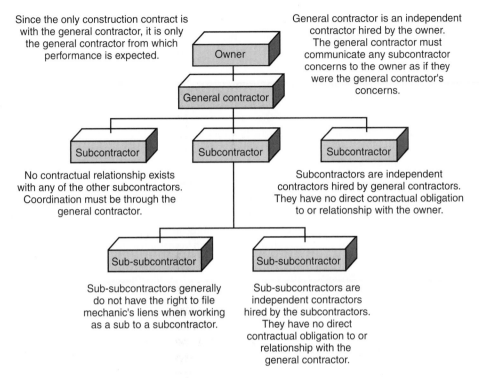

Since the only construction contract is with the general contractor, it is only the general contractor from which performance is expected.

General contractor is an independent contractor hired by the owner. The general contractor must communicate any subcontractor concerns to the owner as if they were the general contractor's concerns.

Owner

General contractor

Subcontractor

Subcontractor

Subcontractor

No contractual relationship exists with any of the other subcontractors. Coordination must be through the general contractor.

Subcontractors are independent contractors hired by general contractors. They have no direct contractual obligation to or relationship with the owner.

Sub-subcontractor

Sub-subcontractor

Sub-subcontractors generally do not have the right to file mechanic's liens when working as a sub to a subcontractor.

Sub-subcontractors are independent contractors hired by the subcontractors. They have no direct contractual obligation to or relationship with the general contractor.

FIGURE 18.1
Typical subcontractor relationships.

project completion. The owner does not typically possess the managerial skills to coordinate the various contracting parties (figure 18.1). If the owner assumed this role, the owner would be held accountable for the successful completion of the project, and this is specifically what the owner wants to avoid in most instances. With the general contractor as the only party with whom the owner has an agreement, the owner can place a stronger reliance on the general contractor for project completion. Since the subcontractors have their contracts with the general contractor, this arrangement effectively insulates the subcontractors from the owner.

Although the subcontract agreement is specifically between the contractor and the subcontractor, the owner typically will reserve the right to reject the use of a specific subcontractor or material supplier. While this option may seldom be exercised, it is apparent that the contractor could be at some risk if the second subcontractor's bid was considerably above the one that was originally selected by the contractor. The ConsensusDOCS address this issue as follows:

> As soon after the award of this Agreement as possible, the Contractor shall provide the Owner and if directed, the Architect/Engineer with a written list of the proposed Subcontractors and significant material suppliers. If the Owner has a reasonable objection to any proposed Subcontractor or material suppliers, the Owner shall notify the Contractor in writing. Failure to promptly object shall constitute acceptance.

Subcontractors are in a somewhat precarious position in that they are contractually bound to answer to the general contractor, but at the same time must perform their work to the satisfaction of an owner with whom they have no direct contractual relationship. Although the subcontractor has no agreement with the owner, a privity relationship can be created if the owner's representative directs the actions of the subcontractor. This could happen if the owner's representative inadvertently directed the work of the subcontractor. Because of the implications of this, many owners are careful to fully indoctrinate their representatives about the pitfalls of getting too closely involved in the work of the general contractor and the subcontractors.

As was mentioned earlier, the general contractor is hired for the most part to provide managerial services. A considerable effort is required to monitor job progress and keep the various parties informed about the need for their services on specific items of work. Despite the strong managerial role played by general contractors, it is generally accepted that the general contractor should not function solely as a broker. That is, the general contractor should not subcontract all the work. If a project is brokered, the general contractor is assumed to be in a position to benefit most in a financial sense if his or her costs are kept to a minimum. The owner does not want the general contractor to simply subcontract 100 percent of the project, and then step back to let the various subcontractors determine how to complete the project. Since the subcontractors have no direct contractual link with each other, it would be very difficult to establish harmonious agreement between all parties. Essentially, one party, such as the general contractor, with contractual links to all the subcontractors, is needed to coordinate the overall project. The direct involvement of the general contractor in day-to-day coordination is assured to a greater degree when he or she has a vested interest in the expedient completion of a project. Consequently, some owners stipulate in the notice to bidders that a minimum, such as 20 percent, of the project must be performed directly by the forces of the general contractor. The following example provision might be included to assure direct involvement by the general contractor in the construction of the project:

> The Contractor shall not sublet, sell, transfer, assign, or otherwise dispose of the contract or any portion thereof, or entitlement to the contract, without written consent of the Owner. With the Owner's consent the Contractor will be permitted to sublet a portion of the work but shall perform with in-house forces work which amounts to not less than 50 percent of the total contract amount.

OWNER-SUBCONTRACTOR RELATIONSHIP

Contractually, the subcontractor will enter into an agreement with the general contractor. In the subcontract, the general contractor will establish a relationship with the subcontractor so that the subcontractor is an independent contractor. This contractual association does not include the owner. Therefore, the subcontractor has a direct responsibility to the general contractor but not to the owner. At the same time, the work of the subcontractor must be approved by a representative of the owner. In effect, the work of the subcontractor is presented to the owner as if it were the work of the general contractor. When payment is made for the work performed by the subcontractor, the payment is made by the owner to the general contractor. Thus, the subcontractor has no direct link to the owner in matters of acceptance of work or the receipt of payment. This arrangement tends to be undesirable at times. Informal communication links may develop at the project level to improve the flow of information. Because of the contractual relationships, however, rigid adherence to contract requirements is often advised.

The contractual distance placed between the subcontractor and the owner may appear to be a disadvantage in many situations, but sometimes it is a blessing. For example, the independence of the subcontractor may bar suit by the owner. This was shown in *Manor Junior College v. Kaller's Inc. and John J. Spencer, Roofing, Inc.* (507 A.2d 1245). Manor Junior College and Kaller entered into a written contract for the installation of a new roof on one of the college buildings. Kaller then entered into a verbal subcontract agreement with Spencer in which Kaller would provide all the materials and Spencer would provide most of the labor to replace the roof. After the old roof had been removed, but before the replacement roof was installed, a rainstorm caused considerable damage to the interior of the building. The college felt that insufficient preventive measures had been taken. It ordered Kaller off the job and hired another roofer to finish the roof. The college then brought suit against both Kaller and Spencer for breach of contract. The suit against Kaller was for negligence; the suit against Spencer was for breach of implied warranty and for "services rendered in an unworkmanlike and negligent fashion." The court ruled that Kaller was liable to the college and would have to pay damages. Spencer, by contrast, was not liable for any damages. The court agreed with Spencer's argument that the college was not a third-party beneficiary of the oral contract between the contractor and the subcontractor, that an implied warranty of reasonable workmanship does not apply to a person who renders construction services but does not sell a product, and that the college never stated "negligence" as a cause of action against Spencer, only "negligent work."

WHAT WORK IS SUBCONTRACTED?

It is understandable that the general contractor will subcontract those portions of the project for which no in-house capability exists. However, there are other reasons why the general contractor may award a subcontract. For example, the general contractor may have the in-house expertise to perform masonry work. If several of the firm's current projects entail a considerable amount of masonry work, it may be desirable to award a subcontract for masonry on a particular project to avoid being overextended. If the masonry portion of a project presents a higher risk because the work must be performed above a major traffic artery, the job may be subcontracted as a simple means of reducing the risk. Work involving blasting is often subcontracted for this reason. Asbestos removal similarly presents a unique risk that most general contractors deal with by obtaining the services of subcontractors. A contractor may undertake a project in an unfamiliar labor market in which a subcontract with a local specialty contractor may be an expedient solution. The validity of the warranty on the equipment being supplied may be conditioned on proper installation of that equipment. The vendor may even stipulate that the warranty is valid only if the vendor installs the equipment. In some cases, notably on public works projects, the owner may require that a specified percentage of the work be awarded to minority-owned businesses and/or women-owned businesses.

HOW ARE SUBCONTRACTORS SELECTED?

On public works projects it is customary for the general contractor to receive bids from a variety of specialty contractors. The names of the general contractors who have obtained the bid documents for a particular project are generally made available to the public on most public works projects. Thus, it is common for the general contractor to receive price quotations or bids from a variety of different specialty contractors. Many of the bids may be from firms with which the general contractor is not familiar. When this occurs, the general contractor must exercise some judgment. For example, if the lowest price from an unfamiliar firm is considerably below the prices received from known firms, the general contractor may decide to use a price submitted by a familiar firm. This is not an easy decision to make, as other general contractors are probably receiving the same low price. For a major work item, using the second lowest price may keep the general contractor from submitting the lowest bid. However, the primary objective should not be to submit the lowest bid, but to submit the lowest price at which a reasonable profit can be made. Often a contractor will try to bid lower than the competition. This is contrary to the primary objective, unless the contractor can bid lower than the competitors and still satisfy the criterion of making a reasonable profit.

If the general contractor enters into an agreement with a specialty contractor with whom a previous working relationship was positive, there is greater assurance that the undertaking will be successful. If the general contractor has not previously worked with a specialty contractor, it is common to conduct a reference check before entering into a binding agreement. Such reference checks will be made with general contractors with whom the specialty contractor has completed similar projects. These general contractors can attest to the ease with which work was coordinated and the general responsiveness of the subcontractor to the needs of the project. An owner or the owner's representative may be contacted to ascertain the standards of quality typically attained by the firm. Surety underwriters and bankers may also provide valuable information regarding the soundness of a company. In some cases these checks are performed as a means of prequalifying specialty contractors. Such reference checks require a considerable amount of time. Since many subcontractors submit their bids shortly before the bid submittal deadline, the general contractor probably will not be able to make any reference checks before finalizing the bid.

SUBMITTING A BID FOR A SUBCONTRACT

During the bidding phase problems can develop from which the project may later suffer. A specialty contractor may be desperate to be awarded a particular contract. In fact, the firm may be so eager to get the job that it is unwilling to simply submit its best price. The firm wants to submit its price because it knows that its bid is lower than the competing bids. To know this with certainty, the firm must obtain the bid prices of its competitors. Obviously, this is not public information. This information about competing bids may be obtained as inside information divulged by an employee of a general contractor. The information that is given is typically of a general nature, such as "Your bid is second low at this point" or "Your bid needs to come down by more than $1,000." In any case, the specialty contractor is given sufficient information to know, at least in rough terms, the relative competitiveness of the bid. Obviously, if the bid is not the lowest bid, it will be reassessed and a reduced quote may be submitted. Needless to say, this is regarded as an unethical practice. Such prebid bid shopping requires cooperation between a specialty contractor and the employee of a general contractor. The practice is often initiated by the specialty contractor; however, the information can also be voluntarily divulged by a general contractor who hopes to be the sole recipient of a reduced price as the deadline for bid submittal approaches. This practice results in the overall reduction of the bid prices submitted to the owner; that is, the owner benefits from prebid bid shopping through the receipt of lower prices. The end benefit to the owner is debatable, because the quality of work may suffer if profits are sufficiently lowered by the practice.

Prebid bid shopping is known to occur in the construction industry. This does not mean that the practice occurs on every bid. The actual frequency of occurrence is difficult to assess, as this type of information is not freely disseminated among firms. A general contractor may suspect that a second bid submitted by a firm is the direct result of that firm obtaining information about the bids of competitors, but the bid may also simply be an honest recalculation.

Because of prebid bid shopping, many quotations to general contractors are delayed until a short time before the deadline for bid submittal. Thus, decisions to shave prices are made under extremely stressful conditions. Subcontractors can easily make errors under these circumstances. Although a subcontractor may obtain the subcontract award as a result of bid shopping, ultimately the parties that suffer to the greatest extent from prebid bid shopping are the specialty contractors. Firms that aggressively engage in this practice will realize small profit margins and may feel compelled to cut corners to compensate for the concessions made to win the contract.

General contractors do not benefit from prebid bid shopping as a rule. The reduced prices are invariably submitted to all known general contractors who are bidding on a particular project, resulting in common use of the reduced prices. Since competing firms will probably use them in their own bids, general contractors often feel compelled to use these lower prices, even though it may mean working with a firm that will try to preserve its profits by compromising its job performance.

If bid shopping occurs prior to bidding, the owner receives the benefit of the practice, unless subsequent quality standards are reduced to accommodate the decreased payments. The general contractor may or may not benefit from the practice, depending on the number of other competing bidders who receive the same reduced quotations. The subcontractor who is selected for the project as a result of submitting the lowest price is the apparent beneficiary of bid shopping; however, this practice is not in the best long-term interests of subcontractors.

Some attempts have been made to reduce prebid bid shopping but have not had widespread success. One of these procedures is to use bid depositories. Bid depositories have been set up in which specialty contractors submit their bids to the depository, and the bids are held in confidence until the day of bid submittal. This procedure prevents subcontractors from submitting successively lower bids as information about other bid amounts becomes known. The use of depositories has had limited success, however, and many have been determined to be in violation of antitrust laws.

Another form of bid shopping is postbid bid shopping. This is bid shopping that is initiated by a general contractor who is awarded a contract. A general contractor who indulges in this practice will contact a particular subcontractor and try to get the subcontractor to lower the submitted bid or accept the reduced price offered by the general contractor. Any reduction in the subcontractor's price will directly increase the general contractor's profit. Postbid bid shopping is not a desirable practice from the point of view of most contracting parties. Owners, whose projects may suffer as a result of postbid bid shopping, may try to curtail the practice by having general contractors include with their bids a list of the major subcontractors that will be utilized on their projects. Although having all subcontractors listed would virtually eliminate all postbid bid shopping, errors would invariably become more common on bids involving numerous subcontractors. Many subcontractors are not identified until minutes prior to bid submittal. As a result, most owners who use a subcontractor listing restrict the list to the major subcontractors, such as mechanical, electrical, steel erection, and so forth. The names of the firms appearing on the list of subcontractors are generally regarded as those specialty contractors with whom the general contractor must contract for the stipulated work (figure 18.2). Thus, the general contractor will not be in a position to negotiate for lower prices from these firms once it has been established that they were named on the list of subcontractors.

When the subcontractors who will be employed on a contract are identified in a general contractor's bid, the general contractor has little recourse but to use them. This was shown in *Ruck Construction Co., Inc. v. The City of Tucson* (570 P.2d 220). In its instructions to bidders for the construction of a police and fire building, the city of Tucson specified that the bidders had to submit "a list naming the subcontractors who will be used in performing the work" with their

---

CLAY COUNTY CIVIC CENTER

List of Subcontractors

March 9, 2011

This list is an integral part of the bid submitted by:

RST Constructors
230 Beltway
Charleston, SC

For the Construction of:

Clay County Civic Center
3499 Stonewall Jackson Way
Augusta, GA

The undersigned, hereinafter called the "Bidder," lists below the names of the Subcontractors who will perform the phases of the work indicated:

TYPE OF WORK   NAME / ADDRESS OF SUBCONTRACTORS

1. Sitework   *ACE Earthmoving, 311 Industrial Way, Waycross, GA*
2. Masonry   *Smith Masonry, 900 Rose Court, Charleston, GA*
3. Mechanical   *Weitz Mechanical, 3443 Edgewood Ave., Savannah, GA*
4. Electrical   *Isaac Hill Electrical, 705 Aurora Road, Bristol, GA*

Respectfully submitted,

Edward J. Billings

By: Edward J. Billings, President
      RST Constructors
      230 Beltway
      Charleston, SC

FIGURE 18.2
Example of a form for the listing of subcontractors.

bids. It also stated that "a subcontractor not named in the 'list of subcontractors' may not be employed on the City's project without express written permission of the City." The low bid was submitted by Ruck Construction with a list of subcontractors that included a mechanical subcontractor, Glover-Miller. On April 8, 1975, Ruck was notified that it was the low bidder. On the same day Ruck sent a letter to the city requesting permission to use SAYCO as the mechanical contractor instead of Glover-Miller. Ruck stated that a clerical error had been made and that SAYCO should have been listed on the original bid. The request was verbally denied by the city on April 11. Three days later Ruck signed the construction contract with the city. A written denial of the substitution request was received by Ruck on May 8. Ruck filed suit against the city, claiming that it lost $5,300 (the alleged difference between the quotes of Glover-Miller and SAYCO) because of the city's refusal to allow the substitution. Ruck argued that both subcontractors were equally competent to perform mechanical work and that the city did not exercise proper discretion in evaluating the bidders. The city stated that the instructions were clear and definite: The subcontractor list was to be submitted for the city's use in determining the "competency and capability of those who work on the City's project." The court agreed with the city, finding that state law required the city to evaluate subcontractors only before awarding public contracts. Once the contract was awarded, the city had no obligation to consider substitutions. Ruck had signed the contract after the city had verbally denied the request, thereby agreeing to its terms. The terms of a contract cannot be changed without the assent of both parties. When subcontractor listing is required, contractors should accept the fact that they generally will be required to contract with the named subcontractors.

Not all subcontracts are awarded through a competitive bidding process. A general contractor may have a strong working relationship with a specialty contractor and not solicit bids from other firms. The relationship may be such that the specialty firm submits a bid only to one general contractor, or a lower price may be submitted to the favored contractor than to the other competing general contractors. On cost-plus contracts, which are common in the private sector, the general contractor may derive greater benefits from working with subcontractors who are team members rather than selecting firms solely on the basis of price. On such projects the specialty firms may be asked to submit quotations that are then subject to negotiation. Since public policy is not an issue, specialty firms use a variety of means to submit bid quotations.

AWARDING SUBCONTRACTS

When a general contractor submits a bid on a public works project, there is some assurance that the lowest responsible bidder with a conforming bid will be awarded the contract. Although such assurances are not guaranteed on private work, the reputation of the owner's representative may be adversely affected if a similar approach is not taken. Thus, most general contracts that are awarded through the competitive bid process will be awarded by means of procedures that are common on public works projects. The awarding of subcontracts is not done as consistently. Even on public works projects, the awarding of subcontracts is not dictated by law or public policy. Thus, subcontractors place considerable reliance on the integrity of the general contractor.

Once the bids of specialty contractors have been incorporated in contractor bids and opened by the owner and the contract has been awarded to the general contractor, the general contractor must begin to assemble the construction team. This phase of entering into firm agreements with the subcontractors and suppliers is referred to as "buying out" the contract. Purchase agreements are signed with suppliers of major materials and equipment to be incorporated in the project, and subcontract agreements are entered. Before entering into the subcontract agreements, the general contractor will carefully review the bids submitted by the subcontractors. The bids of most subcontractors will have been received by means of quotations given to the general contractor through a telephone call or facsimile machine quotation.

Most quotations are submitted by means of facsimile (fax) machines. Quotations received by fax may be advantageous in that more detailed information can be provided than is customarily given via a telephone call. Unfortunately, the use of the fax means that there is no dialogue concerning specific issues. Items of work presumably to be done by a particular specialty contractor may not appear on the fax message. The specialty firm may later claim that since an item was not listed on the fax message, it was not included in the firm's bid. Through a telephone conversation, this type of issue could be quickly resolved if the party receiving the call knew that certain work items were to be included in a particular firm's price. This problem can be minimized if faxed bid quotations are thorough. For example, suppose a firm submits a price through a telephone conversation. It is common to state whether the applicable tax has been included. The bidder may mistakenly have assumed that the owner was tax-exempt and included no allowance for tax in the bid. A fax quote form used to submit bids should have a checkoff box to indicate whether tax is included in the price. Since the fax form did not indicate that tax was included, a quick allowance could be included for it, or a quick telephone call could be made to resolve the matter prior to incorporating the price in the general contractor's bid submittal. At any rate, inconsistencies or errors are best caught before the general contractor submits the bid to the owner.

Once all specialty contractors' prices have been reviewed, the general contractor can award the subcontracts. In general, this review of bids will be restricted to the quotations used in the preparation of the final successful bid. Thus, if there are no disagreements in the work items covered by the quotations of the various firms, the identity of the successful subcontractors is essentially known to the general contractor when the final bid is submitted. Since many public projects are awarded as lump sum contracts, the prices of potential subcontractors are not known to the owner. Even the specialty contractors do not know when the general contractor has used their prices. Only after the general contractor has received the contract award will the general contractor notify the successful specialty contractors of the use of their specific quotations.

If the subcontractor made an error in the preparation of a quotation submitted to the general contractor, and if the general contractor used the quoted amount in finalizing the bid, the subcontractor will generally be required to honor that price. The outcome is different when the general contractor makes a mistake of fact. In that case the general contractor is rarely held to the bid. If the subcontractor were permitted to withdraw the bid, the general contractor would suffer a direct loss. Since the general contractor relied on the price provided by the subcontractor, and since there would be a financial loss if the subcontractor withdrew the bid, the general contractor has an enforceable agreement with the subcontractor under the principle of estoppel.

Contractors should not use the quotes of subcontractors if there is reason to suspect an error. Although the courts often require subcontractors to honor their bids, there have been exceptions, as shown in *Edward Joy Co. v. Noise Control*

Products, Inc. and Cook and Reid, Inc. (443 N.Y.S.2d 361). During bid preparations for a construction contract for the Syracuse University Carrier Dome athletic facility, Joy received a bid of $32,545 from Noise Control for the installation of a sound attenuator. The quote was submitted by Cook and Reid, a firm that was acting as an agent for Noise Control. Since the quote was $24,000 below any other quotes for this work, Joy contacted the agent to verify the numbers. Without consulting Noise Control, Cook and Reid stated that the quote was accurate. Joy allegedly used the quote from Noise Control in its bid. Joy was awarded the contract and promptly notified the agent of its intent to award a subcontract to Noise Control.

When Cook and Reid relayed the information to Noise Control, a mathematical error was discovered that had caused the quote to be low by more than $20,000. Noise Control advised Joy of the error and provided an offer of $54,850 to do the work. This offer was rejected by Joy, and a different subcontractor was hired. Joy then sued Noise Control, claiming that promissory estoppel should force Noise Control to honor its bid. The claim was that subcontractors should perform work at the price they give to the general contractor. Noise Control contended that it should not be forced to honor its bid since it had made an innocent mathematical mistake and that the general contractor should have known that the bid was in error. The error was of sufficient magnitude that fraud on the part of Noise Control could not be claimed. The court agreed with Noise Control and stated that promissory estoppel did not override the fact that the bid, which had been arrived at through a mathematical mistake, was much lower than any other quote. The general contractor should have sensed the mistake and used another quote.

Since the general contractor alone is privy to the means by which the final bid was formulated, the subcontractor must rely on the general contractor to be fair in awarding the subcontracts. Some general contractors cannot resist the temptation posed by the unique position they are in once they have been awarded the contract. The temptation is to get some of the specialty contractors to reduce their prices in order to bolster the contractor's profits on the project. The general contractor may approach the second lowest bidder on an item of work with an offer such as "You were not the lowest bidder, but I really would like to work with you on this project. Could you reevaluate your bid and see if you could do it for $410,000?" The general contractor may fabricate the prices of other firms and even approach the lowest bidder with such a proposal. The firm being asked to reassess its price must then decide if it wants the job for the amount proposed by the general contractor. The firm put into this situation is able to determine its own fate by entering into the subcontract agreement or walking away from the project. Technically, the general contractor has rejected the offer of the specialty contractor when the counteroffer is made. Often the specialty firm will view the project in the sense of a bird in the hand and acquiesce to the general contractor's new price. This practice of general contractors, which is referred to as postbid bid shopping or bid peddling, is regarded as unethical.

Once the general contractor identifies all the subcontractors to be employed on a project, the matter of entering into formal agreements with the various firms can begin. This procedure is best carried out by individually discussing the aspects of the project that pertain to each subcontractor. The general contractor will want to verify the price that was quoted and the exact work items that the subcontractor will be expected to perform. Discrepancies should be resolved prior to signing the subcontract agreement. The general contractor will also want to go over major issues that appear in the general conditions of the subcontract agreement, the general conditions of the general contract agreement, and any special provisions. These issues include each subcontractor's responsibility for safety, cleanup, scheduling, supervision, and so on. The general contractor may also clarify the conditions under which the subcontractor can use the general contractor's trash dumpster, the general contractor's forklift or other equipment, the job office telephone, and the like. The general contractor and the subcontractors can work together most effectively when all parties have a clear understanding of how the project will be organized and the general nature of the work procedures.

For an offer by a subcontractor to be accepted, the acceptance must be absolute and unqualified. The case of *Western Contracting Corporation v. Sooner Construction Company* (256 F. Supp. 163) showed the impact that a qualified acceptance can have. Western Contracting, a general contractor, sought bids for the paving on a runway project. Sooner submitted an asphalt paving quote of $8.32 per ton, minus $0.50 per ton for payments made by the 10th of the month. Western was awarded the contract and on March 25, 1963, received a letter from Sooner confirming its quote. On July 16 Western sent a subcontract to Sooner for a signature. The subcontract showed that Sooner would provide the asphalt at $7.82 per ton. Sooner refused to sign the agreement. For the next two months, Sooner and Western continued to try to reach an agreement. During that period Sooner did some test mixes and performed other administrative duties under the assumption that an agreement would be reached. On September 30 Western terminated negotiations with Sooner and awarded the contract to another firm for $8.53 per ton. Western then sued Sooner for the difference in unit prices, plus interest, overhead, profit, and other expenses. In the absence of a written contract, Western argued that an oral contract had been formed on July 16, and even if it had not been, an explicit contract had been formed when Sooner did the test mixes and performed other duties. Thus the contract had been breached. Sooner contended that no contract had been formed because the original terms of the contract had been altered. Western's version of the contract was a counteroffer which Sooner did not accept. The court ruled that Western's amendment of Sooner's offer voided the original contract. Sooner was left with the option of refusing or accepting the counteroffer presented by Western.

INSURANCE REQUIREMENTS FOR SUBCONTRACTORS

A general contractor usually requires subcontractors to provide and maintain insurance coverage. This coverage is usually comparable with that held by the prime contractor in relation to the owner. It is important that subcontractors have adequate coverage. If a subcontractor's insurance is faulty or inadequate, the responsibility will probably revert to the prime contractor. Workers' compensation is required by law, and if it is not provided, the prime may well be liable. Usually the owner's contract with the general contractor states that if the subcontractor does not have insurance, the prime contractor will obtain it for the sub and charge the sub for the incurred costs. This may adversely affect the experience modification rating of the general contractor if the subcontractor incurs heavy losses.

SUBCONTRACT PROVISIONS

Various construction-related organizations have developed standard forms to be used for subcontracts. Perhaps the most frequently utilized form is similar to AGC Document No. 650 (1998 edition), which was developed by the Associated General Contractors. Various versions of the AGC document have been developed as these forms have evolved. Many general contractors used the AGC subcontract agreement and then modified it to suit their specific needs. The American Institute of Architects (AIA) also has developed a standard subcontract agreement (AIA Document A401). Although the various standard subcontract agreements have differences, some provisions tend to be similar in most subcontracts. Note that some of the provisions that are described below are not necessarily found in either the AGC or the AIA documents.

Ambiguity in the Plans and Specifications

When a subcontract is awarded, the general contractor wants some assurance that each subcontractor fully understands the nature of the work being contracted. For this reason, a provision may be included in which the issue of ambiguity is addressed. This specifically concerns matters in which other trades or subcontractors may be involved. An example of such a provision is as follows:

> If details are ambiguous, then a workable solution will be arranged with the necessary trades and the Contractor, at no extra cost to the Contractor with all costs allocated to the Subcontractors as determined by the Contractor.

Note that the general contractor judges how the costs will be allocated. Some subcontractors may anticipate less than fair treatment if one of the "other trades" consists of work to be performed by the general contractor's own forces. If all trades involved in the ambiguous issue are subcontractors, how will the general contractor render a fair or equitable decision? In addition, will the general contractor argue with the owner on behalf of the subcontractors about ambiguous matters when there is no direct gain for the general contractor? If the ambiguity relates to work that all the subcontractors assumed was to be performed by others, at least one subcontractor will have an increase in the scope of work to be performed without a commensurate increase in compensation.

Scheduling of the Subcontractor's Work

Subcontractors, even relatively small firms, typically undertake numerous projects to maintain a steady flow of work. If several projects are concurrently in progress, the subcontractor is required to pay careful attention to the efficient dispatching of workers and equipment to the various projects. This may not be possible if the following provision is used:

> The Subcontractor is to promptly begin said work as soon as notified by the Contractor and complete the above work as follows: start work within 48 hours after notification and continue with sufficient workers so as not to delay the progress of the job.

While it may be reasonable to expect the subcontractor not to interfere with the overall progress of the project, it may not be realistic to expect every subcontractor to be fully mobilized within a two-day period. Adequate advance scheduling is strongly advised so that the subcontractor can plan well ahead for each project.

A related provision that has also been used concerns overtime and shift work. The provision states, "If overtime and/or shift work is required to meet project schedule requirements, all costs associated with such work shall be included in the subcontract price, and this shall be made clear to Contractor at the time of Subcontract execution so that provisions can be made to accommodate such operations." It is not clear from this provision whether the overtime requirements apply even when the start of the subcontractor's work has been delayed by others. The provision does add, "In the event that Contractor directs additional overtime or shift work, additional compensation shall be limited to the premium portion only." There is an element of risk borne by the subcontractor, in that overtime or shift work may be required simply because the subcontractor was not permitted to begin work at an earlier date, as previously anticipated. In addition, productivity will generally be reduced on work that is performed in shifts or on overtime.

The shift of risk is more apparent in provisions that state, "The Subcontractor shall promptly increase its work force, accelerate its performance, work overtime, work Saturdays, Sundays, and holidays, all without additional compensation, if, in the opinion of the Contractor, such work is necessary to maintain proper progress. The Subcontractor shall conform to the Contractor's hours of work. No premium time will be acknowledged or paid unless pursuant to a written authorization by the Contractor." In this provision, the subcontractor is asked to estimate the likelihood of the need for shift work and overtime work and to include an appropriate amount to cover those costs in the bid. A subcontractor may not be inclined to include much about such uncertain costs in the bid, particularly if the project is one on which the subcontractor would like to be the low bidder. The question will always remain, How much are the competing subcontractors including for this provision, or are they ignoring the provision altogether?

Subcontractor's Payment Conditional on Contractor's Payment

A provision that affects many subcontractors on almost all projects is one in which the subcontractor's payment by the general contractor is conditioned on the general contractor's having been paid by the owner. For example, one such provision stated that periodic payments would be made to the subcontractor when such funds had been "paid to Contractor by the Owner . . . within 5 days of receipt thereof from the Owner." Others are more generic and state that "payments are conditioned on and subject to receipt of payment by Contractor from the Owner" and that the "Subcontractor specifically waives any claims against the Contractor for delays in payment by the Owner." Even when the general contractor has been paid, the subcontractor may not be aware of the payment. An additional risk is assumed when the subcontract does not specify the timing of such payment to the subcontractor, other than that it will be after the owner has made payment to the general contractor.

The "pay when paid" provisions, also known as the "pay if paid" provisions, pose a considerable risk to subcontractors. These have been attacked by subcontractors as being unfair and, in some cases, their enforceability has been challenged. One such challenge occurred in *Galloway Corp. v. S. B. Ballard Construction Co., et al.* (464 S.E.2d 349). Galloway Corporation contracted with Rowe Properties to construct a 14-story office complex for $10,960,000 in Norfolk, VA. Galloway's subcontract agreement with the subcontractors included a paragraph stating, "The Contractor shall pay the Subcontractor each progress payment within three working days after the Contractor receives payment from the Owner." Thus, the subcontractors were to be paid under the condition that the owner had first paid Galloway for the work. During construction, Rowe Properties went into default and was unable to make further payments to Galloway. When Galloway did not pay them, several subcontractors filed suit. The disposition of the subcontractors in this suit clarifies how the courts view the pay when paid provisions in some jurisdictions.

In the Galloway case, S. B. Ballard Construction Company had a different arrangement than the other subcontractors. Steven Ballard, the company president, stated that he understood the implications of the "pay when paid" clauses. In this case, he was relying on a separate "scope of work" agreement that was entered with Galloway prior to the contract. In this agreement, the parties agreed to a discount "to expedite the payment from the contractor to [Ballard] without worrying about the payment from the owner to the contractor." This constituted an early payment discount. At the time of the owner's default, Ballard had already received 12 payments that were not linked to the owner's payment of Galloway. From this, the court construed that Galloway and Ballard were interpreting the contract to permit Galloway only a reasonable amount of time in which to make payments to Ballard.

For the other subcontractors in the Galloway case, no secondary agreements were in existence. The court had to decide on the intent of the contract. Even the subcontractors testified that they knew the risk of pay when paid clauses, but they felt they had to accept such clauses as a condition of being awarded the contract. The court felt compelled to enforce the pay when paid clause. One of the subcontract agreements had not been signed by the subcontractor, but this too gave the subcontractor no relief. The court ruled that by performing the work, the subcontractor had essentially accepted the terms of the subcontract. The court ruled that the credit risk of the owner's insolvency had been shifted from the general contractor to the subcontractors. Only a few jurisdictions have a standard policy of not allowing such a shift of risk from the general contractor to the subcontractor.

Some subcontract provisions give at least some recourse if periodic payments are not made to the subcontractor because of failure by the owner to pay the general contractor. One such provision states that "the Subcontractor shall notify the Contractor of failure to receive payments per the contract, and after 7 days written notice may stop work on the subcontract without penalty until receipt of full progress payments owed by the Contractor. If work is stopped for 30 days, the Subcontractor can take actions to terminate the contract, with additional written notice."

Consider what occurs when the owner's solvency is an issue. This is also addressed in some subcontracts. One such provision states, "The Subcontractor acknowledges the fact that there is the risk that the Owner may not make payment, and as a party to this project agrees to share in this risk with the Contractor." This is open to interpretation in regard to the extent to which risk is actually being shared. Another provision states, "The Subcontractor agrees to assume the credit risk of the solvency of the Owner or any risk for the Owner's inability or refusal to fulfill its obligations under the Main Contract." In these provisions it is generally presumed that the subcontractor has no way to enforce payment from the general contractor if the owner is unable to make payments to the general contractor.

Payment for materials delivered but not installed may also be conditioned on payment by the owner. One such provision states, "Subcontractor may be paid for materials, not incorporated in the work, but delivered and suitably stored at the site, or at some other location agreed upon in writing, only to the extent Contractor receives payment from the Owner."

The amount retained by the contractor from the subcontractor's periodic payments may be different from the retainage withheld by the owner. A provision may state that the contractor will "make partial payments to the Subcontractor in an amount equal to 90 precent of the estimated value of work and materials incorporated in the construction." While many owners may withhold only 5 percent of the periodic payment from the general contractor, it is obvious under these terms that the general contractor is in a position to finance a portion of the project with the subcontractor's funds. A more equitable provision states that the "rate of retainage shall be equal to the percentage retained from the Contractor's payment by the Owner . . . provided that the Subcontractor provides a bond to the Contractor." However, failure to provide a bond may result in a higher stated rate of retainage for the subcontractor.

One type of provision not commonly found in subcontract agreements states that the money withheld from the subcontractor as retainage will be "held in an interest-bearing account, with a proportionate amount of the interest to be paid to the Subcontractor."

Another provision found in some subcontracts states that the "Contractor may deduct from amounts due or to become due to the Subcontractor pursuant to this Subcontract, any sums due or to become due to the Contractor from the Subcontractor whether or not said sums are in any way related to this Subcontract or project." Obviously, the general contractor has the ability under this provision to deny payment to the subcontractor if that payment is used to satisfy an existing indebtedness to the general contractor, even if it is not related to the current subcontract.

A provision may be included that gives the general contractor the right to issue joint checks to the subcontractor. An example of such a provision states, "Contractor reserves the right to make any payment to Subcontractor, including payments due hereunder, through the medium of a check made payable to the joint order of Subcontractor and such of Subcontractor's workers, suppliers, or subcontractors, or any of Subcontractor's creditors having potential lien rights against the work."

As in policies related to periodic payments, final payment to the subcontractor is tied to the owner's payment to the general contractor, with such payments commonly "being provided within 30 days of such payment of the Contractor by the Owner." One version of the final payment provision that is not conditioned on the general contractor's receipt of payment by the owner simply states that final payment will be made "upon acceptance of the work by the Contractor and Owner, and by showing evidence of fulfillment of the contract."

The consolidated cases of *Peacock Construction Co., Inc. v. Modern Air Conditioning, Inc.* and *Peacock Construction Co., Inc. v. Overly Manufacturing Co.* (353 S.2d 840) show how one jurisdiction interprets pay when paid clauses as they pertain to final payment. In its subcontracts, Peacock agreed to make final payment to the subcontractors "within 30 days after completion of the work included in the subcontract, written acceptance by the Architect and full payment therefor by the Owner." On a condominium project in Lee County, Florida, Modern Air Conditioning and Overly Manufacturing completed the work as specified. No deficiencies were found in their work. Despite this, Peacock refused to issue final payments on the grounds that the owner had not paid Peacock for the work performed by these subcontractors. The Florida Supreme Court found for the subcontractors and stated that the pay when paid clause was not absolute but "constitute[d] absolute promises to pay, fixing payment by the owner as a reasonable time for when payment to the subcontractor [was] to be made." While not permitted in all jurisdictions, some subcontracts will specifically state that the subcontractor assumes the credit risk of the owner's insolvency. This strengthens the pay when paid provision from the general contractor's perspective.

Subcontractor Is Bound by the Terms of the General Contract

One common provision of subcontracts is to have the subcontractor bound by the terms of the general contract. The following provision is an example:

> In consideration therefor, the Subcontractor agrees to be bound to the Contractor by the terms of the said Main Contract and to assume toward the Contractor all the obligations and responsibilities that the Contractor, by these documents, assumes toward the Owner (including every part of and all the General Provisions, General and Special Conditions, Drawings, Specifications, and Addenda), in any way applicable to this Subcontract, and also to be bound by the Subcontract General Provisions and the Subcontract Special Conditions attached hereto, which are hereby referred to and made part of this Subcontract.

The Construction Management Association of America (CMAA) stipulates in its General Conditions of the Construction Contract Between Owner and Contractor (CMAA Document A-3) that "Each subcontractor shall be required by the party with whom it contracts to agree to comply with these General Conditions and any other applicable Contract Documents." It is clear in its provisions that this extends equally to sub-subcontractors.

Although the specific wording found in various documents varies, the intent is the same: The subcontractor is bound to the terms of the general contract in addition to those of the subcontract. It is imperative that the subcontractor review and examine all the terms by which a contractual obligation is created. The main contract may contain provisions that contain undesirable clauses. These terms should be communicated prior to bidding, particularly if unique or unusual terms are included. Subcontractors may be asked to sign such subcontract agreements without examining the main contract. This practice is to be avoided. If standard form agreements are used for the general contract, the subcontractor will probably be familiar with most of the terms and will have to examine only the special provisions.

If a conflict is noted between the terms of the subcontract agreement and the terms of the general contract agreement, the means of resolving the conflict should be well understood. In most cases, the more specific document will be assumed to govern. In this case, the subcontract agreement will often be interpreted as taking precedence in the event of a conflict. To avoid confusion, subcontract agreements may include statements such as, "In case of conflict between the terms of the obligations and responsibilities of the parties of this Subcontract and the Main Contract, this subcontract shall control."

It is a simple matter to include in the subcontract an obligation for the subcontractor to be bound by the terms of the general contract. The case of *Sime Construction Co. v. Washington Power Plant Supply System* (621 P.2d 1299) is a good example. Sime was a sub-subcontractor on Hanford Project No. 2, a nuclear power plant being constructed for Washington Power Plant Supply System (WPPSS), who was to perform excavation and foundation work. Marley, the general contractor, subcontracted a portion of the work to Ragnar Benson, Inc., with a portion being sub-subcontracted to Sime. Sime started work in June 1978 but was delayed when critical drawings were not delivered, causing a disruption in the sequence of work. When Sime finished the work, compensation was sought for damages caused by the late drawings. The general contract stated that any claims had to be made to the owner within 15 days of the "cause for action." Since the subcontract made reference to the prime or general contract ("Subcontract documents include all the below listed items . . . the contract between the Owner and the Contractor . . . and conditions thereof. . . ."), WPPSS contended that the reference to the prime contract was clear and that the claim was not valid since it was not made within the mandatory 15 days. Sime argued that the reference to the general contract was intended only to define the scope of Sime's work. The court did not agree with Sime and stated that the prime contract was not referenced for a special purpose, meaning that the entire document did apply to Sime.

Subcontracts are often written so that the subcontractor is bound by the terms of the general contract. The case of *Longview Construction and Development v. Loggins Construction Co.* (523 S.W.2d 771) demonstrates why this is a prudent practice. Longview, a subcontractor, entered into an agreement with Loggins to clear, excavate, and develop the slopes on a football field and parking lot. The terms of the general contract were not referenced in the subcontract agreement. Longview determined that the subcontract had been completed and informed Loggins of this assessment. Loggins inspected the site and found, as confirmed by two surveyors, that the slopes were not at the proper grade. Longview felt that the job was done and did not return to the project when requested to correct the grades. Loggins hired Traylor and Son to finish the job, which took an additional 5 1/2 months. Loggins then sued Longview for $63,976.58, the amount required to pay Traylor to finish the project and pay the liquidated damages assessed against Loggins. Longview contended that its last payment, as certified by the architect, showed that Longview's work was 98 percent complete. Consequently, all that could be deducted from the subcontract sum was the remaining unpaid 2 percent. In regard to the liquidated damages assessment, Longview argued that this had never been communicated and that Longview thus was not bound by the provision.

On the first issue, the court ruled that nothing in the subcontract stated that the determination of the architect was binding, conclusive, or final and that it therefore was not actual proof of the work accomplished. On the second issue, the court stated that Loggins had never shown that it had informed Longview of the liquidated damages provision in the general contract agreement. Thus, the court concluded that the liquidated damages assessed against the general contractor could not be charged to the subcontractor's account. The conditions of the general contract were simply not incorporated in the subcontract agreement.

Backcharges

General contractors generally want assurances that their subcontractors will perform satisfactorily. This is generally done by entering into subcontract agreements with only those firms that have historically performed well on their projects and through carefully crafted subcontract provisions. The following is an example of a provision that is designed to encourage the subcontractor to be responsive to the needs of the project:

> Should Subcontractor fail to satisfy contractual deficiencies within three (3) working days from receipt of Contractor's written notice, then Contractor, without prejudice to any right or remedies, shall have the right to take whatever steps it deems necessary to correct said deficiencies and charge the cost thereof to Subcontractor, who shall be liable for payment of same, including reasonable overhead, profit and attorneys fees.

The provision is broadly stated so that a wide variety of issues might be included. For example, this provision might be invoked if the subcontractor fails to furnish sufficient workers on the project to maintain the project schedule. The provision might also be applicable if the subcontractor fails to keep the work area clean or to promptly address punch list items. Such a failure might cause the general contractor to perform the work in the subcontractor's stead and thereby require compensation from the subcontractor. Such compensation might be in the form of reducing the amount that would otherwise be paid to the subcontractor for other work performed. This potential backcharge against the subcontractor's account is a strong incentive for the subcontractor to promptly perform the work. Needless to say, the term backcharge evokes negative emotions among subcontractors.

Changes and Extra Work

Just as the owner typically is empowered to make changes in the work, the general contractor will typically include a changes provision similar to the following in the subcontract:

> The Contractor may at any time by written order of Contractor's authorized representative, and without notice to the Subcontractor's sureties, make changes in, additions to, and deletions from the work to be performed under this Subcontract, and subcontractor shall promptly proceed with the performance of this Subcontract as so changed. The Contractor and Subcontractor shall attempt in good faith to reach agreement in writing as to any increases or decreases of the Subcontract price or time resulting from such change or extra work, and if agreement is not possible, then the amount of additional time or change in compensation shall be determined as provided in the Disputes Clause of this Agreement.

Another typical changes clause is worded as follows:

> Subcontractor hereby agrees to make any and all changes, furnish the materials and perform the work that the contract may require, without nullifying this agreement, at a reasonable addition to, or reduction from, the contract price. . . . Under no conditions shall the Subcontractor make any changes either as additions or deductions without the written order of the Contractor.

While this provision may appear to be reasonable, the issue of compensation when an agreement is not reached may present problems. This is particularly true if the subcontract contains a provision such as, "If the Subcontractor and Contractor are unable to agree upon the compensation for changes or extra work ordered in writing by the Contractor, the Subcontractor shall nevertheless proceed with the changes and/or extra work, and the compensation to be paid shall be determined by the Architect, whose decision will be final." Of course the subcontractor has reason to question whether fair treatment is assured. Does the architect know about the pricing of the subcontractor's work, and can the architect set a price that is fair when the owner, who is the architect's client, wants to keep costs down?

A harsher provision for when an agreement on the price for extra work cannot be reached is, "If Contractor is not satisfied with the price quoted by Subcontractor with respect to any additional work, Contractor shall have the right to terminate this agreement and to contract with any other person or entity to perform such work."

Delays

Delays on construction projects can be devastating for some contractors and subcontractors. This issue is often addressed in subcontracts. The following wording is typical:

> Should the Subcontractor be delayed in the prosecution or completion of the work by the act, neglect, or default of the Owner, of the Architect, or of the Contractor, or should the Subcontractor be delayed waiting for materials, if required by this contract to be furnished by the Contractor, or by damage caused by fire or other casualty for which the Subcontractor is not responsible . . . the work shall be extended the number of days that said Subcontractor has been thus delayed, but no allowance or extension shall be made unless a claim therefor is presented in writing to the Contractor within 48 hours of the occurrence of such delay, but under no circumstances shall the time of extension exceed the time that the Owner grants the Contractor.

In this provision it is clear that the general contractor may delay the subcontractor, but such a delay will not result in an extension of time for the work of the subcontractor unless the owner grants a similar extension to the general contractor. The risks go beyond the matter of time extensions, as payment for delays can also be a serious matter. The following provision shows one way in which payment for delays is addressed:

> No claims for additional compensation for delays, whether in the furnishing of material by the Contractor or delays by other Subcontractors or the Owner, will be allowed by the Contractor unless they are specifically agreed upon at the time such delays occur and such additional compensation is granted by the Owner with the necessary extension of time. The Subcontractor shall not be allowed any additional overhead expenses by reason of such delays unless, within 48 hours thereafter, it presents evidence to support any actual loss or expense caused by such delay.

The payment implications are obvious. The general contractor assumes no liability for delays, and the subcontractor is not to expect compensation for a delay unless such compensation is granted by the owner. This provision does not specifically address delays caused by the contractor. Presumably, no compensation is permitted for such delays either. This is specifically stated in such provisions as, "If Subcontractor is delayed by an act or omission of the Owner, Contractor, or other subcontractors, Subcontractor shall not be entitled to additional compensation on account of such delay, suspension, or termination, nor for increased costs resulting therefrom." Such no-damage-for-delay provisions can have harsh implications for subcontractors.

Provisions may address delays caused by acts of God. For such delays, if the contractor is penalized by the owner, "then Subcontractor shall be responsible for such portion of the assessment as may be directly attributable to it, regardless of the cause of the delay." According to this provision, if the delay (an act of God) directly affects the subcontractor's operations, the subcontractor will be responsible for any damages or penalties.

No Subcontractor Claims Paid without Owner Payment

Claims may arise as a result of numerous incidents. Some provisions may address this in a general manner. The following provision may seem innocent upon first reading, but closer examination reveals that the subcontractor is placed at considerable risk. This provision reads as follows:

> No interruption, cessation, postponement, or delay . . . shall . . . give rise to any right to damages or additional compensation from the Contractor *except to the extent that reimbursement is received from the Owner by the Contractor* . . . and the Subcontractor hereby expressly waives and releases any other or further right to damages or additional compensation.

It should be obvious that if the general contractor made an error in scoping the work to be performed by two different subcontractors, a dispute could arise as to which party included or should have included a specific work item. For example, an electrical thermostat might have been shown on the mechanical plans and been overlooked by the electrical contractor. The mechanical contractor dismissed the thermostat since it is an electrical device. The error is one that will eventually be paid for by one or both of the subcontractors. According to the above provision, the general contractor is not liable for any payments without being first compensated by the owner. Naturally, the owner will not be inclined to make an additional payment on the thermostat since this was covered in the contract with the general contractor. Thus, the general contractor's error in properly scoping the work can result in a loss that is paid directly by one or more subcontractors. Errors of this kind on costly items can be devastating to an unsuspecting subcontractor.

Responsibility for Liquidated Damages plus Other Damages

Like the general contractor, the subcontractor is expected to perform the required work within a stated time frame. Failure to complete the work in the time required will generally result in consequences that are outlined in the subcontract. The wording of these provisions should be studied. The following provision is of particular interest:

> In the event of any failure of Subcontractor to complete his or her work within the required time . . . the Subcontractor hereby agrees to reimburse the Contractor for any and all liquidated damages . . . collected from the Contractor by the Owner,

which are *directly or indirectly* attributable to . . . the Subcontractor's failure to *comply fully* . . . and further, *whether or not liquidated damages* are so assessed, Subcontractor hereby agrees to pay . . . additional damages as the Contractor may sustain by reason of any such delay directly or indirectly . . . caused by the Subcontractor.

While it seems fair to have the subcontractor bear the burden of liquidated damages that are directly attributed to delays caused by that subcontractor's work, it is harsh to have the subcontractor bear the responsibility for delays that are only indirectly linked to the subcontractor's work. What constitutes *indirect?* Can the subcontractor be asked to pay liquidated damages when the cause of the delay is quite remote? In addition, note that this provision does not limit the amount to be extracted from the subcontractor to the amount of the liquidated damages. This is similar to another provision which states that the "Contractor's recovery against Subcontractor for delays, interferences, impacts, etc., for which Subcontractor is responsible shall not be limited to any liquidated damages assessments." While such a provision may not be enforceable, its presence is still disconcerting. A fairer treatment of the allocation of liquidated damages, as found in many subcontracts, is for the subcontractor to be responsible for the amount of liquidated damages that represents the proportionate share of that subcontractor's responsibility for the delay.

Control over Subcontractor Employees

While it may be reasonable to expect all subcontractors to employ top-quality workers and supervisors, the involvement of the general contractor in the subcontractor's employer-employee relations may be a controversy in disguise. Consider the following provision: "The Subcontractor shall not employ any worker whose employment on the project may be objected to by any of the other Subcontractors or their employees or the Contractor." While this is seemingly an innocent provision that provides a means to assure that objectionable workers will not be employed, there are no specific criteria by which this assessment can be made. Could personal grievances, personality conflicts, or outright discrimination become contractual grounds for removing a worker? Although the intent of the provision is not at issue, its interpretation can be troublesome.

Resolution of Disputes

It is common for dispute resolution to be addressed in subcontracts. For example, a provision may state that "all claims, disputes, and other matters . . . shall be decided by Arbitration in accordance with the Construction Industry Arbitration Rules of the American Arbitration Association." Such provisions are relatively standard and provide some assurance that a fair resolution of all disputes is possible. However, sometimes the provision states that "claims or disputes may, upon mutual agreement of Contractor and Subcontractor, be submitted to arbitration." Note that with this version, either party is in a position to veto the arbitration option. Thus, a formal lawsuit may be the only recourse if the general contractor or the subcontractor refuses arbitration.

Indemnification

Many subcontract agreements now require subcontractors to indemnify the general contractors and the owners. Although this often means that the subcontractor waives protection under workers' compensation, state law will dictate whether the provision is enforceable. In many states, such provisions are enforced. A typical provision of indemnification is as follows: "The Subcontractor hereby releases the Contractor and Owner of all liability on account of any accidents during performance of work in this subcontract." Another provision related to the waiver of workers' compensation protection states, "Subcontractor specifically and expressly waives any immunity that may be granted it under the Worker Compensation Statutes."

Typical indemnification provisions state that the only exception to the requirement to "indemnify and save harmless the Contractor, Owner, and Architect" occurs when the injury or death arose from "the sole negligence of the Contractor or Contractor's agent or employees." An atypical indemnification provision requires the subcontractor to "indemnify and save the Contractor, Owner, and Architect harmless" for all claims arising out of injuries or losses caused by an unsafe work environment *unless* "the Contractor shall have been given written notice of the unsafe condition prior to any accident caused or alleged to have been caused by such unsafe place to work." This provision essentially makes the subcontractor responsible for the safety conditions on an entire work site. This is an area in which the subcontractor can be at considerable risk if a worker is seriously injured.

The case of *Webb v. Lawson-Avila Construction, Inc.* (558 S.2d 433) demonstrates the strength of the indemnity provisions. On a school building project, Lawson-Avila (general contractor) entered into a subcontract agreement with Palmer Steel to provide structural steel for the building. During construction, a crane and a bundle of steel joists fell over, striking two of Palmer's employees, killing one worker and injuring another. The families of both workers sued Lawson-Avila. As the employer, Palmer was immune from suit due to the workers' compensation statute. The jury found that the general contractor was grossly negligent. The jury awarded $500,000 in actual damages and $1,500,000 in punitive damages. The case that emerged was whether the indemnification clause in Palmer's subcontract agreement extended to gross negligence by the general contractor. It was not contested that the indemnification clause covered the actual damages, but the party responsible for punitive damages had to be determined. The court ruled that the issue of negligence in the indemnity clause extended to all shades of negligence, including gross negligence. Thus, the subcontractor became liable for the entire amount of the damages awarded. Note that while workers' compensation provides immunity from suit for the employer, the indemnification clauses for subcontractors essentially mean that the subcontractor is ultimately not immune from paying for the damages.

Indemnification clauses do not always provide the intended protection. This was shown in the case of *Glendale Construction v. Accurate Air Systems* (902 S.W.2d 536). Glendale was the general contractor and Accurate Air Systems was awarded a subcontract for the heating, ventilation, and air conditioning (HVAC) portion of a building in Houston, Texas. Accurate Air Systems agreed to indemnify Glendale for claims arising out of the performance of its work. One of the employees of Accurate Air Systems, Donald Brooks, was electrocuted when the metal duct he was installing became electrically charged. Brooks's widow accepted recovery from workers' compensation for the negligence of Accurate Air Systems and sued Glendale for its contributory negligence. Glendale then filed suit against Accurate Air Systems for contributory negligence, indemnity, and its legal fees. The court interpreted the indemnification provision as not protecting Glendale for its own negligence. In addition, the court determined that Accurate Air Systems was also not obligated to pay for Glendale's legal fees.

Scope of Work

When the subcontractor submits a bid, there is undoubtedly a clear idea in the subcontractor's mind of what work items are included in the bid. The subcontractor should make this scope of work being covered very clear in the bid. Furthermore, the subcontractor should carefully examine the subcontract agreement before signing it to determine whether the scope has been changed from what was originally bid. Various types of subcontract provisions exist on this matter.

The subcontractor shall perform "any items of work normally performed by such Subcontractor in association with its work, including such items which may be specified in other parts of the plans and technical specifications. Unless specifically provided for herein, Subcontractor shall also be responsible for any items which may be included in the specifications but excluded by the Subcontractor from his or her bid, and any items included in Subcontractor's bid but not included in the designated specification sections." Such a provision necessitates that the subcontractor fully review the drawings and specifications. Omission of information in the bid documents could, by interpretation, become the responsibility of the subcontractor.

The full range of the subcontractor's obligations is not always clear. For example, a provision might state, "The Subcontractor is to take proper care of all construction materials on the grounds, and take full responsibility for all materials and equipment provided to the Subcontractor whether in Subcontractor's possession or on the project." This provision may be construed as broadening the responsibilities of the subcontractor. Is the subcontractor responsible for equipment used by the subcontractor's workers even though such equipment has been returned to the general contractor? The wording appears to be deliberately broad.

A provision may enumerate the obligation of the subcontractor to protect the work performed. For example, the clause may state, "Subcontractor specifically agrees that it is responsible for the protection of its work until final completion and acceptance thereof by the Owner and it will make good or replace, at no additional expense to others, any damage to its work which occurs prior to said final acceptance." The implications of this provision are of greater concern when another provision in the subcontract states, "Whenever it may be useful or necessary for the Contractor to do so, the Contractor or Owner shall be permitted to occupy and/or use any portion of the work which has been either partially or fully completed by the Subcontractor before final inspection and acceptance thereof by the Owner, but such use and/or occupation shall not relieve the Subcontractor of its guarantee of said work nor of its obligation to make good at its own expense any defect in material and/or workmanship which may occur or develop prior to Contractor's release from responsibility to the Owner." It appears that the subcontractor is asked to accept responsibility for any damage that the subcontractor's work may sustain regardless of who caused the damage. Subcontractors should carefully assess the implications of such provisions.

Changed (Differing Site) Conditions

It is typical to include in subcontracts a provision that states that the subcontractor has an obligation to visit the site when preparing the estimate. Such provisions are included to provide assurance that the subcontractors are fully apprised of the actual site conditions. By submitting a bid, the subcontractor is typically presumed to have made a site investigation of the proposed project. If unusual subsurface conditions are subsequently encountered, the subcontract may place full responsibility on the subcontractor by including a provision that states, "Subcontractor will complete the work for the compensation stated in this Subcontract and assume full and complete responsibility for the conditions (including subsurface) existing at the site and its surroundings."

Another provision may state, "All work subject to this Subcontract shall be performed to the complete satisfaction of the Contractor, the Owner, and/or the representative of the Owner authorized to interpret and judge the performance of the work."

The subcontractor will find it difficult to interpret these provisions. Phrases such as "to the Engineer's satisfaction" and "to the Contractor's satisfaction" cannot be interpreted in a manner that allows the costs associated with the provisions to be accurately quantified. Such provisions should be avoided by contractors and viewed with skepticism by subcontractors.

Termination of the Subcontract

Like general contractors, subcontractors are subject to having their projects canceled by owners. While it would not be realistic to assume that the subcontract cannot be canceled when the project is canceled, an examination of subcontract provisions shows that some subcontracts can be terminated for less substantial reasons. Consider the following provisions:

> In the event that the Contractor shall at any time be of the good faith opinion, after consultation with the Subcontractor, that the Subcontractor is not proceeding with diligence and in such a manner as to satisfactorily complete the work within the required time . . . Contractor shall have the right, after a seventy-two-hour notice confirmed in writing, to take over the work.

> In the event that Subcontractor fails to comply, becomes disabled from complying, or fails to furnish, where requested, written assurance of its ability to comply with the provisions herein as to character or time of performance, and the failure is not corrected within 5 calendar days after written notice by the Contractor to the Subcontractor, the Contractor may, without prejudice to any other right or remedy against Subcontractor or its surety, take over and complete the performance of this subcontract.

The terms under which the subcontract agreement can be terminated should be reviewed carefully. Many differing versions of these provisions exist. One is as follows:

> If at any time, in the Contractor's opinion, the Subcontractor is not proceeding with diligence and in such a manner as to satisfactorily complete the work within the time scheduled for this work, or if Subcontractor shall fail to immediately correct defective work or replace unsatisfactory materials, then and in that event, Contractor shall have the right, after giving two (2) working days notice, confirmed in writing, to take over the work, or any portion thereof, and to complete, correct, or replace the same at the expense of Subcontractor without prejudice to Contractor's other rights or remedies for any loss or damage sustained. If such action is necessary, Subcontractor shall be in default of this Subcontract.

Some provisions state that the general contractor will give a 72-hour notice. Another gives the general contractor the right to terminate a subcontract with a four-hour notice, and one states that the general contractor can take over the sub-contractor's work "after reasonable notice." These provisions must be given careful consideration as the time allotted for notification can have a dramatic impact on the subcontractor's ability to respond in sufficient time.

When the subcontractor is deemed to be in default, additional rights may be contractually conferred to the general contractor. Several subcontracts have provisions similar to the following:

> The Contractor may take possession of the Subcontractor's materials, supplies, machines, tools, equipment, and plant which may be located at the site of the work or en route to the site, as may be necessary to prosecute the work hereunder to completion, all without liability on the part of the Contractor for any damage, wear or tear, depreciation, theft, action of the elements, acts of God, fire, flood, vandalism or for any other injury or damage to such materials, tools and equipment.

The risks to a subcontractor deemed to be in default are considerable.

Also included in some subcontracts is a limit to the remedies available to the subcontractors. For example, a provision may state that if the general contractor deems it necessary to take over the subcontractor's work, "the Subcontractor will accept the terms and penalties and not file any liens against the project or file suit to recover any compensation felt to be due." Note that the potential for filing a lien is at the heart of the subcontractor's security. Giving up lien rights reduces the subcontractor's power considerably. Such provisions are not honored or enforced in all states.

The definition of default in order to terminate a subcontract cannot be determined solely by the general contractor. This was illustrated in *Ned Paduano v. J. C. Boespflug Construction Company* (403 P.2d 841). Boespflug, a general contractor on a building project, awarded a subcontract to Paduano for clearing, grubbing, backfilling, and respreading of the topsoil. Paduano began work on the specified date and continued working until his initial duties were completed. The only remaining work for Paduano consisted of landscaping that would take place near project completion. Later, when requested, Paduano returned to the project to finish the work. When Paduano's work was not completed after 5 1/2 weeks, Boespflug declared Paduano to be in default on the subcontract. Paduano requested payment for the work he had done, but Boespflug denied the request. Paduano then filed suit against Boespflug. Boespflug contended that he was simply enforcing the subcontract, which stated in part, "If the subcontractor shall fail to . . . prosecute said work continuously with sufficient workmen and equipment to insure its completion within the time herein specified . . . the contractor may elect to give notice in writing of such default. . . ." Paduano argued that he was doing all that could be reasonably expected of him to complete his work. The court ruled that Boespflug was "arbitrary and capricious" in declaring Paduano in default. Payment for the work performed and the release of the retainage were awarded to Paduano.

Termination of a subcontract by the subcontractor is rare. The right of a subcontractor to terminate an agreement was tested in *Vermont Marble Company v. Baltimore Contractors, Inc.* (520 F. Supp. 922). Baltimore Contractors, a general contractor, was awarded a contract to construct a 6-story addition to an existing office building. Vermont Marble entered into a subcontract agreement to perform all the stone and masonry work on the project. The agreement stated that time was of the essence. Furthermore, it stated that Vermont Marble would start its work within 12 months and finish within the following 6 months. After the project was under construction, numerous delays were caused by the owner and general contractor. Vermont Marble was unable to begin its work until 18 months had passed. After starting work on the project, Vermont Marble recognized that the major portion of its work could not be started for at least another 6 months. At that point it formally notified Baltimore Contractors that it was rescinding its subcontract. When Baltimore Contractors refused to grant the withdrawal, Vermont Marble sued to be released of its subcontract and receive payment for the work it had done. Vermont Marble argued that since the subcontract had stated that "time is of the essence," it was implied that delays would be kept to a minimum, and that the delays that did occur amounted to a material breach of the subcontract. This breach gave it the right to terminate the agreement. Baltimore Contractors recognized the inconvenience caused by the delays, but contended that this did not constitute a breach of contract. In fact, it presented a clause in the subcontract that specifically gave it the right to suspend or delay work done by Vermont Marble with just cause. It was also argued that since Vermont Marble had started work, it waived any rights to rescind the contract. The court found no element of the subcontract that clearly defined an "unreasonable delay." The court stated, "We conclude that even unreasonable delays are not material breaches of this subcontract. Rather, a material breach might arise only if BCI refused to pay a properly presented claim." Thus, the subcontractor could not terminate the subcontract for unanticipated delays.

Knowing and Understanding the Terms of a Subcontract

The terms of a subcontract form the legal document by which the relationship between the general contractor and subcontractor is defined. Inferences should not be made about the validity of matters that do not conform to the specific terms of the subcontract. While some terms may be waived, as evidenced by past actions, this should not be the assumption in most cases. The following cases bear testimony to the importance of adhering to the terms of the subcontract.

As is true of most contracts, the terms of the subcontract should be understood and followed. Failure to follow these terms may preclude recovery if a claim is made. This was illustrated in *Collins v. Vlesko and Post* (362 P.2d 325), a case concerning E. Vlesko and Claude Post, who were awarded a general contract by the state of Oregon to construct the main building of a state correctional institution near Salem. A subcontract was awarded to Collins Plumbing and Heating for the construction of a tunnel and steam line that would run from the main building to several residences of the institution's personnel. After Collins completed the steam line in August 1958, but before the project was accepted by the owner, the steam line was damaged by another contractor performing work on the residences in January 1959. As verbally requested by Vlesko and Post, Collins repaired the damaged lines and subsequently requested additional payment, claiming that the repair work was not within the scope of the original subcontract. Collins filed suit when the payment request was denied, claiming that there was an implied promise to pay.

The case was decided by the Oregon Supreme Court, which ruled in favor of Vlesko and Post. The court stated that the work performed by Collins was work that was required to fulfill the obligations of the original contract, enforcing a subcontract provision in which Collins agreed "to be bound with the Contractor by all the terms of the contract. . . ." Thus, the repair work did not constitute a change to the original subcontract. The court also enforced the provision stating that "no extra work will be recognized or paid for unless done pursuant to written instructions from the Contractor." Collins had provided no evidence that showed that Vlesko and Post promised to pay for the work and that urging Collins to perform the repair work did not automatically imply they would pay for it. Although the damage to the steam line was caused by an independent contractor, the responsibility for the repair of the damage was placed on Collins, since the project had not yet been accepted by the owner.

Subcontractors, like other parties, must fully understand all the terms of a contract. The case of *Hensel Phelps Construction Co. v. King County* (787 P.2d 58) shows why this is true. Phoenix Painting Company was awarded a subcontract for painting the new King County Jail. Hensel Phelps was the general contractor. Several problems developed that affected the work of Phoenix. Because of job delays caused by numerous factors in the early stages of construction, the original work schedule was accelerated. The painter was affected by a requirement that the original schedule of 45 days per floor was reduced to 19 days per floor. This accelerated work schedule meant that in some cases the painters would have to work alongside other contractors. In addition to making work in the congested areas more difficult, on some occasions the other workers would mark or scratch the paint during their daily work routine. In some cases painting could not continue until other crafts had vacated the premises. Even the design of the building made painting difficult. In the 18-story jail, to make inmate escape more difficult, each consecutive floor had its staircase on the opposite end of the building, making it difficult to carry up paint and painting supplies. For the added costs, Phoenix sought additional compensation. A suit was filed when Hensel Phelps refused to pay.

The court decided in favor of Hensel Phelps, concluding that the plans on which the painting price was based clearly showed the staircase locations and that appropriate allowances should have been made for the decreased efficiency. Regarding the acceleration of the schedule, the subcontract agreement gave Hensel Phelps the power to accelerate the schedule in order to complete the project on time. Finally, it was determined that Phoenix had failed to follow the appropriate procedures for filing for added compensation. This procedure was described in the subcontract agreement, and Phoenix had received nearly $120,000 in added compensation on other claims from Hensel Phelps by using these procedures. Since this demonstrated that Phoenix was familiar with the appropriate means for requesting reimbursement, the court stated that this procedure should have been followed.

The case of *Del Guzzi Construction Co., Inc. v. Global Northwest, Ltd., Inc. and Balboa Insurance Co.* (719 P.2d 120) also shows the need to follow the terms of a subcontract. Del Guzzi was the low bidder and the recipient of a contract for the construction of a sewer interceptor in Clallam County, Washington. Soon after the contract award, Del Guzzi discovered that the native material would not meet the compaction specifications. When Del Guzzi asked the county to issue a change order that would cost $400,000, the county stated that the contract was substantially altered by this change and canceled it. The county redrafted the contract documents to reflect the additional work and readvertised the project for bidding. Del Guzzi was again the low bidder. To satisfy a federal funding requirement calling for minority participation, Del Guzzi awarded a subcontract to Global Northwest, a minority contractor, for the backfilling and compaction work. Global also encountered problems with the compaction, but the project was successfully completed in December 1976. After project completion, Global submitted a request for additional payment for the additional work required to satisfy the contract. Del Guzzi and the county approved a change order request. However, since federal funds were applied to the project, this disbursement of funds was contingent on U.S. Environmental Protection Agency (EPA) approval. In the meantime, Del Guzzi sued Global for damages it claimed were incurred in completing a portion of Global's contract. The EPA rejected the change order request. In April 1980 Global countered by suing Del Guzzi, the county, and the designer of the project for breach of contract. Global's claim was based on Del Guzzi's failure to inform Global of the soil conditions (Del Guzzi had encountered the same compaction problems on its earlier contract with the county), the county's knowledge of Global's inexperience, and the implied warranty of the adequacy of the specifications.

The appellate court dismissed Global's claim against Del Guzzi, stating that Del Guzzi did not owe any special consideration to Global since no such contractual obligation existed. The Washington Supreme Court ruled that Global was not a third-party beneficiary of the contract between Clallam County and Del Guzzi. The claim against the county was also denied since Global did not file its claim within the statutory 3-year period following subcontract completion. Although various reasons were given for denying Global's claim, the failure to file the claim in a timely fashion was most damaging to its case against the county.

Job-related injuries are often the source of large court settlements. In *Barry L. Husfloen, et al. v. MTA Construction, Inc. et al.* (794 P.2d 859), Bill's Plumbing, the general contractor and owner of the site, awarded a subcontract to MTA to build the foundation for the project. MTA then sub-subcontracted with Pumpcrete for placing concrete in forms around the job site. On the morning of February 18, 1987, Barry Husfloen, an employee of Pumpcrete, arrived at the site. He parked the concrete-pumping boom truck in a driveway directly underneath overhead power lines. Husfloen did most of the concrete placement by means of a remote control device which hung from his neck. This device was connected to the truck by a cable. The concrete was placed successfully, but when the cleanup operation began, Husfloen was injured. Having forgotten about the electric lines, Husfloen had extended the boom vertically above the truck. When the boom made contact with the 7200-volt wire, the truck was energized and the electric current was transmitted to Husfloen through the cable attached to the remote control device. Husfloen brought suit against MTA and Bill's Plumbing, alleging that they allowed him to perform the work in an unsafe fashion, failed to have the power line disconnected, and failed to assist in folding up the boom on the pump truck. The court agreed with Husfloen and found that MTA and Bill's Plumbing had a duty to comply with the safety regulations. Note that another party is not immune from suit simply because someone elects to work in an unsafe manner.

The pay when paid provisions generally are interpreted as placing the burden of owner insolvency on the contractor. In *Thos. J. Dyer v. Bishop International Engineering Co.,* the court refused to enforce a pay when paid clause. In this case, the general contractor (Dyer) on a project was not paid by the owner who declared bankruptcy. As a result, the general contractor did not pay for work that had been done by a subcontractor. The court ruled that the conditions of payment are enforceable if the conditions are clearly described. The contract stated that the contractor was to pay subcontractors "five days after the owner shall have paid the contractor." This was not clear. The court had to consider the intent of the provision and that the general contractor generally bears the risk of nonpayment because of owner insolvency. The court ruled that the contract wording was only to delay payment, but that the owner's insolvency would not erase the contractor's obligation to the subcontractor. Thus, some courts interpret the pay when paid provisions as merely influencing the timing of payments, but not creating an absolute condition that must be satisfied before the contractor is obligated to pay the subcontractors. It should be noted, however, that there are many court cases on this subject and the decisions do not always favor the subcontractors.

The pay if paid provisions are interpreted by some courts as being enforceable, and that they effectively shift the burden of owner insolvency to the subcontractors. These provisions state that the subcontractors are to be paid only if the general contractor is paid, or that the subcontractors will not be paid if the general contractor does not receive payment from the owner. Some pay if paid provisions will add such words as "the subcontractor assumes the risk of nonpayment due to owner insolvency" to further clarify the intent. Courts tend to enforce these provisions because of their clarity of intent. There are exceptions to this, as some courts have ruled that such provisions violate public policy and are not enforceable. For example, rulings that refuse to enforce the paid if paid provisions have occurred in California and New York. The litigation on subcontractor payment by the general contractor will doubtless continue.

REVIEW QUESTIONS

1. What are the essential differences between prebid and postbid bid shopping? Who benefits most from each type of bid shopping? What means can be employed to minimize bid shopping?
2. What is the role of the subcontractor in relation to the owner of a project?
3. What are the implications to the subcontractor of the "pay when paid" policy in regard to the timing of payments to the subcontractor from the general contractor?
4. What type of work is generally subcontracted by general contractors?
5. On what grounds are subcontractors generally obligated to honor their bids in spite of the fact that a material error of fact may have been made in their preparation?
6. Examine the AIA documents and the ConsensusDOCS that are included in the Appendix and determine the similarities and differences in the provisions related to the approval of subcontractors and material suppliers. Specifically examine AIA provisions §5.2.1 and the applicable ConsensusDOCS provisions.

19

INTERNATIONAL CONSTRUCTION CONTRACTS

WITH THE BREAKTHROUGHS in technology that have enhanced communications around the globe, international contracts have become commonplace. While firms from a few countries dominated the international construction market three decades ago, the field of competitors now includes many firms from numerous countries. In fact, the United States has ceased to be the leader in this area. International construction projects require the skills of numerous types of field-workers and design professionals. Projects may include water and wastewater treatment facilities, processing plants, petrochemical plants, housing projects, airports, and pipelines. If a country can supply its own skilled labor, an international contract may be awarded for design services only. If that country does not have the available labor to complete a project, international contracts will be awarded for the labor. International firms that can efficiently provide design services may not be competitive in providing skilled labor. Thus, different countries have emerged as leaders in supplying either design services or skilled labor. U.S. firms have been most successful in providing the design services, while the most competitive firms providing skilled labor have generally come from countries with lower hourly wage rates for field-workers. In some cases countries hire workers directly from other countries.

Although the U.S. construction volume in 1999 was approximately $700 billion, this is only 20 percent of the global construction market of approximately $3.2 trillion. Over 70 percent of this volume is spent by 10 countries, namely United States, Japan, Germany, China, United Kingdom, Brazil, France, Italy, South Korea, and Canada. Figure 19.1 shows how the construction volume is distributed among the different continents.

Because of the different ways countries have developed, differences are common in the way construction contracts are let. Few countries have a protocol for awarding public sector contracts that is spelled out as succinctly as the one used in the United States. The practice of awarding contracts in the United States has evolved because of a strong sense that fairness should exist when contracts are

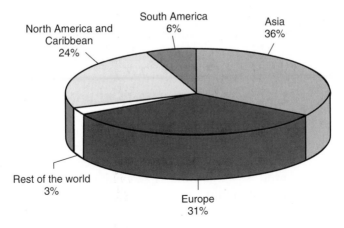

FIGURE 19.1
Construction volume by global region.

awarded. The Sherman Antitrust Act is perhaps the most important law that has helped formulate the bidding policies enforced today. Few countries have similar laws. This is not to imply that practices in other countries are less pure than those in the United States, but that they may be different and probably are not as stringently outlined.

When considering the pursuit of work in other countries, there will very likely be some significant differences from conditions encountered in the United States. Firms venturing into international markets will be wise to ask some fundamental questions, including the following:

How stable is the government?
What are the assurances of being paid?
To what extent is English spoken and understood?
What unique customs or traditions exist?
Will family members of employees be welcomed and secure?
Will disputes be handled fairly?
How qualified are the local craft workers?
What is the level of technological know-how?

The answers to these and other questions may be instrumental in deciding on whether to pursue projects in a particular country.

TYPES OF CONTRACTS

The types of contracts awarded depend to a large degree on the type of projects being undertaken. For process plant or petrochemical plant construction, the contracts will generally be some version of a cost-plus contract. Such projects may also require a considerable amount of design work, and so a fixed-price contract would be inappropriate. On smaller, well-defined projects, fixed-price contracts will be used. Projects involving earthwork often use unit price contracts.

CONTRACT AWARDS IN THE PUBLIC SECTOR

Contractors wishing to enter into construction contracts in the international market frequently find significant differences in the means by which contracts are awarded in the public sector. Few countries award the preponderance of their contracts solely to the lowest responsible bidder submitting a regular bid. In general, most countries do not have antitrust laws as stringent as those in the United States. In addition, few countries experience the everyday concern for litigation that is common for many U.S. contractors. In short, the rules are typically less defined in most foreign countries.

The contracting procedures of public agencies in many countries are probably more closely related to contracting practices in the private sector in the United States. For example, a country or a government agency within a country may open bidding only to those on a select bidders list. There may be a public advertisement for construction firms to submit application materials that attest to their qualifications. Although 20 firms may apply for prequalification and 15 firms may satisfy the essential criteria, the actual number of firms permitted to bid may be only 6 or 7. When the bids are received, the contract award may not necessarily go to the lowest bidder, but to the firm that will provide the best-quality project for the funds being expended. Under such a system, which is known to be practiced to some degree in European countries, the contract award is not made to the lowest bidder or to the one submitting a regular or conforming bid, but to the most qualified contractor for the funds to be expended. Another approach used in some countries is to open the bids, throw out the highest bid, throw out the lowest bid, and then award the contract to the bid that is closest to the average of the remaining bids. The strategy is then not to be the lowest bidder, but rather to give a realistic bid and rely to some extent on luck.

SURETY BONDS AND GUARANTEES

It is common in the international market for owners to require contractors to provide assurance that satisfactory performance will occur. The contractor who is the successful bidder is expected to obtain the required performance bond or guarantee after receiving notification that the offer or bid has been accepted. Contrary to the practice of using performance bonds in the United States, many owners in the international arena require a demand instrument, such as a letter of credit, issued by a bank. Since payment and performance bonds are difficult to obtain on international projects, alternatives must be sought. It is common for such guarantees to consist of letters of credit for 10 percent of the contract amount. Larger percentages may be required by some owners. The owner will generally reserve the right to approve the terms of the guarantee instrument, as well as the firm, bank, surety, insurance company, individual, or federal agency providing the guarantee. The owner should stipulate specific terms related to the guarantee, including the period for which the instrument is valid, the forfeiture procedure, the means of executing its release, and the type of currency to be used. In some cases, the owner may insist that the guarantee be issued by a bank in the owner's country.

The guarantees may be conditional or unconditional. A *conditional guarantee* is one in which the owner must show or demonstrate that the contractor has defaulted before the guarantee is forfeited. It is desirable for such guarantees to stipulate that notification be provided to the contractor whenever the contractor is deemed to be in default. Furthermore, such notification should allow a specified number of days (e.g., 30 or 40 days) in which the contractor can remedy the default. An *unconditional guarantee* is one in which the owner may foreclose on the guarantee without having to show proof of contractor default. The bank or party issuing the guarantee may simply be bound to issue payment to the owner without the benefit of due process. Unconditional guarantees are particularly worrisome to contractors who want to negotiate major disputes with the owner, because the owner can simply request payment on the guarantee. A trusting relationship must exist between the contractor and the government of the host country, as litigation is seldom a viable alternative.

To reduce the risk to the contractor, the host country may also provide a guaranteed letter of credit or irrevocable letter of credit to the contractor for 100 percent of the contract value. If the letter of credit is issued by an American bank, contractors have a strong assurance of being paid. A similar letter may be required of the contractor. The fee for a letter of credit is in the range of 1.5 percent of the credit limit amount per year, comparable to the fee associated with payment and performance bonds. If one party defaults, the other party can make a claim against the letter of credit upon approval within a stated period of time, such as 30 days.

While payment and performance bonds are quite common in the United States, they are only beginning to be utilized in many countries. A form of bond that is more common in many countries than in the United States is the owner's bond. The owner's bond is one that the owner provides to the contractor and assures the contractor that the owner will pay for the construction services provided. The surety essentially provides assurances that the owner has the ability to pay for the construction work that is performed. This reduces the contractor's risk of nonpayment to a considerable extent.

PERIODIC CONSTRUCTION PAYMENTS

Payments for construction services are similar to those regularly utilized in the United States. Only a few differences may be anticipated. Since many projects include extensive engineering and design work prior to construction, it is desirable to keep this effort independent of the construction effort. Thus, upon completion of the engineering and design work, payment will be made in full for that work.

Periodic payments are commonly made for construction that is put in place on a monthly basis. A retainage is typically withheld, as in the United States. Some contractors may be successful in negotiating for "up-front" money, such as 10 percent of the contract. If this money is provided, the retainage percentage is adjusted to reflect that. For example, if the up-front payment is 10 percent, the retainage may be boosted from 10 to 20 percent. Thus, the owner will be withholding 10 percent at project completion.

Partial payment (possibly as high as 75 percent) for materials and equipment, often with long lead times for manufacture, may be made when the contractor presents evidence that the materials have been purchased and have been properly warehoused in the United States. When the materials arrive at the job site, the remainder of the payment is made. The contract may also stipulate that the contractor provide an array of spare parts for major pieces of equipment. Training for the operation and maintenance of the equipment may be contractually provided. In many instances the training is provided in the United States, where the facilities may be more effective.

In some cases the owner may elect to handle the shipping of materials and equipment. This may be prudent, particularly if the goods will move more rapidly through customs if shipped by the host government. In addition to the possible savings in shipping time, the contractor does not need to include a duty fee in the bid if the owner elects to handle shipping. This also avoids the need to consider providing bribes (a common procedure in some countries) to hasten the processing of goods through customs. The important point is to have the government handle the material or equipment receipt if customs presents a potential problem.

TIME PROVISIONS

As on most construction projects, time is usually an important aspect of international construction contracts. It is common for liquidated damages provisions to be included. The liquidated damages may consist of a stipulated amount for each day of late completion, with the total allocation not to exceed 10 percent of the contract amount. If a project is delivered late, the owner may simply withhold the retainage. Some contractors acknowledge that if legitimate time extensions will be difficult to obtain, an allowance for liquidated damages (up to 10 percent of the contract) is added to the contract amount. It must be remembered that litigation is not common in many countries. For example, in contracts with royalty or a monarch, some contractors feel that any effort to win a court battle will be in vain.

OWNER ACCEPTANCE

The definition of substantial completion is almost universal. The last periodic payment is generally authorized upon substantial completion. After final acceptance, the retainage is released, typically within 30 to 60 days. The contractor is generally required to warranty the project for one year; again, this is similar to the practice in the United States.

SAFETY

Safety legislation for the protection of construction workers is not as stringent in most third world countries as in the United States. Worker safety is largely in the hands of the contractor. In short, the contractor must initiate most safety programs for the construction project. The absence of stringent safety legislation does not mean that construction firms need not be concerned about worker safety. Some countries have laws that make an injured employee the responsibility of the employer. Such laws are not without clout. Unfortunately, laws in some countries make employers responsible for worker safety, but there is no legal system to enforce compliance with safety regulations. Companies must simply police themselves to avoid the consequences of noncompliance. In addition, much goodwill can be generated by a firm that is a strong proponent of safety.

INSURANCE

Many of the types of insurance that are common on construction projects in the United States are also common on international projects. The only difference is that the coverages are provided by firms that do business in international markets. Coverages for builder's risk and liability are similar to the policies provided for U.S. contractors. Workers' compensation or a similar counterpart is not as common. Socialized medicine and health care constitutes one expenditure that may be higher than is customary in the United States. One type of policy not found in the United States that may be obtained for work in some countries is political risk insurance. *Political risk insurance* is best obtained on projects in countries whose stability is in question. If the government is overthrown, the dependents of employees in the country will continue to be paid by the insurance company. Essentially, this type of insurance guarantees a paycheck to dependents while employees are incarcerated. These payments are based on the payment schedules for the workers. For the financial security that is provided, political risk insurance is relatively inexpensive.

FOREIGN CORRUPT PRACTICES ACT

Construction firms contemplating work in other world markets should be familiar with the Foreign Corrupt Practices Act of 1977. This law makes it a felony for a United States firm to make corrupt disbursements to foreign officials or to candidates for foreign political office. This law was passed to curtail the practice of making payments to individuals in foreign countries for the purpose of having that official's influence assist a firm in obtaining or retaining business in that country. Since other countries may not have comparable laws to prevent these types of "payoffs," many U.S. firms have contended that this act disables American companies and prevents them from competing successfully in some markets. Regardless of the arguments against the Foreign Corrupt Practices Act, firms must be mindful of the requirements, as violations can readily result in criminal prosecution.

UNIQUE CONSIDERATIONS IN OTHER COUNTRIES

The many countries in which construction contracts may be sought include some that vary only slightly in regard to the procedures used for contracting in the United States. However, some countries have specific constraints or requirements unlike any typically encountered in the United States.

Owner Financing

The finances of some owners, particularly in underdeveloped countries or politically unstable countries, are a major concern to some contractors. The contractor may question whether the owner has the necessary funds to finance the project. The contractor may wonder if the contract will be secure if the country's leadership is overthrown or changes. Although construction contracts with such countries may be relatively rare, the concern is a serious one. If the country's finances or political climate are unsure, unique arrangements may have to be established to entice a construction firm to undertake a project. Such arrangements may include the establishment of an escrow account, possibly in another country, from which periodic payments will be made. This account should provide adequate assurance to the contractor that payments for work accomplished will be made, regardless of any changes in the financial or political picture in the host country.

Funding is more secure for U.S. Agency for International Development (AID) projects. This funding is used for the construction of facilities related to food production, food storage, water treatment, sewage treatment, and similar projects. These funds are handled through the World Bank and are generally regarded as being very secure. These projects may have as much as 90 to 100 percent external support. A small portion, covered by the host country, may be allocated for local services, including food, housing, and pay for expatriates. The contract may also stipulate that only a portion of the payments to the contractor will be in U.S. dollars, to assure that a portion of the funds will be expended locally to bolster the local economy.

Some countries wish to avoid the negative balance of trade that accompanies most construction projects. This can be done if the construction firm is asked to undertake the entire project as a profit venture. This has been done successfully by a number of firms undertaking projects in the People's Republic of China.

Essentially, the Chinese government gives limited lease-free use of land to developers. The developer is then permitted to use the land and the constructed project for a stipulated period of time, typically 25 to 30 years. At the end of the lease period, the entire project reverts to the Chinese government. Hotels and similar ventures have been constructed under such arrangements. Such projects require that the construction firm, or its backers, have the necessary funds to finance the entire project for an extended period of time. Furthermore, the project must turn a profit before the end of the lease period.

Rate of Inflation

Although a country may be politically stable, the strength of the local economy may be subject to sudden changes. Any number of events can take place that may alter the value of money in a country. Although U.S. contractors were shocked by a domestic inflation rate of nearly 20 percent in the early 1980s, some countries have had inflation rates above 1000 percent.

Use of Local Labor

Some construction projects are huge undertakings that require the expenditure of considerable funds. If all the construction funds were paid to international firms, none of the funds would remain in the country. Thus, large projects can adversely affect the economies of some countries. If the economy in a country is weak, a requirement may be imposed on the construction firm concerning the percentage of local hire. Requirements on local hires will bolster the local economy, or at least minimize the impact of the project on the country's balance of trade. Furthermore, such requirements help develop the skills of local workers.

Subcontracting

Subcontracting is a common practice in most countries. The specific practices, as implemented in a particular country, should be carefully examined to ensure that no surprises exist. For example, owners may play a stronger role in the area of subcontracting. In some countries, it is becoming more common for the owner to select one or more of the subcontractors on the project. Such subcontractors are called nominated subcontractors, but the general contractor is still expected to enter into a subcontract agreement with the selected specialty contractor. The owner may "nominate" or select the elevator supplier and installer in order that there can be input from the elevator supplier during the design phase. Nominating electrical contractors, mechanical contractors, or pool contractors may also suit the owner's need to have design input from these parties. By nominating certain subcontractors, owners can utilize their skills to make their designs more efficient.

Transfer of Technology

Countries may have the funds to finance large construction projects but lack the necessary design capability. Thus, such countries may seek international design-build or design-construct firms to assume overall responsibility for projects. However, some countries recognize that purchasing design and construction expertise abroad does not improve their own capacity to undertake similar projects. This can be remedied to some extent by incorporating native engineers in the design and construction phases. As with the local hire of a percentage of the field-workers, a requirement may be imposed in which local engineers are integrated into the process. In effect, the local engineers become trainees. The ultimate intent of this arrangement is for the skills of the country's technical personnel to become enhanced, so that eventually projects can be undertaken solely by locals. This may involve an iterative process in which the expertise leading to self-sufficiency is gained only after several projects have been completed. As each subsequent project is completed, the degree of mandatory local involvement will be increased.

Mandated Use of SI Units

The United States is one of only a few countries still commonly using the English system of units on construction projects; the International System of Units (SI units), also called the metric system, has been almost universally adopted. Although U.S. firms can convert to SI units while doing most of their work with the English system, additional efforts will be required in training personnel and double-checking construction documents. This may give a slight competitive edge to firms that do all their work in SI units. Even a firm that has its historical cost data in English units can minimize conversion problems by using one of several software packages that convert English units to SI units.

Differences in Language and Culture

Language differences present an obvious challenge to international firms. However, English is a universal language that poses no serious problems in the contracting arena. Greater difficulty may be expected among workers who do not understand English. This is often resolved by communicating to them through bilingual supervisors on the site.

When control systems are installed on a project, the contract may require that the instructions for the operation of the system be written in English and in the language of the host country. Written documents related to the installation of materials will generally be only in English. Only when the end user must be able to understand specific instructions related to the assembly, disassembly, or operation of an item may another language be required.

What may be underestimated are the differences in culture. Statements, actions, or gestures that are innocuous in one culture may have specific implications in another. This can lead to uncomfortable situations if these differences are not fully understood. Training related to the culture of a society can help alleviate problems in this area. When multinational forces are involved in a project, the training effort may have to be considerable.

Differences in Ethics

Without discussing what constitutes good ethical practices and poor ethical practices, it should be recognized that there are wide differences in these definitions. Even if an owner adheres closely to the accepted practices of U.S. firms, other contacts within the host country may not be as accommodating. For example, customs officials may expect a payoff in exchange for efficient processing. Other officials, such as police or regulatory personnel, may also insist on some form of payment. Examples have occurred where payoffs had to be made to get telephone lines installed, obtain smooth passage of supplies through customs, or move household effects. Some firms have indicated that mail was opened before being delivered. It would be unfair to suggest that such practices are inevitable in an international project, but it would also be inappropriate to state that such occurrences are rare. In certain countries such practices can be anticipated. Firms have hired employees on some large projects who were employed solely for the purpose of making payoffs. U.S. contractors must be careful to avoid such practices, because U.S. laws forbid the use of influence money.

Historically, practices have varied in regard to the means by which contract awards are made. In some cases the practices employed in one country may be readily accepted, while in another country the same practices may be regarded as being blatantly unethical. For example, an agent representing a country to secure construction services has considerable clout and may be unduly influenced by an "under the table" payment. Firms seeking such contracts may simply regard the additional payment as a business expense. Another firm may be unsuccessful in its bid for the contract simply because it regards the practice as unethical and refuses to issue such payments.

In some countries, the securing of international contracts may be made quite difficult by traditional contracting procedures that exclude international firms. The *dango* system practiced in Japan for generations is such a system. It is somewhat of a "good old boys" system in which the firms permitted to compete are preselected. Historically, a firm had to be able to show that it had successfully undertaken similar projects in the past. These projects were often restricted to Japanese firms, effectively eliminating any newcomers from contention. The *dango* system, despite its established history, is rapidly eroding.

REVIEW QUESTIONS

1. What mechanism is used on many international contracts to demonstrate the solvency of the contractor? That is, what is used instead of performance bonds?

2. What ethical concerns may arise when some international construction contracts are awarded?

3. How do some contractors respond to the use of liquidated damages provisions for late completion on international contracts?

4. What mechanisms may be used to give contractors assurances of being paid for construction contracts in politically volatile or economically depressed countries?

5. What laws may apply in some countries, in lieu of workers' compensation laws, that provide for worker safety?

6. What can be done when a firm undertakes a construction project on which many non-English-speaking people with different cultures are employed?

20

METHODS OF DISPUTE RESOLUTION

CONSTRUCTION PROJECTS ARE complex undertakings. They invariably are described by a unique set of drawings and specifications and are generally performed by a general contractor and numerous subcontractors, many of whom have not worked together previously. The unique aspects of each project and the unique constitution of each construction team are common reasons for disagreements to occur. Since each project is unique, not all of its aspects can be anticipated by the designer. This results in omissions and in some cases contradictions within the contract documents. Even though materials may have been addressed by the designer in the contract documents, the various parties involved in the construction process may not agree on the interpretation of that information. Whatever the source of the disagreement, disputes in the construction industry are common. Many construction professionals feel that the frequency of construction disputes has risen, along with the costs of resolving them.

Because conflicts are commonplace on construction projects, the potential disputing parties are well advised to outline the appropriate procedures to follow during disputes. This should be done before starting construction and can be best accomplished through the inclusion of a claims clause. A claims clause permits the contractor to present disputes to the owner without having to resort to litigation as a first step. Claims clauses typically require the contractor to notify the owner of any work items that are in dispute and state that this notification must precede the performance of the disputed work. Failure to give the proper notification is generally construed as a waiver of entitlement to added compensation on disputed items of work. A typical provision is as follows:

> If the Contractor deems additional compensation is warranted for work or materials not covered in the contract or not ordered by the owner as "extra work," according to Section . . . the Contractor shall, prior to beginning work on which the claim will be based, notify the Owner in writing of the intent to make claim and the basis for such claim for additional compensation. If the basis for the claim does not become apparent

until the Contractor has proceeded with the work and it is not feasible to stop the work, the Contractor shall immediately notify the Owner that work is continuing and that written notification of the intent to make a claim will be submitted within 10 calendar days. Failure of the Contractor to give required notification and to provide the Owner with proper facilities and assistance in keeping strict account of actual costs will constitute a waiver of claim for additional compensation in connection with the work already performed. Notification of a claim, and the fact that the Engineer has kept account of the costs involved, shall not be construed as proving or substantiating the claim's validity. All claims by the Contractor for additional compensation shall be submitted in writing within 120 days after completion of the work on which the claim is based.

Once the disputed work is completed, the contractor is generally given a stated period of time in which to present the amount of the claim to the owner. This time period may be as little as 10 days after work completion to 4 months or even longer. The primary aspect of the claims clause is that it establishes a mechanism by which the contractor can request additional compensation. The notification portion of the clause is very important. The notification, as one provision stated, "need not detail the amount of the claim but shall state the facts surrounding the claim in sufficient detail to identify the claim, together with its character and scope." Essentially, the notification places both parties on an equal footing in identifying the disputed work. Thus, both parties can independently monitor the effort required to perform the disputed work.

Some provisions stipulate that when a claim arises, the party making the claim must notify the other party in writing within a specified number of days after the occurrence of the event, or within a specified period after recognizing the condition giving rise to the claim, whichever is later. When such stipulations exist, it is imperative that the contractor file all claims within the stated periods, or all rights for such claims may be lost.

Despite the existence of a dispute, it is to the benefit of the contracting parties for the other aspects of the construction project to be completed without being compromised or adversely affected by the dispute. The following provision specifically states that the owner and contractor will not let disputes impede job progress:

The Owner and the Contractor shall attempt to resolve all disputes by good faith negotiations. If, after good faith negotiations, disputes shall remain between the parties, resolution of such dispute shall be by litigation. Notwithstanding any dispute or litigation between the Contractor and the Owner, the Contractor shall proceed diligently with the performance of the work required by the contract as directed by the Owner.

Disputes can be resolved in many ways, including negotiation, traditional litigation in court, arbitration, mediation, dispute review boards, or other novel means. If the means of resolving disputes is stipulated in the contract with the owner, it is imperative that the general contractor be fully aware of and comply with the specific requirements in order to retain a viable means of pursuing any claim. Furthermore, whatever dispute resolution is stipulated in the contract with the owner, similar means of resolving disputes should be included in the subcontract agreements. If the resolution of disputes is not addressed in the contract, the disputing parties will generally rely on the courts to resolve the claim.

NEGOTIATION

Most disputes in construction have small beginnings. It is at this point that the disagreement can best be resolved—before the opposing parties have gotten embroiled in an emotional conflict. If both parties calmly discuss an issue and listen carefully to each other's comments, a good opportunity will exist for the satisfactory resolution of the matter. They must also recognize that they have a common objective and that they are members of the same team. It is during periods of conflict that this is most difficult to accomplish.

There are many styles of negotiation. Regardless of the particular approach being used, it is important that the point of disagreement be communicated to the opposing party or parties as soon as the matter has been identified as a topic that needs to be resolved. This is when parties have not yet formulated strong positions and will be more likely to try to resolve the matter on its own merits. One party may even see the legitimacy in the other party's point of view and concede the argument. The parties may see that each has a legitimate perspective, and some type of barter agreement may be achieved. For example, the owner may not want to pay for an alleged changed condition, and so a trade-off is made in regard to the scope of work on another part of the project, such as the owner accepting a substitute material that does not fully satisfy the specifications. Of course, different companies have differing philosophies on these types of trades.

As the unresolved dispute becomes more advanced, each disputing party has a greater interest in trying to prevail. In some cases the position of each party becomes more unyielding as each focuses on winning, rather than on resolving the dispute. In the advanced stages of the dispute, the parties may try various techniques to weaken the other party's position, including the threat of a lawsuit.

Some disputes cannot be readily resolved even when the parties negotiate in good faith; that is, failure at resolving a dispute does not necessarily mean the parties have negotiated improperly. It is important, however, that when negotiations fail, the parties not position themselves so that the successful completion of the project is jeopardized. The heated emotions that can result from conflict must be confined to the dispute, and not be permitted to adversely affect the remaining portion of the project.

LITIGATION

When negotiations fail to resolve a dispute and there is no contractual guidance for dispute resolution, the parties will generally find themselves in a lawsuit. Although the normal progression of circumstances would imply that the services of a lawyer are obtained when negotiations break down, this does not necessarily constitute the most effective use of legal services. While most people have reservations, about calling a lawyer until a dispute has fully matured, the prudent move in many instances would be to obtain a legal opinion about an issue in the early stages. Legal counsel can serve a party well in developing a case early to help negotiations, or in pointing out that there may be no foundation for the dispute. In any case, the drafting of the lawsuit should not necessarily coincide with the initial involvement of legal counsel.

The process for filing claims is often specifically addressed in the contract documents. The claims provision often includes a time period in which claims must be made and the appropriate procedure for initiating claims. Failure to follow these requirements may bar a contractor from successfully pursuing a claim against the owner. The following provision illustrates some key elements that may be included in a claims clause:

> If, in any case, the Contractor deems that additional compensation is due for work or material not clearly covered in the contract or not ordered by the Owner as extra work, the Contractor shall notify the Owner in writing of his or her intention to make a claim for such additional compensation before he or she begins the work on which the claim is based. If such notification is not given and the Owner is not granted an opportunity to keep strict account of actual costs as required, then the Contractor hereby agrees to waive any claim for such additional compensation. Proper notification of a potential claim shall not be construed as substantiating the validity of a claim. If the claim is considered valid by the Owner, the claim will be paid as extra work.

If the parties do not agree on the validity of the claim, the contractor may wish to pursue litigation. For this to be successful, the contract should again be examined to determine whether any specific provisions address this subject. The contractor may be required to submit to the owner a formal document that may be called the notice of potential claim or another similar document. This is essentially a means of giving the owner fair notice of the claim potential, thereby granting the owner an equal opportunity to acquire relevant information and develop a case.

There are essentially two court systems: the federal system and the state system. Most cases are resolved in the state system. The federal system typically is used if one of the contracting parties is the federal government, or if the disputing parties reside in different states and the amount in dispute exceeds $10,000. Most construction cases resolved in the federal court system are those in which the decision of the contracting officer is appealed directly to the U.S. Claims Court or appealed from the Board of Contract Appeals.

In essence, the filing of a lawsuit is simply one of the more advanced milestones of the negotiation process. It differs from negotiations in that the costs for legal services will be considerably higher and the end result will be left in the hands of the court. Once the lawsuit is filed, direct communications between the disputing parties are essentially brought to an end, with dialogue being confined to the lawyers.

Once a lawsuit is filed, a number of discovery procedures will occur before the matter is brought to court. This is the fact-finding stage of the case, and will assure both parties that no surprises are going to be presented in the courtroom. The most time-consuming elements consist of interrogatories, depositions, and the review of various records and documents. *Interrogatories* are questions that one party submits to the opposing party; the opposing party is given a specified time frame in which to respond. A series of these interrogatories may be submitted for responses from various parties.

Depositions also consist of questions; however, these questions are generally asked in person. Theoretically, a lawyer could request depositions from all the witnesses supporting the opponent. Through depositions, one party obtains information about the strength of the opposing party's position. A deposition with one individual can take from a few minutes to several days. The parties typically involved in a deposition session include the party being deposed, legal counsel (representatives of each of the disputing parties), and a court reporter. Depositions are conducted with the deposed party under oath. The testimony is in essence a pretrial substitute for testimony from the witness stand. Witnesses must therefore be very careful when answering questions during a deposition. Statements made from the witness stand that contradict comments made in a deposition can compromise the credibility of a witness.

Part of the discovery period may be devoted to reviewing various documents that the other party possesses. Other than material related to the strategy one party plans to follow in preparation for litigation, often called work product, material that is germane to the case must be made available. The information need not be volunteered by one party, but must be presented if specifically requested.

While all these procedures are taking place, the parties never relinquish the option to resolve the dispute themselves. One of the objectives of interrogatories and depositions is to assess the strength of the opposing party's case. In fact, it has been estimated that fewer than 5 percent of the lawsuits filed ever make it to the courtroom. Even after a case is started in the courtroom, the parties may elect to settle the issue on their own. In some instances, one of the parties may begin to see the strength of the other party's case and wish to cut losses by negotiating an out-of-court settlement. In other instances, one party's bluff may be successful in getting the opposing party to settle the case. It may even occur to the parties that the matter can best be resolved mutually while they themselves still control the outcome. Probably over 95 percent of these civil cases are actually resolved outside of the courtroom.

It must also be remembered that most juries are not well versed in the nature of the construction industry. The contractual arrangements on construction projects are unlike those encountered in the rest of the business world. Unfortunately, most construction disputes are complicated, and it would take a unique jury to fully understand all the concepts and render a fully-informed decision. With this kept in perspective, it becomes clear that it is difficult to predict the verdict to be made by a jury. Of course, a lawyer well versed in construction terminology can help a great deal in informing the jury. The careful selection of technical consultants or expert witnesses can also help to clarify key points.

Litigation is very time-consuming with the resolution of the dispute often occurring long after the project is completed. Many courts are backlogged with cases, so a speedy court decision is seldom realized. In addition, the cost of a court settlement is generally quite high. With a heavy investment of time by attorneys, it is common for attorney and court fees to consume 25 to 50 percent or even more of the claim. Once the matter is decided by the court, the issue is essentially closed. The exception is when a case is appealed, and this can occur only when procedural errors have been made by a lower court. Thus, few cases are appealed.

ALTERNATIVE DISPUTE RESOLUTION (ADR) TECHNIQUES

It is common when disputing parties reach a stalemate for the phrase "See you in court!" to be used. This is often said at the culmination of an emotional debate.

Unfortunately, the use of the court system to resolve disputes is often extremely inefficient. The cost of litigation can also be quite high. It is often said that the lawyers are the only parties who win in litigation cases. The resolution of disputes through the judicial system is often very time-consuming. When a lawsuit is filed, the court date may be set two or three years in the future. Much time and energy will be devoted to litigating the case, and this will constitute a drain on the firm's resources. These are resources above and beyond the legal fees, which invariably are significant. In the past two decades, the cost of litigation has become so high that much attention has been focused on alternative methods of dispute resolution. These methods have been developed to accelerate the resolution process and to keep the legal expenses under control. Of the various types that have been employed, arbitration has been used most often. Other methods that have been introduced to resolve construction-related disputes include partnering, mediation, disputes review boards, and minitrials. Their increased use may spur revised versions of these methods or experimentation with other methods.

Partnering

In the late 1980s a new form of dispute resolution emerged that has dramatically changed the way many projects are constructed. This method is called partnering and it is essentially an attempt to change the mind-sets of the parties involved in the construction effort. Partnering is a voluntary approach to establishing teamwork among the contracting parties. With partnering, disputes are ideally resolved at the lowest managerial level. This means that disputes are resolved quickly and often without costly claims. Proponents of partnering contend that partnering reduces the exposure of the contracting parties to claims, lowers the risk of cost overruns, results in better quality projects, fosters open communication, decreases administrative costs, and generally improves project performance. With the improved communications, decisions are made more quickly, and innovations are embraced more than on nonpartnered projects.

Partnering was developed to alter the manner of resolving problems that arise on construction projects. Changes in the original plans, specifications, or contract are seen as sources of added costs to the contractor, owner, or both. Differing site conditions, delays, and change orders are often viewed as opportunities for contractors to make up for lost profit from an improper bid, or losses in other parts of the project. This is clearly a breeding ground for disputes. When disputes cannot be resolved effectively, the owner and the contractor may end up losing. Partnering attempts to change the lose-lose situation into a win-win situation. Partnering attempts to solve disputes early and at the lowest levels of project management.

The partnering process typically begins as soon as the contracting parties have been identified for a project. Participation in partnering must be voluntary. The partnering process generally begins with an invitation to the key parties on a construction project, including the owner, designers, general contractor, key subcontractors, key suppliers, and possibly others. The process continues generally with a facilitated session consisting of organized workshops attended by the key parties to the contract. The workshop sessions will initially focus on team building, group awareness, and conflict awareness. This workshop is generally led by an outside facilitator who helps in getting the parties acquainted with one another, and in developing a cooperative attitude and commitment toward the partnering objective. At the conclusion of the workshop, generally two or three days, a partnering agreement is drafted. The agreement, mission statement, or partnering charter will state the common goals of the participants, including the communication objectives, the framework for resolving conflict, and the performance objectives (avoid claims, reduce costs, complete on time, eliminate delays, zero injuries, positive public relations, etc.) for the project. All participants then sign the charter. If all stakeholders do not "buy into" the concept of partnering, it will generally not succeed. Fundamental to the success of partnering is the fair and equitable sharing of risk.

Fundamental to partnering is changing the view of the parties involved. Under partnering, a problem for one party becomes a problem for both parties. Both parties use their resources and experience to solve problems and keep the project moving toward a successful completion. The personnel involved in solving the problem or dispute are the ones working on-site and most familiar with the project. This is in sharp contrast to the traditional approach to resolving problems.

The use of partnering is more than just a change in contract administration; rather it is the use of good common sense. It consists of getting along with people and doing the work at hand in a mature and honorable way. While there are several definitions of partnering, they all have the same focus. They stress changing the traditional adversarial owner-contractor relationship to one of cooperation and the achievement of mutual goals. It is important that all members of the partnerships stay in continual contact with each other, that all matters of the contract be discussed as issues come up, and that issues be resolved at the earliest time and at the lowest possible level.

There are several keys that make a successful partnering relationship: trust, commitment, and a shared vision. In private construction, partnering seeks to be a long-term relationship. In public works, where the low bidder gets the contract, the partnering arrangement can be developed only after awarding the contract; the partnering process terminates with the completion of the job. In any case, participation in partnering is voluntary.

The successes of partnering have resulted in many public agencies implementing partnering arrangements on many of their construction projects. For most of these agencies, partnering is considered only on the larger projects, namely, those in the vicinity of $1 million or more. For these agencies, the invitation to partner is offered soon after the low bidder is identified. Some public agencies have reported resounding successes. In some cases, projects were completed in time periods far shorter than originally expected. Savings through value engineering have been highly praised and attributed to partnering.

There are some detractors who are not fond of partnering. Some owners contend that the owner's representatives try to "keep the peace" at the expense of project quality, or that the inspectors simply do not enforce the specifications as rigorously on partnered projects. This compromising response to partnering is allegedly to keep from provoking a claim. Nonetheless there are many proponents of partnering, and the concept will undoubtedly continue to be used on many projects.

There are numerous benefits to partnering a project. Most of the results of partnering are difficult to quantify, but they are generally perceived by the partners as being beneficial. One quantified benefit is the decrease in litigation and the number of unresolved conflicts at project completion. The open communications and teamwork approach solves problems as they develop. The problems are solved by on-site personnel who can make informed decisions. This has eliminated escalating the problem to higher management and evolving the problem into an "us against them" approach. Partnering must begin at project start. Partnering will generally not be able to be successfully implemented after disputes have developed at the project level.

Mediation

Mediation is a nonbinding method of dispute resolution that contains elements of negotiation and arbitration. A mediator might be regarded as a third party who tries to force or persuade the disputing parties to agree on an appropriate settlement of an issue. A mediator must have strong skills of negotiation and should be able to grasp the technical aspects of construction disputes. The procedures followed for mediation are not as formalized as those for arbitration. Different mediators may use very different approaches. Mediation is the most popular of the recent ADR methods.

If a dispute occurs on a construction project, the parties may agree to try to resolve the matter through mediation. A mediator who is mutually agreed upon by both parties is contacted, and a date (usually one day only) is set for the mediation. Depending on the nature of the dispute, relevant documentation may be sent to the mediator for review. This information is generally restricted to crucial information that can be reviewed by an expert in a fairly short time.

The choice of a mediator consists primarily of locating an expert whose primary livelihood is serving as a mediator. Mediators tend to be very good at what they do. The fee is generally fixed, and can be as high as $3,000 or $4,000 for a one-day mediation. Mediators sell their services as not being one-sided; that is, their strength is in being able to get both parties to agree on a common solution. The fee applies regardless of the success in reaching a resolution of the dispute. The best mediators keep statistics on the percentage of disputes for which settlements have been obtained through their efforts. These mediators try to maintain their records by using their expertise to resolve every dispute they undertake. If the mediation is successful, much has been gained because the dispute resolution is quick (one day) and the cost is limited to the mediator's fee, which is paid by both disputing parties.

As has already been mentioned, prior to the mediation day the mediator may be apprised in general terms of the nature of the dispute. Then, on the day of mediation, the mediator meets with both disputing parties. Both parties then are given a short period of time, possibly a half hour, in which to present their cases to the mediator. The parties may also be requested to prepare a one-page summary of their respective arguments. This period is used primarily to orient the mediator to the root causes of the problem and the basic differences in the perspectives of the disputing parties.

After the mediator has been oriented, representatives of the disputing parties are assigned to different rooms. These rooms are generally close to each other, as the mediator will shuttle between them for the remainder of the day. Once assigned to separate rooms, the parties will not be assembled together again during the mediation unless a resolution is obtained. Essentially, the mediator will address one of the disputing parties and try to point out reasons why that party should try to moderate its demands. The mediator will then go to the other party, possibly with counterproposals, and try to get that party to give in to some extent. During this shuttling between the disputing parties, the mediator's skills are tested to the greatest extent. Some of these skills involve negotiation, and may relate very little to the merits of either party's case. For example, if the parties' demands are close, with neither party wishing to budge, the negotiator may stress that failure to accept the given terms will mean litigation and all the uncertainty, delay, and cost associated with it. Many negotiation skills will be brought to bear to get the parties to resolve their differences. The technical skills of the mediator may also prove to be a compelling asset. For example, the mediator may point out particular weaknesses in an argument made by one of the parties. The mediator may use technical skills to get one of the parties to soften its argument, while using negotiation skills to move the other party from its position.

The mediator will continue to work with the parties throughout the day in order to resolve the dispute. Once the parties have agreed on a common resolution, they are brought together and the issue is resolved. If the day ends without a settlement, the parties are left to their own devices to resolve the dispute. In one instance where the case was not resolved by the end of the day but the mediator felt that the parties were close to an agreement, the mediator volunteered to return for a second day at no expense. This was done in part because the parties were close to converging on a settlement, and the mediator wanted every opportunity to keep her success record intact.

If a case is not resolved through mediation and subsequently goes to court, the information brought forth during mediation will have little value. No records are kept of the summations of the parties or of the comments of the mediator. Furthermore, the mediator is never asked to formulate an opinion on the proper resolution of the case, as an arbitrator might do.

Mediation has some strong points in resolving disputes, particularly claims that involve tens of thousands of dollars. The resolution is fast compared with other methods, typically one day. The cost is simply the cost of obtaining the services of a mediator. Throughout the mediation process the resolution is controlled solely by the disputing parties; that is, either party can stop the procedure at any time. The mediator is simply a vehicle for the parties to seek a compromise or a solution. The mediator has no authority to decide the outcome of a dispute. The disputing parties must therefore enter the mediation process in good faith, and mutually work toward an acceptable resolution of the dispute. When the disputing parties have a true desire to reach a resolution, mediation will generally be successful.

Arbitration

Arbitration has traditionally been the most popular alternative to litigation. The use of arbitration in the construction industry dates back to the nineteenth century. It became particularly popular as a means of resolving construction disputes in the 1960s. Many construction contracts now require that disputes be resolved by binding arbitration, particularly construction contracts in the private sector. Binding arbitration means that the disputing parties will agree to adhere to the decision reached. Without the contractual requirement for arbitration, litigation is the only option unless the parties agree to another means of resolving the dispute. State law may preclude the use of arbitration in some instances. Construction contracts that state that disputes will be resolved by binding arbitration often stipulate that the procedures of the American Arbitration Association be used. The following provision is typical of provisions that mandate the use of arbitration to resolve disputes:

> Claims (demands for monetary compensation or damages) arising under or related to performance of the contract shall be resolved by arbitration unless the Owner and the Contractor agree in writing, after the claim has arisen, to waive arbitration and to have the claim litigated in a court of competent jurisdiction. The arbitration decision shall be decided under and in accordance with the law of this State, supported by substantial evidence and, in writing, contain the basis for the decision, findings of fact, and conclusions of law. Arbitration shall be initiated by a Demand for Arbitration made in compliance with the requirements of said regulations. A Demand for Arbitration by the Contractor shall be made not later than 180 days after the date of the cause of action.

Provisions mandating binding arbitration for resolving disputes are quite powerful. The signatories do not have any alternatives unless they agree on another method.

The American Arbitration Association maintains a slate of prequalified potential arbitrators, including lawyers, contractors, claims consultants, architects, and engineers, who can be hired to resolve disputes. These arbitrators are knowledgeable about the construction industry, understand the vocabulary, and can readily grasp the nature of a dispute involving technical matters. This is one aspect of arbitration that represents a particular advantage compared with litigation. Jurors rarely can fully appreciate or understand all the technical details germane to a case. With the ability of the arbitrators to quickly grasp the nature of the problem, the arbitration process often can resolve disputes in a matter of days, while a court trial might take several weeks. The decisions of the arbitrators are generally not one-sided, as is commonly the case in court decisions. Since they can more easily visualize the various points of view, a more moderate resolution will often be the result. There will still be winners and losers, but the actual settlement costs for the losing parties will not be as high as they often are in court decisions.

While a single arbitrator may be utilized to resolve a small dispute, it is common to use three arbitrators for larger disputes. The selection process may vary. One way to organize the panel is for each disputing party to select one individual from the list of arbitrators, with the third arbitrator being selected by the arbitration association, jointly by the disputing parties, or by the arbitrators already selected. The disputing parties may each be given a list of potential arbitrators, and each party will indicate, perhaps by means of a ranking system, which individuals are deemed acceptable. The arbitrator or arbitrators for the dispute will be chosen from among the potential arbitrators selected by both parties. The arbitrators may also be selected by the association. If there is a single arbitrator, it is common for the association to play a significant role in the selection process.

Although the arbitration process may begin almost immediately after a dispute is presented, the arbitration hearings may be held somewhat sporadically. These hearings are essentially held at the convenience of the arbitrators, many of whom are professionals with a variety of other commitments.

Arbitration differs from litigation in that the rules of evidence are eased and the discovery proceedings are not formally defined. The hearings could be classified as being informal. Information that might be ruled inadmissible in court may form the basis for the ruling of the arbitrators. Case law and precedent have little influence on arbitration decisions. Thus, arbitration hearings occasionally include surprises as new information is presented. These surprises are essentially nonexistent in litigation, where discovery proceedings are carefully conducted.

When the arbitration panel rules on a particular dispute, the decision is binding or final. That is, the disputing parties cannot appeal the decision unless they can prove fraud, lack of impartiality, conflict of interest, or bad faith, or show that the scope of authority of the arbitrators was exceeded. Unlike litigation, the rationale for an arbitration decision does not become public record. This aspect of confidentiality appeals to many firms.

The advantages of arbitration compared with litigation are that it is less time-consuming and less expensive. The following example demonstrates this point. A contractor constructed a new building for a convenience store. Shortly after project completion, the owner of the store complained to the contractor that water was entering the basement after every heavy rainfall. The matter was not resolved between the store owner and the contractor, and so they agreed to have it resolved by means of arbitration. A single arbitrator was assigned to resolve the dispute. Shortly afterward, the arbitrator met the contractor and the owner at the store in which the water problem existed. The arbitrator reviewed the plans and evaluated the site conditions. Several holes were dug by hand to establish the location of the drainpipe on the uphill portion of the building. It was quickly discovered that the drainpipe existed, but that the discharge end had been inadvertently buried, preventing the escape of trapped water. Thus, in the company of the disputing parties, the arbitrator found the source of the problem. The arbitrator ruled that the contractor should correctly install the drainpipe and pay for the damage caused by the water in the basement. The dispute was resolved in a matter of hours. The cost of the arbitration consisted of the filing fee and a modest fee for the arbitrator. A court resolution of the dispute would have taken several months.

Not all cases can be resolved quickly. Many cases are very involved and technically complex. Much time will be required to establish all the facts, and this can be costly. Arbitrators are paid for their time, with little or no costs paid by the taxpayers. On complex cases, three arbitrators are commonly used to constitute the panel. The fees for arbitrators vary. The fee for an arbitrator on a small case may be fixed at $50. On other cases, the arbitrators may each be paid over $100 per hour. Although the arbitrators are trying to resolve the dispute, each disputing party will generally continue to have legal counsel involved in all the hearings, particularly in larger disputes. In short, the advantages of arbitration in terms of time and money are minor or nonexistent in some disputes.

Arbitration is generally restricted to disputes between two parties; third parties are specifically excluded. For example, the general contractor may have a dispute with the owner and present the case for arbitration. The contractor wants to claim a differing site condition for the discovery of asbestos on a renovation project. The owner wants to implicate the architect for drafting faulty drawings and specifications. The architect in turn contends that the error lies with the consulting engineer who inspected the building. Arbitration rules do not let these third parties—the architect and consulting engineer—become parties to the arbitration decision. While they may be included in typical litigation, third parties are always excluded from arbitration. In an incident like the one above, arbitration appears to be inefficient.

If arbitration is deemed desirable to resolve disputes on a project, a contractual requirement for binding arbitration should be included. This provision should address the scope of arbitration and the procedures to be employed. A requirement that the procedures of the American Arbitration Association be used for arbitrating disputes will simplify dispute resolution.

Agreements can be entered into in which arbitration is the agreed-upon means of resolving disputes. This was shown in *Pro Tech Industries v. URS Corp* (377 F.3d 868). In this case, URS Corporation was the general contractor on an environmental reclamation project in New Mexico, and Pro Tech Industries, Inc., was a subcontractor for the pipe work. According to the subcontract agreement, the parties agreed to arbitrate all disputes. A dispute arose when Pro Tech removed its personnel and equipment from the site before completing its work. Pro Tech then filed suit against URS in state court for additional work that it claimed it had performed. The case was moved to federal court jurisdiction since the project was operated by the National Aeronautics and Space Administration's Johnson Space Center. URS tried to compel Pro Tech to enter into arbitration in accord with the subcontract agreement. Pro Tech argued against arbitration, since it did not have the financial resources to enter arbitration. The court relied heavily on the subcontract agreement that called for disputes to be resolved via arbitration. The court stated that the status of a company's financial resources was not a valid issue to bar arbitration.

Disputes Review Board

In the attempt to come up with alternatives to litigation and arbitration, another method has emerged: the use of *disputes review boards.* This type of board generally consists of three individuals who meet whenever one of the contracting parties on a project desires a hearing on an issue of conflict. Since the board is assembled early in the life of a construction project, disputes can be resolved quickly. Although various formats exist, perhaps the most widely accepted approach is detailed in a set of guidelines prepared under the joint sponsorship of the American Society of Civil Engineers (Construction Division) and the American Institute of Mining Engineers.

The board is assembled shortly after the contract award. Each contracting party nominates a person to serve on the board. Each nominee must then be approved by the other contracting party. After two board members have been selected, they jointly decide on the third member, normally the chair for the board. The compensation for the board members is split between the owner and the contractor. The owner is typically solely responsible for absorbing any other operating costs of the board.

The procedures by which the board operates are typically determined by the board members. These procedures are generally fairly informal. The board will be expected to stay informed about progress on the construction project through regular meetings. This will provide assurance that the board can grasp the nature of any conflicts or disputes that arise.

The board will review any disputes that are presented to it at its regular meetings. This aspect of presenting a case to the board can be unilateral; the two disputing parties need not both present the case to the board. However, the owner and the contractor should try to negotiate their own resolution of disputes. Cases brought to the board should be those for which the owner and contractor have exhausted other means within their organizations to resolve the dispute. The owner and the contractor are expected to have representation whenever a dispute is considered by the board. The board will act alone, however, when deliberations take place. Construction progress is generally expected to continue as the board reviews a dispute.

The board is expected to prepare a set of recommendations for each dispute it reviews. Ideally, the decisions of the board should be unanimous. Decisions that are not unanimous should have the dissenting member's opinions or views expressed in the recommendations. The contractor and the owner are given a period of time, such as two weeks, to consider the board's recommendations, which can be accepted or rejected. If either party does not accept the recommendations, the dispute can be appealed to the board for another review, or other methods of dispute resolution may be pursued. The decisions of the board, unlike many arbitration decisions, are not binding; that is, litigation is still an option if one of the parties is not satisfied with the decision. Of course, appeals are generally made reluctantly, since the board is made up of carefully selected professionals or experts. In addition, it can be contractually stipulated that all records and recommendations of the board will be admissible in court if litigation ensues. This will be a further deterrent to litigating a board decision.

The 2007 American Institute of Architects (AIA) documents (A201) permit the contracting parties to choose the means of resolving disputes, whether by mediation, arbitration, litigation, or a variation of the dispute resolution board. The dispute resolution board provision carries with it an option of using a board/panel or an Initial Decision Maker (IDM). The IDM is an outside third party who can be selected by the disputing contractor and owner; if none is selected, the architect will become the IDM by default. This is a change from the previous version where the architect was designated as the initial arbiter for disputes. The new documents recognize that the architect might not always be the most impartial party in a dispute between the owner and the contractor. Architects might be biased in a dispute when one of the disputing parties is actually compensating the architect. The dispute might also involve design errors or omissions in which impartiality by the architect will also be questioned.

The IDM is charged with reviewing the circumstances related to the dispute, making preliminary assessments of the arguments, and facilitating communications between the disputing parties. At the start of a working relationship, the parties to a contract should decide which mode of dispute resolution will be used. If the disputes review board or IDM will be employed, the selection of the board or IDM should be made early, so that these third parties can remain informed about the project as it evolves.

According to the revised American Institute of Architects contract documents, arbitration is an option that the contracting parties can select, but it is no longer a requirement. Of the different dispute resolution techniques, the contracting parties will be asked to simply select the desired method. If no method is selected, the formal process of litigation will become the default option.

Minitrials

Another alternative method of resolving disputes that has emerged is the *minitrial,* also called private litigation. As the name implies, this method has some of the features of the courtroom, but it involves less than a full-blown courtroom procedure. There is no current model by which all minitrials are conducted. Instead, the parties to a dispute are free to draft their own procedures. Many versions of minitrials may be used.

Minitrials are not commonly addressed in construction contract documents, and so the parties must agree to a minitrial as an appropriate means to settle a dispute after the dispute has materialized. The parties must have similar objectives in order for an agreement to be reached on how the minitrial will be conducted. The disputing parties may decide beforehand if the resolution of the dispute by minitrial will be binding or not. The paramount objective should be to resolve the dispute quickly and at minimal cost. The following description involves one set of rules upon which the parties might agree.

The parties must first agree on the party who will hear the case and render a decision. It is possible that more than one party will be selected. These individuals will then act as a jury. If one person is selected, the option most often taken, that individual will act as the judge. This person must be agreed upon by both parties and should be very familiar with the construction industry. Above all, this individual must have a reputation for impartiality. Candidates who may act as the "judge pro tem" may be selected from among former judges, construction attorneys, construction claims consultants, university professors specializing in construction, or others who are knowledgeable about construction.

The minitrial rules must also be established. They may be very specific or may be outlined in general terms. A major decision to be made by the parties is whether the judgment rendered through the minitrial will be binding. Those placing great trust in minitrials advocate that the judgment will be binding upon both parties. Thus, the parties will know from the start that whatever the outcome, the decision of the minitrial will be final. This will provide a good incentive for each party to present the best possible case.

The procedures for conducting the minitrial can then be established. One possible procedure is to set a time limit for the minitrial. Although the time limitations may vary depending on the nature, scope, or complexity of the dispute, the parties should be able to establish time limitations that are satisfactory to both. With the realization that a court case can involve weeks of testimony, the minitrial may be limited to three days. For example, on the first day one party will be given six hours to present its case. As the case is being presented, questions may be asked by the judge pro tem or by opposing counsel. After this, the other party will be given two hours in which to present a rebuttal. On the second day the roles will be reversed; the other party will have six hours to present its case, and the first party will have two hours to provide a rebuttal. On the third day each party will have one hour to give any additional information, one hour to provide closing comments, and one hour to give a final rebuttal. The sequence and time limitations can be altered as the parties deem appropriate. The important point is that each party has an equal amount of time to present information regarding the dispute.

Other procedural issues may also be agreed upon by the parties, such as the type of evidence that is admissible, the number of witnesses to be presented, and the extent of adherence to discovery procedures. Naturally, it might appear reasonable for the parties to relax the rules on what is admissible in a minitrial. The parties may also agree on the limits of any settlements, and whether these limits will be presented to the judge prior to the judgment. The parties may ask the judge to provide a summary statement regarding the judgment. A court reporter may record the proceedings, and these can be utilized in a court of law if a resolution is not accepted by both parties.

Since minitrial procedures are not formalized, different versions will be utilized. One approach is to simply utilize a neutral advisor who listens to both sides of the dispute. The key issues are presented to the neutral advisor in the presence of executives or principals from each party. The principals discuss the potential settlement with each other with expert insight being provided by the neutral advisor. Thus, the disputing parties have some advance insight about how the dispute might be resolved if it were to go to court. Under this scenario, the disputing parties attempt to negotiate a settlement.

With successful minitrial procedures, the parties will be able to put the dispute behind them within three days, and the results are confidential. The legal costs will be much lower than they would be if the dispute were litigated. Minitrials may be agreed upon after the parties have unsuccessfully negotiated a settlement, after an arbitration decision that was not binding, or even after unsuccessful mediation. With an impartial judge who understands construction disputes, a quick and fair decision is possible in many instances.

THE IMPORTANCE OF DOCUMENTATION

Once the construction contract has been signed or the notice to proceed has been given to the contractor, construction operations can begin. These activities will be undertaken with guidance provided by the contract documents. The drawings and the technical specifications will be the primary resource for directing the contractor's operations. The general conditions will serve as a guide for establishing procedures to be followed when the contractor and owner interact in a wide variety of areas. If the drawings and technical specifications were "perfect," perhaps the only interaction of note would relate to payments made to the contractor. Unfortunately, this is not an ideal world; conflicts and problems are invariably encountered on a construction project. Many of these problems (changes, differing site conditions, delays, and other disputed issues) are addressed in the general conditions, but disputes are still common.

It would not be healthy for the contractor or the owner to enter into an agreement with the intent of subsequently being involved in litigation. However, the possibility of litigation cannot be ignored. Disputes can arise from a wide variety of sources. Many disputes could be eliminated if there was clear communication between all parties, because misunderstandings often cause disputes. These misunderstandings are frequently caused by verbal communications that may be clear when they take place, but subsequently deteriorate through distortion and memory loss. In fact, some conversations take place in which only seemingly unimportant matters are discussed. Under such circumstances, no great value is placed on the interchange. Unfortunately, the information shared and the decisions made at such interchanges often become the crucial subject matter of major disputes. For this reason, it is important that all these communications be documented.

Documentation of information does not mean that one anticipates a subsequent dispute, but it is impossible to predict which subjects will become sources of dispute. Therefore, all communications should be well documented. If there is a large disparity between the disputing parties in the amount or quality of information that is documented, a clear advantage will be enjoyed by the party exhibiting the greatest detail of documentation, whether in the form of telephone logs, internal memoranda, information on conversations, and so on. Not only will this documentation be valuable if a dispute arises, but the probability of a dispute will be diminished if information is carefully recorded. Perhaps the best way to avoid disputes is to treat all information as if it might become the subject of a dispute. In other words, the best way to avoid disputes is to always be prepared for them.

It should be obvious that the purpose of documentation is to avoid disputes. For example, if a discussion takes place between the owner and the contractor in which a seemingly minor detail is resolved, it is prudent to put this in writing and share the written documentation with the other party. Upon receipt of the written description of the resolution, the recipient is in a position to take issue with the information if it is in error, or to file it if it appears accurate. If the information is filed without any corrections, the subject will probably not be the source of any disputes.

Since it is never known which items or topics will result in a dispute, the proper approach is for all parties to document all communications. In some cases the documentation is done merely so that specific items can be more easily retrieved when needed. The types of documentation that are maintained include logs of incoming correspondence, logs of all letters sent, logs of photos, records of conversations and meetings, records of safety meetings, and daily job diaries. A wide assortment of additional records should also be kept. Figures 20.1 through 20.8 show some forms that can be used to document valuable information.

LOG OF INCOMING CORRESPONDENCE
Contract Title:_____
Contract No.:_____

SER. #	DATE	FROM	SUBJECT	ROUTED TO	FILE LOCATION	COMMENTS

Log of in

FIGURE 20.1

LOG OF OUTGOING CORRESPONDENCE
Contract Title:_____
Contract No.:_____

SER. #	DATE	PREPARED BY	SENT TO	SUBJECT	FILE LOCATION	COMMENTS

Log of out

FIGURE 20.2

CE & M Constructors, Inc. **PROJECT PHOTO LOG**

Project Title: _____

Project Number: _____

Roll #	Photo #	Date Taken	Taken by	Location and Subject of Photo	Devel. Dates		Files	
					Sent	Return	When?	Where?

Instructions:
1. Number rolls and photos sequentially in the order taken
2. Record the following information on the back of each photo prior to filing:
   a. Project number and title
   b. When, where and by whom the picture was taken
   c. A brief description of what the picture is supposed to show

Photo Log

FIGURE 20.3

CE & M Constructors, Inc.
5725 Olympic Avenue
Seattle, Washington

REPORT OF MEETING
OR PHONE CONVERSATION   ☐ ☐ (check one)

DATE _____ START TIME/FINISH TIME _____

JOB NAME _____ LOCATION _____

JOB NO. _____

SUBJECT _____

PARTICIPANTS:

_____   _____
_____   _____
_____   _____
_____   _____

SUMMARY
Note all decisions made: Specify who will do any actions required and when those actions must be completed.

_____
_____
_____
_____
_____
_____
_____
_____
_____
_____
_____
_____

CC TO

_____   _____
_____   _____
_____   _____
_____   _____

PREPARED BY _____

Report of meeting

FIGURE 20.4

CE & M Constructors, Inc.
5725 Olympic Avenue
Seattle, Washington

Weekly Safety Meeting

Project: _____ Date: _____

Presided By: _____

Attendees:

_____   _____   _____
_____   _____   _____
_____   _____   _____
_____   _____   _____
_____   _____   _____
_____   _____   _____

Topics Presented:

_____
_____
_____
_____
_____
_____
_____
_____
_____
_____

Comments:

_____
_____
_____
_____
_____
_____
_____
_____

Supervisor's Signature: _____

Safety Meeting

FIGURE 20.5

CE & M Constructors, Inc.
5725 Olympic Avenue
Seattle, Washington

DAILY JOB DIARY

DATE _____   PROJECT NAME _____

PREPARED BY _____   PROJECT NO. _____

CE & M PERSONNEL ON SITE	SUBCONTRACTORS ON SITE	MATERIAL DELIVERED
SUPERINTENDENT _____	EXCAVATORS _____	_____
CARPENTER FOREMAN _____	PLUMBERS _____	_____
CARPENTERS _____	ELECTRICIANS _____	_____
CARPENTER APPRENTICE _____	PIPEFITTERS _____	_____
LABOR FOREMAN _____	CONCRETE _____	_____
LABORERS _____	HVAC _____	_____
CEMENT MASON FOREMAN _____	FRAMING _____	_____
CEMENT MASON _____	CABINET _____	_____
IRON WORKER FOREMAN _____	ROOFING _____	_____
IRON WORKERS _____	MASONRY _____	_____
OPERATING ENGINEERS _____	GLASS _____	**EQUIPMENT RECEIVED**
TEAMSTERS _____	INSULATION _____	_____
_____ _____	ACOUSTICAL _____	_____
_____ _____	PAINTING _____	_____
_____ _____	GYP. BOARD _____	_____
	FLOOR COVERING _____	_____
	ASPHALT _____	_____
	_____ _____	_____
	_____ _____	_____
	_____ _____	_____

WEATHER	VISITORS		
	NAME	FIRM	PURPOSE
CLEAR _____ FOG _____			
WARM _____ COLD _____	_____		
RAIN:LIGHT _____ OVR CST _____	_____		
(SNOW)INTER _____ WINDY _____	_____		
HEAVY _____	_____		
CONT _____	_____		
APPROX. TEMP. AM ___ PM ___			

DAILY ACTIVITY PROGRESS

COST CODE	ACTIVITY DESCRIPTION

SAFETY NOTES: _____

COMMENTS: _____

Daily diary

FIGURE 20.6

CE & M Constructors, Inc.
5725 Olympic Avenue
Seattle, Washington

☐ MATERIAL SUPPLIER
☐ SUBCONTRACT

Telephone Quotation

Project _____ Date _____

_____ Time Received _____

Firm _____ Taken by _____

Address _____

Talked to _____ Phone _____

Tax Included _____ Add Tax _____ FOB Jobsite _____ Material Only _____

Bond Included _____ Add Bond_____ Discount _____ Installed (mat'l & labor) _____

Delivery Date _____ Quote Valid Until_____ Addendas Noted _____

MBE _____ WBE _____ Specifications Section(s) _____

ITEM NO.	QUANTITY	UNIT	DESCRIPTION	UNIT PRICE	TOTAL
				TOTAL BID	

ADDITIONS OR DELETIONS/REMARKS

	Total Adjustment	
	Adjusted Total Bid	

Tele Quote

FIGURE 20.7

CE & M Constructors, Inc.
5725 Olympic Avenue
Seattle, Washington

MEMO

DATE: _____

ACTION REQUIRED:
- ☐ REPLY
- ☐ FOR YOUR INFORMATION
- ☐ OTHER (SPECIFY) _____

SUBJECT: _____

TO _____

_____   _____

_____   _____

_____   _____

(Fold Here For Windowed Envelope)

MEMO: _____

_____

_____

_____

_____

_____

_____

_____

_____

_____

_____

SIGNED: _____

ENCLOSURES: _____

REPLY: _____

_____

_____

_____

_____

_____

_____

_____

_____

_____

_____

CC: Original to addressee (to be returned if reply is required)
1st copy to addressee (to retain)
2nd copy retained by sender (for file)
Other copies to: _____

Memo

FIGURE 20.8

REVIEW QUESTIONS

1. What are some of the most frequently cited disadvantages of resolving disputes through court decisions?
2. How does arbitration differ from conventional resolution of disputes through court decisions?
3. What are some advantages of mediation compared with arbitration?
4. What are some disadvantages of mediation compared with arbitration?
5. What are some advantages of using disputes review boards instead of litigation, arbitration, or mediation?
6. What are the disadvantages of using disputes review boards to resolve construction disputes?
7. What are some advantages of using minitrials for resolving disputes?
8. What are some disadvantages of using minitrials for resolving disputes?
9. Compare the various means of resolving disputes in regard to time and cost.
10. How does partnering assist in resolving disputes?

21

PROFESSIONAL ETHICS

MOST SOCIETIES RELY on laws to control or guide behavior. These laws consist of common-law practices and statutes. *Common law* need not be written; it is defined through traditional usage and custom. As a result, common law may differ between societies or between states within a country. *Statute law* is established by legislative action. Statute laws may be enacted by a variety of governments: municipal, county, state, or federal. Laws can also be created or defined through judicial review. If two laws are in conflict, the law enacted by the higher governing body will control or dominate. If one law is more stringent than another, the more stringent law will govern.

Although laws have been enacted to provide guidance for seemingly all forms of conduct, laws are still considered to be inadequate. It must be recognized that individuals can inflict pain or cause harm to others through actions that may be legal. Under such circumstances, the laws fall short of their full intent. When laws are deemed to be inadequate, codes of ethics and standards of moral conduct are formulated. Morals and codes of ethics are very closely related. *Morals* tend to relate to norms in a society, while *ethics* often relate to practices within a profession. Morals are rules of conduct with reference to standards of right and wrong. Ethics are rules or standards governing the conduct of the members of a profession. Even without the formal drafting of these codes, professionals as a group aspire to higher ideals of behavior. Professionals are often held in high regard by society, and consequently, they want to formalize their code of behavior to help their associates live up to that level of esteem. The role of ethics is shown conceptually in figure 21.1.

The definition of ethics is by no means absolute. This is exemplified by several quotes by famous individuals regarding morals and ethics. (*Note:* All quotes as reported in the *Dictionary of Quotable Definitions,* edited by Eugene E. Brussell, Prentice-Hall, Inc., Englewood Cliffs, New Jersey, 1970.) These quotes reveal the breadth of the definition of this term.

Legal behavior   Illegal behavior

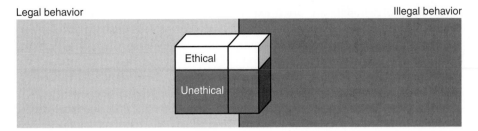

FIGURE 21.1
Conceptualization of ethical and unethical behavior in relation to legal and illegal behavior.

Henry Adams: "a private and costly luxury"

Karl Barth: "a terribly thin covering of ice over a sea of primitive barbarity"

Ambrose Bierce: "conforming to a local and mutable [inconsistent] standard of right"

J. V. Casserley: "not only the way in which we behave towards our neighbors, but also the way in which we cling to the integrity of our own thinking"

Gilbert Keith Chesterton: "drawing the line somewhere"

Sigmund Freud: "feeling temptation but resisting it"

Elbert Hubbard: "the line of conduct that pays"

Thomas Henry Huxley: "to have done, once and for all, with lying"

Edgar Lee Masters: "a hollow tooth which must be propped with gold"

G. E. Moore: "merely statements that certain kinds of actions will have good effects"

Friedrich Nietzsche: "the best of all devices for leading mankind by the nose"

Charles Peirce: "to obey the traditional maxims of your community without hesitation or discussion"

Leon Roth: "the meeting-place between the human and divine"

Herbert Spencer: "the regulation of conduct in such a way that pain shall not be inflicted"

Francis Thompson: "the act of defining your principles to oppose your practices"

Oliver Wendell Holmes: "a body of imperfect social generalizations expressed in terms of emotions"

Reginald A. Rogers: "the science which investigates the general principles for determining the true worth of the ultimate ends of human conduct"

Albert Schweitzer: "obeying the compulsion to help all life which one is able, while shrinking from injuring anything that lives"

The definition of morals or ethics is not concise. There are frequent references to right and wrong, but these too are not absolute terms. Behavior that is right in one culture may be considered wrong in another. Even within the same culture the interpretations may be vague and even contradictory. Contributing to this vagueness is the fact that individuals often make their own interpretations of proper moral and ethical behavior.

Since professionals are often expected to have standards that are more re-stricted than what the law permits, it is common for professionals to develop for-mal codes of ethical practice. Even among professionals in an industry who do not belong to a society that has drafted a code of ethics, many behavioral standards are established through practice. On an individual basis, these standards can be ac-cepted as is, or modified to become guidelines for an individual's actions.

To a large extent, questions of whether an action is ethical or unethical can be answered by whether the action does any harm to anyone, takes advantage of any-one, or gives one party an unfair advantage. It ends up basically being a question of whether an action is good or bad. The distinction between good and bad is not al-ways clear, particularly for people placed in situations where the rewards for their actions are so significant that they cannot be objective. Some public figures state that it is not just a matter of good versus bad, but that actions must also be tempered by how they appear. A public figure may be perfectly innocent in having lunch with a contractor who happens to be a neighbor, yet some public figures avoid appearing in public with individuals to whom contracts might be awarded, simply because of the way in which that action may be interpreted. Some people argue that the distinction might be valid if contracts, such as those for design, are awarded through a proce-dure other than competitive bidding. It is contended that if awards are based on com-petitive bidding, the friendly association of a public official with a contractor is not in and of itself unethical. Other people retort that even on competitively bid projects, one party's judgment may be compromised by this type of association. For example, is the public figure more inclined to concede to a request for a time extension or a change order if a friendly association exists between the contractor and the public figure? Clearly, ethics comes into play when one party's judgment becomes influ-enced by an association with another party. It is also clear that some discretion may be warranted to avoid the suggestion that an improper action is taking place.

CODES OF ETHICS

Because of individual variations in the interpretations of proper behavior for particu-lar groups, professional groups commonly draft a formal document to define the desired behavior of the profession. This *code of ethics* is then used by all members of the profession. However, variations may exist among the members as to the appli-cation of the code to professional practice. Thus, the strength of the code is based on the extent to which the members of the profession abide by it in practice.

Construction associations commonly develop codes of ethics to provide mini-mal standards of professional behavior for their members. One such association that has had a code of ethics for many years is the American Institute of Constructors (AIC). The AIC Code of Ethics (Figure 21.2) prescribes the expecta-tions of the individual members in the conduct of their professional activities. The provisions in this code of ethics embody such issues as always being honest, not being deceptive or unfair, and not performing in a way that might unjustly place others at a disadvantage/advantage or that might harm them in some way.

AIC Code of Ethics

The "AIC" designation that our constructor members use after their names is a symbol of their dedication to the Code of Ethics and the elevation of the profession through individual excellence.

I. A member shall have full regard to the public interest in fulfilling his or her responsibilities to the employer or client.

II. A member shall not engage in any deceptive practice, or in any practice which creates an unfair advantage for the member or another.

III. A member shall not maliciously or recklessly injure or attempt to injure, whether directly or indirectly, the professional reputation of others.

IV. A member shall ensure that when providing a service which includes advice, such advice shall be fair and unbiased.

V. A member shall not divulge to any person, firm, or company, information of a confidential nature acquired during the course of professional activities.

VI. A member shall carry out responsibilities in accordance with current professional practice, so far as it lies within his or her power.

VII. A member shall keep informed of new thought and development in the construction process appropriate to the type and level of his or her responsibilities and shall support research and the educational processes associated with the construction profession.

FIGURE 21.2
Code of Ethics of the American Institute of Constructors (AIC) and the Certified Professional Contractor (CPC), *reprinted with permission of the AIC.*

Another construction industry association with a code of ethics is the Construction Management Association of America (CMAA). The overall intent, like the AIC Code of Ethics, has some common objectives including honesty, fairness, integrity, and so on. (Figure 21.3). The CMAA ethics code includes statements that are quite specific about some of the activities of a construction manager, and include examples of how ethical behavior is to be maintained.

The Associated Builders and Contractors, Inc., (ABC) is an association that represents many general contractors and specialty contractors. The ABC has a code of ethics that consists of eight basic ethical professional behaviors that are to be followed (Figure 21.4). Note that this code of ethics is directed to both the activities of the member companies, as well as, it can be construed, to the individual employees of the member companies.

A code of ethics is effective in guiding the conduct of only those who subscribe to it. Most professions have adopted codes of ethics to further define behavior for their members. Unfortunately, the courts have begun to evaluate the relevance of codes of ethics in relation to the law. For example, some court

Construction Management Association of America (CMAA)

Code of Professional Ethics for the Construction and Program Manager

All members of the Construction Management Association of America commit to conduct themselves and their practice of Construction and Program Management in accordance with the Code of Professional Ethics of the Construction Manager.

As a professional engaged in the business of providing construction and program management services, and as a member of CMAA, I agree to conduct myself and my business in accordance with the following:

1. **Client Service.** I will serve my clients with honesty, integrity, candor, and objectivity. I will provide my services with competence, using reasonable care, skill and diligence consistent with the interests of my client and the applicable standard of care.

2. **Representation of Qualifications and Availability.** I will only accept assignments for which I am qualified by my education, training, professional experience and technical competence, and I will assign staff to projects in accordance with their qualifications and commensurate with the services to be provided, and I will only make representations concerning my qualifications and availability which are truthful and accurate.

3. **Standards of Practice.** I will furnish my services in a manner consistent with the established and accepted standards of the profession and with the laws and regulations which govern its practice.

4. **Fair Competition.** I will represent my project experience accurately to my prospective clients and offer services and staff that I am capable of delivering. I will develop my professional reputation on the basis of my direct experience and service provided, and I will only engage in fair competition for assignments.

5. **Conflicts of Interest.** I will endeavor to avoid conflicts of interest; and will disclose conflicts which in my opinion may impair my objectivity or integrity.

6. **Fair Compensation.** I will negotiate fairly and openly with my clients in establishing a basis for compensation, and I will charge fees and expenses that are reasonable and commensurate with the services to be provided and the responsibilities and risks to be assumed.

7. **Release of Information.** I will only make statements that are truthful, and I will keep information and records confidential when appropriate and protect the proprietary interests of my clients and professional colleagues.

8. **Public Welfare.** I will not discriminate in the performance of my Services on the basis of race, religion, national origin, age, disability, or sexual orientation. I will not knowingly violate any law, statute, or regulation in the performance of my professional services.

9. **Professional Development.** I will continue to develop my professional knowledge and competency as Construction Manager, and I will contribute to the advancement of the construction and program management practice as a profession by fostering research and education and through the encouragement of fellow practitioners.

10. **Integrity of the Profession.** I will avoid actions which promote my own self-interest at the expense of the profession, and I will uphold the standards of the construction management profession with honor and dignity.

FIGURE 21.3

Code of Ethics of the Construction Management Association of America (CMAA), *reprinted with permission from the Construction Management Association of America®.*

Associated Builders and Contractors, Inc. (ABC)

ABC Code of Ethics

To maintain a standard of performance consistent with the owner's best interest.

To quote only realistic prices and completion dates and perform accordingly.

To fully cooperate with the architect and other agents of the owner toward fulfillment of the contract undertaken.

To solicit and accept bids and/or quotations only from firms with whom we are willing to do business.

To make all payments promptly within the terms of the contract.

To observe and foster the highest standards of safety and working conditions for employees.

To establish fair wage schedules for employees commensurate with their ability and their industry.

To actively participate in the training of skilled tradespersons for the future welfare of the construction industry.

FIGURE 21.4
Code of Ethics of the Associated Builders and Contractors, Inc. (ABC), *reprinted with permission of the Associated Builders and Contractors.*

decisions have stated that codes of ethics infringe on competition. As a result of such a court decision, the AIA code of ethics is now voluntary. That is, an AIA member may no longer be denied AIA membership because of unethical behavior. It might be implied that this applies to all codes of ethics. The ASCE's code of ethics includes some of the following provisions regarding the behavior of an engineer. An engineer should do the following:

- (3) A "be realistic and honest in all estimates, reports, statements and testimony."
  There can be a direct conflict of interest if an engineer's fee is based on a percentage of the project costs. The engineer has an incentive to inflate the costs so that the fee will be higher. Testimony may be given in favor of the client simply because of the source of payment. It also may not be in the engineer's best interest for the client to decide against doing a project.
- (10) A "inform his client or employer of any business connections, interest, or circumstances which may be deemed as influencing his judgement or quality of his services to his client or employer."
  The engineer may select a site for a project where he or she owns a portion of the property or otherwise has a vested interest in it. The engineer could have a ready-mix plant (the only one in town) and include extra concrete in the project requirements when steel or wood might be more economical.
- (11) B "not undertake work at a fee or salary below the accepted standards of the profession in the area."

Consulting work performed at a cost below the standard for the area can lead to rate cutting and cause some firms to be put out of business. Thus, an engineer who decides to moonlight by accepting small consulting jobs at a level below scale is acting counter to this provision.

- (13) F "not use equipment, supplies, laboratory, or office facilities of his employer to carry on outside private practice without consent."

An engineer may perform outside consulting work while employed by a firm, and in effect run a second business from the office. This engineer is clearly taking advantage of the employment with another firm to undertake additional work. The engineer is taking advantage of the overhead being paid for by the employer. Note that this is unethical only if it is not fully communicated to and sanctioned by the employer.

What gifts are acceptable for an engineer to accept from a client? There is no clear-cut answer. One might feel comfortable accepting a pad of paper, a bumper sticker, a decal, or other inexpensive items, but at what stage does the expense of the gift become sufficient to influence the recipient's judgment? In many instances this must be answered on an individual basis. Can a box of apples be accepted as a Christmas gift from a client? Four tickets to a professional ball game? Season tickets? Obviously, at some point the value of the gift exceeds what would be expected of someone who is giving a token gift.

Codes of ethics are commonly considered to be voluntary codes of conduct for businesses. On some federal construction projects, the question might arise of whether ethical practice is voluntary or compulsory. The Federal Acquisitions Regulations (FAR) now includes a policy statement that requires all contractors and subcontractors to have a written code of business ethics on domestic projects that exceed $5 million and that exceed 120 days in duration. The written code of business ethics and conduct is to be provided within 30 days of contract award. This code is to be provided to each employee working on the project covered by this requirement. This must be communicated to employees via training, and the contractors are also to ensure that prompt corrective actions are taken when infractions are noted. There are no small business criteria that will exempt any firms.

ETHICS SCENARIOS

One of the best ways to develop a greater appreciation for the role of ethics in a profession is to consider specific examples of how questions of ethics arise in different settings. Following are several scenarios in which a question of ethical conduct is raised. These scenarios have not been written to show specifically what is ethical and what is unethical. In fact, in some of them the intent is to show situations in which a clear distinction cannot be made between ethical conduct and unethical conduct. In reviewing these scenarios, consider how the following questions relate to the specific circumstances:

- Does one party have, as a result of a particular action, an advantage not enjoyed by others?
- Does another party incur an unfair disadvantage as a result of an action?

- Is the action dishonest or deceptive?
- Is someone harmed or damaged as a result of an action?
- Is someone's judgment likely to be altered by an action when that party should remain impartial?

A positive response to any one of the above questions may be a clear indication that an action is or could be unethical. These questions may be used for guidance in examining the following scenarios. Other questions may be generated. As these scenarios are considered, remember that full agreement on the ethical merits of each case will not necessarily occur.

In some situations an action may appear unethical, but in fact is not. Appearances can be deceiving and could lead to individuals making the wrong conclusions. An action that simply appears to be unethical, is not inherently unethical. Most practitioners would advise that such actions are to be avoided, even though an action is not unethical. If an action has the potential of being misconstrued by others, it probably will be.

Ethics Scenarios Involving University Students

1. George, a senior in civil engineering, is about to graduate with a bachelor's degree. Various recruiters are on campus during the semester before he graduates. Students are granted interviews by signing up on a recruiter interview schedule. George has worked for the past three summers for Ace Engineers and has already accepted a generous offer from them to start full-time employment upon graduation. George intends to work for Ace, but all his friends anxiously await their opportunities to be interviewed by the many companies that will be visiting the campus. George gets caught up in the excitement of interviewing and decides to schedule a few interviews. Beatriz, one of his friends, confronts George and asks why he is signing up for job interviews when he already has a job sewn up. George responds that he just wants to see what else is out there, and that he really would give serious consideration to a good offer from another company. Beatriz then suggests that perhaps this is not ethical, because Ace Engineers, a small firm, will not be attempting to interview any other seniors since the company intends to hire just one person. Are these concerns valid? Is ethics an issue in what George is doing? What should George do? What should he not do?

2. When the various companies came to do interviews on her Illinois campus, Joan got interviews with six of the best companies. Since the interviewers were often on campus for only one or two days, students who procrastinated in signing up for interviews did not have much success in getting interviews with the companies that were rated highly by the other students. Joan was prompt in signing up for interviews. In addition, because of her academic record and the impression she made during the interviews, Joan was offered plant trips by each of the companies. One of the firms, Bayou Canal-Builders, is based in New Orleans, and the 30 positions that are available with the firm will be filled in New Orleans. This presents a problem as Joan has already firmed up her plans to get married. In addition, her fiancé has received a fellowship to attend graduate school in Chicago. Joan arranges

to make the plant trip to New Orleans, where she will be staying with a cousin. Riley, a friend of Joan, asked rather candidly if Joan would seriously consider taking a job that would require her to move to New Orleans. Joan's response was that at this stage she had major doubts, but that she would like to at least become acquainted with the firm. She added that after her fiancé finished graduate school, she would welcome the opportunity to go to New Orleans. Riley suggested that perhaps her intentions were not compatible with the expectations of the company. Joan stated that with so many positions available, the cost of one plant trip was not significant, and besides, she might not get an offer. Joan also reminded Riley that she would be staying with relatives for the entire visit, and that since she was staying over Saturday night, the airfare was actually quite low. Riley felt that Joan should be honest about her inability or unwillingness to relocate to New Orleans upon graduation. Is Joan being unethical? What should she do?

3. One day Professor Nealy was visited in her office by John, a senior who was about to enter the work world. John was troubled. He gave this account: "This past semester I have been interviewing with a lot of companies that have come to visit our campus. I went to 11 interviews and did not have any firm job offers. Since I thought I would not get any job offers through the interviews I had on campus, I decided to knock on doors. So I went to Houston and personally went to see several engineering firms. Even that was frustrating, but I felt that the Henderson Group was interested in me. Three days after I returned to campus following spring break, I got a letter from them expressing an interest in me. Shortly after that, I got a telephone call from them in which they made an offer to me. They also said they would have to know my answer within two days. Since I did not have any other offers and I was getting desperate, I accepted the job. This acceptance occurred two days ago. This morning I got a letter from Grit Contractors that included a nice job offer. That letter was held up in the dormitory because of spring break. Anyway, I should have received it over a week ago, but I didn't get it until today. I have already accepted the job with Henderson. Can I still take this job with Grit? What should I do?" The professor thought about it for a while before she responded. She said that there were actually several things to consider: "It is noteworthy that the late receipt of the offer from Grit is not the fault of either you or Grit. The real issue is, Are you now bound, from an ethical point of view, to the job you accepted with Henderson? If the job with Henderson is not that attractive to you, you will probably be doing them a service by telling them the circumstances of the job offers, and explaining that you would really prefer to accept the offer from Grit. Perhaps they would simply understand and be grateful for your honesty. It would be quite costly for them to employ you for six months or a year and then have you quit the job. I suggest you give them a call." Did John get good advice? Are there other concerns that should be addressed?

4. Mike and Joe were engineering seniors who were frantically studying for a major examination in thermodynamics. They were both members of a social fraternity that kept exam files for a large number of courses. As a strategy for studying for the exam, these students were reviewing a thermodynamics exam file that contained six exams, and working every problem. They finished all the problems and verified their answers with the master solutions that had been prepared. The

next day Mike and Joe went to class to take the exam. To their surprise, the examination was identical to one of the ones they had studied. Mike immediately felt uncomfortable about taking an exam for which he had an unfair advantage, even though he could have used a good grade. Before continuing with the exam, Mike approached the professor, and told her about the coincidence of having studied the exact exam in the test file. She told Mike that she was grateful for his honesty and said that she would waive his taking the exam. Instead of dropping the low exam, she would simply base his grade on the average of his grades on the other exams. In the meantime, Joe continued working on the exam, delighted at his good fortune. After the exam, Joe told Mike, "Look, if you and I had made up some problems and one of the problems was just like one being asked on the exam, would you tell the professor, or would you simply work the problem? It is not our fault if we have access to the test file and the professor elects to use the same exam over and over again. Isn't that right?" Was Joe unethical for taking the exam, or was it simply a stroke of good luck for him? Is the professor's treatment of Mike just, or is he being unduly punished for expressing his integrity? Should the exam problems have been changed by the professor?

5. A half hour before hydrology class started, Tom, Doug, and Ann were already in the classroom. These students were individually reviewing their homework problems. After a long silence, Tom went over to Doug's desk and asked, "What did you get as an answer for the third problem?" When they compared answers, they realized that they had made different assumptions when solving the problem. They discussed this at length and could not agree on the correct approach. Finally, Tom said, "Hey, Ann, how did you do the third problem?" Ann was noticeably uncomfortable about being asked this question, so Tom said, "I don't want to copy your solution; I just want to compare your assumptions with the one we made. There's nothing wrong with that." Ann was not convinced and responded, "Well, I thought we were supposed to work alone on this homework. The homework does count for 30 percent of the course grade, so I just assumed we were supposed to work independently." Tom countered, "I understand your concern about actually working the problems. All I want to do is check some assumptions. I think we actually learn a lot by working the problems by ourselves and them comparing answers. It's not like I'm making a photocopy of your homework." This was unresolved by the time class started. Was Tom's action unethical? Was Ann being too strict in her interpretation of appropriate behavior?

6. Marcella was taking an independent study class in which she was examining the laws related to the transfer of ownership in family-owned companies. She had this interest because her father had plans to give her the business once she was able to run it. To demonstrate her independent work for the class, Marcella submitted a written report to her advising professor. The next semester Marcella took a class in business management. In this class she was asked to submit a major report on a topic in which she had some interest. Marcella requested and had approved a topic related to the transfer of ownership in family-owned companies. Harry, Marcella's friend, knew about the report Marcella had done the previous semester and asked how the reports would differ. Marcella responded, "Well, actually they are one and the same. Since I have it all on a word processor, all I have to do is change

the cover page." Harry then asked, "Is it really okay to do that? Is the business pro-fessor expecting you to start from scratch on this topic and then write the report?" Marcella said, "There were no restrictions placed on the report. All he said was that the report was to be my own work, and that it will most assuredly be." Harry contended that Marcella was plagiarizing her own paper by not telling the business professor what she was really doing. Is Marcella doing something she should not do? What should she do?

7. At the fall membership drive of the ASCE, Bob joined the student chapter. After four meetings had passed, Sandra, the chapter president, noticed that Bob had never attended a meeting. Since she had a class with him, she asked Bob if anything was wrong, and if he had been informed about the meetings. Bob said, "Well, I guess I have heard about most of them, since they post the announcements and some faculty even announce it in their classes. But actually, I don't take much interest in being involved in any extracurricular activities at this time. To be honest with you, I joined so I could put this on my résumé. Since most civil engineers are members of ASCE, I thought it would look good on my résumé if I was listed as a member. Anyway, the student dues are not that much." Sandra did not respond to Bob, but she could not help but wonder if this was unethical. She thought to her-self that most people would agree that it is wrong to list information that is not true on a résumé. She knew it is wrong to list membership in a society to which one does not belong. Now she was thinking that simply paying the dues to be able to list the membership on the résumé is not much different. Has Bob stretched honesty too far?

8. Jaime, a graduate student in construction engineering, was conducting a re-search study related to aggregate size and concrete strength in pumped concrete. Part of his research consisted of doing an extensive literature search. In the engi-neering library he found numerous papers and articles on the topic. He also found a good resource in a master's nonthesis report done two years earlier that was essentially an extensive literature study. After Jaime completed his search on cam-pus, his adviser suggested that he also look at references at another university li-brary not far away. Jaime did this and found a few other references, which he checked out. After he brought them home and began to study them, he noticed a striking resemblance between the nonthesis report he had previously read, and a master's thesis done at the other university. In fact, he noticed that the nonthesis report had been done before the master's thesis. Jaime went to his adviser to ask her if he should do anything about his new information. It was obvious to Jaime that plagiarism had occurred, but he was concerned whether he had any obligation to notify anyone about this blantant copying of someone else's work without the benefit of citation. What should the adviser suggest?

9. Henry and George were old college buddies who worked for different con-struction firms. Henry worked for a construction firm based in St. Louis, and George worked for a firm based in Memphis. Each firm pursued projects primarily in their own metropolitan areas, so these were not considered by Henry or George to be competing firms. One fall weekend Henry and George got together at their alma mater's homecoming football game. Henry and George spent a considerable amount of time together, and after the home team's victory, they proceeded to

celebrate like in the "olden days." After several drinks, Henry was clearly the more intoxicated and became rather talkative. He started to tell George about an upcoming bid that he was working on for a lucrative project to be built in Paducah. He said there were only three bidders expected on the project. George asked about additional details regarding the project, and Henry readily shared the information. After a late night, the friends parted company with plans to meet again at a future football game. The Paducah project bid opening was three weeks after the homecoming, and when the bids were opened, George's firm submitted the low bid. Henry did not see George at the bid opening, so he called him at his office to discuss why George did not mention his company's interest in the Paducah project. When Henry told the receptionist his identity, the response was that George was unavailable. Henry felt that George's company's interest in the project developed as a result of Henry's comments, and that George should have told him of his intention before Henry shared details about the project. When Henry eventually was able to confront George about his concerns, George said "It's only business." Henry contended that George was unethical in his actions. Was George unethical in this scenario, or was he just being entrepreneurial?

10. Walter was a licensed professional engineer who had worked for a wood truss manufacturing firm for twelve years. He was the chief engineer in the company. Eventually, one of the company's clients recognized Walter's strengths and offered him a position that constituted a substantial pay increase. He was hired as the chief of operations for a small commercial development company. Although he did not do any engineering design work, he enjoyed the challenges and the variety offered by the new position. A month after starting to work for the company, Walter's boss, Josh, came to Walter with an unusual request. He said, "Hey Walt, I know you are a licensed engineer. One of our superintendents did some design work on a wooden mezzanine, and now the owner insists that this design be stamped by a PE. This is just standard construction, so all we need is for you to put your seal on these drawings so we can submit them." Walter said, "I will be happy to look these over and then put my stamp on them." At this point, Josh replied, "Walt, this is an eleventh hour kind of thing. The design is fine. We just need your stamp on it, and we need it now." Walter insisted that he would not seal the drawings without first reviewing the design, which would take at least a day to complete. Josh told Walter that his review was not needed, since the structure was not really complicated. Josh grew more and more agitated by Walter's reluctance to put his seal on the drawings without a review. He said, "Walt, we did not hire you to do design work for us. We don't need you to do this review. Just put you seal on the drawings." Should Walter stand firm on his initial stance or should he appease his boss?

Ethics Scenarios Involving University Faculty

1. At a local professional society MEETING, Jill sat next to Mary, a friend who had obtained a graduate degree the previous year. Jill said to Mary, "I have been meaning to talk to you. Last Tuesday I received my journal for this month and

noticed that your graduate research work was written up. But you aren't listed as an author of the article, and no acknowledgment is made that you did any of the work. Tell me, was that your work mentioned in the article or not?" Mary responded, "If you compared the article to my thesis, you'd find that it is my work being described. You probably remember that Professor Davidwood was my thesis adviser and the sole author of the article. I called him last Thursday to see why I was not mentioned. He had a ready answer for me. He said that the research I did was originally his idea; he obtained the funding to do the research; I was paid to do the research; he wrote the paper by himself; and under such conditions, he felt justified in being listed as the sole author. He did say that he should have added an acknowledgment mentioning my contribution, and for this omission he apologized. Well, he was right, I suppose. I was paid for my work. He did write the article by himself, as I checked this and determined that he did not copy any sentences I had written. Even the figures, which showed the same information, were redrawn. I did make unique contributions to the research, but it was all under his guidance. So I guess it is okay to do what he did." Do you agree with Mary's conclusion? Would the circumstances be different if Mary had not been paid?

2. One day in the civil engineering office, Frank and Debbie, two graduate students, came by to pick up the new departmental brochures. While they were waiting, Debbie asked Frank if he was funded for any graduate work. Frank told her that he had a research assistantship and would be fully funded for the academic year. At that point Beverly, the office secretary, interrupted Frank and said, "Frank, you do not have a research assistantship." Frank responded to her, "Oh, yes, I do. In fact, I have already received two payments." Then Beverly added, "Maybe I wasn't supposed to tell you this, but you have a fellowship. You do not have a research assistantship. I've done the paperwork for this myself." Then Frank turned to Debbie and asked, "Why would Professor Tieman tell me I had a research assistantship when I actually had a fellowship?" Debbie said, "It's just a guess on my part, but I'd say that Professor Tieman needs somebody on that research project real bad, but doesn't have the money to pay for it." Frank said, "You know, I would not be able to go to graduate school without some type of support. I was happy to get the letter from Tieman about the research assistantship. Now I feel I've been lied to. What do you think I should do?" What advice would you offer Frank? Was Professor Tieman's behavior unethical?

3. One day in the faculty lounge two professors were having a friendly conversation. One asked the other, "Whatever happened to your research proposal on developing a manual on construction equipment safety?" The response was surprising to the faculty members who overheard it. The professor replied, "Well, I really wanted to get this project. I first heard that the research funds had run out. Then I heard that my proposal was 'given' to our neighboring research institute, and that they are now doing what I had proposed." The other professor said, "You mean that they submitted the same type of proposal you did?" The first professor clarified his point by saying, "No. The funding agency literally took my proposal to the institute and asked them if they could do the research. They accepted it, even though their researchers are not well versed on the subject. I know this for a fact. The institute people visited me shortly after that

happened. They wanted to hire me as a consultant on that project. I turned them down. Their research will probably be a bust, but I couldn't see me working at arm's length on a project that was my brainchild." Discuss the ethical issues involved in these circumstances.

4. Two faculty members were discussing their success at getting research funding. One of the professors offered this as a means of improving one's chances: "As a known researcher, I am asked to review research proposals submitted by others. If the proposal is to an agency to which I have recently submitted a proposal, I try my best to improve the chances of my proposal, by really cutting down the proposals that compete with mine. Mind you, I don't just say that a proposal is worthless. I carefully read them, and then I start to pick them apart. No proposal is perfect. Once I find a flaw, I just exploit it. I'm sure everybody else does the same thing. Anyway, I've had pretty good success with it." Discuss the ethical ramifications of this professor's actions.

Ethics Scenarios Involving Construction Contractors

1. Jerry, a graduating senior in civil engineering, entered his faculty adviser's office one day. He could hardly contain himself. His adviser said, "Now, just start from the beginning and tell me what is going on." Jerry began, "As you know, I will be graduating in two weeks. You will recall that I have been looking for a position with a firm specializing in harbor structures. I worked eight years in that area before pursuing a university degree. Well, through some of the leads you gave me, I found what I thought were excellent job prospects with two firms in Baltimore. I interviewed with Hatch Contractors and with Pier Constructors. Both looked just excellent to me, and they were both interested in me. Because of my experience, I got great job offers from both firms. I accepted the position with Hatch about six weeks ago, and I also notified Pier Constructors of my decision. Everything seemed to be going so well. Then, this morning, I got a short letter from Hatch stating that they would not be able to honor the job offer. I don't know if their contract work is dropping, like they say, or if they are just making room for one of the children of the owner. Anyway, I am frantic. I called Hatch to see if this was possibly a mistake, but my call never got transferred beyond the receptionist. I called Pier Constructors to see if they still had a position, and they said it had been filled. I know that I should be able to get another job, but all the interviewing on campus has already come to an end for this semester. So I will graduate in two weeks without a job. I wouldn't mind so much if I were single. But my wife and I were so elated over the job and never considered that the offer would be withdrawn, so we bought a house. Is there something I can do?" Discuss only the ethical aspects of Hatch's behavior. What should Hatch have done? What should Hatch do now?

2. Margaret confided in her friend about a dilemma a potential employer had created for her. She began, "You know, I was always looking forward to this semester, when I would be interviewing with all these companies and making a decision about the job I would accept. I had identified about eight companies that

I wanted to interview with over a 6-week period. Two weeks ago I had my first interview, and last week I had two more interviews. They all went really well as far as I could tell. Well, this morning I got a letter from TXR Constructors stating that they were impressed with my interview and academic record. They also extended an offer to me. The clincher here is that they want an answer from me by next Friday. I have only one interview lined up between now and then. So if I accept, I will not be able to interview with half the firms I had planned. Actually, TXR was one of the firms I was most excited about interviewing with, since they do the types of projects I want to be involved with. But I would rather be able to say that I accepted the offer with TXR because it was the best firm among all those interviewed, rather than admit that I was forced to make a decision before any other firms had a chance to make an offer to me. What should I do?" As a friend of Margaret, what would you suggest? Is TXR's behavior unethical, or is it simply a good business practice that provides assurance that if one offer is not accepted, the position may still be filled, since another offer can be extended to someone else?

3. Bud and Monte were the sole owners of a successful firm specializing in industrial renovation, or "turnaround," projects. These were typically short-duration projects with quick buildups of large work forces. This Monday morning Bud, who was in charge of the field operations, had a special meeting with Monte, who ran most of the administrative portions of the business. Bud was concerned about Steve, the project manager on one of the projects. Bud said to Monte, "We have a real problem with Steve. I just learned that he fired Helen, the office manager. Apparently, Steve was the person who originally hired Helen. Word has it that Helen was never much of an asset on the project. Steve hired her because of some kind of a fling he had going with her. Helen told one of the other guys at the job that as long as she did what Steve wanted after work hours, she didn't have to do much actual work on the job." Then Monte asked, "Are you suggesting that Steve hired Helen as a playmate? I thought he was married." Bud continued, "Yes, he is married, and he has three kids at home. I thought his marriage was solid, but apparently one woman isn't enough for him. I think we need to fire Steve. We cannot tolerate such wanton behavior, but—" Monte interjected, "I see where you are heading now. Steve has the best safety record of all the project managers, and in fact, that was how we persuaded the owner on this job to award the contract to us. Do we compromise our job by dismissing the project manager in the middle of a major work force buildup on a job?" Bud continued, "You do get the picture, all right. My concern is that Steve was obviously using this woman to satisfy some personal needs. She was not hired for her job skills. I'm not concerned as to whether she should have been fired or not. My concern is that we have an employee of such low moral character representing us on a job." Then Monte said, "I don't like that kind of behavior one bit, but don't we owe it to the workers to keep Steve on the job? Won't the safety of the job be jeopardized if we try to change project managers during such a crucial part of the project?" Offer suggestions about appropriate actions that could be taken by Bud and Monte. Should the safety of the workers be a consideration?

4. Vernon, one of the company's superintendents, was called into the main office by Barbara, the owner. Vernon sat down in Barbara's office, and she immediately got

to the point. "Vernon, I have it on good authority that you accepted a microwave as a Christmas gift from the Detter Corporation, the owners of your project. I know we have never talked about professional ethics as far as gifts go, but you have stepped over the line by accepting this gift. I want you to return it." Vernon responded, "Now, I know what this might look like to you, but you don't know the whole story. My project consists of 165 apartment units, and each is furnished with a microwave. When they were installing these microwave units, the installers noticed that one was damaged during shipment. There was a chrome piece missing, and part of the metal casing was severely dented in front. They asked me if I wanted it, because it was too much hassle to send it back. You have to realize, Barb, that they only pay about fifty bucks for those things. I admit that it works great, but it wouldn't look right in those apartments. Anyway, one of the other supers got a roasting turkey for Christmas, and I understand that is acceptable. Well, they can cost about thirty bucks. The way I figure it, my microwave wouldn't buy a turkey with the cosmetic damage on it." Barb considered his comments and said, "Well, maybe I see your point. But do be aware that accepting a gift like that doesn't look good." Do you agree with Barbara's initial assessment that the gift should be returned, or with her final assessment that taking the gift was acceptable?

5. Kevin was a new employee. His first assignment was as a field engineer on a concrete-lined canal. Ned was the superintendent from whom Kevin sought advice, directions, and job assignments. They got along very well. One day Kevin asked Ned, "I have noticed on every Friday that Aaron, the field inspector, brings his own car to the job, and on all other days he has a government car. Last week I noticed that he parked his car real close to our fuel tanks. When I asked him why he parked there, he just walked away. Do you think that he is doing something wrong?" Ned tried to calm his friend down. "You will soon realize that this is the real world. Everything does not go by the book. I don't know the history behind this all, but it is an arrangement that Rob, the project manager, made with Aaron when he first came to this job. For some reason Aaron was promised a full tank of gas every week as long as he is assigned to the project. I personally don't like it. I even confronted Rob about it. I told him we don't need to bribe anyone. My work will stand up to any inspector's eye. If my work isn't up to spec, I don't want it accepted. I told Rob all this. But he thinks things just run smoother if we have what he called 'friends in the owner's camp.' Then he brought up the fact that I am no different, because I spring for lunch for Aaron twice a month. And it is true that I do buy him lunch. But if I didn't buy him lunch, I wouldn't have a good way to get his undivided attention without the disruptions that we have here on the job. You have to realize that by lunch I mean a hamburger. I'm not trying to influence Aaron by buying him lunch, but simply buying his time. For a three-dollar lunch to get an hour of Aaron's time, it is worth it. I don't think buying an inspector an inexpensive business lunch, with the specific purpose of getting together to discuss project affairs, constitutes a bribe or anything that is remotely unethical. But I draw the line when it comes to buying a tank of gas for someone, for no other reason than to keep him in our back pocket. What do you think?" Kevin did not know what to say. Is Rob's practice unethical? Is Ned's practice unethical?

6. Molly had been employed by Xert Contractors for almost 10 years. She had spent virtually all that time in the field. Later, when she wanted to have a family, she requested a position in the main office that would not entail constant uprooting from project to project. In her new position, she quickly became assistant chief estimator. One day, as she was beginning to review the bidding documents on a project, she got a call from a competitor. The call was from Hank, the chief estimator for Viking Erectors, and it involved an unusual request. Hank said, "I was wondering if Xert was going to bid on the Bristal Hotel project in Rabstown." Molly said, "Yes, Hank. In fact, I was just starting to review the plans this morning. I expect you will be sharpening your pencil on this one, too." Hank responded, "That's why I'm calling. We just got that school job in Spring and won't be able to take on any more work." Molly then said, "That should make you pleased. I wish we had been so fortunate. So what's the problem?" Then Hank asked, "I was wondering if I could get a courtesy bid or complimentary bid price from you." Molly inquired, "Hank, you'll have to explain this. I haven't been estimating that long. What is a complimentary bid?" Then Hank said, "You probably know that we have done several jobs in the past for the Bristal hotels in our region of the country. They are an excellent company to work for. And we had to work hard to get on the select bidder's list. They are real particular about their contractors. But once you start to work with them, they want you to stay with them. If they don't receive a bid from us on this project, Viking will most likely be taken off the bidder's list. So I really want to put in a bid on this job, but I don't want to get the job. What I would like to get from you, on bid day, is a price that looks serious but that is well above your price. Can I count on you to give me a price?" Molly started to think, Is this being honest to the Bristal hotels? However, Molly realized that this would give Xert a bit of an advantage, since one of the bidders would be under her control. Discuss the ethics of submitting a courtesy or complimentary bid. Discuss the ethics of providing a price for another company to use in a complimentary bid. What do you recommend that Molly do?

7. Todd, who works for a building contractor, was finalizing a bid. Much of the work of the firm is subcontracted, so many subcontractor prices were received on the fax and by telephone. On bid day things are always hectic. This bid day was no exception. One of the subcontract items that caught Todd's attention was masonry. He had received four bids, which were as follows: $98,000, $121,000, $124,000, and $132,000. The low bid was from Gnu Masonry, a subcontractor not familiar to him. The second was from Carla Masonry, a subcontracting firm that had successfully finished several jobs for Todd in the past. Todd reviewed the bids and felt uncomfortable about the bid from Gnu. The bid had to be submitted at 2 P.M., and the bid from Gnu was received at 1:30 P.M. Todd would have liked to have checked the references Gnu had listed on the fax, but he did not have the time to make telephone calls at that stage of the bidding process. The dilemma raced through Todd's mind: "If the competitors use the low bid and I use the second low bid, they will have a considerable edge on my bid. If I use Gnu's bid and they don't work out, then I have over $20,000 to make up if I later contract with Carla." Todd made several telephone calls to Gnu, but the line was always busy. Then he decided that he would split the difference. He used $110,000 as his price for masonry. Todd did

submit the low bid. When it came time to award the subcontracts, Todd called the references Gnu had listed. Through these calls he was also able to identify other contractors who had worked with Gnu. The results had not been good. The record of liens from unpaid workers and deficient work was enough to convince Todd that Gnu could not do the job. Todd then called Carla at Carla Masonry. He said, "Carla, I have a real dilemma on this job. I used a price of $110,000, which was more than the price given by the low masonry bidder, but now I am feeling uncomfortable about using that firm. If you could take on this job for $110,000, the job is yours." Carla must now make a serious decision. She may even resent having had her bid shopped. She may think that Todd is bluffing about the low bid price. Regardless of her response, discuss the ethics of Todd's request.

8. Ed is the owner of a ready-mix project in town. He acquired the plant after several years of successful construction contracting. Now he sells concrete to himself whenever he needs it and also sells on the commercial market. One night Ed attended a local contractor's association banquet to which various city engineers from the area were also invited. During the cocktail hour, Ed was approached in the hallway by Paula, the city engineer for one of the local municipalities. Paula said, "Boy, you really pulled a fast one last week." Ed inquired, "What are you talking about?" Paula continued, "Well, I was at the two bid openings in Shelby. It was pretty clear what was going on." Ed became indignant. "Why don't you tell me what was going on?" Then Paula said, "I couldn't help but notice that you and Ace Concrete were the only bidders on both jobs. With your good luck, you were the low bidder on both jobs. Word on the street now has it that you are subcontracting the concrete to Ace. It kind of looks like there was really only one serious bidder. What kind of markup did you use, anyway?" Ed began to feel uneasy since others in the hallway had heard the accusation. He said, "You know darn well that there are only two ready-mix plants within 50 miles of Shelby. I submitted competitive bids on each project, and I assume Ace did the same. The first job, using roller-compacted concrete, will require about 400 cubic yards of concrete a day for over three months. That is approaching my total daily concrete production capacity of 600 cubic yards. The second job will need about 300 cubic yards of concrete a day for nearly six months. You don't need to be a rocket scientist to figure it out. I can't supply concrete for both jobs. It is unusual that we have two large concrete jobs at the same time, but I can't help it. I was going to pull my bid on the second job, but the owner talked me out of it. If the owner had received only Ace's bid, the entire project would have had to have been rebid. I knew I could provide the construction services with my personnel, but I can't provide the concrete, at least not most of it. What I did was not unethical. It is just one of those things. The people at Ace should tell you the same thing, because I didn't talk to them until after I realized I needed a concrete price from them on the second job. Once I was the low bidder on the first job, Ace knew, or should have known, that they would be providing the concrete on the second job. But I can tell you this: Their price looked fair to me. Don't accuse me of any wrongdoing unless you are prepared to answer to a lawyer for defamation of character." Have ethical boundaries been overstepped by Ed? Are there any serious ethical concerns in this type of situation?

9. An electrical contractor put together an estimate on an addition to a country club. The bid that was submitted to the general contractor on this cost-reimbursable contract was $2,160,000 that represented a combination of $960,000 in labor and $1,200,000 in material costs. Before a formal subcontract agreement was signed, the project got delayed due to financing complications. The project was delayed for nearly a year. Later, the general contractor asked the electrical contractor if the firm was still interested in working on the project. The response was that the firm was interested, but that the costs of materials had risen by $80,000 and that the bid should be commensurately increased. This was accepted by the general contractor, who indicated that a formal subcontract agreement for $2,240,000 would be forthcoming in the mail. Harry was the estimator for the electrical contractor who negotiated this contract with the general contractor. Harry was approached by Aaron, a fellow estimator, who was concerned about the contract pricing arrangement. Aaron said that he had reexamined the estimate and discovered that the labor for one part of the project had been counted twice, resulting in an extra $60,000 in the bid. Harry said that this was just fortuitous for the company. The general contractor had accepted the original bid with the extra $60,000 in it and the new contract is no different. Aaron was suggesting that the $80,000 increase in material costs should be balanced against the $60,000 inflated amount in the labor costs. Harry did not agree. Is it unethical for the company to reap the benefits of its own estimating error?

10. A large warehouse was being built to house materials for a sporting goods store. The foundation was in the process of being installed by Semke Foundations, Inc. The block walls for the structure were to be erected by A1 Masonry. Kelly, the foreman for A1 Masonry, was on the site to become familiar with the layout and to finalize the job-sequencing plan. One day he was concerned about what was going on, and so he decided to call a friend, Norma, to obtain advice on an appropriate action response on his part. He told Norma, "I can't believe what Semke Foundations is doing on this job. The structural drawings for the slab in the warehouse call for #9 reinforcing throughout. In about one-third of the slab, they put in #7 rebar. The #7 rebar is in that part of the warehouse where heavy equipment is not likely to be operating. In fact, the area in the vicinity of the loading dock is exactly according to the drawings. The rebar is all in place and, somehow, it passed inspection. Maybe it is acceptable. I just think the owner should know what is going on. But if I cause this job to grind to a halt, the company will lose money, the people at Semke Foundations will hate me, and the architect will look bad. It seems more bad things can happen if I mention this to someone than if I just let it be." Norma listened intently to the scenario. What should her advice be? Should Kelly take some action? If so, what action would be appropriate?

11. Douglas graduated six years ago from a prestigious university with a strong construction/engineering program. He has worked for the same firm in Louisville since graduation and has progressed to a position with considerable responsibility. Through one of his colleagues he was introduced to Janice. Douglas spent much of his spare time in the past two years with Janice. They decided to get married, but this is complicated by the fact that she would soon be moving to Chicago where she would be employed by her parents' business, which she hopes

to run someday. Douglas decided to seek employment in Chicago. Douglas contacted Stareen Developers in Chicago about possible employment with them. Stareen checked with a few references Douglas had provided and they promptly asked for an interview, sent him an airline ticket, and arranged hotel accommodations for him. Douglas went to the interview and felt that Stareen would be a good match for him. He then asked Stareen if he could stay a few extra days to become more familiar with Chicago. Stareen agreed. In the next three days, Douglas contacted other potential employers and went to two additional interviews. Douglas felt confident that he would probably receive offers from each firm he interviewed. What are the ethical implications of the trip Douglas took to Chicago?

12. Sharon's firm specializes in long-span roof systems. Sharon was asked to assist the Robison Group in making a design-build presentation on a new museum in Dallas. Sharon worked with the Robison Group and helped make an impressive presentation to the museum board of directors. As Sharon was leaving the room after the presentation, she was greeted by Gilbert, a principal with a leading competitor, TY Trinity Builders. A few days after the presentation, Sharon received a call from Gilbert, who informed her that TY Trinity was selected for the museum project and that the Robison Group placed second. Gilbert then asked Sharon if her firm would assist TY Trinity Builders in the museum project. Are there any ethical implications in Sharon's decision? What should she do?

REVIEW QUESTIONS

1. What distinguishes morals from ethics?
2. Why do professionals sense the need for a code of ethics?
3. What guidelines may be used to determine whether a particular action is ethical?
4. Describe procedures that could be adopted in the construction industry that would help reduce the occurrence of at least some unethical practices.
5. What guidelines might be followed regarding the acceptance of gifts offered to public officials? To private citizens?
6. In general, to whom do codes of ethics apply?
7. Examine the codes of ethics of the AIC, CMAA, and ABC. What are the similarities and the primary differences between these codes of ethics?

22

CONSTRUCTION SAFETY

In the 1960s, the number of occupational injuries that occurred annually was staggering. On each working day, approximately 40 to 45 workers died, and 6500 were disabled as a result of a work-related injury. Ralph Nader and Jerome Gordon publicized these statistics in 1968 (June 15 issue of *The New Republic*, Vol. 158, No. 25, pp. 23–25) in an attempt to encourage Congress to pass national safety legislation. They referred to these statistics as "grisly evidence" and echoed Labor Secretary Wirtz's plea to Congress to "stop the carnage." They felt that legislation would eliminate deaths and injuries being suffered by employees of business. The costs of accidents are difficult to ascertain, but they were estimated to be $29.5 billion in 1972, a tragic waste of our resources. Current estimates of the annual costs of injuries exceed $50 billion.

The occurrence of fatalities and injuries in the construction industry is disproportionate compared with all industries. Over 1200 construction workers die each year as a result of work-related injuries, and over 400,000 incur disabling injuries. These are alarming statistics, since construction workers constitute less than 6 percent of the industrial workforce, but account for 20 percent of the fatalities and 11 percent of the disabling injuries.

The construction industry is among those industries with the worst fatality records. In recent years, the U.S. construction industry fatality statistics have been surpassed only by mining and agriculture. The logging industry also has high fatality rates, but it is closely aligned with the construction industry. When compared to other industrialized countries, the U.S. construction industry's record continues to be among the worst as well. Among construction deaths, the primary cause in 2004–06 was noted to be falls from elevation, followed by equipment (struck by and caught in or between) and electrical shock (primarily powerline contacts). These do not include transportation fatalities that did not occur on the work sites. A disproportionate number of the falls were sustained

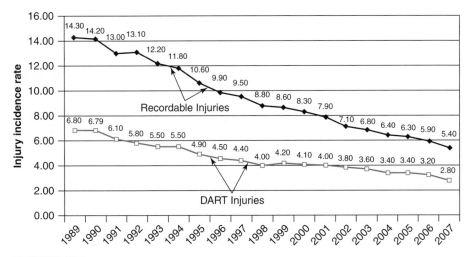

FIGURE 22.1

Construction injury incidence rates (Source: Bureau of Labor Statistics).

by ironworkers and roofers. Different trades are exposed to differing hazards, and the causes of death are often quite varied between trades.

Safety performance is commonly measured in terms of the number of injuries incurred per 200,000 hours of worker exposure. This equates roughly to the number of injuries incurred per 100 workers employed full-time for one year. There are two broadly used measures. One relates to the recordable injury rate. Recordable injuries are primarily those injuries that are treated by a physician, so first aid injuries are not included. The other injuries, known as DART injuries, pertain to those where the worker could not show up for work following the day of the injury (Days Away), the injury was such that the worker could not continue to perform the same work that had been performed prior to the injury (Restricted), or the worker had to be assigned to different work tasks due to the injury (Transferred). The construction industry safety performance over nearly two decades is shown in figure 22.1. It is clear from the figure that the injury rates have steadily declined or improved, but it is widely accepted that the current level of safety performance is still quite unsatisfactory.

THE OCCUPATIONAL SAFETY AND HEALTH ACT

Nader's efforts and those of other safety advocates resulted in several pieces of safety legislation in the late 1960s. One was in the form of an amendment on August 9, 1969, to the Federal Contract Work Hours Standards Act, called the Construction Safety Act of 1969. This act applied only to federal and federally assisted construction projects. While these efforts yielded some legislation, the impact was not significant. There were clear indicators that more had to be done for worker safety, setting the stage for what was to come. In 1970 Congress passed the Williams-Steiger Act, which is more commonly known as the Occupational Safety and Health Act of 1970 (OSHAct). This major piece of legislation became effective on April 28, 1971, and it has had a significant impact on the construction industry.

The essence of the OSHAct is that every worker should be provided with a safe place to work. This guarantee is stated in Section 5 of the OSHAct: "(a) Each employer (1) shall furnish to each of his employees employment and a place of employment which are free from recognized hazards that are causing or are likely to cause death or serious physical harm to his employees; (2) shall comply with occupational safety and health standards promulgated under this chapter." This portion of the legislation is known as the general duty clause, stating that the employer must comply with the OSHA standards, and that the safety of employees is to be ensured, even if specific safety regulations do not address certain conditions or actions.

Virtually every business is affected. OSHAct is meant to cover every employer who engages in interstate commerce. This can be loosely interpreted as including any employer who uses tools, equipment, materials, or devices made in other states. It has also been stated that anyone using the U.S. mail to conduct business, or placing long-distance telephone calls as part of doing business, can be construed as being involved in interstate commerce. Thus more than 7.5 million businesses employing more than 90 million workers are covered by this law. Federal, state, and local governments are not covered by OSHAct.

OSHAct designated the construction industry as one of its target industries. This designation implied that construction would be monitored more closely since the industry has had a disproportionate number of injuries.

OSHAct consisted of (1) the Construction Safety Act, (2) 70 pages of the *Federal Register* on new safety regulations issued on April 17, 1971, (3) 248 pages of additional regulations, and (4) for reference, various established federal standards and national consensus standards. The initial reaction of employers to the legislation was negative, primarily because of the large volume of material included by reference (one estimate stated that stacked together it would be 17 feet high). These reference materials were regarded as complex and incomprehensible.

OSHAct set up three different agencies to carry out the intent of the law. The most visible agency is the Occupational Safety and Health Administration (OSHA), which has the primary responsibility for promulgating standards, inspecting workplaces to enforce compliance, performing short-term training and education, and developing injury and illness statistics.

The second agency created by the OSHAct is the National Institute for Occupation Safety and Health (NIOSH). It is charged with developing criteria for standards, research and development, professional training and education, and with performing special health surveys. NIOSH is the research arm of OSHA and helps to initiate standards.

The third agency is the Occupational Safety and Health Review Commission (OSHRC). This is the judicial branch (consisting of three members appointed by the President) of OSHA in that it hears cases that are contested when violations of the regulations are found. Employers can contest a citation for an alleged violation,

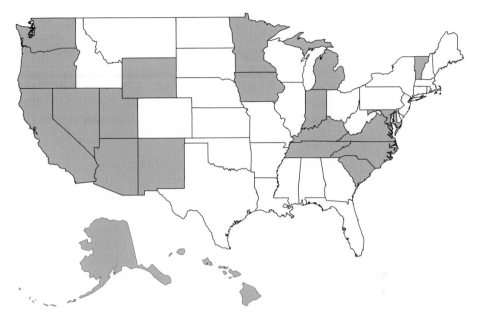

FIGURE 22.2
Distribution of OSHA state-plan states. State-plan states are shaded. (Note: Connecticut, New Jersey, and New York have state plans for public sector employees only.)

the abatement period, or the proposed penalty. OSHRC will issue orders affirming, modifying, or denying the proposals of OSHA compliance officers.

When OSHAct was passed, a provision was included by which the individual states could carry out the wishes of the legislation while being monitored by the federal OSHA, which is organized into 10 geographic regions. The federal government would subsidize the states taking this role. Approximately half the states have adopted plans for carrying out the intent of OSHA (figure 22.2). However, most state plans are closely tailored to the federal legislation, with many being little more than carbon copies of the federal regulations. States could generate additional standards or modify existing ones, but they had to be at least as stringent as the federal law. Federal OSHA maintains an office within each region to carry out OSHA's mission, including the monitoring of the state-plan states.

OSHA INSPECTIONS

The successful enforcement of OSHAct hinges on OSHA inspections. OSHA makes inspections (unannounced in most cases) of work sites to ascertain the extent of compliance with the regulations. Inspections can be made at a place of work under five criteria:

1. Random or scheduled inspections.
2. Reports of major accidents (fatalities or multiple-employee incidents).

3. Employee complaints. Employee complaints must be in writing, must state the basis for the complaint, and must be signed. The nature of the complaint is explained to the employer, but the complaining employee is not identified.
4. Referrals by police, field building inspectors, media, etc.
5. Follow-up inspections.

Inspections are made by OSHA compliance officers. Under the law, these officers must identify themselves before inspecting a job site. They must show their credentials and indicate the purpose and scope of the visit. The compliance officer will probably first ask to see the OSHA postings. These include the posting of information about OSHA and the right of workers to refuse to work in unsafe conditions without fearing retaliation. Information about injuries on the job site should also be available. A summary of the prior year's injuries must be posted during the months of February, March, and April. After the required postings have been examined, the compliance officer will ask permission to make an inspection of the work site or a particular part of it. The compliance officer can ask questions of the workers and can take pictures and make notes. The employer is allowed to accompany the compliance officer and is well advised to do so.

The employer has an obligation to provide a place of employment that is free from recognized hazards and meets the safety and health standards. The employees must comply with the rules, regulations, and standards applicable to their type of work. However, if an employee blatantly refuses to comply with the OSHA regulations, the only recourse available to the employer is to dismiss or suspend that employee. Employees cannot be fined by an OSHA compliance officer.

As the regulations were originally interpreted, OSHA compliance officers could enter an establishment or work site at a reasonable time and in a reasonable manner to inspect, investigate, and question employers and employees. This was successfully challenged in court (*Marshall v. Barlow's Inc.*, 436 U.S. 307), meaning that an employer could deny site entry to a compliance officer. Such denial of entry would mean that a court order or warrant had to be obtained first. Although initially regarded as a setback for OSHA inspections, this impact was small. Although a compliance officer may be denied access to a work site, the compliance officer can generally obtain a warrant or court order permitting entrance. After having been denied access to a work site, a compliance officer need only show probable cause to get a federal warrant to gain legal access. Then the possibly irate compliance officer may be more thorough in the inspection.

After conducting a workplace inspection, the OSHA compliance officer will discuss the results of the visit in a closing conference. The inspector does not levy fines or penalties; this is done by the area director of OSHA. However, the compliance officer may issue citations and can post a notice of imminent danger if a life-threatening condition is observed. The employer is notified of any citations that could bring a penalty. The particular section or sections of the standards violated will be noted. A reasonable time for abatement will be determined.

OSHA FINES AND PENALTIES

Historically, OSHA fines have not been severe, but this has begun to change in recent years. Heavy penalties have been levied against many employers for failing to provide safe workplaces. Previously, OSHA fines had averaged less than $100 per violation, but recent statistics show that the average fine is more than $1,000 per violation. Not all violations bring fines, but all infractions must be corrected. The correction of unsafe conditions must occur within a designated abatement period. Failure to correct a violation will probably result in an additional fine. If the infraction is serious, the fine may be quite high, up to $7,000 for each day it remains uncorrected.

For a workplace condition to be considered an imminent danger, workers in the area must be at immediate serious risk of death or serious physical harm. Such conditions could include a vertical trench wall without shoring, fall hazards without appropriate fall protection, or exposure of roofing workers to contact with an overhead power line. If an OSHA compliance officer determines that an imminent danger exists, the officer is to request that the hazard be corrected immediately, and that all employees be removed from the danger area. The inspector must inform affected employees; a recommendation would then be made that OSHA take steps to stop the imminent danger. OSHA will conduct a follow-up inspection to ensure that the condition has been addressed. If the employer is not responsive to the request, OSHA may post an "Imminent Danger" notice, and seek a temporary restraining order from the nearest federal district court (federal court order) requiring the employer to remove employees from exposure to the danger, or to eliminate the imminent danger.

After an OSHA inspection, notification by certified mail is sent to the employer, listing the standards allegedly violated and the corresponding penalty. For citations of serious violations, the imposition of a penalty is mandatory. The employer has 15 federal working days from receipt of the letter to decide to contest any of the allegations or penalties (violations, penalties, and abatement periods). If anything will be contested, written notice must be sent to the OSHA area director, clearly stating the intentions of the contest. This matter will be referred to the OSHRC, and a hearing date will be set. If the employer does not challenge the results of an inspection within 15 federal working days, the allegations become final.

If the violations are contested in good faith, the abatement period is automatically delayed. No action will be required until after a decision has been rendered by the commission. If only the penalty or the abatement period is contested, the employer must still initiate appropriate action to correct the violation, or additional penalties can be levied. If any of the options under the law are not clear to the employer, a meeting can be scheduled with the area director. This should be done within the 15-day period so that appropriate actions can still be taken regarding a contest procedure.

The amounts of the fines can vary considerably. It was once alleged that OSHA fines were initially kept deliberately low to minimize the number of citations that were contested. Such allegations are no longer common. Violations that are "other-than-serious" (physical harm not likely to cause death) and violations that are serious (substantial probability that death could result) can result in penalties of up to $7,000 for each offence. *Serious* refers to life threatening, and the term is certainly open to interpretation. Fines can be more severe if, for example, an infraction is not corrected ($7,000 per day) or if a serious violation, called a willful and repeat violation, occurs (up to $70,000). A serious penalty can be assessed against persons giving a tip to employers of an upcoming OSHA inspection ($1,000 and/or one year in jail).

Although the fines may be high, they can be reduced under several criteria. For example, if a nonserious violation is corrected immediately, it is possible that no fine will be issued at all. Fine reductions may be made when the following criteria may apply:

- Reduction for demonstrated good faith.
  - 25% reduction in the fine for having a comprehensive written safety and health program.
  - 15% reduction for providing documented evidence that the safety and health program is being implemented.
  - 0% reduction if major discrepancies are noted in the safety and health program, or if there is no program.
- 10% reduction for having a past history of compliance.
- Proposed penalties will be reduced by the following percentages in considering employer size:
  - 60% penalty reduction may be applied if an employer has 25 employees or fewer.
  - 40% penalty reduction if the employer has 26–100 employees.
  - 20% penalty reduction if the employer has 101–250 employees.

In recent years OSHA has become more aggressive in levying heavy penalties for noncompliance. Fines of more than $1 million have been assessed against some employers.

RECORDKEEPING REQUIREMENTS

Employers of more than 11 employees in covered industries, including construction, must maintain an up-to-date record of job-related injuries. These records must be kept a minimum of five years. These records include the following:

Official OSHA notice informing employees about their rights under the act.

OSHA No. 300: A continuous log of each recordable occupational injury or illness. This includes all lost workday cases, cases where a worker cannot return to the same task or operation, and cases where medical treatment is required. Entries must be made within seven days of the employer's receipt of information about the incident.

OSHA No. 301: Supplementary record containing additional detailed information about every injury and illness. This is similar to the insurance form known as the first report of injury. The first report can be used in lieu of No. 301 if it is approved by OSHA. This form need not be posted, but it must be available upon request by an OSHA compliance officer.

OSHA No. 301A: A summary of occupational injuries and illnesses that is posted after the year is completed. It is posted beginning no later than

February 1 and kept in place until April 30. This must be available upon request by a compliance officer at any time.

Citations of OSHA violations must also be posted.

Falsification of a required report may result in serious fines. One company that falsified the information in its OSHA forms was recently fined in excess of $5 million. Falsification of documentation can also result in imprisonment. These recordkeeping requirements are waived for establishments that have 10 or fewer employees, unless the small employer has been requested to participate in a BLS or OSHA survey. Violations of the posting requirements may result in penalties of up to $7,000 per incident.

If a serious accident—a fatality or a multiple-injury incident (five or more workers)—occurs on the job, OSHA must be notified within eight hours.

TYPICAL SAFETY STANDARDS

Although all industries fall under the jurisdiction of the OSHA regulations, some of these regulations were promulgated specifically for the construction industry. The construction standards are codified in 29 Code of Federal Regulations (CFR) Part 1926. The general industry standards, which also can be applied to construction, are codified in 29 CFR Part 1910. An example of a general industry standard is the one related to respirators. Workers subjected to toxic environments must have appropriate protection. Compliance with the respirator standard is required in all places of work. This has special implications for construction workers who work with or around chemicals and volatile substances.

The following construction safety standards might apply on a typical construction project. They are among the standards most often violated.

1926.501 Guardrails, handrails, covers
.451 Scaffolding
.1050 Ladders
.350 Gas welding and cutting
.401 Grounding and bonding
.550 Cranes and derricks
.250 Housekeeping
.152 Flammable and combustible liquids
.400 General electric
.402 Electrical equipment—installation and maintenance
.150 Fire protection
.652 Trenching
.601 Motor vehicles
.100 Head protection
.552 Material hoists, personnel hoists, elevators
.500 Medical services, first aid
.510 Drinking water

These areas would constitute a good guideline for self-inspections on most construction projects.

OSHA also affected equipment design. Rollover protection structures (ROPS) are required on all equipment manufactured after 1972 at an added estimated average cost of $4,315. In some cases the fuel tank had to be moved to accommodate the protection. Fenders were required on scrapers at an added cost of $870 per axle. Cantilever-type canopies were required for front-end loaders at an added cost of $2,700. Load-moment indicators were required to retrofit some cranes (costs are unknown, but some firms reportedly went out of business). Many required changes were costly. Some have been eased or rescinded.

Initial estimates of some skeptics were that the OSHA regulations would add about 30 percent to the costs of construction. Also, many people felt that some of the added costs would not make the workplace any safer. In fact, some individuals alleged that the added requirements would merely reduce the efficiency of machinery and workers. Despite these allegations, the frequency rates of injuries, expressed as the number of injuries per 200,000 hours of worker exposure, appear to have declined since the OSHA regulations became effective.

Although the OSHA regulations cover most types of work, not all are covered in detail. For such conditions, the general duty clause of the regulations should be deemed to apply. This clause states that employers should "furnish to their employees, employment and places of employment which are free from recognized hazards that are causing or are likely to cause death or serious physical harm to their employees." This clause is effective in addressing areas not specifically dealt with elsewhere in the OSHA regulations.

OSHA COMPLIANCE

OSHA has not always enjoyed a healthy public image, partly as a result of the compliance officers who have represented OSHA. The initial complaints included the following:

"They are nitpicky."

"There is no uniformity among inspectors."

"Some want to be authoritarian, while others are nice. They should see the law as being important."

While these complaints had some merit in the past, employers now appear to be less critical of OSHA and OSHA compliance officers. Many "nitpicky" rules were eliminated in the first decade of OSHA to correct some of the problems associated with compliance. Increased acceptance of OSHA may also be due in part to the greater awareness of the need to be safe that resulted from the escalation of workers' compensation insurance costs. An increase in the number of liability suits filed by injured construction workers may have contributed to this awareness.

OSHA CONSULTATION SERVICES

To contractors, OSHA is often viewed as a threat, and many contractors would never consider calling OSHA for advice. Despite the thoughts that the term "OSHA" conjures up in the minds of many employers, OSHA can be a valuable source of assistance in promoting worker safety. This is done through consultation services that are available in each state. These services are provided by a state agency that operates independent of OSHA, but is funded in part by OSHA. Employers need not fear citations as a result of a consultation, as penalties are not proposed for hazards identified by OSHA consultants. Consultation advice may be obtained through a telephone conversation or through an extensive field visit of a construction site. The consultant will conduct a closing conference with the employer, followed by a written report to the employer of the consultant's findings and recommendations. These services are available to employers with fewer than 250 employees nationally. In addition to providing an appraisal of the safety and health conditions of the workplace, the employer may receive training and education services. The identity of the employer is not routinely revealed to the OSHA enforcement staff or compliance officers. Consultation services are provided only to employers who request them. The best part of the advice is that it is provided at no cost to the employer.

SHOULD CITATIONS BE APPEALED?

There are two points of view on whether an OSHA citation should be appealed. Of course, this question is more relevant if the employer feels that there is a good case. One view is that the appeal of a citation will antagonize the compliance officer and that the compliance officer will come back to harass the employer. The second view is that good cases should be taken through an appeal. "Nitpicky" citations will decrease if more appeals are successfully made. Another point is that a successful appeal wipes the slate clean. If a more serious citation is not vacated, a subsequent similar circumstance may be regarded as a willful and repeat violation. Such violations can carry very harsh penalties.

NEW OSHA INITIATIVES

OSHA has initiated several new programs to increase its effectiveness in reducing worker injuries. One such program is the *voluntary protection program* (VPP), which is designed to recognize outstanding achievement of firms that have successfully incorporated comprehensive safety and health programs in their overall management system. Firms can be designated as Star, Merit, or Demonstration. The VPP designation is granted when a firm has established a cooperative relationship with its employees and OSHA. The benefits of this program have shown that participants regularly experience injury rates that are 60 percent to 80 percent below the industry average. Employers who participate in one of these proactive programs are not scheduled for OSHA programmed inspections. Employers interested in participating in a VPP should contact their local OSHA office.

OSHA created a *focused inspection program* for the construction industry which is designed to streamline the inspection process, and essentially to emphasize inspections on those job sites that are most perilous to workers. With focused inspections, OSHA will limit its time spent on job sites with effective safety programs; it concentrate its efforts on those projects that do not have effective safety programs. The inspection process will begin in the normal manner, but the compliance officer will be particularly interested in seeing the written safety program and in discussing the program with the competent person in charge of administering and enforcing it. A project "walkaround" will be conducted to verify that the safety program is being fully implemented. On the job tour, the compliance officer will focus on the leading hazards that cause 90 percent of the deaths and injuries in construction, namely: falls, stuck by, caught in/between, and electrical hazards. The compliance officer will meet with employees to determine their knowledge of the safety and health program. If it is determined that the program is being effectively implemented, the inspection will be terminated at that point. Otherwise, a comprehensive inspection of the entire project will result. Note that a VPP inspection is designed to be completed in two or three hours, while a comprehensive inspection could take days to complete.

CONTRACTUAL SAFETY REQUIREMENTS

While OSHA mandates that employers provide for the safety and well-being of their employees, construction contracts are increasingly being drafted whereby additional emphasis is placed on the need for performing work in a safe manner. This is evident in the Construction Management Association of America (CMAA) Document A-3 that states, (Before beginning the Work, the Contractor shall prepare and submit to the CM the Contractor's safety program that provides for the implementation of all of the Contractor's safety responsibilities in connection with the Work at the site and the coordination of that program and its associated procedures and precautions with the safety programs, precautions and procedures of each of the other contractors performing the Work at the site. The Contractor shall be solely responsible for initiating, maintaining, monitoring and supervising all safety programs, precautions and procedures in connection with the Work and for coordinating its programs, precautions and procedures with those other contractors performing the Work at the site).

The provisions continue by stipulating that the safety efforts are to include the safety of "all employees on the Work; employees of all subcontractors, and other persons and organizations who may be affected thereby." With the continuing increases in health care costs and the continuation of the litigious environment in construction, safety emphasis on construction projects can be expected to increase.

REVIEW QUESTIONS

1. What is the approximate death and injury toll among construction workers each year?
2. What are the three agencies established by the Williams-Steiger Act, and what are the essential functions of each one?
3. How can OSHA fines against a contractor be reduced?
4. What are the reasons for contesting an OSHA fine instead of simply paying the fine?
5. Discuss the need for an agency such as OSHA, when the costs of injuries and insurance are sufficient to warrant concern by construction firms.

23

LABOR RELATIONS IN CONSTRUCTION

EVERY MANAGER IN the construction community should have some knowledge of labor relations. This is particularly true of managers of construction projects in which at least some of the workers are union members. This chapter is organized into three sections. The first contains a list of common labor relations terms accompanied by their definitions. The second consists of typical labor agreement provisions that should be reviewed prior to working with members of a craft union. The third describes major legislation that has had a significant impact on the construction industry.

LABOR RELATIONS TERMS

The topic of labor relations is of major importance in the construction industry. The complexity of this area is increased by the unique organization of labor in most sectors of the construction industry. In most industries, the workers are organized in one union. In construction, it is not uncommon for as many as 18 different labor unions to be represented on a project. To provide a view of labor relations in the construction industry, selected terms are defined and a brief description of each one is given.

Employer Organization Terms

agency shop: a business that formally agrees with a union that all employees must pay union dues whether or not the workers are union members. The dues are meant to compensate the union for services which benefit all the workers.

closed shop: a business that formally agrees with a union local to hire only union members. Closed shops were outlawed in 1947 by the Taft-Hartley Act.

double-breasted operation: an arrangement by which a union shop company and an open shop company are owned by the same parent firm. Common ownership of such firms can be legitimate, but care must be exercised to ensure that one firm is not established as an interim means of absorbing the other.

front: a business, ostensibly a minority or women's business enterprise, in which the owners of record are controlled or manipulated by dominant group members (see *minority business enterprise* and *women's business enterprise*).

merit shop: a business whose labor relations are not governed by a labor agreement. The employer, rather than a negotiated labor agreement, dictates how labor relations decisions are made (see *open shop*).

minority business enterprise (MBE): a business that is at least 50 percent owned and managed by minority group members (blacks, Hispanics, Native Americans, et al.).

open shop: a business whose labor relations are not governed by a labor agreement. Employees are not required to be or become union members as a condition of employment.

right-to-work law: a state law that prohibits prehire agreements. States that have enacted such laws are known as right-to-work states. These laws are found only in states in which union strength has been diminished (figure 23.1).

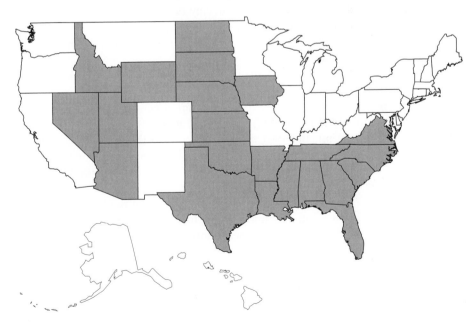

FIGURE 23.1
Distribution of right-to-work states. Right-to-work states are shaded.

set-aside requirement: a requirement in a construction contract in which a specified percentage of the work is established as a goal for participation by minority business enterprises and/or women's business enterprises.

union shop: a business that enters into an agreement with a union local by which all workers who are hired must join the union in order to remain employed. Newly hired workers generally have a stipulated period of time, such as 30 or 60 days, in which to join.

women's business enterprise (WBE): a business that is at least 50 percent owned and managed by women.

Labor-Management Relations Terms

boycott: an organized effort in which no business is conducted with a particular establishment and in which other workers are encouraged to refrain from doing business with that firm. In labor relations a *primary boycott* is one involving a union and an employer, while a *secondary boycott* is a boycott of companies that are doing business with an employer with whom the union has a grievance.

collective bargaining: the process by which representatives of employers and employees hammer out a labor agreement. The representatives of employers negotiate with the representatives of a union about wages, hours of employment, work practices, safety, and the like.

featherbedding: a work rule provision in labor agreements that dictates an inefficient and often costly way of carrying out specified tasks. An example would be a project where two workers are required to perform a task that could easily be performed by one. Another example occurs when the maximum productivity level for a task is prescribed.

good faith: adherence to standards that are honest and fair when involved in bargaining.

injunction: a judge's order to a person or group of persons (such as a union or employer) to refrain from doing a particular act.

job targeting: the practice of a union's subsidizing the labor costs of union contractors on selected projects as a means of helping union contractors underbid their open shop competitors. This is a way to help ensure that a project will be built by union workers. The program is financed by payroll deductions from union employees on all projects. The funds are pooled and are selectively distributed to union contractors and subcontractors.

jurisdiction: the extent of control or authority a union has in administrating the rules governing its members and the work they do. In labor unions it is related to the range of work performed by a particular craft. Other crafts are expected to honor the "ownership" rights of another craft over such work. Exclusive control is assumed over the content of work performed by the members of the local in a given geographic territory.

jurisdictional dispute: a work assignment dispute in which two unions claim the same work as a "property right." This occurs when the boundary between two unions is not clearly defined. Technological advances in certain product lines are often responsible for such disputes. Disputes may arise when a union tries to assure employment security, or maintain job ownership, or when a disagreement over the geographic territory covered by two similar craft unions occurs.

labor dispute: a disagreement or controversy arising between a labor union and an employer or group of employers. Disagreements commonly concern wages, safety, work conditions, working hours, and the like.

lockout: a refusal on the part of an employer to permit workers to work at a work site. This tactic may be used to weaken a union when there is a labor dispute.

national agreement: a labor agreement that is established between a particular employer and a particular craft union. The agreement can be applied to all portions of the country. Disputes that a craft union may have with other employers in a local area will generally not influence the union's association with a firm that is a signatory to a national agreement.

picketing: as related to labor disputes, a gathering of union members near the entrance to a place of employment of an employer with whom the union has a grievance. The purpose may be to interfere with the normal operation of the business establishment (construction project) or to inform others about the nature of the labor dispute. Other union members will generally honor a picketed entrance by not passing through it.

prehire agreement: an agreement between a contractor and a union in which the contractor agrees to hire workers who will become members of the union if they are not already members. This agreement is made with the union, without any input by the workers who are to be hired.

project agreement: a labor agreement that is negotiated solely for a particular project. The advantages include assurances that agreements will exist with all labor unions for the full duration of the project, and that there will be greater consistency in the work practices and work rules of all the crafts.

salting: the practice of sending union-paid professional union organizers to work for a selected nonunion contractor for the purpose of organizing the workers employed by the firm. The practice is legal and is regarded as a first amendment right of workers to exercise their right of free speech. The "salt" cannot be discriminated against simply because the right of free speech is exercised in the organizing effort.

secondary boycott: the blacklisting by a union of an employer who is doing business with a firm with which the union has a dispute. This is an indirect means of applying pressure.

situs (or common situs) picketing: pickets set up to discourage union workers, as well as others, from entering a place of employment (such as a construction project), as a means of gaining demands made of an employer. It may also be a means of communicating the grievances a union has with a particular employer. This technique has been deemed a secondary boycott, which in some cases constitutes a violation of the National Labor Relations Act.

slowdown: an organized and systematic reduction in productivity at a job site, generally to force the employer to make certain concessions.

strike: the organized refusal of workers to work at a particular place of employment as a means of gaining demands made of the employer. This is also known as a walkout.

sweetheart agreement: derogatory term referring to an arrangement between a union and an employer whereby concessions are made from one party to the other. Generally the intent is to keep a rival union out of an establishment, but the term may apply more broadly to permit either or both parties to gain in some way.

unfair labor practice: an action by a union or employer that is prohibited by law or court decision.

union security clause: a contract provision in a labor agreement that establishes and clarifies the union's status as the bargaining agent for the employees.

wildcat strike: an unauthorized action in which the workers refuse to work without first gaining the consent of the union for a strike, or in direct opposition to the directives of union leaders. These strikes are often spontaneous.

work rules: work provisions included in labor agreements that clearly define appropriate procedures to be followed in performing certain work tasks. Examples include work hours per day, jurisdiction of work, multiple shifts, overtime, holidays, apprentices, prohibition of piecework, paydays, reporting time, supervisor crew size, safety provisions, safety devices, job stewards, and drinking water.

yellow-dog contract: an agreement by which a job applicant, as a condition of being hired, formally agrees not to join a labor union or participate in organizing a union. Such agreements are not legally binding and constitute an unfair labor practice. They are now illegal.

Organized Labor Terms

business agent (BA): the representative of the union local. The BA has general responsibility for promoting the best interests of the union. The BA, who is elected by the union membership, generally wields considerable power.

craft union: a labor union in which all the members perform the same type of work (carpentry, masonry, pipefitting, electrical, operating engineering, etc.).

hiring hall: the home base for union members. The hiring of workers often is done directly through the hiring hall. Employers who want a worker simply call the hall, and an available worker will be dispatched.

hot cargo agreement: an agreement among union workers that they will not work with or handle materials that were prepared or prefabricated by nonunion workers.

industrial union: a labor union whose members may have different skills but all work for the same employer or type of industry (automotive, textiles, etc.).

job steward: the union representative on a construction project to whom grievances, such as safety infractions, are directed. The steward may be designated by the business agent, may be elected by the workers, or may be the first worker in a craft union to be assigned to the project. Not all crafts have a job steward. The job steward is often the first craftsperson hired and the last fired.

scabs: a derogatory term used by union workers for nonunion tradespeople who work at jobs from which union workers were displaced as a result of a strike or lockout.

union local: the members of a particular craft who entrust their governance to elected leaders. A local has jurisdiction over union members in a particular craft in a prescribed geographic area.

LABOR AGREEMENT PROVISIONS

Whenever workers are represented by a labor union on a construction site, the primary source of guidance for the relationship between the workers and the employer is the labor agreement. In most cases, these are pocket-sized booklets that contain all the provisions agreed upon by the labor union and the contractor. In many cases, several contractors will negotiate with the union as a unit; however, some contractors elect to negotiate their own agreements. Since these agreements are the basis for understanding the rights and obligations of the unions and the employers, it is important to appreciate the impact of different provisions. Since not all provisions are the same within a region or within a craft, it is important to recognize how differences in the provisions may affect the relationship between the union members and the employers. Strong craft unions generally gain more concessions from management than do weaker unions. In addition, labor agreements in right-to-work states are generally more favorable to management than are those in other states. Some of the provisions to consider when reviewing a labor agreement are as follows:

coffee break provisions: provisions that allow work stoppages by an entire work crew to drink coffee or another nonalcoholic beverage. While such provisions can still be found in labor agreements, they are not common. In some instances, area practices may allow coffee breaks. If such an area practice already exists, unions do not feel compelled to negotiate for the privilege.

contract duration: a labor agreement is negotiated for a specified period. While these periods may range from one to five years, durations of three years are the most common; very few agreements are for four or five years. A new agreement is generally negotiated at the end of the contract duration.

cost-of-living-allowance provisions: used when wages do not keep pace with other costs during periods of high inflation. To safeguard against these relative losses, cost-of-living-allowance (COLA) provisions may be desirable for a union. Such provisions tie future wage increases to a price index, such as the U.S. All Cities Consumer Price Index for Urban Wage Earners and Clerical Workers. Employers do not favor such provisions, as they make it impossible to estimate the exact wage rates to apply to a project over a given time period. COLA provisions are very rare in right-to-work states and are not common in union shop states.

expiration date: the date on which a labor agreement ends. This is important for a contractor, as it designates when a new agreement must be negotiated. Expiration dates are generally set so that negotiations will take place when the demand for the services of the craft union is high. In construction, most labor agreements expire in the spring months, when construction activity runs high: March, April, May, and June. Some crafts, such as plasterers, may have a later expiration date because their services are generally needed later in most projects.

no-strike, no-lockout provision: an agreement by employers to refrain from locking out employees in exchange for a union agreeing not to strike in a labor dispute. Lockouts and strikes are generally regarded as harsh maneuvers and result in economic losses for both sides. Resolving disputes without interrupting work is beneficial to both sides. Almost all labor agreements now contain such provisions.

overtime provisions: provisions that stipulate the nature of the premium to be paid when the number of hours worked per day typically exceeds 8 hours or the weekly total exceeds 40 hours. The most common method is to pay time and a half for all work over 8 hours, Monday through Friday, and for work performed on Saturdays. Double time may be stipulated for work performed on Sundays and holidays. Some provisions increase the premium to double time when the daily total exceeds 10 hours and the Saturday total exceeds 8 hours. Many variations of premiums will be noted in labor agreements. The extreme, which is found in only a few agreements, would be for all overtime to be paid at double time. Note that Saturday makeup provisions and provisions that allow work to be performed in four 10-hour days may alter the premium pay schedule.

Saturday makeup provisions: provisions that permit work to be performed on Saturday at straight-time pay, when conditions out of the employer's control precluded the possibility of working 40 hours during the regular workweek. Such provisions acknowledge that weather conditions may halt construction during the workweek, and that progress may be regained to some extent by working on Saturdays. Since most projects are estimated with the assumption of straight pay, Saturday makeup work at straight pay is desirable for a contractor. With union strength having been diminished in the past decade, such provisions have become more common, particularly in right-to-work states.

shift provisions: provisions that define the nature of the premium that is paid when work is performed in shifts. Most labor agreements pay for the shift premium by reducing the number of hours actually worked while paying for 8 hours. For example, the second shift will work 7 1/2 hours while being paid for 8, and the third shift (night shift) will work 7 hours while being paid for 8. Less common provisions provide for a premium wage to be paid for work performed on the second and third shifts, or a premium wage to be paid in addition to a reduction in the number of hours worked.

show-up-time provision: a payment made to regular workers who report to work but cannot be assigned to a job. Rain may prevent workers from being able to work. Payment is typically in the form of two to four hours of straight-time pay for showing up.

subcontractor provisions: provisions that are enforced only if one or more open shop subcontractors are expected on a construction project. If such a provision is in the labor agreement, open shop subcontractors are required to become signatories to the labor agreement for the duration of the project or, at the very least, are required to abide by all the provisions. Most agreements in union shop states have subcontractor provisions, while these provisions are less likely to be included in labor agreements in right-to-work states.

travel pay provisions: provisions that provide compensation for workers who are required to travel considerable distances to construction sites. Compensation may be in the form of a predetermined amount to be paid per mile of travel (usually measured from a designated central location), a stipulated sum per day (such as an extra hour of pay), or reimbursement for actual expenses. Many variations of such provisions are employed. Many labor agreements contain travel pay provisions, with slightly fewer of these provisions occurring in right-to-work states.

wage rate provisions: provisions that stipulate the rates of pay to be received by various worker classifications, including apprentice, journeyman, supervisor, and general supervisor. Premiums may also be included for work that is performed in more hazardous conditions, such as above the ground level.

work through lunch provision: a stipulation in a labor agreement that provides added compensation for workers who are asked to work through the normal lunch period. This does not permit an employer to have workers forgo lunch, only to delay it. Workers who have lunch delayed as a result of being asked to work at the regularly scheduled lunchtime are typically compensated for the inconvenience. The compensation may be in the form of time and a half pay for the work performed until the lunch period is taken, or straight-time pay for the lunch period itself. Crafts involved in concrete finishing are particularly likely to have these provisions in their agreements. Such provisions are most common in union shop states.

FEDERAL LAWS RELATED TO CONSTRUCTION

Just as labor relations are unique in the construction industry, so too are several laws that have had a great influence on the construction industry. Many federal laws dictate or define appropriate behavior for construction managers and crafts-people. Note that these laws may not apply only to the construction industry. They are discussed below in chronological order.

Sherman Antitrust Act of 1890: the first law that made all contracts, trusts, and conspiracies that restrain trade or commerce illegal. Although the law did not specifically mention labor unions, court decisions determined that unions were covered.

Clayton Act of 1914: a law that specifically removed unions from falling under the Sherman Antitrust Act. Under this law, it was specifically stated that people are not a commodity and that the Sherman Act should not be construed as forbidding the existence or operation of unions. Thus, unions are not regarded as being in restraint of trade. Under this law, no restraining order could be granted simply because a dispute concerned the terms of employment, unless it was intended to prevent personal injury or property loss. In other words, no injunctions could be placed against unions for striking. Court decisions did, however, define activities by workers other than district union members as being in restraint of trade. Consequently, the actions of a sympathetic union, such as a boycott, could result in an injunction being sought by an employer. Previously, such injunctions could be obtained only by a U.S. district attorney. In effect, only the employees of a company could legally boycott that company. The courts defined peaceful picketing as consisting of a single picket, with greater numbers warranting an injunction.

Davis Bacon Act of 1931: a law that states the prevailing wages for an area shall be paid on all federally funded and federally assisted projects valued above $2,000. Under this act, workers are paid on a weekly basis. The primary purpose is to protect the local wage rates and economy of each community. This law has the effect of putting union and nonunion contractors on a nearly equal competitive footing in bidding on federal work. Many states and municipalities have enacted similar requirements known as Little Davis Bacon laws. This law is frequently under attack, and attempts to repeal it are common. Critics state that the law is inflationary, while others contend that repeal of the law would result in more work going to open shop contractors, because the prevailing wage is often defined as the union wage. Minority groups feel that repeal of the law would result in increased exploitation of workers.

Norris-LaGuardia Act of 1932: an act whose primary purpose was to prevent federal courts from issuing injunctions against union activities occurring in the context of labor disputes. This anti-injunction movement made it very difficult for an employer to obtain an injunction against union activities. Today, injunctions can generally be obtained only if the activities of a union are illegal or imperil national health or safety. An employer cannot get an injunction for a strike resulting from a breach of contract. The act also outlawed yellow-dog contracts.

Buy America Act of 1933: a law that requires the use of U.S.-produced or U.S.-mined goods unless a local shortage exists.

National Labor Relations Act (Wagner Act) of 1935: a law that was designed to protect union organizing activities and foster collective bargaining.

Yellow-dog contracts were made illegal. The act states that employers must bargain in good faith with properly chosen representatives of their workers. It also forbids employers from practicing discrimination against employees simply because the employees are involved in labor activities. Furthermore, employers cannot attempt to influence an employee's membership in any labor organization. This act created the National Labor Relations Board, which is empowered to enforce it by overseeing collective bargaining procedures, union certification elections, allegations of unfair labor practices, and so on. No comparable prohibitions were imposed on employees in dealing with employers. This act gave considerable strength to unions, with union membership increasing sharply from 1937 to 1945.

Miller Act of 1935: this law (last amended in 1999) prescribes the requirements of performance and payment bonds used in conjunction with federal construction projects valued over $200,000. Generally, the performance bond will be written for 100 percent of the contract amount and the payment bond (protecting persons supplying labor and materials) must be written for an amount that is at least equal to the performance bond. The act guarantees that workers and suppliers will be paid within 90 days after the date on which the last labor was done or the last materials were furnished. If they are not paid, a suit can follow the 90-day period, but must not be filed more than one year after the labor or materials were provided.

Walsh-Healy Public Contracts Act of 1936: a law that requires all persons employed by a contractor to be paid no less than the prevailing minimum wage for industry in the locality for any contract involving the manufacture or furnishing of materials valued at over $10,000 to any U.S. government agency. An employer cannot force workers to work more than 8 hours a day or more than 40 hours a week without a collective bargaining agreement. This law also eliminated child and convict labor.

Fair Labor Standards Act (Wage and Hour Law) of 1938: a law that established the minimum wage, overtime pay, equal pay standards, and child labor standards. It covers workers whose employment is related to interstate commerce or consists of producing goods for interstate commerce. The minimum wage has been increased a number of times. This law does not require that premium pay be offered for work done on weekends or holidays, but it states that time and a half must be paid for more than 40 hours of work in one week.

Hobbs Act (Antiracketeering Act) as redrafted in 1946: a law that made it a felony to obstruct, delay, or affect interstate commerce by committing robbery or extortion, or by attempting or conspiring to do so. Extortion consists of obtaining property with consent by the wrongful use of actual or threatened force, violence, or fear. The act was aimed at stopping payments to union officials by employers under the guise of recompense for services rendered. Examples of extortion include payments to union leaders to avoid union trouble—usually disguised as gifts, commissions, equipment rentals, or services. This act applies even if no violence has occurred. It is sufficient to have a threat of economic loss, personal injury, or damage to equipment.

Taft-Hartley Act (Labor Management Relations Act) of 1947: essentially an amendment to the Wagner Act that included provisions about unfair labor practices by unions. This was the first federal law imposing controls on the activities of organized labor. It established the basic right of every worker to participate in union activities or refrain from them, subject to authorized agreements requiring membership in a union as a condition of employment. It defined unfair labor practices for both employers and labor organizations. It outlawed secondary boycotts, excessive entrance fees for union membership, and featherbedding. Special constraints were established on strikes that would affect or imperil national health or safety.

Copeland Act (Antikickback Law) as amended in 1948: a law specifically focused on contractors on federally funded and federally assisted projects. This act makes it illegal for an employer to deprive anyone working on federal construction projects, or projects financed wholly or in part by federal funds, of any portion of the compensation to which he or she is entitled. The employer can take out or withhold only the standard deductions. Violators can be punished by a fine and/or imprisonment. For example, suppose a worker is employed on a private project for which the wages are $12 per hour. The worker may then be assigned to a public (federally funded) project in which the Davis Bacon wages are $19 per hour. In exchange for this assignment, the employer may stipulate that the worker return $1 or $2 for each hour that is worked at the higher-paying position. Such a practice would constitute a violation of the act.

Landrum-Griffin Act (Labor Management Reporting and Disclosure Act) of 1959: a law that was designed to eliminate improper labor practices by labor organizations, employees of a union, and others. It imposes stringent controls on internal union affairs. It was designed to protect the rights of the individual union members by requiring democratic elections, combating corruption and racketeering in unions, and protecting the public and innocent parties against unscrupulous union tactics. It requires reports to be submitted on the finances, activities, and policies of unions, union officials, union employees, labor relations consultants, and union trusteeships.

Equal Pay Act of 1963: a law that prohibits discrimination on the basis of sex. An employer cannot pay workers of one sex a different wage when equal work is performed. Pay can differ as a result of seniority, merit, a piecework pay system, or other systems that are not based on sex. The minimum age of workers remained at 16, but the age was raised to 18 for hazardous work such as truck driving, wrecking and demolition, and power tool operation.

Civil Rights Act of 1964: a law that made it illegal to discriminate on the basis of race, sex, or creed. It was enacted to boost the economic conditions of groups that had historically experienced discrimination. It became legal for a company to pass over whites with seniority in order to help disadvantaged parties. This was done in part to make up for past inequities in business practices. Court cases have found that reverse discrimination may not be upheld if the practice is contested.

Occupational Safety and Health Act (Williams-Steiger Act) of 1970: a law on safety that states that every employee is entitled to a safe and healthy place in which to work.

Local Public Works Law of 1977: a stipulation added to construction projects supported in part by federal funds that required 10 percent of the funds be diverted to minority business enterprises (MBEs). Initially, MBEs were broadly defined to include minorities and women. Similar requirements are in common use on all public works projects, but the percentages may be different for MBEs and WBEs. The federal government currently refers to both groups as disadvantaged business enterprises (DBEs). These requirements are referred to as set-asides and are continuing to experience considerable change.

REVIEW QUESTIONS

1. What are the essential differences between agency shops, closed shops, open shops, and union shops?
2. What is the difference between a lockout and a strike?
3. Give an example of a jurisdictional dispute.
4. Discuss the implications of union strength for each of the following labor agreement provisions: COLA provisions, Saturday makeup provisions, show-up-time provisions, subcontractor provisions, travel pay provisions, and work through lunch provisions.
5. What is the impact of Davis Bacon requirements on a construction project?
6. What are the primary benefits of the Miller Act requirements?
7. Which federal law imposed considerable constraints on union practices?

APPENDIX

▓AIA® Document A101™ – 2007

Standard Form of Agreement Between Owner and Contractor *where the basis of payment is a Stipulated Sum*

AGREEMENT made as of the   day of
in the year
(In words, indicate day, month and year)

BETWEEN the Owner:
(Name, address and other information)

> This document has important legal consequences. Consultation with an attorney is encouraged with respect to its completion or modification.
>
> AIA Document A201™–2007, General Conditions of the Contract for Construction, is adopted in this document by reference. Do not use with other general conditions unless this document is modified.

and the Contractor:
(Name, address and other information)

for the following Project:
(Name, location, and detailed description)

The Architect:
(Name, address and other information)

The Owner and Contractor agree as follows.

TABLE OF ARTICLES

ARTICLE 1 THE CONTRACT DOCUMENTS
The Contract Documents consist of this Agreement, Conditions of the Contract (General, Supplementary and other Conditions), Drawings, Specifications, Addenda issued prior to execution of this Agreement, other documents listed in this Agreement and Modifications issued after execution of this Agreement, all of which form the Contract, and are as fully a part of the Contract as if attached to this Agreement or repeated herein. The Contract represents the entire and integrated agreement between the parties hereto and supersedes prior negotiations, representations or agreements, either written or oral. An enumeration of the Contract Documents, other than a Modification, appears in Article 9.

ARTICLE 2 THE WORK OF THIS CONTRACT
The Contractor shall fully execute the Work described in the Contract Documents, except as specifically indicated in the Contract Documents to be the responsibility of others.

ARTICLE 3 DATE OF COMMENCEMENT AND SUBSTANTIAL COMPLETION
§ 3.1 The date of commencement of the Work shall be the date of this Agreement unless a different date is stated below or provision is made for the date to be fixed in a notice to proceed issued by the Owner.
(Insert the date of commencement if it differs from the date of this Agreement or, if applicable, state that the date will be fixed in a notice to proceed.)

If, prior to the commencement of the Work, the Owner requires time to file mortgages and other security interests, the Owner's time requirement shall be as follows:

§ 3.2 The Contract Time shall be measured from the date of commencement.

§ 3.3 The Contractor shall achieve Substantial Completion of the entire Work not later than
(   ) days from the date of commencement, or as follows:
(Insert number of calendar days. Alternatively, a calendar date may be used when coordinated with the date of commencement. If appropriate, insert requirements for earlier Substantial Completion of certain portions of the Work.)

, subject to adjustments of this Contract Time as provided in the Contract Documents.
(Insert provisions, if any, for liquidated damages relating to failure to achieve Substantial Completion on time or for bonus payments for early completion of the Work.)

ARTICLE 4   CONTRACT SUM
§ 4.1 The Owner shall pay the Contractor the Contract Sum in current funds for the Contractor's performance of the Contract. The Contract Sum shall be
Dollars ($), subject to additions and deductions as provided in the Contract Documents.

§ 4.2 The Contract Sum is based upon the following alternates, if any, which are described in the Contract Documents and are hereby accepted by the Owner:
(State the numbers or other identification of accepted alternates. If the bidding or proposal documents permit the Owner to accept other alternates subsequent to the execution of this Agreement, attach a schedule of such other alternates showing the amount for each and the date when that amount expires.)

§ 4.3 Unit prices, if any:
(Identify and state the unit price; state quantity limitations, if any, to which the unit price will be applicable.)

Item	Units and Limitations	Price Per Unit

§ 4.4 Allowances included in the Contract Sum, if any:
(Identify allowance and state exclusions, if any, from the allowance price.)

Item	Price

ARTICLE 5   PAYMENTS
§ 5.1 PROGRESS PAYMENTS
§ 5.1.1 Based upon Applications for Payment submitted to the Architect by the Contractor and Certificates for Payment issued by the Architect, the Owner shall make progress payments on account of the Contract Sum to the Contractor as provided below and elsewhere in the Contract Documents.

§ 5.1.2 The period covered by each Application for Payment shall be one calendar month ending on the last day of the month, or as follows:

§ 5.1.3 Provided that an Application for Payment is received by the Architect not later than the ( ) day of a month, the Owner shall make payment of the certified amount to the Contractor not later than the ( ) day of the ( ) month. If an Application for Payment is received by the Architect after the application date fixed above, payment shall be made by the Owner not later than ( ) days after the Architect receives the Application for Payment.
(Federal, state or local laws may require payment within a certain period of time.)

§ 5.1.4 Each Application for Payment shall be based on the most recent schedule of values submitted by the Contractor in accordance with the Contract Documents. The schedule of values shall allocate the entire Contract Sum among the various portions of the Work. The schedule of values shall be prepared in such form and supported by such data to substantiate its accuracy as the Architect may require. This schedule, unless objected to by the Architect, shall be used as a basis for reviewing the Contractor's Applications for Payment.

§ 5.1.5 Applications for Payment shall show the percentage of completion of each portion of the Work as of the end of the period covered by the Application for Payment.

§ 5.1.6 Subject to other provisions of the Contract Documents, the amount of each progress payment shall be computed as follows:

.1 Take that portion of the Contract Sum properly allocable to completed Work as determined by multiplying the percentage completion of each portion of the Work by the share of the Contract Sum allocated to that portion of the Work in the schedule of values, less retainage of percent (%). Pending final determination of cost to the Owner of changes in the Work, amounts not in dispute shall be included as provided in Section 7.3.9 of AIA Document A201™–2007, General Conditions of the Contract for Construction;

.2 Add that portion of the Contract Sum properly allocable to materials and equipment delivered and suitably stored at the site for subsequent incorporation in the completed construction (or, if approved in advance by the Owner, suitably stored off the site at a location agreed upon in writing), less retainage of percent (%);

.3 Subtract the aggregate of previous payments made by the Owner; and

.4 Subtract amounts, if any, for which the Architect has withheld or nullified a Certificate for Payment as provided in Section 9.5 of AIA Document A201–2007.

§ 5.1.7 The progress payment amount determined in accordance with Section 5.1.6 shall be further modified under the following circumstances:

.1 Add, upon Substantial Completion of the Work, a sum sufficient to increase the total payments to the full amount of the Contract Sum, less such amounts as the Architect shall determine for incomplete Work, retainage applicable to such work and unsettled claims; and
(Section 9.8.5 of AIA Document A201–2007 requires release of applicable retainage upon Substantial Completion of Work with consent of surety, if any.)

.2 Add, if final completion of the Work is thereafter materially delayed through no fault of the Contractor, any additional amounts payable in accordance with Section 9.10.3 of AIA Document A201–2007.

§ 5.1.8 Reduction or limitation of retainage, if any, shall be as follows:
(If it is intended, prior to Substantial Completion of the entire Work, to reduce or limit the retainage resulting from the percentages inserted in Sections 5.1.6.1 and 5.1.6.2 above, and this is not explained elsewhere in the Contract Documents, insert here provisions for such reduction or limitation.)

§ 5.1.9 Except with the Owner's prior approval, the Contractor shall not make advance payments to suppliers for materials or equipment which have not been delivered and stored at the site.

§ 5.2 FINAL PAYMENT

§ 5.2.1 Final payment, constituting the entire unpaid balance of the Contract Sum, shall be made by the Owner to the Contractor when

    .1 the Contractor has fully performed the Contract except for the Contractor's responsibility to correct Work as provided in Section 12.2.2 of AIA Document A201–2007, and to satisfy other requirements, if any, which extend beyond final payment; and

    .2 a final Certificate for Payment has been issued by the Architect.

§ 5.2.2 The Owner's final payment to the Contractor shall be made no later than 30 days after the issuance of the Architect's final Certificate for Payment, or as follows:

ARTICLE 6 DISPUTE RESOLUTION

§ 6.1 INITIAL DECISION MAKER

The Architect will serve as Initial Decision Maker pursuant to Section 15.2 of AIA Document A201–2007, unless the parties appoint below another individual, not a party to this Agreement, to serve as Initial Decision Maker.
(If the parties mutually agree, insert the name, address and other contact information of the Initial Decision Maker, if other than the Architect.)

§ 6.2 BINDING DISPUTE RESOLUTION

For any Claim subject to, but not resolved by, mediation pursuant to Section 15.3 of AIA Document A201–2007, the method of binding dispute resolution shall be as follows:
(Check the appropriate box. If the Owner and Contractor do not select a method of binding dispute resolution below, or do not subsequently agree in writing to a binding dispute resolution method other than litigation, Claims will be resolved by litigation in a court of competent jurisdiction.)

    ☐   Arbitration pursuant to Section 15.4 of AIA Document A201–2007

    ☐   Litigation in a court of competent jurisdiction

    ☐   Other *(Specify)*

ARTICLE 7 TERMINATION OR SUSPENSION

§ 7.1 The Contract may be terminated by the Owner or the Contractor as provided in Article 14 of AIA Document A201–2007.

§ 7.2 The Work may be suspended by the Owner as provided in Article 14 of AIA Document A201–2007.

ARTICLE 8   MISCELLANEOUS PROVISIONS

§ 8.1 Where reference is made in this Agreement to a provision of AIA Document A201–2007 or another Contract Document, the reference refers to that provision as amended or supplemented by other provisions of the Contract Documents.

§ 8.2 Payments due and unpaid under the Contract shall bear interest from the date payment is due at the rate stated below, or in the absence thereof, at the legal rate prevailing from time to time at the place where the Project is located.
(Insert rate of interest agreed upon, if any.)

§ 8.3 The Owner's representative:
(Name, address and other information)

§ 8.4 The Contractor's representative:
(Name, address and other information)

§ 8.5 Neither the Owner's nor the Contractor's representative shall be changed without ten days written notice to the other party.

§ 8.6 Other provisions:

ARTICLE 9   ENUMERATION OF CONTRACT DOCUMENTS

§ 9.1 The Contract Documents, except for Modifications issued after execution of this Agreement, are enumerated in the sections below.

§ 9.1.1 The Agreement is this executed AIA Document A101–2007, Standard Form of Agreement Between Owner and Contractor.

§ 9.1.2 The General Conditions are AIA Document A201–2007, General Conditions of the Contract for Construction.

§ 9.1.3 The Supplementary and other Conditions of the Contract:

Document	Title	Date	Pages

§ 9.1.4 The Specifications:
(Either list the Specifications here or refer to an exhibit attached to this Agreement.)

Section	Title	Date	Pages

§ 9.1.5 The Drawings:
(Either list the Drawings here or refer to an exhibit attached to this Agreement.)

Number	Title	Date

§ 9.1.6 The Addenda, if any:

Number	Date	Pages

Portions of Addenda relating to bidding requirements are not part of the Contract Documents unless the bidding requirements are also enumerated in this Article 9.

§ 9.1.7 Additional documents, if any, forming part of the Contract Documents:
    .1   AIA Document E201™–2007, Digital Data Protocol Exhibit, if completed by the parties, or the following:

    .2   Other documents, if any, listed below:
(List here any additional documents that are intended to form part of the Contract Documents. AIA Document A201–2007 provides that bidding requirements such as advertisement or invitation to bid, Instructions to Bidders, sample forms and the Contractor's bid are not part of the Contract Documents unless enumerated in this Agreement. They should be listed here only if intended to be part of the Contract Documents.)

ARTICLE 10   INSURANCE AND BONDS
The Contractor shall purchase and maintain insurance and provide bonds as set forth in Article 11 of AIA Document A201–2007.
(State bonding requirements, if any, and limits of liability for insurance required in Article 11 of AIA Document A201–2007.)

This Agreement entered into as of the day and year first written above.

_____   _____
OWNER *(Signature)*   **CONTRACTOR** *(Signature)*

_____   _____
(Printed name and title)   *(Printed name and title)*

CAUTION: You should sign an original AIA Contract Document, on which this text appears in RED. An original assures that changes will not be obscured.

▲AIA® Document A201™ – 2007

General Conditions of the Contract for Construction

for the following PROJECT:
(Name and location or address)

This document has important legal consequences. Consultation with an attorney is encouraged with respect to its completion or modification.

THE OWNER:
(Name and address)

THE ARCHITECT:
(Name and address)

TABLE OF ARTICLES

Init.

/

1

ARTICLE 1 GENERAL PROVISIONS
§ 1.1 BASIC DEFINITIONS
§ 1.1.1 THE CONTRACT DOCUMENTS

The Contract Documents are enumerated in the Agreement between the Owner and Contractor (hereinafter the Agreement) and consist of the Agreement, Conditions of the Contract (General, Supplementary and other Conditions), Drawings, Specifications, Addenda issued prior to execution of the Contract, other documents listed in the Agreement and Modifications issued after execution of the Contract. A Modification is (1) a written amendment to the Contract signed by both parties, (2) a Change Order, (3) a Construction Change Directive or (4) a written order for a minor change in the Work issued by the Architect. Unless specifically enumerated in the Agreement, the Contract Documents do not include the advertisement or invitation to bid, Instructions to Bidders, sample forms, other information furnished by the Owner in anticipation of receiving bids or proposals, the Contractor's bid or proposal, or portions of Addenda relating to bidding requirements.

§ 1.1.2 THE CONTRACT

The Contract Documents form the Contract for Construction. The Contract represents the entire and integrated agreement between the parties hereto and supersedes prior negotiations, representations or agreements, either written or oral. The Contract may be amended or modified only by a Modification. The Contract Documents shall not be construed to create a contractual relationship of any kind (1) between the Contractor and the Architect or the Architect's consultants, (2) between the Owner and a Subcontractor or a Sub-subcontractor, (3) between the Owner and the Architect or the Architect's consultants or (4) between any persons or entities other than the Owner and the Contractor. The Architect shall, however, be entitled to performance and enforcement of obligations under the Contract intended to facilitate performance of the Architect's duties.

§ 1.1.3 THE WORK

The term "Work" means the construction and services required by the Contract Documents, whether completed or partially completed, and includes all other labor, materials, equipment and services provided or to be provided by the Contractor to fulfill the Contractor's obligations. The Work may constitute the whole or a part of the Project.

§ 1.1.4 THE PROJECT

The Project is the total construction of which the Work performed under the Contract Documents may be the whole or a part and which may include construction by the Owner and by separate contractors.

§ 1.1.5 THE DRAWINGS

The Drawings are the graphic and pictorial portions of the Contract Documents showing the design, location and dimensions of the Work, generally including plans, elevations, sections, details, schedules and diagrams.

§ 1.1.6 THE SPECIFICATIONS

The Specifications are that portion of the Contract Documents consisting of the written requirements for materials, equipment, systems, standards and workmanship for the Work, and performance of related services.

§ 1.1.7 INSTRUMENTS OF SERVICE

Instruments of Service are representations, in any medium of expression now known or later developed, of the tangible and intangible creative work performed by the Architect and the Architect's consultants under their respective professional services agreements. Instruments of Service may include, without limitation, studies, surveys, models, sketches, drawings, specifications, and other similar materials.

§ 1.1.8 INITIAL DECISION MAKER

The Initial Decision Maker is the person identified in the Agreement to render initial decisions on Claims in accordance with Section 15.2 and certify termination of the Agreement under Section 14.2.2.

§ 1.2 CORRELATION AND INTENT OF THE CONTRACT DOCUMENTS

§ 1.2.1 The intent of the Contract Documents is to include all items necessary for the proper execution and completion of the Work by the Contractor. The Contract Documents are complementary, and what is required by one shall be as binding as if required by all; performance by the Contractor shall be required only to the extent consistent with the Contract Documents and reasonably inferable from them as being necessary to produce the indicated results.

§ 1.2.2 Organization of the Specifications into divisions, sections and articles, and arrangement of Drawings shall not control the Contractor in dividing the Work among Subcontractors or in establishing the extent of Work to be performed by any trade.

§ 1.2.3 Unless otherwise stated in the Contract Documents, words that have well-known technical or construction industry meanings are used in the Contract Documents in accordance with such recognized meanings.

§ 1.3 CAPITALIZATION
Terms capitalized in these General Conditions include those that are (1) specifically defined, (2) the titles of numbered articles or (3) the titles of other documents published by the American Institute of Architects.

§ 1.4 INTERPRETATION
In the interest of brevity the Contract Documents frequently omit modifying words such as "all" and "any" and articles such as "the" and "an," but the fact that a modifier or an article is absent from one statement and appears in another is not intended to affect the interpretation of either statement.

§ 1.5 OWNERSHIP AND USE OF DRAWINGS, SPECIFICATIONS AND OTHER INSTRUMENTS OF SERVICE
§ 1.5.1 The Architect and the Architect's consultants shall be deemed the authors and owners of their respective Instruments of Service, including the Drawings and Specifications, and will retain all common law, statutory and other reserved rights, including copyrights. The Contractor, Subcontractors, Sub-subcontractors, and material or equipment suppliers shall not own or claim a copyright in the Instruments of Service. Submittal or distribution to meet official regulatory requirements or for other purposes in connection with this Project is not to be construed as publication in derogation of the Architect's or Architect's consultants' reserved rights.

§ 1.5.2 The Contractor, Subcontractors, Sub-subcontractors and material or equipment suppliers are authorized to use and reproduce the Instruments of Service provided to them solely and exclusively for execution of the Work. All copies made under this authorization shall bear the copyright notice, if any, shown on the Instruments of Service. The Contractor, Subcontractors, Sub-subcontractors, and material or equipment suppliers may not use the Instruments of Service on other projects or for additions to this Project outside the scope of the Work without the specific written consent of the Owner, Architect and the Architect's consultants.

§ 1.6 TRANSMISSION OF DATA IN DIGITAL FORM
If the parties intend to transmit Instruments of Service or any other information or documentation in digital form, they shall endeavor to establish necessary protocols governing such transmissions, unless otherwise already provided in the Agreement or the Contract Documents.

ARTICLE 2 OWNER
§ 2.1 GENERAL
§ 2.1.1 The Owner is the person or entity identified as such in the Agreement and is referred to throughout the Contract Documents as if singular in number. The Owner shall designate in writing a representative who shall have express authority to bind the Owner with respect to all matters requiring the Owner's approval or authorization. Except as otherwise provided in Section 4.2.1, the Architect does not have such authority. The term "Owner" means the Owner or the Owner's authorized representative.

§ 2.1.2 The Owner shall furnish to the Contractor within fifteen days after receipt of a written request, information necessary and relevant for the Contractor to evaluate, give notice of or enforce mechanic's lien rights. Such information shall include a correct statement of the record legal title to the property on which the Project is located, usually referred to as the site, and the Owner's interest therein.

§ 2.2 INFORMATION AND SERVICES REQUIRED OF THE OWNER
§ 2.2.1 Prior to commencement of the Work, the Contractor may request in writing that the Owner provide reasonable evidence that the Owner has made financial arrangements to fulfill the Owner's obligations under the Contract. Thereafter, the Contractor may only request such evidence if (1) the Owner fails to make payments to the Contractor as the Contract Documents require; (2) a change in the Work materially changes the Contract Sum; or (3) the Contractor identifies in writing a reasonable concern regarding the Owner's ability to make payment when due. The Owner shall furnish such evidence as a condition precedent to commencement or continuation of the Work or the portion of the Work affected by a material change. After the Owner furnishes the evidence, the Owner shall not materially vary such financial arrangements without prior notice to the Contractor.

§ **2.2.2** Except for permits and fees that are the responsibility of the Contractor under the Contract Documents, including those required under Section 3.7.1, the Owner shall secure and pay for necessary approvals, easements, assessments and charges required for construction, use or occupancy of permanent structures or for permanent changes in existing facilities.

§ **2.2.3** The Owner shall furnish surveys describing physical characteristics, legal limitations and utility locations for the site of the Project, and a legal description of the site. The Contractor shall be entitled to rely on the accuracy of information furnished by the Owner but shall exercise proper precautions relating to the safe performance of the Work.

§ **2.2.4** The Owner shall furnish information or services required of the Owner by the Contract Documents with reasonable promptness. The Owner shall also furnish any other information or services under the Owner's control and relevant to the Contractor's performance of the Work with reasonable promptness after receiving the Contractor's written request for such information or services.

§ **2.2.5** Unless otherwise provided in the Contract Documents, the Owner shall furnish to the Contractor one copy of the Contract Documents for purposes of making reproductions pursuant to Section 1.5.2.

§ 2.3 OWNER'S RIGHT TO STOP THE WORK
If the Contractor fails to correct Work that is not in accordance with the requirements of the Contract Documents as required by Section 12.2 or repeatedly fails to carry out Work in accordance with the Contract Documents, the Owner may issue a written order to the Contractor to stop the Work, or any portion thereof, until the cause for such order has been eliminated; however, the right of the Owner to stop the Work shall not give rise to a duty on the part of the Owner to exercise this right for the benefit of the Contractor or any other person or entity, except to the extent required by Section 6.1.3.

§ 2.4 OWNER'S RIGHT TO CARRY OUT THE WORK
If the Contractor defaults or neglects to carry out the Work in accordance with the Contract Documents and fails within a ten-day period after receipt of written notice from the Owner to commence and continue correction of such default or neglect with diligence and promptness, the Owner may, without prejudice to other remedies the Owner may have, correct such deficiencies. In such case an appropriate Change Order shall be issued deducting from payments then or thereafter due the Contractor the reasonable cost of correcting such deficiencies, including Owner's expenses and compensation for the Architect's additional services made necessary by such default, neglect or failure. Such action by the Owner and amounts charged to the Contractor are both subject to prior approval of the Architect. If payments then or thereafter due the Contractor are not sufficient to cover such amounts, the Contractor shall pay the difference to the Owner.

ARTICLE 3 CONTRACTOR
§ 3.1 GENERAL
§ **3.1.1** The Contractor is the person or entity identified as such in the Agreement and is referred to throughout the Contract Documents as if singular in number. The Contractor shall be lawfully licensed, if required in the jurisdiction where the Project is located. The Contractor shall designate in writing a representative who shall have express authority to bind the Contractor with respect to all matters under this Contract. The term "Contractor" means the Contractor or the Contractor's authorized representative.

§ **3.1.2** The Contractor shall perform the Work in accordance with the Contract Documents.

§ **3.1.3** The Contractor shall not be relieved of obligations to perform the Work in accordance with the Contract Documents either by activities or duties of the Architect in the Architect's administration of the Contract, or by tests, inspections or approvals required or performed by persons or entities other than the Contractor.

§ 3.2 REVIEW OF CONTRACT DOCUMENTS AND FIELD CONDITIONS BY CONTRACTOR
§ **3.2.1** Execution of the Contract by the Contractor is a representation that the Contractor has visited the site, become generally familiar with local conditions under which the Work is to be performed and correlated personal observations with requirements of the Contract Documents.

§ 3.2.2 Because the Contract Documents are complementary, the Contractor shall, before starting each portion of the Work, carefully study and compare the various Contract Documents relative to that portion of the Work, as well as the information furnished by the Owner pursuant to Section 2.2.3, shall take field measurements of any existing conditions related to that portion of the Work, and shall observe any conditions at the site affecting it. These obligations are for the purpose of facilitating coordination and construction by the Contractor and are not for the purpose of discovering errors, omissions, or inconsistencies in the Contract Documents; however, the Contractor shall promptly report to the Architect any errors, inconsistencies or omissions discovered by or made known to the Contractor as a request for information in such form as the Architect may require. It is recognized that the Contractor's review is made in the Contractor's capacity as a contractor and not as a licensed design professional, unless otherwise specifically provided in the Contract Documents.

§ 3.2.3 The Contractor is not required to ascertain that the Contract Documents are in accordance with applicable laws, statutes, ordinances, codes, rules and regulations, or lawful orders of public authorities, but the Contractor shall promptly report to the Architect any nonconformity discovered by or made known to the Contractor as a request for information in such form as the Architect may require.

§ 3.2.4 If the Contractor believes that additional cost or time is involved because of clarifications or instructions the Architect issues in response to the Contractor's notices or requests for information pursuant to Sections 3.2.2 or 3.2.3, the Contractor shall make Claims as provided in Article 15. If the Contractor fails to perform the obligations of Sections 3.2.2 or 3.2.3, the Contractor shall pay such costs and damages to the Owner as would have been avoided if the Contractor had performed such obligations. If the Contractor performs those obligations, the Contractor shall not be liable to the Owner or Architect for damages resulting from errors, inconsistencies or omissions in the Contract Documents, for differences between field measurements or conditions and the Contract Documents, or for nonconformities of the Contract Documents to applicable laws, statutes, ordinances, codes, rules and regulations, and lawful orders of public authorities.

§ 3.3 SUPERVISION AND CONSTRUCTION PROCEDURES

§ 3.3.1 The Contractor shall supervise and direct the Work, using the Contractor's best skill and attention. The Contractor shall be solely responsible for, and have control over, construction means, methods, techniques, sequences and procedures and for coordinating all portions of the Work under the Contract, unless the Contract Documents give other specific instructions concerning these matters. If the Contract Documents give specific instructions concerning construction means, methods, techniques, sequences or procedures, the Contractor shall evaluate the jobsite safety thereof and, except as stated below, shall be fully and solely responsible for the jobsite safety of such means, methods, techniques, sequences or procedures. If the Contractor determines that such means, methods, techniques, sequences or procedures may not be safe, the Contractor shall give timely written notice to the Owner and Architect and shall not proceed with that portion of the Work without further written instructions from the Architect. If the Contractor is then instructed to proceed with the required means, methods, techniques, sequences or procedures without acceptance of changes proposed by the Contractor, the Owner shall be solely responsible for any loss or damage arising solely from those Owner-required means, methods, techniques, sequences or procedures.

§ 3.3.2 The Contractor shall be responsible to the Owner for acts and omissions of the Contractor's employees, Subcontractors and their agents and employees, and other persons or entities performing portions of the Work for, or on behalf of, the Contractor or any of its Subcontractors.

§ 3.3.3 The Contractor shall be responsible for inspection of portions of Work already performed to determine that such portions are in proper condition to receive subsequent Work.

§ 3.4 LABOR AND MATERIALS

§ 3.4.1 Unless otherwise provided in the Contract Documents, the Contractor shall provide and pay for labor, materials, equipment, tools, construction equipment and machinery, water, heat, utilities, transportation, and other facilities and services necessary for proper execution and completion of the Work, whether temporary or permanent and whether or not incorporated or to be incorporated in the Work.

§ 3.4.2 Except in the case of minor changes in the Work authorized by the Architect in accordance with Sections 3.12.8 or 7.4, the Contractor may make substitutions only with the consent of the Owner, after evaluation by the Architect and in accordance with a Change Order or Construction Change Directive.

§ 3.4.3 The Contractor shall enforce strict discipline and good order among the Contractor's employees and other persons carrying out the Work. The Contractor shall not permit employment of unfit persons or persons not properly skilled in tasks assigned to them.

§ 3.5 WARRANTY

The Contractor warrants to the Owner and Architect that materials and equipment furnished under the Contract will be of good quality and new unless the Contract Documents require or permit otherwise. The Contractor further warrants that the Work will conform to the requirements of the Contract Documents and will be free from defects, except for those inherent in the quality of the Work the Contract Documents require or permit. Work, materials, or equipment not conforming to these requirements may be considered defective. The Contractor's warranty excludes remedy for damage or defect caused by abuse, alterations to the Work not executed by the Contractor, improper or insufficient maintenance, improper operation, or normal wear and tear and normal usage. If required by the Architect, the Contractor shall furnish satisfactory evidence as to the kind and quality of materials and equipment.

§ 3.6 TAXES

The Contractor shall pay sales, consumer, use and similar taxes for the Work provided by the Contractor that are legally enacted when bids are received or negotiations concluded, whether or not yet effective or merely scheduled to go into effect.

§ 3.7 PERMITS, FEES, NOTICES, AND COMPLIANCE WITH LAWS

§ 3.7.1 Unless otherwise provided in the Contract Documents, the Contractor shall secure and pay for the building permit as well as for other permits, fees, licenses, and inspections by government agencies necessary for proper execution and completion of the Work that are customarily secured after execution of the Contract and legally required at the time bids are received or negotiations concluded.

§ 3.7.2 The Contractor shall comply with and give notices required by applicable laws, statutes, ordinances, codes, rules and regulations, and lawful orders of public authorities applicable to performance of the Work.

§ 3.7.3 If the Contractor performs Work knowing it to be contrary to applicable laws, statutes, ordinances, codes, rules and regulations, or lawful orders of public authorities, the Contractor shall assume appropriate responsibility for such Work and shall bear the costs attributable to correction.

§ 3.7.4 Concealed or Unknown Conditions. If the Contractor encounters conditions at the site that are (1) subsurface or otherwise concealed physical conditions that differ materially from those indicated in the Contract Documents or (2) unknown physical conditions of an unusual nature that differ materially from those ordinarily found to exist and generally recognized as inherent in construction activities of the character provided for in the Contract Documents, the Contractor shall promptly provide notice to the Owner and the Architect before conditions are disturbed and in no event later than 21 days after first observance of the conditions. The Architect will promptly investigate such conditions and, if the Architect determines that they differ materially and cause an increase or decrease in the Contractor's cost of, or time required for, performance of any part of the Work, will recommend an equitable adjustment in the Contract Sum or Contract Time, or both. If the Architect determines that the conditions at the site are not materially different from those indicated in the Contract Documents and that no change in the terms of the Contract is justified, the Architect shall promptly notify the Owner and Contractor in writing, stating the reasons. If either party disputes the Architect's determination or recommendation, that party may proceed as provided in Article 15.

§ 3.7.5 If, in the course of the Work, the Contractor encounters human remains or recognizes the existence of burial markers, archaeological sites or wetlands not indicated in the Contract Documents, the Contractor shall immediately suspend any operations that would affect them and shall notify the Owner and Architect. Upon receipt of such notice, the Owner shall promptly take any action necessary to obtain governmental authorization required to resume the operations. The Contractor shall continue to suspend such operations until otherwise instructed by the Owner but shall continue with all other operations that do not affect those remains or features. Requests for adjustments in the Contract Sum and Contract Time arising from the existence of such remains or features may be made as provided in Article 15.

§ 3.8 ALLOWANCES

§ 3.8.1 The Contractor shall include in the Contract Sum all allowances stated in the Contract Documents. Items covered by allowances shall be supplied for such amounts and by such persons or entities as the Owner may direct, but the Contractor shall not be required to employ persons or entities to whom the Contractor has reasonable objection.

§ 3.8.2 Unless otherwise provided in the Contract Documents,

.1   allowances shall cover the cost to the Contractor of materials and equipment delivered at the site and all required taxes, less applicable trade discounts;

.2   Contractor's costs for unloading and handling at the site, labor, installation costs, overhead, profit and other expenses contemplated for stated allowance amounts shall be included in the Contract Sum but not in the allowances; and

.3   whenever costs are more than or less than allowances, the Contract Sum shall be adjusted accordingly by Change Order. The amount of the Change Order shall reflect (1) the difference between actual costs and the allowances under Section 3.8.2.1 and (2) changes in Contractor's costs under Section 3.8.2.2.

§ 3.8.3 Materials and equipment under an allowance shall be selected by the Owner with reasonable promptness.

§ 3.9 SUPERINTENDENT

§ 3.9.1 The Contractor shall employ a competent superintendent and necessary assistants who shall be in attendance at the Project site during performance of the Work. The superintendent shall represent the Contractor, and communications given to the superintendent shall be as binding as if given to the Contractor.

§ 3.9.2 The Contractor, as soon as practicable after award of the Contract, shall furnish in writing to the Owner through the Architect the name and qualifications of a proposed superintendent. The Architect may reply within 14 days to the Contractor in writing stating (1) whether the Owner or the Architect has reasonable objection to the proposed superintendent or (2) that the Architect requires additional time to review. Failure of the Architect to reply within the 14 day period shall constitute notice of no reasonable objection.

§ 3.9.3 The Contractor shall not employ a proposed superintendent to whom the Owner or Architect has made reasonable and timely objection. The Contractor shall not change the superintendent without the Owner's consent, which shall not unreasonably be withheld or delayed.

§ 3.10 CONTRACTOR'S CONSTRUCTION SCHEDULES

§ 3.10.1 The Contractor, promptly after being awarded the Contract, shall prepare and submit for the Owner's and Architect's information a Contractor's construction schedule for the Work. The schedule shall not exceed time limits current under the Contract Documents, shall be revised at appropriate intervals as required by the conditions of the Work and Project, shall be related to the entire Project to the extent required by the Contract Documents, and shall provide for expeditious and practicable execution of the Work.

§ 3.10.2 The Contractor shall prepare a submittal schedule, promptly after being awarded the Contract and thereafter as necessary to maintain a current submittal schedule, and shall submit the schedule(s) for the Architect's approval. The Architect's approval shall not unreasonably be delayed or withheld. The submittal schedule shall (1) be coordinated with the Contractor's construction schedule, and (2) allow the Architect reasonable time to review submittals. If the Contractor fails to submit a submittal schedule, the Contractor shall not be entitled to any increase in Contract Sum or extension of Contract Time based on the time required for review of submittals.

§ 3.10.3 The Contractor shall perform the Work in general accordance with the most recent schedules submitted to the Owner and Architect.

§ 3.11 DOCUMENTS AND SAMPLES AT THE SITE

The Contractor shall maintain at the site for the Owner one copy of the Drawings, Specifications, Addenda, Change Orders and other Modifications, in good order and marked currently to indicate field changes and selections made during construction, and one copy of approved Shop Drawings, Product Data, Samples and similar required submittals. These shall be available to the Architect and shall be delivered to the Architect for submittal to the Owner upon completion of the Work as a record of the Work as constructed.

§ 3.12 SHOP DRAWINGS, PRODUCT DATA AND SAMPLES

§ 3.12.1 Shop Drawings are drawings, diagrams, schedules and other data specially prepared for the Work by the Contractor or a Subcontractor, Sub-subcontractor, manufacturer, supplier or distributor to illustrate some portion of the Work.

§ 3.12.2 Product Data are illustrations, standard schedules, performance charts, instructions, brochures, diagrams and other information furnished by the Contractor to illustrate materials or equipment for some portion of the Work.

§ 3.12.3 Samples are physical examples that illustrate materials, equipment or workmanship and establish standards by which the Work will be judged.

§ 3.12.4 Shop Drawings, Product Data, Samples and similar submittals are not Contract Documents. Their purpose is to demonstrate the way by which the Contractor proposes to conform to the information given and the design concept expressed in the Contract Documents for those portions of the Work for which the Contract Documents require submittals. Review by the Architect is subject to the limitations of Section 4.2.7. Informational submittals upon which the Architect is not expected to take responsive action may be so identified in the Contract Documents. Submittals that are not required by the Contract Documents may be returned by the Architect without action.

§ 3.12.5 The Contractor shall review for compliance with the Contract Documents, approve and submit to the Architect Shop Drawings, Product Data, Samples and similar submittals required by the Contract Documents in accordance with the submittal schedule approved by the Architect or, in the absence of an approved submittal schedule, with reasonable promptness and in such sequence as to cause no delay in the Work or in the activities of the Owner or of separate contractors.

§ 3.12.6 By submitting Shop Drawings, Product Data, Samples and similar submittals, the Contractor represents to the Owner and Architect that the Contractor has (1) reviewed and approved them, (2) determined and verified materials, field measurements and field construction criteria related thereto, or will do so and (3) checked and coordinated the information contained within such submittals with the requirements of the Work and of the Contract Documents.

§ 3.12.7 The Contractor shall perform no portion of the Work for which the Contract Documents require submittal and review of Shop Drawings, Product Data, Samples or similar submittals until the respective submittal has been approved by the Architect.

§ 3.12.8 The Work shall be in accordance with approved submittals except that the Contractor shall not be relieved of responsibility for deviations from requirements of the Contract Documents by the Architect's approval of Shop Drawings, Product Data, Samples or similar submittals unless the Contractor has specifically informed the Architect in writing of such deviation at the time of submittal and (1) the Architect has given written approval to the specific deviation as a minor change in the Work, or (2) a Change Order or Construction Change Directive has been issued authorizing the deviation. The Contractor shall not be relieved of responsibility for errors or omissions in Shop Drawings, Product Data, Samples or similar submittals by the Architect's approval thereof.

§ 3.12.9 The Contractor shall direct specific attention, in writing or on resubmitted Shop Drawings, Product Data, Samples or similar submittals, to revisions other than those requested by the Architect on previous submittals. In the absence of such written notice, the Architect's approval of a resubmission shall not apply to such revisions.

§ 3.12.10 The Contractor shall not be required to provide professional services that constitute the practice of architecture or engineering unless such services are specifically required by the Contract Documents for a portion of the Work or unless the Contractor needs to provide such services in order to carry out the Contractor's responsibilities for construction means, methods, techniques, sequences and procedures. The Contractor shall not be required to provide professional services in violation of applicable law. If professional design services or certifications by a design professional related to systems, materials or equipment are specifically required of the Contractor by the Contract Documents, the Owner and the Architect will specify all performance and design criteria that such services must satisfy. The Contractor shall cause such services or certifications to be provided by a properly licensed design professional, whose signature and seal shall appear on all drawings, calculations, specifications, certifications, Shop Drawings and other submittals prepared by such professional. Shop Drawings and other submittals related to the Work designed or certified by such professional, if prepared by others, shall bear such professional's written approval when submitted to the Architect. The Owner and the Architect shall be entitled to rely upon the adequacy, accuracy and completeness of the services, certifications and approvals performed or provided by such design professionals, provided the Owner and Architect have specified to the Contractor all performance and design criteria that such services must satisfy. Pursuant to this Section 3.12.10, the Architect will review, approve or take other appropriate action on submittals only for the limited purpose of checking for conformance with information given and the design concept expressed in the Contract Documents. The Contractor shall not be responsible for the adequacy of the performance and design criteria specified in the Contract Documents.

§ 3.13 USE OF SITE

The Contractor shall confine operations at the site to areas permitted by applicable laws, statutes, ordinances, codes, rules and regulations, and lawful orders of public authorities and the Contract Documents and shall not unreasonably encumber the site with materials or equipment.

§ 3.14 CUTTING AND PATCHING

§ 3.14.1 The Contractor shall be responsible for cutting, fitting or patching required to complete the Work or to make its parts fit together properly. All areas requiring cutting, fitting and patching shall be restored to the condition existing prior to the cutting, fitting and patching, unless otherwise required by the Contract Documents.

§ 3.14.2 The Contractor shall not damage or endanger a portion of the Work or fully or partially completed construction of the Owner or separate contractors by cutting, patching or otherwise altering such construction, or by excavation. The Contractor shall not cut or otherwise alter such construction by the Owner or a separate contractor except with written consent of the Owner and of such separate contractor; such consent shall not be unreasonably withheld. The Contractor shall not unreasonably withhold from the Owner or a separate contractor the Contractor's consent to cutting or otherwise altering the Work.

§ 3.15 CLEANING UP

§ 3.15.1 The Contractor shall keep the premises and surrounding area free from accumulation of waste materials or rubbish caused by operations under the Contract. At completion of the Work, the Contractor shall remove waste materials, rubbish, the Contractor's tools, construction equipment, machinery and surplus materials from and about the Project.

§ 3.15.2 If the Contractor fails to clean up as provided in the Contract Documents, the Owner may do so and Owner shall be entitled to reimbursement from the Contractor.

§ 3.16 ACCESS TO WORK

The Contractor shall provide the Owner and Architect access to the Work in preparation and progress wherever located.

§ 3.17 ROYALTIES, PATENTS AND COPYRIGHTS

The Contractor shall pay all royalties and license fees. The Contractor shall defend suits or claims for infringement of copyrights and patent rights and shall hold the Owner and Architect harmless from loss on account thereof, but shall not be responsible for such defense or loss when a particular design, process or product of a particular manufacturer or manufacturers is required by the Contract Documents, or where the copyright violations are contained in Drawings, Specifications or other documents prepared by the Owner or Architect. However, if the Contractor has reason to believe that the required design, process or product is an infringement of a copyright or a patent, the Contractor shall be responsible for such loss unless such information is promptly furnished to the Architect.

§ 3.18 INDEMNIFICATION

§ 3.18.1 To the fullest extent permitted by law the Contractor shall indemnify and hold harmless the Owner, Architect, Architect's consultants, and agents and employees of any of them from and against claims, damages, losses and expenses, including but not limited to attorneys' fees, arising out of or resulting from performance of the Work, provided that such claim, damage, loss or expense is attributable to bodily injury, sickness, disease or death, or to injury to or destruction of tangible property (other than the Work itself), but only to the extent caused by the negligent acts or omissions of the Contractor, a Subcontractor, anyone directly or indirectly employed by them or anyone for whose acts they may be liable, regardless of whether or not such claim, damage, loss or expense is caused in part by a party indemnified hereunder. Such obligation shall not be construed to negate, abridge, or reduce other rights or obligations of indemnity that would otherwise exist as to a party or person described in this Section 3.18.

§ **3.18.2** In claims against any person or entity indemnified under this Section 3.18 by an employee of the Contractor, a Subcontractor, anyone directly or indirectly employed by them or anyone for whose acts they may be liable, the indemnification obligation under Section 3.18.1 shall not be limited by a limitation on amount or type of damages, compensation or benefits payable by or for the Contractor or a Subcontractor under workers' compensation acts, disability benefit acts or other employee benefit acts.

ARTICLE 4   ARCHITECT
§ 4.1 GENERAL
§ **4.1.1** The Owner shall retain an architect lawfully licensed to practice architecture or an entity lawfully practicing architecture in the jurisdiction where the Project is located. That person or entity is identified as the Architect in the Agreement and is referred to throughout the Contract Documents as if singular in number.

§ **4.1.2** Duties, responsibilities and limitations of authority of the Architect as set forth in the Contract Documents shall not be restricted, modified or extended without written consent of the Owner, Contractor and Architect. Consent shall not be unreasonably withheld.

§ **4.1.3** If the employment of the Architect is terminated, the Owner shall employ a successor architect as to whom the Contractor has no reasonable objection and whose status under the Contract Documents shall be that of the Architect.

§ 4.2 ADMINISTRATION OF THE CONTRACT
§ **4.2.1** The Architect will provide administration of the Contract as described in the Contract Documents and will be an Owner's representative during construction until the date the Architect issues the final Certificate For Payment. The Architect will have authority to act on behalf of the Owner only to the extent provided in the Contract Documents.

§ **4.2.2** The Architect will visit the site at intervals appropriate to the stage of construction, or as otherwise agreed with the Owner, to become generally familiar with the progress and quality of the portion of the Work completed, and to determine in general if the Work observed is being performed in a manner indicating that the Work, when fully completed, will be in accordance with the Contract Documents. However, the Architect will not be required to make exhaustive or continuous on-site inspections to check the quality or quantity of the Work. The Architect will not have control over, charge of, or responsibility for, the construction means, methods, techniques, sequences or procedures, or for the safety precautions and programs in connection with the Work, since these are solely the Contractor's rights and responsibilities under the Contract Documents, except as provided in Section 3.3.1.

§ **4.2.3** On the basis of the site visits, the Architect will keep the Owner reasonably informed about the progress and quality of the portion of the Work completed, and report to the Owner (1) known deviations from the Contract Documents and from the most recent construction schedule submitted by the Contractor, and (2) defects and deficiencies observed in the Work. The Architect will not be responsible for the Contractor's failure to perform the Work in accordance with the requirements of the Contract Documents. The Architect will not have control over or charge of and will not be responsible for acts or omissions of the Contractor, Subcontractors, or their agents or employees, or any other persons or entities performing portions of the Work.

§ 4.2.4 COMMUNICATIONS FACILITATING CONTRACT ADMINISTRATION
Except as otherwise provided in the Contract Documents or when direct communications have been specially authorized, the Owner and Contractor shall endeavor to communicate with each other through the Architect about matters arising out of or relating to the Contract. Communications by and with the Architect's consultants shall be through the Architect. Communications by and with Subcontractors and material suppliers shall be through the Contractor. Communications by and with separate contractors shall be through the Owner.

§ **4.2.5** Based on the Architect's evaluations of the Contractor's Applications for Payment, the Architect will review and certify the amounts due the Contractor and will issue Certificates for Payment in such amounts.

§ **4.2.6** The Architect has authority to reject Work that does not conform to the Contract Documents. Whenever the Architect considers it necessary or advisable, the Architect will have authority to require inspection or testing of the

Work in accordance with Sections 13.5.2 and 13.5.3, whether or not such Work is fabricated, installed or completed. However, neither this authority of the Architect nor a decision made in good faith either to exercise or not to exercise such authority shall give rise to a duty or responsibility of the Architect to the Contractor, Subcontractors, material and equipment suppliers, their agents or employees, or other persons or entities performing portions of the Work.

§ 4.2.7 The Architect will review and approve, or take other appropriate action upon, the Contractor's submittals such as Shop Drawings, Product Data and Samples, but only for the limited purpose of checking for conformance with information given and the design concept expressed in the Contract Documents. The Architect's action will be taken in accordance with the submittal schedule approved by the Architect or, in the absence of an approved submittal schedule, with reasonable promptness while allowing sufficient time in the Architect's professional judgment to permit adequate review. Review of such submittals is not conducted for the purpose of determining the accuracy and completeness of other details such as dimensions and quantities, or for substantiating instructions for installation or performance of equipment or systems, all of which remain the responsibility of the Contractor as required by the Contract Documents. The Architect's review of the Contractor's submittals shall not relieve the Contractor of the obligations under Sections 3.3, 3.5 and 3.12. The Architect's review shall not constitute approval of safety precautions or, unless otherwise specifically stated by the Architect, of any construction means, methods, techniques, sequences or procedures. The Architect's approval of a specific item shall not indicate approval of an assembly of which the item is a component.

§ 4.2.8 The Architect will prepare Change Orders and Construction Change Directives, and may authorize minor changes in the Work as provided in Section 7.4. The Architect will investigate and make determinations and recommendations regarding concealed and unknown conditions as provided in Section 3.7.4.

§ 4.2.9 The Architect will conduct inspections to determine the date or dates of Substantial Completion and the date of final completion; issue Certificates of Substantial Completion pursuant to Section 9.8; receive and forward to the Owner, for the Owner's review and records, written warranties and related documents required by the Contract and assembled by the Contractor pursuant to Section 9.10; and issue a final Certificate for Payment pursuant to Section 9.10.

§ 4.2.10 If the Owner and Architect agree, the Architect will provide one or more project representatives to assist in carrying out the Architect's responsibilities at the site. The duties, responsibilities and limitations of authority of such project representatives shall be as set forth in an exhibit to be incorporated in the Contract Documents.

§ 4.2.11 The Architect will interpret and decide matters concerning performance under, and requirements of, the Contract Documents on written request of either the Owner or Contractor. The Architect's response to such requests will be made in writing within any time limits agreed upon or otherwise with reasonable promptness.

§ 4.2.12 Interpretations and decisions of the Architect will be consistent with the intent of, and reasonably inferable from, the Contract Documents and will be in writing or in the form of drawings. When making such interpretations and decisions, the Architect will endeavor to secure faithful performance by both Owner and Contractor, will not show partiality to either and will not be liable for results of interpretations or decisions rendered in good faith.

§ 4.2.13 The Architect's decisions on matters relating to aesthetic effect will be final if consistent with the intent expressed in the Contract Documents.

§ 4.2.14 The Architect will review and respond to requests for information about the Contract Documents. The Architect's response to such requests will be made in writing within any time limits agreed upon or otherwise with reasonable promptness. If appropriate, the Architect will prepare and issue supplemental Drawings and Specifications in response to the requests for information.

ARTICLE 5 SUBCONTRACTORS
§ 5.1 DEFINITIONS

§ 5.1.1 A Subcontractor is a person or entity who has a direct contract with the Contractor to perform a portion of the Work at the site. The term "Subcontractor" is referred to throughout the Contract Documents as if singular in number and means a Subcontractor or an authorized representative of the Subcontractor. The term "Subcontractor" does not include a separate contractor or subcontractors of a separate contractor.

§ 5.1.2 A Sub-subcontractor is a person or entity who has a direct or indirect contract with a Subcontractor to perform a portion of the Work at the site. The term "Sub-subcontractor" is referred to throughout the Contract Documents as if singular in number and means a Sub-subcontractor or an authorized representative of the Sub-subcontractor.

§ 5.2 AWARD OF SUBCONTRACTS AND OTHER CONTRACTS FOR PORTIONS OF THE WORK

§ 5.2.1 Unless otherwise stated in the Contract Documents or the bidding requirements, the Contractor, as soon as practicable after award of the Contract, shall furnish in writing to the Owner through the Architect the names of persons or entities (including those who are to furnish materials or equipment fabricated to a special design) proposed for each principal portion of the Work. The Architect may reply within 14 days to the Contractor in writing stating (1) whether the Owner or the Architect has reasonable objection to any such proposed person or entity or (2) that the Architect requires additional time for review. Failure of the Owner or Architect to reply within the 14-day period shall constitute notice of no reasonable objection.

§ 5.2.2 The Contractor shall not contract with a proposed person or entity to whom the Owner or Architect has made reasonable and timely objection. The Contractor shall not be required to contract with anyone to whom the Contractor has made reasonable objection.

§ 5.2.3 If the Owner or Architect has reasonable objection to a person or entity proposed by the Contractor, the Contractor shall propose another to whom the Owner or Architect has no reasonable objection. If the proposed but rejected Subcontractor was reasonably capable of performing the Work, the Contract Sum and Contract Time shall be increased or decreased by the difference, if any, occasioned by such change, and an appropriate Change Order shall be issued before commencement of the substitute Subcontractor's Work. However, no increase in the Contract Sum or Contract Time shall be allowed for such change unless the Contractor has acted promptly and responsively in submitting names as required.

§ 5.2.4 The Contractor shall not substitute a Subcontractor, person or entity previously selected if the Owner or Architect makes reasonable objection to such substitution.

§ 5.3 SUBCONTRACTUAL RELATIONS

By appropriate agreement, written where legally required for validity, the Contractor shall require each Subcontractor, to the extent of the Work to be performed by the Subcontractor, to be bound to the Contractor by terms of the Contract Documents, and to assume toward the Contractor all the obligations and responsibilities, including the responsibility for safety of the Subcontractor's Work, which the Contractor, by these Documents, assumes toward the Owner and Architect. Each subcontract agreement shall preserve and protect the rights of the Owner and Architect under the Contract Documents with respect to the Work to be performed by the Subcontractor so that subcontracting thereof will not prejudice such rights, and shall allow to the Subcontractor, unless specifically provided otherwise in the subcontract agreement, the benefit of all rights, remedies and redress against the Contractor that the Contractor, by the Contract Documents, has against the Owner. Where appropriate, the Contractor shall require each Subcontractor to enter into similar agreements with Sub-subcontractors. The Contractor shall make available to each proposed Subcontractor, prior to the execution of the subcontract agreement, copies of the Contract Documents to which the Subcontractor will be bound, and, upon written request of the Subcontractor, identify to the Subcontractor terms and conditions of the proposed subcontract agreement that may be at variance with the Contract Documents. Subcontractors will similarly make copies of applicable portions of such documents available to their respective proposed Sub-subcontractors.

§ 5.4 CONTINGENT ASSIGNMENT OF SUBCONTRACTS

§ 5.4.1 Each subcontract agreement for a portion of the Work is assigned by the Contractor to the Owner, provided that

.1 assignment is effective only after termination of the Contract by the Owner for cause pursuant to Section 14.2 and only for those subcontract agreements that the Owner accepts by notifying the Subcontractor and Contractor in writing; and

.2 assignment is subject to the prior rights of the surety, if any, obligated under bond relating to the Contract.

When the Owner accepts the assignment of a subcontract agreement, the Owner assumes the Contractor's rights and obligations under the subcontract.

§ 5.4.2 Upon such assignment, if the Work has been suspended for more than 30 days, the Subcontractor's compensation shall be equitably adjusted for increases in cost resulting from the suspension.

§ 5.4.3 Upon such assignment to the Owner under this Section 5.4, the Owner may further assign the subcontract to a successor contractor or other entity. If the Owner assigns the subcontract to a successor contractor or other entity, the Owner shall nevertheless remain legally responsible for all of the successor contractor's obligations under the subcontract.

ARTICLE 6   CONSTRUCTION BY OWNER OR BY SEPARATE CONTRACTORS
§ 6.1 OWNER'S RIGHT TO PERFORM CONSTRUCTION AND TO AWARD SEPARATE CONTRACTS

§ 6.1.1 The Owner reserves the right to perform construction or operations related to the Project with the Owner's own forces, and to award separate contracts in connection with other portions of the Project or other construction or operations on the site under Conditions of the Contract identical or substantially similar to these including those portions related to insurance and waiver of subrogation. If the Contractor claims that delay or additional cost is involved because of such action by the Owner, the Contractor shall make such Claim as provided in Article 15.

§ 6.1.2 When separate contracts are awarded for different portions of the Project or other construction or operations on the site, the term "Contractor" in the Contract Documents in each case shall mean the Contractor who executes each separate Owner-Contractor Agreement.

§ 6.1.3 The Owner shall provide for coordination of the activities of the Owner's own forces and of each separate contractor with the Work of the Contractor, who shall cooperate with them. The Contractor shall participate with other separate contractors and the Owner in reviewing their construction schedules. The Contractor shall make any revisions to the construction schedule deemed necessary after a joint review and mutual agreement. The construction schedules shall then constitute the schedules to be used by the Contractor, separate contractors and the Owner until subsequently revised.

§ 6.1.4 Unless otherwise provided in the Contract Documents, when the Owner performs construction or operations related to the Project with the Owner's own forces, the Owner shall be deemed to be subject to the same obligations and to have the same rights that apply to the Contractor under the Conditions of the Contract, including, without excluding others, those stated in Article 3, this Article 6 and Articles 10, 11 and 12.

§ 6.2 MUTUAL RESPONSIBILITY

§ 6.2.1 The Contractor shall afford the Owner and separate contractors reasonable opportunity for introduction and storage of their materials and equipment and performance of their activities, and shall connect and coordinate the Contractor's construction and operations with theirs as required by the Contract Documents.

§ 6.2.2 If part of the Contractor's Work depends for proper execution or results upon construction or operations by the Owner or a separate contractor, the Contractor shall, prior to proceeding with that portion of the Work, promptly report to the Architect apparent discrepancies or defects in such other construction that would render it unsuitable for such proper execution and results. Failure of the Contractor so to report shall constitute an acknowledgment that the Owner's or separate contractor's completed or partially completed construction is fit and proper to receive the Contractor's Work, except as to defects not then reasonably discoverable.

§ 6.2.3 The Contractor shall reimburse the Owner for costs the Owner incurs that are payable to a separate contractor because of the Contractor's delays, improperly timed activities or defective construction. The Owner shall be responsible to the Contractor for costs the Contractor incurs because of a separate contractor's delays, improperly timed activities, damage to the Work or defective construction.

§ 6.2.4 The Contractor shall promptly remedy damage the Contractor wrongfully causes to completed or partially completed construction or to property of the Owner, separate contractors as provided in Section 10.2.5.

§ 6.2.5 The Owner and each separate contractor shall have the same responsibilities for cutting and patching as are described for the Contractor in Section 3.14.

§ 6.3 OWNER'S RIGHT TO CLEAN UP

If a dispute arises among the Contractor, separate contractors and the Owner as to the responsibility under their respective contracts for maintaining the premises and surrounding area free from waste materials and rubbish, the Owner may clean up and the Architect will allocate the cost among those responsible.

ARTICLE 7 CHANGES IN THE WORK
§ 7.1 GENERAL

§ 7.1.1 Changes in the Work may be accomplished after execution of the Contract, and without invalidating the Contract, by Change Order, Construction Change Directive or order for a minor change in the Work, subject to the limitations stated in this Article 7 and elsewhere in the Contract Documents.

§ 7.1.2 A Change Order shall be based upon agreement among the Owner, Contractor and Architect; a Construction Change Directive requires agreement by the Owner and Architect and may or may not be agreed to by the Contractor; an order for a minor change in the Work may be issued by the Architect alone.

§ 7.1.3 Changes in the Work shall be performed under applicable provisions of the Contract Documents, and the Contractor shall proceed promptly, unless otherwise provided in the Change Order, Construction Change Directive or order for a minor change in the Work.

§ 7.2 CHANGE ORDERS

§ 7.2.1 A Change Order is a written instrument prepared by the Architect and signed by the Owner, Contractor and Architect stating their agreement upon all of the following:

    .1   The change in the Work;
    .2   The amount of the adjustment, if any, in the Contract Sum; and
    .3   The extent of the adjustment, if any, in the Contract Time.

§ 7.3 CONSTRUCTION CHANGE DIRECTIVES

§ 7.3.1 A Construction Change Directive is a written order prepared by the Architect and signed by the Owner and Architect, directing a change in the Work prior to agreement on adjustment, if any, in the Contract Sum or Contract Time, or both. The Owner may by Construction Change Directive, without invalidating the Contract, order changes in the Work within the general scope of the Contract consisting of additions, deletions or other revisions, the Contract Sum and Contract Time being adjusted accordingly.

§ 7.3.2 A Construction Change Directive shall be used in the absence of total agreement on the terms of a Change Order.

§ 7.3.3 If the Construction Change Directive provides for an adjustment to the Contract Sum, the adjustment shall be based on one of the following methods:

    .1   Mutual acceptance of a lump sum properly itemized and supported by sufficient substantiating data to permit evaluation;
    .2   Unit prices stated in the Contract Documents or subsequently agreed upon;
    .3   Cost to be determined in a manner agreed upon by the parties and a mutually acceptable fixed or percentage fee; or
    .4   As provided in Section 7.3.7.

§ 7.3.4 If unit prices are stated in the Contract Documents or subsequently agreed upon, and if quantities originally contemplated are materially changed in a proposed Change Order or Construction Change Directive so that application of such unit prices to quantities of Work proposed will cause substantial inequity to the Owner or Contractor, the applicable unit prices shall be equitably adjusted.

§ 7.3.5 Upon receipt of a Construction Change Directive, the Contractor shall promptly proceed with the change in the Work involved and advise the Architect of the Contractor's agreement or disagreement with the method, if any, provided in the Construction Change Directive for determining the proposed adjustment in the Contract Sum or Contract Time.

§ 7.3.6 A Construction Change Directive signed by the Contractor indicates the Contractor's agreement therewith, including adjustment in Contract Sum and Contract Time or the method for determining them. Such agreement shall be effective immediately and shall be recorded as a Change Order.

§ 7.3.7 If the Contractor does not respond promptly or disagrees with the method for adjustment in the Contract Sum, the Architect shall determine the method and the adjustment on the basis of reasonable expenditures and savings of those performing the Work attributable to the change, including, in case of an increase in the Contract Sum, an amount for overhead and profit as set forth in the Agreement, or if no such amount is set forth in the Agreement, a reasonable amount. In such case, and also under Section 7.3.3.3, the Contractor shall keep and present, in such form as the Architect may prescribe, an itemized accounting together with appropriate supporting data. Unless otherwise provided in the Contract Documents, costs for the purposes of this Section 7.3.7 shall be limited to the following:

 .1 Costs of labor, including social security, old age and unemployment insurance, fringe benefits required by agreement or custom, and workers' compensation insurance;

 .2 Costs of materials, supplies and equipment, including cost of transportation, whether incorporated or consumed;

 .3 Rental costs of machinery and equipment, exclusive of hand tools, whether rented from the Contractor or others;

 .4 Costs of premiums for all bonds and insurance, permit fees, and sales, use or similar taxes related to the Work; and

 .5 Additional costs of supervision and field office personnel directly attributable to the change.

§ 7.3.8 The amount of credit to be allowed by the Contractor to the Owner for a deletion or change that results in a net decrease in the Contract Sum shall be actual net cost as confirmed by the Architect. When both additions and credits covering related Work or substitutions are involved in a change, the allowance for overhead and profit shall be figured on the basis of net increase, if any, with respect to that change.

§ 7.3.9 Pending final determination of the total cost of a Construction Change Directive to the Owner, the Contractor may request payment for Work completed under the Construction Change Directive in Applications for Payment. The Architect will make an interim determination for purposes of monthly certification for payment for those costs and certify for payment the amount that the Architect determines, in the Architect's professional judgment, to be reasonably justified. The Architect's interim determination of cost shall adjust the Contract Sum on the same basis as a Change Order, subject to the right of either party to disagree and assert a Claim in accordance with Article 15.

§ 7.3.10 When the Owner and Contractor agree with a determination made by the Architect concerning the adjustments in the Contract Sum and Contract Time, or otherwise reach agreement upon the adjustments, such agreement shall be effective immediately and the Architect will prepare a Change Order. Change Orders may be issued for all or any part of a Construction Change Directive.

§ 7.4 MINOR CHANGES IN THE WORK
The Architect has authority to order minor changes in the Work not involving adjustment in the Contract Sum or extension of the Contract Time and not inconsistent with the intent of the Contract Documents. Such changes will be effected by written order signed by the Architect and shall be binding on the Owner and Contractor.

ARTICLE 8 TIME
§ 8.1 DEFINITIONS
§ 8.1.1 Unless otherwise provided, Contract Time is the period of time, including authorized adjustments, allotted in the Contract Documents for Substantial Completion of the Work.

§ 8.1.2 The date of commencement of the Work is the date established in the Agreement.

§ 8.1.3 The date of Substantial Completion is the date certified by the Architect in accordance with Section 9.8.

§ 8.1.4 The term "day" as used in the Contract Documents shall mean calendar day unless otherwise specifically defined.

§ 8.2 PROGRESS AND COMPLETION
§ 8.2.1 Time limits stated in the Contract Documents are of the essence of the Contract. By executing the Agreement the Contractor confirms that the Contract Time is a reasonable period for performing the Work.

§ 8.2.2 The Contractor shall not knowingly, except by agreement or instruction of the Owner in writing, prematurely commence operations on the site or elsewhere prior to the effective date of insurance required by Article 11 to be furnished by the Contractor and Owner. The date of commencement of the Work shall not be changed by the effective date of such insurance.

§ 8.2.3 The Contractor shall proceed expeditiously with adequate forces and shall achieve Substantial Completion within the Contract Time.

§ 8.3 DELAYS AND EXTENSIONS OF TIME

§ 8.3.1 If the Contractor is delayed at any time in the commencement or progress of the Work by an act or neglect of the Owner or Architect, or of an employee of either, or of a separate contractor employed by the Owner; or by changes ordered in the Work; or by labor disputes, fire, unusual delay in deliveries, unavoidable casualties or other causes beyond the Contractor's control; or by delay authorized by the Owner pending mediation and arbitration; or by other causes that the Architect determines may justify delay, then the Contract Time shall be extended by Change Order for such reasonable time as the Architect may determine.

§ 8.3.2 Claims relating to time shall be made in accordance with applicable provisions of Article 15.

§ 8.3.3 This Section 8.3 does not preclude recovery of damages for delay by either party under other provisions of the Contract Documents.

ARTICLE 9   PAYMENTS AND COMPLETION
§ 9.1 CONTRACT SUM

The Contract Sum is stated in the Agreement and, including authorized adjustments, is the total amount payable by the Owner to the Contractor for performance of the Work under the Contract Documents.

§ 9.2 SCHEDULE OF VALUES

Where the Contract is based on a stipulated sum or Guaranteed Maximum Price, the Contractor shall submit to the Architect, before the first Application for Payment, a schedule of values allocating the entire Contract Sum to the various portions of the Work and prepared in such form and supported by such data to substantiate its accuracy as the Architect may require. This schedule, unless objected to by the Architect, shall be used as a basis for reviewing the Contractor's Applications for Payment.

§ 9.3 APPLICATIONS FOR PAYMENT

§ 9.3.1 At least ten days before the date established for each progress payment, the Contractor shall submit to the Architect an itemized Application for Payment prepared in accordance with the schedule of values, if required under Section 9.2., for completed portions of the Work. Such application shall be notarized, if required, and supported by such data substantiating the Contractor's right to payment as the Owner or Architect may require, such as copies of requisitions from Subcontractors and material suppliers, and shall reflect retainage if provided for in the Contract Documents.

§ 9.3.1.1 As provided in Section 7.3.9, such applications may include requests for payment on account of changes in the Work that have been properly authorized by Construction Change Directives, or by interim determinations of the Architect, but not yet included in Change Orders.

§ 9.3.1.2 Applications for Payment shall not include requests for payment for portions of the Work for which the Contractor does not intend to pay a Subcontractor or material supplier, unless such Work has been performed by others whom the Contractor intends to pay.

§ 9.3.2 Unless otherwise provided in the Contract Documents, payments shall be made on account of materials and equipment delivered and suitably stored at the site for subsequent incorporation in the Work. If approved in advance by the Owner, payment may similarly be made for materials and equipment suitably stored off the site at a location agreed upon in writing. Payment for materials and equipment stored on or off the site shall be conditioned upon compliance by the Contractor with procedures satisfactory to the Owner to establish the Owner's title to such materials and equipment or otherwise protect the Owner's interest, and shall include the costs of applicable insurance, storage and transportation to the site for such materials and equipment stored off the site.

§ 9.3.3 The Contractor warrants that title to all Work covered by an Application for Payment will pass to the Owner no later than the time of payment. The Contractor further warrants that upon submittal of an Application for Payment all Work for which Certificates for Payment have been previously issued and payments received from the

Owner shall, to the best of the Contractor's knowledge, information and belief, be free and clear of liens, claims, security interests or encumbrances in favor of the Contractor, Subcontractors, material suppliers, or other persons or entities making a claim by reason of having provided labor, materials and equipment relating to the Work.

§ 9.4 CERTIFICATES FOR PAYMENT

§ 9.4.1 The Architect will, within seven days after receipt of the Contractor's Application for Payment, either issue to the Owner a Certificate for Payment, with a copy to the Contractor, for such amount as the Architect determines is properly due, or notify the Contractor and Owner in writing of the Architect's reasons for withholding certification in whole or in part as provided in Section 9.5.1.

§ 9.4.2 The issuance of a Certificate for Payment will constitute a representation by the Architect to the Owner, based on the Architect's evaluation of the Work and the data comprising the Application for Payment, that, to the best of the Architect's knowledge, information and belief, the Work has progressed to the point indicated and that the quality of the Work is in accordance with the Contract Documents. The foregoing representations are subject to an evaluation of the Work for conformance with the Contract Documents upon Substantial Completion, to results of subsequent tests and inspections, to correction of minor deviations from the Contract Documents prior to completion and to specific qualifications expressed by the Architect. The issuance of a Certificate for Payment will further constitute a representation that the Contractor is entitled to payment in the amount certified. However, the issuance of a Certificate for Payment will not be a representation that the Architect has (1) made exhaustive or continuous on-site inspections to check the quality or quantity of the Work, (2) reviewed construction means, methods, techniques, sequences or procedures, (3) reviewed copies of requisitions received from Subcontractors and material suppliers and other data requested by the Owner to substantiate the Contractor's right to payment, or (4) made examination to ascertain how or for what purpose the Contractor has used money previously paid on account of the Contract Sum.

§ 9.5 DECISIONS TO WITHHOLD CERTIFICATION

§ 9.5.1 The Architect may withhold a Certificate for Payment in whole or in part, to the extent reasonably necessary to protect the Owner, if in the Architect's opinion the representations to the Owner required by Section 9.4.2 cannot be made. If the Architect is unable to certify payment in the amount of the Application, the Architect will notify the Contractor and Owner as provided in Section 9.4.1. If the Contractor and Architect cannot agree on a revised amount, the Architect will promptly issue a Certificate for Payment for the amount for which the Architect is able to make such representations to the Owner. The Architect may also withhold a Certificate for Payment or, because of subsequently discovered evidence, may nullify the whole or a part of a Certificate for Payment previously issued, to such extent as may be necessary in the Architect's opinion to protect the Owner from loss for which the Contractor is responsible, including loss resulting from acts and omissions described in Section 3.3.2, because of

   .1   defective Work not remedied;

   .2   third party claims filed or reasonable evidence indicating probable filing of such claims unless security acceptable to the Owner is provided by the Contractor;

   .3   failure of the Contractor to make payments properly to Subcontractors or for labor, materials or equipment;

   .4   reasonable evidence that the Work cannot be completed for the unpaid balance of the Contract Sum;

   .5   damage to the Owner or a separate contractor;

   .6   reasonable evidence that the Work will not be completed within the Contract Time, and that the unpaid balance would not be adequate to cover actual or liquidated damages for the anticipated delay; or

   .7   repeated failure to carry out the Work in accordance with the Contract Documents.

§ 9.5.2 When the above reasons for withholding certification are removed, certification will be made for amounts previously withheld.

§ 9.5.3 If the Architect withholds certification for payment under Section 9.5.1.3, the Owner may, at its sole option, issue joint checks to the Contractor and to any Subcontractor or material or equipment suppliers to whom the Contractor failed to make payment for Work properly performed or material or equipment suitably delivered. If the Owner makes payments by joint check, the Owner shall notify the Architect and the Architect will reflect such payment on the next Certificate for Payment.

§ 9.6 PROGRESS PAYMENTS

§ 9.6.1 After the Architect has issued a Certificate for Payment, the Owner shall make payment in the manner and within the time provided in the Contract Documents, and shall so notify the Architect.

§ 9.6.2 The Contractor shall pay each Subcontractor no later than seven days after receipt of payment from the Owner the amount to which the Subcontractor is entitled, reflecting percentages actually retained from payments to the Contractor on account of the Subcontractor's portion of the Work. The Contractor shall, by appropriate agreement with each Subcontractor, require each Subcontractor to make payments to Sub-subcontractors in a similar manner.

§ 9.6.3 The Architect will, on request, furnish to a Subcontractor, if practicable, information regarding percentages of completion or amounts applied for by the Contractor and action taken thereon by the Architect and Owner on account of portions of the Work done by such Subcontractor.

§ 9.6.4 The Owner has the right to request written evidence from the Contractor that the Contractor has properly paid Subcontractors and material and equipment suppliers amounts paid by the Owner to the Contractor for subcontracted Work. If the Contractor fails to furnish such evidence within seven days, the Owner shall have the right to contact Subcontractors to ascertain whether they have been properly paid. Neither the Owner nor Architect shall have an obligation to pay or to see to the payment of money to a Subcontractor, except as may otherwise be required by law.

§ 9.6.5 Contractor payments to material and equipment suppliers shall be treated in a manner similar to that provided in Sections 9.6.2, 9.6.3 and 9.6.4.

§ 9.6.6 A Certificate for Payment, a progress payment, or partial or entire use or occupancy of the Project by the Owner shall not constitute acceptance of Work not in accordance with the Contract Documents.

§ 9.6.7 Unless the Contractor provides the Owner with a payment bond in the full penal sum of the Contract Sum, payments received by the Contractor for Work properly performed by Subcontractors and suppliers shall be held by the Contractor for those Subcontractors or suppliers who performed Work or furnished materials, or both, under contract with the Contractor for which payment was made by the Owner. Nothing contained herein shall require money to be placed in a separate account and not commingled with money of the Contractor, shall create any fiduciary liability or tort liability on the part of the Contractor for breach of trust or shall entitle any person or entity to an award of punitive damages against the Contractor for breach of the requirements of this provision.

§ 9.7 FAILURE OF PAYMENT
If the Architect does not issue a Certificate for Payment, through no fault of the Contractor, within seven days after receipt of the Contractor's Application for Payment, or if the Owner does not pay the Contractor within seven days after the date established in the Contract Documents the amount certified by the Architect or awarded by binding dispute resolution, then the Contractor may, upon seven additional days' written notice to the Owner and Architect, stop the Work until payment of the amount owing has been received. The Contract Time shall be extended appropriately and the Contract Sum shall be increased by the amount of the Contractor's reasonable costs of shut-down, delay and start-up, plus interest as provided for in the Contract Documents.

§ 9.8 SUBSTANTIAL COMPLETION
§ 9.8.1 Substantial Completion is the stage in the progress of the Work when the Work or designated portion thereof is sufficiently complete in accordance with the Contract Documents so that the Owner can occupy or utilize the Work for its intended use.

§ 9.8.2 When the Contractor considers that the Work, or a portion thereof which the Owner agrees to accept separately, is substantially complete, the Contractor shall prepare and submit to the Architect a comprehensive list of items to be completed or corrected prior to final payment. Failure to include an item on such list does not alter the responsibility of the Contractor to complete all Work in accordance with the Contract Documents.

§ 9.8.3 Upon receipt of the Contractor's list, the Architect will make an inspection to determine whether the Work or designated portion thereof is substantially complete. If the Architect's inspection discloses any item, whether or not included on the Contractor's list, which is not sufficiently complete in accordance with the Contract Documents so that the Owner can occupy or utilize the Work or designated portion thereof for its intended use, the Contractor shall, before issuance of the Certificate of Substantial Completion, complete or correct such item upon notification by the Architect. In such case, the Contractor shall then submit a request for another inspection by the Architect to determine Substantial Completion.

§ 9.8.4 When the Work or designated portion thereof is substantially complete, the Architect will prepare a Certificate of Substantial Completion that shall establish the date of Substantial Completion, shall establish responsibilities of the Owner and Contractor for security, maintenance, heat, utilities, damage to the Work and insurance, and shall fix the time within which the Contractor shall finish all items on the list accompanying the Certificate. Warranties required by the Contract Documents shall commence on the date of Substantial Completion of the Work or designated portion thereof unless otherwise provided in the Certificate of Substantial Completion.

§ 9.8.5 The Certificate of Substantial Completion shall be submitted to the Owner and Contractor for their written acceptance of responsibilities assigned to them in such Certificate. Upon such acceptance and consent of surety, if any, the Owner shall make payment of retainage applying to such Work or designated portion thereof. Such payment shall be adjusted for Work that is incomplete or not in accordance with the requirements of the Contract Documents.

§ 9.9 PARTIAL OCCUPANCY OR USE

§ 9.9.1 The Owner may occupy or use any completed or partially completed portion of the Work at any stage when such portion is designated by separate agreement with the Contractor, provided such occupancy or use is consented to by the insurer as required under Section 11.3.1.5 and authorized by public authorities having jurisdiction over the Project. Such partial occupancy or use may commence whether or not the portion is substantially complete, provided the Owner and Contractor have accepted in writing the responsibilities assigned to each of them for payments, retainage, if any, security, maintenance, heat, utilities, damage to the Work and insurance, and have agreed in writing concerning the period for correction of the Work and commencement of warranties required by the Contract Documents. When the Contractor considers a portion substantially complete, the Contractor shall prepare and submit a list to the Architect as provided under Section 9.8.2. Consent of the Contractor to partial occupancy or use shall not be unreasonably withheld. The stage of the progress of the Work shall be determined by written agreement between the Owner and Contractor or, if no agreement is reached, by decision of the Architect.

§ 9.9.2 Immediately prior to such partial occupancy or use, the Owner, Contractor and Architect shall jointly inspect the area to be occupied or portion of the Work to be used in order to determine and record the condition of the Work.

§ 9.9.3 Unless otherwise agreed upon, partial occupancy or use of a portion or portions of the Work shall not constitute acceptance of Work not complying with the requirements of the Contract Documents.

§ 9.10 FINAL COMPLETION AND FINAL PAYMENT

§ 9.10.1 Upon receipt of the Contractor's written notice that the Work is ready for final inspection and acceptance and upon receipt of a final Application for Payment, the Architect will promptly make such inspection and, when the Architect finds the Work acceptable under the Contract Documents and the Contract fully performed, the Architect will promptly issue a final Certificate for Payment stating that to the best of the Architect's knowledge, information and belief, and on the basis of the Architect's on-site visits and inspections, the Work has been completed in accordance with terms and conditions of the Contract Documents and that the entire balance found to be due the Contractor and noted in the final Certificate is due and payable. The Architect's final Certificate for Payment will constitute a further representation that conditions listed in Section 9.10.2 as precedent to the Contractor's being entitled to final payment have been fulfilled.

§ 9.10.2 Neither final payment nor any remaining retained percentage shall become due until the Contractor submits to the Architect (1) an affidavit that payrolls, bills for materials and equipment, and other indebtedness connected with the Work for which the Owner or the Owner's property might be responsible or encumbered (less amounts withheld by Owner) have been paid or otherwise satisfied, (2) a certificate evidencing that insurance required by the Contract Documents to remain in force after final payment is currently in effect and will not be canceled or allowed to expire until at least 30 days' prior written notice has been given to the Owner, (3) a written statement that the Contractor knows of no substantial reason that the insurance will not be renewable to cover the period required by the Contract Documents, (4) consent of surety, if any, to final payment and (5), if required by the Owner, other data establishing payment or satisfaction of obligations, such as receipts, releases and waivers of liens, claims, security interests or encumbrances arising out of the Contract, to the extent and in such form as may be designated by the Owner. If a Subcontractor refuses to furnish a release or waiver required by the Owner, the Contractor may furnish a bond satisfactory to the Owner to indemnify the Owner against such lien. If such lien remains unsatisfied after payments are made, the Contractor shall refund to the Owner all money that the Owner may be compelled to pay in discharging such lien, including all costs and reasonable attorneys' fees.

§ 9.10.3 If, after Substantial Completion of the Work, final completion thereof is materially delayed through no fault of the Contractor or by issuance of Change Orders affecting final completion, and the Architect so confirms, the Owner shall, upon application by the Contractor and certification by the Architect, and without terminating the Contract, make payment of the balance due for that portion of the Work fully completed and accepted. If the remaining balance for Work not fully completed or corrected is less than retainage stipulated in the Contract Documents, and if bonds have been furnished, the written consent of surety to payment of the balance due for that portion of the Work fully completed and accepted shall be submitted by the Contractor to the Architect prior to certification of such payment. Such payment shall be made under terms and conditions governing final payment, except that it shall not constitute a waiver of claims.

§ 9.10.4 The making of final payment shall constitute a waiver of Claims by the Owner except those arising from
.1   liens, Claims, security interests or encumbrances arising out of the Contract and unsettled;
.2   failure of the Work to comply with the requirements of the Contract Documents; or
.3   terms of special warranties required by the Contract Documents.

§ 9.10.5 Acceptance of final payment by the Contractor, a Subcontractor or material supplier shall constitute a waiver of claims by that payee except those previously made in writing and identified by that payee as unsettled at the time of final Application for Payment.

ARTICLE 10   PROTECTION OF PERSONS AND PROPERTY
§ 10.1 SAFETY PRECAUTIONS AND PROGRAMS
The Contractor shall be responsible for initiating, maintaining and supervising all safety precautions and programs in connection with the performance of the Contract.

§ 10.2 SAFETY OF PERSONS AND PROPERTY
§ 10.2.1 The Contractor shall take reasonable precautions for safety of, and shall provide reasonable protection to prevent damage, injury or loss to
.1   employees on the Work and other persons who may be affected thereby;
.2   the Work and materials and equipment to be incorporated therein, whether in storage on or off the site, under care, custody or control of the Contractor or the Contractor's Subcontractors or Sub-subcontractors; and
.3   other property at the site or adjacent thereto, such as trees, shrubs, lawns, walks, pavements, roadways, structures and utilities not designated for removal, relocation or replacement in the course of construction.

§ 10.2.2 The Contractor shall comply with and give notices required by applicable laws, statutes, ordinances, codes, rules and regulations, and lawful orders of public authorities bearing on safety of persons or property or their protection from damage, injury or loss.

§ 10.2.3 The Contractor shall erect and maintain, as required by existing conditions and performance of the Contract, reasonable safeguards for safety and protection, including posting danger signs and other warnings against hazards, promulgating safety regulations and notifying owners and users of adjacent sites and utilities.

§ 10.2.4 When use or storage of explosives or other hazardous materials or equipment or unusual methods are necessary for execution of the Work, the Contractor shall exercise utmost care and carry on such activities under supervision of properly qualified personnel.

§ 10.2.5 The Contractor shall promptly remedy damage and loss (other than damage or loss insured under property insurance required by the Contract Documents) to property referred to in Sections 10.2.1.2 and 10.2.1.3 caused in whole or in part by the Contractor, a Subcontractor, a Sub-subcontractor, or anyone directly or indirectly employed by any of them, or by anyone for whose acts they may be liable and for which the Contractor is responsible under Sections 10.2.1.2 and 10.2.1.3, except damage or loss attributable to acts or omissions of the Owner or Architect or anyone directly or indirectly employed by either of them, or by anyone for whose acts either of them may be liable, and not attributable to the fault or negligence of the Contractor. The foregoing obligations of the Contractor are in addition to the Contractor's obligations under Section 3.18.

§ 10.2.6 The Contractor shall designate a responsible member of the Contractor's organization at the site whose duty shall be the prevention of accidents. This person shall be the Contractor's superintendent unless otherwise designated by the Contractor in writing to the Owner and Architect.

§ 10.2.7 The Contractor shall not permit any part of the construction or site to be loaded so as to cause damage or create an unsafe condition.

§ 10.2.8 INJURY OR DAMAGE TO PERSON OR PROPERTY

If either party suffers injury or damage to person or property because of an act or omission of the other party, or of others for whose acts such party is legally responsible, written notice of such injury or damage, whether or not insured, shall be given to the other party within a reasonable time not exceeding 21 days after discovery. The notice shall provide sufficient detail to enable the other party to investigate the matter.

§ 10.3 HAZARDOUS MATERIALS

§ 10.3.1 The Contractor is responsible for compliance with any requirements included in the Contract Documents regarding hazardous materials. If the Contractor encounters a hazardous material or substance not addressed in the Contract Documents and if reasonable precautions will be inadequate to prevent foreseeable bodily injury or death to persons resulting from a material or substance, including but not limited to asbestos or polychlorinated biphenyl (PCB), encountered on the site by the Contractor, the Contractor shall, upon recognizing the condition, immediately stop Work in the affected area and report the condition to the Owner and Architect in writing.

§ 10.3.2 Upon receipt of the Contractor's written notice, the Owner shall obtain the services of a licensed laboratory to verify the presence or absence of the material or substance reported by the Contractor and, in the event such material or substance is found to be present, to cause it to be rendered harmless. Unless otherwise required by the Contract Documents, the Owner shall furnish in writing to the Contractor and Architect the names and qualifications of persons or entities who are to perform tests verifying the presence or absence of such material or substance or who are to perform the task of removal or safe containment of such material or substance. The Contractor and the Architect will promptly reply to the Owner in writing stating whether or not either has reasonable objection to the persons or entities proposed by the Owner. If either the Contractor or Architect has an objection to a person or entity proposed by the Owner, the Owner shall propose another to whom the Contractor and the Architect have no reasonable objection. When the material or substance has been rendered harmless, Work in the affected area shall resume upon written agreement of the Owner and Contractor. By Change Order, the Contract Time shall be extended appropriately and the Contract Sum shall be increased in the amount of the Contractor's reasonable additional costs of shut-down, delay and start-up.

§ 10.3.3 To the fullest extent permitted by law, the Owner shall indemnify and hold harmless the Contractor, Subcontractors, Architect, Architect's consultants and agents and employees of any of them from and against claims, damages, losses and expenses, including but not limited to attorneys' fees, arising out of or resulting from performance of the Work in the affected area if in fact the material or substance presents the risk of bodily injury or death as described in Section 10.3.1 and has not been rendered harmless, provided that such claim, damage, loss or expense is attributable to bodily injury, sickness, disease or death, or to injury to or destruction of tangible property (other than the Work itself), except to the extent that such damage, loss or expense is due to the fault or negligence of the party seeking indemnity.

§ 10.3.4 The Owner shall not be responsible under this Section 10.3 for materials or substances the Contractor brings to the site unless such materials or substances are required by the Contract Documents. The Owner shall be responsible for materials or substances required by the Contract Documents, except to the extent of the Contractor's fault or negligence in the use and handling of such materials or substances.

§ 10.3.5 The Contractor shall indemnify the Owner for the cost and expense the Owner incurs (1) for remediation of a material or substance the Contractor brings to the site and negligently handles, or (2) where the Contractor fails to perform its obligations under Section 10.3.1, except to the extent that the cost and expense are due to the Owner's fault or negligence.

§ 10.3.6 If, without negligence on the part of the Contractor, the Contractor is held liable by a government agency for the cost of remediation of a hazardous material or substance solely by reason of performing Work as required by the Contract Documents, the Owner shall indemnify the Contractor for all cost and expense thereby incurred.

§ 10.4 EMERGENCIES
In an emergency affecting safety of persons or property, the Contractor shall act, at the Contractor's discretion, to prevent threatened damage, injury or loss. Additional compensation or extension of time claimed by the Contractor on account of an emergency shall be determined as provided in Article 15 and Article 7.

ARTICLE 11   INSURANCE AND BONDS
§ 11.1 CONTRACTOR'S LIABILITY INSURANCE
§ 11.1.1 The Contractor shall purchase from and maintain in a company or companies lawfully authorized to do business in the jurisdiction in which the Project is located such insurance as will protect the Contractor from claims set forth below which may arise out of or result from the Contractor's operations and completed operations under the Contract and for which the Contractor may be legally liable, whether such operations be by the Contractor or by a Subcontractor or by anyone directly or indirectly employed by any of them, or by anyone for whose acts any of them may be liable:

.1   Claims under workers' compensation, disability benefit and other similar employee benefit acts that are applicable to the Work to be performed;

.2   Claims for damages because of bodily injury, occupational sickness or disease, or death of the Contractor's employees;

.3   Claims for damages because of bodily injury, sickness or disease, or death of any person other than the Contractor's employees;

.4   Claims for damages insured by usual personal injury liability coverage;

.5   Claims for damages, other than to the Work itself, because of injury to or destruction of tangible property, including loss of use resulting therefrom;

.6   Claims for damages because of bodily injury, death of a person or property damage arising out of ownership, maintenance or use of a motor vehicle;

.7   Claims for bodily injury or property damage arising out of completed operations; and

.8   Claims involving contractual liability insurance applicable to the Contractor's obligations under Section 3.18.

§ 11.1.2 The insurance required by Section 11.1.1 shall be written for not less than limits of liability specified in the Contract Documents or required by law, whichever coverage is greater. Coverages, whether written on an occurrence or claims-made basis, shall be maintained without interruption from the date of commencement of the Work until the date of final payment and termination of any coverage required to be maintained after final payment, and, with respect to the Contractor's completed operations coverage, until the expiration of the period for correction of Work or for such other period for maintenance of completed operations coverage as specified in the Contract Documents.

§ 11.1.3 Certificates of insurance acceptable to the Owner shall be filed with the Owner prior to commencement of the Work and thereafter upon renewal or replacement of each required policy of insurance. These certificates and the insurance policies required by this Section 11.1 shall contain a provision that coverages afforded under the policies will not be canceled or allowed to expire until at least 30 days' prior written notice has been given to the Owner. An additional certificate evidencing continuation of liability coverage, including coverage for completed operations, shall be submitted with the final Application for Payment as required by Section 9.10.2 and thereafter upon renewal or replacement of such coverage until the expiration of the time required by Section 11.1.2. Information concerning reduction of coverage on account of revised limits or claims paid under the General Aggregate, or both, shall be furnished by the Contractor with reasonable promptness.

§ 11.1.4 The Contractor shall cause the commercial liability coverage required by the Contract Documents to include (1) the Owner, the Architect and the Architect's Consultants as additional insureds for claims caused in whole or in part by the Contractor's negligent acts or omissions during the Contractor's operations; and (2) the Owner as an additional insured for claims caused in whole or in part by the Contractor's negligent acts or omissions during the Contractor's completed operations.

§ 11.2 OWNER'S LIABILITY INSURANCE
The Owner shall be responsible for purchasing and maintaining the Owner's usual liability insurance.

§ 11.3 PROPERTY INSURANCE
§ 11.3.1 Unless otherwise provided, the Owner shall purchase and maintain, in a company or companies lawfully authorized to do business in the jurisdiction in which the Project is located, property insurance written on a builder's

risk "all-risk" or equivalent policy form in the amount of the initial Contract Sum, plus value of subsequent Contract Modifications and cost of materials supplied or installed by others, comprising total value for the entire Project at the site on a replacement cost basis without optional deductibles. Such property insurance shall be maintained, unless otherwise provided in the Contract Documents or otherwise agreed in writing by all persons and entities who are beneficiaries of such insurance, until final payment has been made as provided in Section 9.10 or until no person or entity other than the Owner has an insurable interest in the property required by this Section 11.3 to be covered, whichever is later. This insurance shall include interests of the Owner, the Contractor, Subcontractors and Sub-subcontractors in the Project.

§ 11.3.1.1 Property insurance shall be on an "all-risk" or equivalent policy form and shall include, without limitation, insurance against the perils of fire (with extended coverage) and physical loss or damage including, without duplication of coverage, theft, vandalism, malicious mischief, collapse, earthquake, flood, windstorm, falsework, testing and startup, temporary buildings and debris removal including demolition occasioned by enforcement of any applicable legal requirements, and shall cover reasonable compensation for Architect's and Contractor's services and expenses required as a result of such insured loss.

§ 11.3.1.2 If the Owner does not intend to purchase such property insurance required by the Contract and with all of the coverages in the amount described above, the Owner shall so inform the Contractor in writing prior to commencement of the Work. The Contractor may then effect insurance that will protect the interests of the Contractor, Subcontractors and Sub-subcontractors in the Work, and by appropriate Change Order the cost thereof shall be charged to the Owner. If the Contractor is damaged by the failure or neglect of the Owner to purchase or maintain insurance as described above, without so notifying the Contractor in writing, then the Owner shall bear all reasonable costs properly attributable thereto.

§ 11.3.1.3 If the property insurance requires deductibles, the Owner shall pay costs not covered because of such deductibles.

§ 11.3.1.4 This property insurance shall cover portions of the Work stored off the site, and also portions of the Work in transit.

§ 11.3.1.5 Partial occupancy or use in accordance with Section 9.9 shall not commence until the insurance company or companies providing property insurance have consented to such partial occupancy or use by endorsement or otherwise. The Owner and the Contractor shall take reasonable steps to obtain consent of the insurance company or companies and shall, without mutual written consent, take no action with respect to partial occupancy or use that would cause cancellation, lapse or reduction of insurance.

§ 11.3.2 BOILER AND MACHINERY INSURANCE
The Owner shall purchase and maintain boiler and machinery insurance required by the Contract Documents or by law, which shall specifically cover such insured objects during installation and until final acceptance by the Owner; this insurance shall include interests of the Owner, Contractor, Subcontractors and Sub-subcontractors in the Work, and the Owner and Contractor shall be named insureds.

§ 11.3.3 LOSS OF USE INSURANCE
The Owner, at the Owner's option, may purchase and maintain such insurance as will insure the Owner against loss of use of the Owner's property due to fire or other hazards, however caused. The Owner waives all rights of action against the Contractor for loss of use of the Owner's property, including consequential losses due to fire or other hazards however caused.

§ 11.3.4 If the Contractor requests in writing that insurance for risks other than those described herein or other special causes of loss be included in the property insurance policy, the Owner shall, if possible, include such insurance, and the cost thereof shall be charged to the Contractor by appropriate Change Order.

§ 11.3.5 If during the Project construction period the Owner insures properties, real or personal or both, at or adjacent to the site by property insurance under policies separate from those insuring the Project, or if after final payment property insurance is to be provided on the completed Project through a policy or policies other than those insuring the Project during the construction period, the Owner shall waive all rights in accordance with the terms of Section 11.3.7 for damages caused by fire or other causes of loss covered by this separate property insurance. All separate policies shall provide this waiver of subrogation by endorsement or otherwise.

§ 11.3.6 Before an exposure to loss may occur, the Owner shall file with the Contractor a copy of each policy that includes insurance coverages required by this Section 11.3. Each policy shall contain all generally applicable conditions, definitions, exclusions and endorsements related to this Project. Each policy shall contain a provision that the policy will not be canceled or allowed to expire, and that its limits will not be reduced, until at least 30 days' prior written notice has been given to the Contractor.

§ 11.3.7 WAIVERS OF SUBROGATION

The Owner and Contractor waive all rights against (1) each other and any of their subcontractors, sub-subcontractors, agents and employees, each of the other, and (2) the Architect, Architect's consultants, separate contractors described in Article 6, if any, and any of their subcontractors, sub-subcontractors, agents and employees, for damages caused by fire or other causes of loss to the extent covered by property insurance obtained pursuant to this Section 11.3 or other property insurance applicable to the Work, except such rights as they have to proceeds of such insurance held by the Owner as fiduciary. The Owner or Contractor, as appropriate, shall require of the Architect, Architect's consultants, separate contractors described in Article 6, if any, and the subcontractors, sub-subcontractors, agents and employees of any of them, by appropriate agreements, written where legally required for validity, similar waivers each in favor of other parties enumerated herein. The policies shall provide such waivers of subrogation by endorsement or otherwise. A waiver of subrogation shall be effective as to a person or entity even though that person or entity would otherwise have a duty of indemnification, contractual or otherwise, did not pay the insurance premium directly or indirectly, and whether or not the person or entity had an insurable interest in the property damaged.

§ 11.3.8 A loss insured under the Owner's property insurance shall be adjusted by the Owner as fiduciary and made payable to the Owner as fiduciary for the insureds, as their interests may appear, subject to requirements of any applicable mortgagee clause and of Section 11.3.10. The Contractor shall pay Subcontractors their just shares of insurance proceeds received by the Contractor, and by appropriate agreements, written where legally required for validity, shall require Subcontractors to make payments to their Sub-subcontractors in similar manner.

§ 11.3.9 If required in writing by a party in interest, the Owner as fiduciary shall, upon occurrence of an insured loss, give bond for proper performance of the Owner's duties. The cost of required bonds shall be charged against proceeds received as fiduciary. The Owner shall deposit in a separate account proceeds so received, which the Owner shall distribute in accordance with such agreement as the parties in interest may reach, or as determined in accordance with the method of binding dispute resolution selected in the Agreement between the Owner and Contractor. If after such loss no other special agreement is made and unless the Owner terminates the Contract for convenience, replacement of damaged property shall be performed by the Contractor after notification of a Change in the Work in accordance with Article 7.

§ 11.3.10 The Owner as fiduciary shall have power to adjust and settle a loss with insurers unless one of the parties in interest shall object in writing within five days after occurrence of loss to the Owner's exercise of this power; if such objection is made, the dispute shall be resolved in the manner selected by the Owner and Contractor as the method of binding dispute resolution in the Agreement. If the Owner and Contractor have selected arbitration as the method of binding dispute resolution, the Owner as fiduciary shall make settlement with insurers or, in the case of a dispute over distribution of insurance proceeds, in accordance with the directions of the arbitrators.

§ 11.4 PERFORMANCE BOND AND PAYMENT BOND

§ 11.4.1 The Owner shall have the right to require the Contractor to furnish bonds covering faithful performance of the Contract and payment of obligations arising thereunder as stipulated in bidding requirements or specifically required in the Contract Documents on the date of execution of the Contract.

§ 11.4.2 Upon the request of any person or entity appearing to be a potential beneficiary of bonds covering payment of obligations arising under the Contract, the Contractor shall promptly furnish a copy of the bonds or shall authorize a copy to be furnished.

ARTICLE 12   UNCOVERING AND CORRECTION OF WORK
§ 12.1 UNCOVERING OF WORK

§ 12.1.1 If a portion of the Work is covered contrary to the Architect's request or to requirements specifically expressed in the Contract Documents, it must, if requested in writing by the Architect, be uncovered for the Architect's examination and be replaced at the Contractor's expense without change in the Contract Time.

§ 12.1.2 If a portion of the Work has been covered that the Architect has not specifically requested to examine prior to its being covered, the Architect may request to see such Work and it shall be uncovered by the Contractor. If such Work is in accordance with the Contract Documents, costs of uncovering and replacement shall, by appropriate Change Order, be at the Owner's expense. If such Work is not in accordance with the Contract Documents, such costs and the cost of correction shall be at the Contractor's expense unless the condition was caused by the Owner or a separate contractor in which event the Owner shall be responsible for payment of such costs.

§ 12.2 CORRECTION OF WORK
§ 12.2.1 BEFORE OR AFTER SUBSTANTIAL COMPLETION

The Contractor shall promptly correct Work rejected by the Architect or failing to conform to the requirements of the Contract Documents, whether discovered before or after Substantial Completion and whether or not fabricated, installed or completed. Costs of correcting such rejected Work, including additional testing and inspections, the cost of uncovering and replacement, and compensation for the Architect's services and expenses made necessary thereby, shall be at the Contractor's expense.

§ 12.2.2 AFTER SUBSTANTIAL COMPLETION

§ 12.2.2.1 In addition to the Contractor's obligations under Section 3.5, if, within one year after the date of Substantial Completion of the Work or designated portion thereof or after the date for commencement of warranties established under Section 9.9.1, or by terms of an applicable special warranty required by the Contract Documents, any of the Work is found to be not in accordance with the requirements of the Contract Documents, the Contractor shall correct it promptly after receipt of written notice from the Owner to do so unless the Owner has previously given the Contractor a written acceptance of such condition. The Owner shall give such notice promptly after discovery of the condition. During the one-year period for correction of Work, if the Owner fails to notify the Contractor and give the Contractor an opportunity to make the correction, the Owner waives the rights to require correction by the Contractor and to make a claim for breach of warranty. If the Contractor fails to correct nonconforming Work within a reasonable time during that period after receipt of notice from the Owner or Architect, the Owner may correct it in accordance with Section 2.4.

§ 12.2.2.2 The one-year period for correction of Work shall be extended with respect to portions of Work first performed after Substantial Completion by the period of time between Substantial Completion and the actual completion of that portion of the Work.

§ 12.2.2.3 The one-year period for correction of Work shall not be extended by corrective Work performed by the Contractor pursuant to this Section 12.2.

§ 12.2.3 The Contractor shall remove from the site portions of the Work that are not in accordance with the requirements of the Contract Documents and are neither corrected by the Contractor nor accepted by the Owner.

§ 12.2.4 The Contractor shall bear the cost of correcting destroyed or damaged construction, whether completed or partially completed, of the Owner or separate contractors caused by the Contractor's correction or removal of Work that is not in accordance with the requirements of the Contract Documents.

§ 12.2.5 Nothing contained in this Section 12.2 shall be construed to establish a period of limitation with respect to other obligations the Contractor has under the Contract Documents. Establishment of the one-year period for correction of Work as described in Section 12.2.2 relates only to the specific obligation of the Contractor to correct the Work, and has no relationship to the time within which the obligation to comply with the Contract Documents may be sought to be enforced, nor to the time within which proceedings may be commenced to establish the Contractor's liability with respect to the Contractor's obligations other than specifically to correct the Work.

§ 12.3 ACCEPTANCE OF NONCONFORMING WORK

If the Owner prefers to accept Work that is not in accordance with the requirements of the Contract Documents, the Owner may do so instead of requiring its removal and correction, in which case the Contract Sum will be reduced as appropriate and equitable. Such adjustment shall be effected whether or not final payment has been made.

ARTICLE 13 MISCELLANEOUS PROVISIONS
§ 13.1 GOVERNING LAW
The Contract shall be governed by the law of the place where the Project is located except that, if the parties have selected arbitration as the method of binding dispute resolution, the Federal Arbitration Act shall govern Section 15.4.

§ 13.2 SUCCESSORS AND ASSIGNS
§ 13.2.1 The Owner and Contractor respectively bind themselves, their partners, successors, assigns and legal representatives to covenants, agreements and obligations contained in the Contract Documents. Except as provided in Section 13.2.2, neither party to the Contract shall assign the Contract as a whole without written consent of the other. If either party attempts to make such an assignment without such consent, that party shall nevertheless remain legally responsible for all obligations under the Contract.

§ 13.2.2 The Owner may, without consent of the Contractor, assign the Contract to a lender providing construction financing for the Project, if the lender assumes the Owner's rights and obligations under the Contract Documents. The Contractor shall execute all consents reasonably required to facilitate such assignment.

§ 13.3 WRITTEN NOTICE
Written notice shall be deemed to have been duly served if delivered in person to the individual, to a member of the firm or entity, or to an officer of the corporation for which it was intended; or if delivered at, or sent by registered or certified mail or by courier service providing proof of delivery to, the last business address known to the party giving notice.

§ 13.4 RIGHTS AND REMEDIES
§ 13.4.1 Duties and obligations imposed by the Contract Documents and rights and remedies available thereunder shall be in addition to and not a limitation of duties, obligations, rights and remedies otherwise imposed or available by law.

§ 13.4.2 No action or failure to act by the Owner, Architect or Contractor shall constitute a waiver of a right or duty afforded them under the Contract, nor shall such action or failure to act constitute approval of or acquiescence in a breach there under, except as may be specifically agreed in writing.

§ 13.5 TESTS AND INSPECTIONS
§ 13.5.1 Tests, inspections and approvals of portions of the Work shall be made as required by the Contract Documents and by applicable laws, statutes, ordinances, codes, rules and regulations or lawful orders of public authorities. Unless otherwise provided, the Contractor shall make arrangements for such tests, inspections and approvals with an independent testing laboratory or entity acceptable to the Owner, or with the appropriate public authority, and shall bear all related costs of tests, inspections and approvals. The Contractor shall give the Architect timely notice of when and where tests and inspections are to be made so that the Architect may be present for such procedures. The Owner shall bear costs of (1) tests, inspections or approvals that do not become requirements until after bids are received or negotiations concluded, and (2) tests, inspections or approvals where building codes or applicable laws or regulations prohibit the Owner from delegating their cost to the Contractor.

§ 13.5.2 If the Architect, Owner or public authorities having jurisdiction determine that portions of the Work require additional testing, inspection or approval not included under Section 13.5.1, the Architect will, upon written authorization from the Owner, instruct the Contractor to make arrangements for such additional testing, inspection or approval by an entity acceptable to the Owner, and the Contractor shall give timely notice to the Architect of when and where tests and inspections are to be made so that the Architect may be present for such procedures. Such costs, except as provided in Section 13.5.3, shall be at the Owner's expense.

§ 13.5.3 If such procedures for testing, inspection or approval under Sections 13.5.1 and 13.5.2 reveal failure of the portions of the Work to comply with requirements established by the Contract Documents, all costs made necessary by such failure including those of repeated procedures and compensation for the Architect's services and expenses shall be at the Contractor's expense.

§ 13.5.4 Required certificates of testing, inspection or approval shall, unless otherwise required by the Contract Documents, be secured by the Contractor and promptly delivered to the Architect.

§ 13.5.5 If the Architect is to observe tests, inspections or approvals required by the Contract Documents, the Architect will do so promptly and, where practicable, at the normal place of testing.

§ 13.5.6 Tests or inspections conducted pursuant to the Contract Documents shall be made promptly to avoid unreasonable delay in the Work.

§ 13.6 INTEREST
Payments due and unpaid under the Contract Documents shall bear interest from the date payment is due at such rate as the parties may agree upon in writing or, in the absence thereof, at the legal rate prevailing from time to time at the place where the Project is located.

§ 13.7 TIME LIMITS ON CLAIMS
The Owner and Contractor shall commence all claims and causes of action, whether in contract, tort, breach of warranty or otherwise, against the other arising out of or related to the Contract in accordance with the requirements of the final dispute resolution method selected in the Agreement within the time period specified by applicable law, but in any case not more than 10 years after the date of Substantial Completion of the Work. The Owner and Contractor waive all claims and causes of action not commenced in accordance with this Section 13.7.

ARTICLE 14 TERMINATION OR SUSPENSION OF THE CONTRACT
§ 14.1 TERMINATION BY THE CONTRACTOR
§ 14.1.1 The Contractor may terminate the Contract if the Work is stopped for a period of 30 consecutive days through no act or fault of the Contractor or a Subcontractor, Sub-subcontractor or their agents or employees or any other persons or entities performing portions of the Work under direct or indirect contract with the Contractor, for any of the following reasons:

.1 Issuance of an order of a court or other public authority having jurisdiction that requires all Work to be stopped;

.2 An act of government, such as a declaration of national emergency that requires all Work to be stopped;

.3 Because the Architect has not issued a Certificate for Payment and has not notified the Contractor of the reason for withholding certification as provided in Section 9.4.1, or because the Owner has not made payment on a Certificate for Payment within the time stated in the Contract Documents; or

.4 The Owner has failed to furnish to the Contractor promptly, upon the Contractor's request, reasonable evidence as required by Section 2.2.1.

§ 14.1.2 The Contractor may terminate the Contract if, through no act or fault of the Contractor or a Subcontractor, Sub-subcontractor or their agents or employees or any other persons or entities performing portions of the Work under direct or indirect contract with the Contractor, repeated suspensions, delays or interruptions of the entire Work by the Owner as described in Section 14.3 constitute in the aggregate more than 100 percent of the total number of days scheduled for completion, or 120 days in any 365-day period, whichever is less.

§ 14.1.3 If one of the reasons described in Section 14.1.1 or 14.1.2 exists, the Contractor may, upon seven days' written notice to the Owner and Architect, terminate the Contract and recover from the Owner payment for Work executed, including reasonable overhead and profit, costs incurred by reason of such termination, and damages.

§ 14.1.4 If the Work is stopped for a period of 60 consecutive days through no act or fault of the Contractor or a Subcontractor or their agents or employees or any other persons performing portions of the Work under contract with the Contractor because the Owner has repeatedly failed to fulfill the Owner's obligations under the Contract Documents with respect to matters important to the progress of the Work, the Contractor may, upon seven additional days' written notice to the Owner and the Architect, terminate the Contract and recover from the Owner as provided in Section 14.1.3.

§ 14.2 TERMINATION BY THE OWNER FOR CAUSE
§ 14.2.1 The Owner may terminate the Contract if the Contractor

.1 repeatedly refuses or fails to supply enough properly skilled workers or proper materials;

.2 fails to make payment to Subcontractors for materials or labor in accordance with the respective agreements between the Contractor and the Subcontractors;

.3 repeatedly disregards applicable laws, statutes, ordinances, codes, rules and regulations, or lawful orders of a public authority; or

.4 otherwise is guilty of substantial breach of a provision of the Contract Documents.

§ 14.2.2 When any of the above reasons exist, the Owner, upon certification by the Initial Decision Maker that sufficient cause exists to justify such action, may without prejudice to any other rights or remedies of the Owner and after giving the Contractor and the Contractor's surety, if any, seven days' written notice, terminate employment of the Contractor and may, subject to any prior rights of the surety:

.1 Exclude the Contractor from the site and take possession of all materials, equipment, tools, and construction equipment and machinery thereon owned by the Contractor;

.2 Accept assignment of subcontracts pursuant to Section 5.4; and

.3 Finish the Work by whatever reasonable method the Owner may deem expedient. Upon written request of the Contractor, the Owner shall furnish to the Contractor a detailed accounting of the costs incurred by the Owner in finishing the Work.

§ 14.2.3 When the Owner terminates the Contract for one of the reasons stated in Section 14.2.1, the Contractor shall not be entitled to receive further payment until the Work is finished.

§ 14.2.4 If the unpaid balance of the Contract Sum exceeds costs of finishing the Work, including compensation for the Architect's services and expenses made necessary thereby, and other damages incurred by the Owner and not expressly waived, such excess shall be paid to the Contractor. If such costs and damages exceed the unpaid balance, the Contractor shall pay the difference to the Owner. The amount to be paid to the Contractor or Owner, as the case may be, shall be certified by the Initial Decision Maker, upon application, and this obligation for payment shall survive termination of the Contract.

§ 14.3 SUSPENSION BY THE OWNER FOR CONVENIENCE
§ 14.3.1 The Owner may, without cause, order the Contractor in writing to suspend, delay or interrupt the Work in whole or in part for such period of time as the Owner may determine.

§ 14.3.2 The Contract Sum and Contract Time shall be adjusted for increases in the cost and time caused by suspension, delay or interruption as described in Section 14.3.1. Adjustment of the Contract Sum shall include profit. No adjustment shall be made to the extent

.1 that performance is, was or would have been so suspended, delayed or interrupted by another cause for which the Contractor is responsible; or

.2 that an equitable adjustment is made or denied under another provision of the Contract.

§ 14.4 TERMINATION BY THE OWNER FOR CONVENIENCE
§ 14.4.1 The Owner may, at any time, terminate the Contract for the Owner's convenience and without cause.

§ 14.4.2 Upon receipt of written notice from the Owner of such termination for the Owner's convenience, the Contractor shall

.1 cease operations as directed by the Owner in the notice;

.2 take actions necessary, or that the Owner may direct, for the protection and preservation of the Work; and

.3 except for Work directed to be performed prior to the effective date of termination stated in the notice, terminate all existing subcontracts and purchase orders and enter into no further subcontracts and purchase orders.

§ 14.4.3 In case of such termination for the Owner's convenience, the Contractor shall be entitled to receive payment for Work executed, and costs incurred by reason of such termination, along with reasonable overhead and profit on the Work not executed.

ARTICLE 15   CLAIMS AND DISPUTES
§ 15.1 CLAIMS
§ 15.1.1 DEFINITION
A Claim is a demand or assertion by one of the parties seeking, as a matter of right, payment of money, or other relief with respect to the terms of the Contract. The term "Claim" also includes other disputes and matters in question between the Owner and Contractor arising out of or relating to the Contract. The responsibility to substantiate Claims shall rest with the party making the Claim.

§ 15.1.2 NOTICE OF CLAIMS
Claims by either the Owner or Contractor must be initiated by written notice to the other party and to the Initial Decision Maker with a copy sent to the Architect, if the Architect is not serving as the Initial Decision Maker.

Claims by either party must be initiated within 21 days after occurrence of the event giving rise to such Claim or within 21 days after the claimant first recognizes the condition giving rise to the Claim, whichever is later.

§ 15.1.3 CONTINUING CONTRACT PERFORMANCE

Pending final resolution of a Claim, except as otherwise agreed in writing or as provided in Section 9.7 and Article 14, the Contractor shall proceed diligently with performance of the Contract and the Owner shall continue to make payments in accordance with the Contract Documents. The Architect will prepare Change Orders and issue Certificates for Payment in accordance with the decisions of the Initial Decision Maker.

§ 15.1.4 CLAIMS FOR ADDITIONAL COST

If the Contractor wishes to make a Claim for an increase in the Contract Sum, written notice as provided herein shall be given before proceeding to execute the Work. Prior notice is not required for Claims relating to an emergency endangering life or property arising under Section 10.4.

§ 15.1.5 CLAIMS FOR ADDITIONAL TIME

§ 15.1.5.1 If the Contractor wishes to make a Claim for an increase in the Contract Time, written notice as provided herein shall be given. The Contractor's Claim shall include an estimate of cost and of probable effect of delay on progress of the Work. In the case of a continuing delay, only one Claim is necessary.

§ 15.1.5.2 If adverse weather conditions are the basis for a Claim for additional time, such Claim shall be documented by data substantiating that weather conditions were abnormal for the period of time, could not have been reasonably anticipated and had an adverse effect on the scheduled construction.

§ 15.1.6 CLAIMS FOR CONSEQUENTIAL DAMAGES

The Contractor and Owner waive Claims against each other for consequential damages arising out of or relating to this Contract. This mutual waiver includes

> .1 damages incurred by the Owner for rental expenses, for losses of use, income, profit, financing, business and reputation, and for loss of management or employee productivity or of the services of such persons; and
>
> .2 damages incurred by the Contractor for principal office expenses including the compensation of personnel stationed there, for losses of financing, business and reputation, and for loss of profit except anticipated profit arising directly from the Work.

This mutual waiver is applicable, without limitation, to all consequential damages due to either party's termination in accordance with Article 14. Nothing contained in this Section 15.1.6 shall be deemed to preclude an award of liquidated damages, when applicable, in accordance with the requirements of the Contract Documents.

§ 15.2 INITIAL DECISION

§ 15.2.1 Claims, excluding those arising under Sections 10.3, 10.4, 11.3.9, and 11.3.10, shall be referred to the Initial Decision Maker for initial decision. The Architect will serve as the Initial Decision Maker, unless otherwise indicated in the Agreement. Except for those Claims excluded by this Section 15.2.1, an initial decision shall be required as a condition precedent to mediation of any Claim arising prior to the date final payment is due, unless 30 days have passed after the Claim has been referred to the Initial Decision Maker with no decision having been rendered. Unless the Initial Decision Maker and all affected parties agree, the Initial Decision Maker will not decide disputes between the Contractor and persons or entities other than the Owner.

§ 15.2.2 The Initial Decision Maker will review Claims and within ten days of the receipt of a Claim take one or more of the following actions: (1) request additional supporting data from the claimant or a response with supporting data from the other party, (2) reject the Claim in whole or in part, (3) approve the Claim, (4) suggest a compromise, or (5) advise the parties that the Initial Decision Maker is unable to resolve the Claim if the Initial Decision Maker lacks sufficient information to evaluate the merits of the Claim or if the Initial Decision Maker concludes that, in the Initial Decision Maker's sole discretion, it would be inappropriate for the Initial Decision Maker to resolve the Claim.

§ 15.2.3 In evaluating Claims, the Initial Decision Maker may, but shall not be obligated to, consult with or seek information from either party or from persons with special knowledge or expertise who may assist the Initial Decision Maker in rendering a decision. The Initial Decision Maker may request the Owner to authorize retention of such persons at the Owner's expense.

§ 15.2.4 If the Initial Decision Maker requests a party to provide a response to a Claim or to furnish additional supporting data, such party shall respond, within ten days after receipt of such request, and shall either (1) provide a response on the requested supporting data, (2) advise the Initial Decision Maker when the response or supporting data will be furnished or (3) advise the Initial Decision Maker that no supporting data will be furnished. Upon receipt of the response or supporting data, if any, the Initial Decision Maker will either reject or approve the Claim in whole or in part.

§ 15.2.5 The Initial Decision Maker will render an initial decision approving or rejecting the Claim, or indicating that the Initial Decision Maker is unable to resolve the Claim. This initial decision shall (1) be in writing; (2) state the reasons therefor; and (3) notify the parties and the Architect, if the Architect is not serving as the Initial Decision Maker, of any change in the Contract Sum or Contract Time or both. The initial decision shall be final and binding on the parties but subject to mediation and, if the parties fail to resolve their dispute through mediation, to binding dispute resolution.

§ 15.2.6 Either party may file for mediation of an initial decision at any time, subject to the terms of Section 15.2.6.1.

§ 15.2.6.1 Either party may, within 30 days from the date of an initial decision, demand in writing that the other party file for mediation within 60 days of the initial decision. If such a demand is made and the party receiving the demand fails to file for mediation within the time required, then both parties waive their rights to mediate or pursue binding dispute resolution proceedings with respect to the initial decision.

§ 15.2.7 In the event of a Claim against the Contractor, the Owner may, but is not obligated to, notify the surety, if any, of the nature and amount of the Claim. If the Claim relates to a possibility of a Contractor's default, the Owner may, but is not obligated to, notify the surety and request the surety's assistance in resolving the controversy.

§ 15.2.8 If a Claim relates to or is the subject of a mechanic's lien, the party asserting such Claim may proceed in accordance with applicable law to comply with the lien notice or filing deadlines.

§ 15.3 MEDIATION
§ 15.3.1 Claims, disputes, or other matters in controversy arising out of or related to the Contract except those waived as provided for in Sections 9.10.4, 9.10.5, and 15.1.6 shall be subject to mediation as a condition precedent to binding dispute resolution.

§ 15.3.2 The parties shall endeavor to resolve their Claims by mediation which, unless the parties mutually agree otherwise, shall be administered by the American Arbitration Association in accordance with its Construction Industry Mediation Procedures in effect on the date of the Agreement. A request for mediation shall be made in writing, delivered to the other party to the Contract, and filed with the person or entity administering the mediation. The request may be made concurrently with the filing of binding dispute resolution proceedings but, in such event, mediation shall proceed in advance of binding dispute resolution proceedings, which shall be stayed pending mediation for a period of 60 days from the date of filing, unless stayed for a longer period by agreement of the parties or court order. If an arbitration is stayed pursuant to this Section 15.3.2, the parties may nonetheless proceed to the selection of the arbitrator(s) and agree upon a schedule for later proceedings.

§ 15.3.3 The parties shall share the mediator's fee and any filing fees equally. The mediation shall be held in the place where the Project is located, unless another location is mutually agreed upon. Agreements reached in mediation shall be enforceable as settlement agreements in any court having jurisdiction thereof.

§ 15.4 ARBITRATION
§ 15.4.1 If the parties have selected arbitration as the method for binding dispute resolution in the Agreement, any Claim subject to, but not resolved by, mediation shall be subject to arbitration which, unless the parties mutually agree otherwise, shall be administered by the American Arbitration Association in accordance with its Construction Industry Arbitration Rules in effect on the date of the Agreement. A demand for arbitration shall be made in writing, delivered to the other party to the Contract, and filed with the person or entity administering the arbitration. The party filing a notice of demand for arbitration must assert in the demand all Claims then known to that party on which arbitration is permitted to be demanded.

§ 15.4.1.1 A demand for arbitration shall be made no earlier than concurrently with the filing of a request for mediation, but in no event shall it be made after the date when the institution of legal or equitable proceedings based on the Claim would be barred by the applicable statute of limitations. For statute of limitations purposes, receipt of a written demand for arbitration by the person or entity administering the arbitration shall constitute the institution of legal or equitable proceedings based on the Claim.

§ 15.4.2 The award rendered by the arbitrator or arbitrators shall be final, and judgment may be entered upon it in accordance with applicable law in any court having jurisdiction thereof.

§ 15.4.3 The foregoing agreement to arbitrate and other agreements to arbitrate with an additional person or entity duly consented to by parties to the Agreement shall be specifically enforceable under applicable law in any court having jurisdiction thereof.

§ 15.4.4 CONSOLIDATION OR JOINDER

§ 15.4.4.1 Either party, at its sole discretion, may consolidate an arbitration conducted under this Agreement with any other arbitration to which it is a party provided that (1) the arbitration agreement governing the other arbitration permits consolidation, (2) the arbitrations to be consolidated substantially involve common questions of law or fact, and (3) the arbitrations employ materially similar procedural rules and methods for selecting arbitrator(s).

§ 15.4.4.2 Either party, at its sole discretion, may include by joinder persons or entities substantially involved in a common question of law or fact whose presence is required if complete relief is to be accorded in arbitration, provided that the party sought to be joined consents in writing to such joinder. Consent to arbitration involving an additional person or entity shall not constitute consent to arbitration of any claim, dispute or other matter in question not described in the written consent.

§ 15.4.4.3 The Owner and Contractor grant to any person or entity made a party to an arbitration conducted under this Section 15.4, whether by joinder or consolidation, the same rights of joinder and consolidation as the Owner and Contractor under this Agreement.

CONSENSUSDOCS 200

STANDARD AGREEMENT AND GENERAL CONDITIONS BETWEEN OWNER AND CONTRACTOR

(Where the Contract Price is a Lump Sum)

This document was developed through a collaborative effort of entities representing a wide cross-section of the construction industry. The organizations endorsing this document believe it represents a fair and reasonable consensus among the collaborating parties of allocation of risk and responsibilities in an effort to appropriately balance the critical interests and concerns of all project participants.

These endorsing organizations recognize and understand that users of this document must review and adapt this document to meet their particular needs, the specific requirements of the project, and applicable laws. Users are encouraged to consult legal, insurance and surety advisors before modifying or completing this document. Further information on this document and the perspectives of endorsing organizations is available in the ConsensusDOCS Guidebook.

TABLE OF ARTICLES

1. AGREEMENT

2. GENERAL PROVISIONS

3. CONTRACTOR'S RESPONSIBILITIES

4. OWNER'S RESPONSIBILITIES

5. SUBCONTRACTS

6. CONTRACT TIME

7. CONTRACT PRICE

1

8. CHANGES

9. PAYMENT

10. INDEMNITY, INSURANCE, WAIVERS AND BONDS

11. SUSPENSION, NOTICE TO CURE AND TERMINATION OF THE AGREEMENT

12. DISPUTE RESOLUTION

13. MISCELLANEOUS PROVISIONS

14. CONTRACT DOCUMENTS

This Agreement has important legal and insurance consequences. Consultations with an attorney and with insurance and surety consultants are encouraged with respect to its completion or modification. Notes indicate where information is to be inserted to complete this Agreement.

ARTICLE 1

AGREEMENT

This Agreement is made this _____ day of _____ in the year _____,

by and between the

OWNER

and the

CONTRACTOR

for services in connection with the following

PROJECT

Notice to the Parties shall be given at the above addresses.

ARTICLE 2

GENERAL PROVISIONS

2.1 RELATIONSHIP OF PARTIES The Owner and the Contractor agree to proceed with the Project on the basis of mutual trust, good faith and fair dealing.

2.1.1 The Contractor shall furnish construction administration and management services and use the Contractor's diligent efforts to perform the Work in an expeditious manner consistent with the Contract Documents. The Owner and Contractor shall endeavor to promote harmony and cooperation among all Project participants.

2.1.2 The Contractor represents that it is an independent contractor and that in its performance of the Work it shall act as an independent contractor.

2

2.1.3 Neither Contractor nor any of its agents or employees shall act on behalf of or in the name of Owner except as provided in this Agreement or unless authorized in writing by Owner's Representative.

2.1.4 The Owner and the Contractor shall perform their obligations with integrity, ensuring at a minimum that

2.1.4.1 Conflicts of interest shall be avoided or disclosed promptly to the other Party; and

2.1.4.2 The Contractor and the Owner warrant that they have not and shall not pay nor receive any contingent fees or gratuities to or from the other Party, including its agents, officers and employees, Subcontractors or others for whom they may be liable, to secure preferential treatment.

2.2 EXTENT OF AGREEMENT This Agreement is solely for the benefit of the Parties, represents the entire and integrated agreement between the Parties, and supersedes all prior negotiations, representations or agreements, either written or oral. This Agreement and each and every provision is for the exclusive benefit of the Owner and Contractor and not for the benefit of any third party except to the extent expressly provided in this Agreement.

2.3 ARCHITECT/ENGINEER The Owner, through its Architect/Engineer, shall provide all architectural and engineering design services necessary for the completion of the Work, except the following: _____. The Contractor shall not be required to provide professional services which constitute the practice of architecture or engineering except as otherwise provided in Paragraph 3.15.

2.3.1 The Owner shall obtain from the Architect/Engineer either a license for Contractor and Subcontractors to use the design documents prepared by the Architect/Engineer or ownership of the copyrights for such design documents, and shall indemnify and hold harmless the Contractor against any suits or claims of infringement of any copyrights or licenses arising out of the use of the design documents for the project.

2.4 DEFINITIONS

2.4.1 Agreement means this ConsensusDOCS 200 Standard Agreement and General Conditions Between Owner and Contractor (Where the Contract Price is a Lump Sum), as modified by the Parties, and exhibits and attachments made part of this Agreement upon its execution.

2.4.2 Architect/Engineer means the licensed Architect, Architect/Engineer or Engineer and its consultants, retained by Owner to perform design services for the Project. The Owner's Architect/Engineer for the Project is _____.

2.4.3 A Change Order is a written order signed by the Owner and the Contractor after execution of this Agreement, indicating changes in the scope of the Work, the Contract Price or Contract Time, including substitutions proposed by the Contractor and accepted by the Owner.

2.4.4 The Contract Documents consist of this Agreement, the drawings, specifications, addenda issued prior to execution of this Agreement, approved submittals, information furnished by the Owner under Paragraph 4.3, other documents listed in this Agreement and any modifications issued after execution.

2.4.5 The Contract Price is the amount indicated in Paragraph 7.1 of this Agreement.

2.4.6 The Contract Time is the period between the Date of Commencement and Final Completion.

3

See Article 6.

2.4.7 The Contractor is the person or entity identified in Article 1 and includes the Contractor's Representative.

2.4.8 The term Day shall mean calendar day unless otherwise specifically defined.

2.4.9 Final Completion occurs on the date when the Contractor's obligations under this Agreement are complete and accepted by the Owner and final payment becomes due and payable, as established in Article 6. This date shall be confirmed by a Certificate of Final Completion signed by the Owner and the Contractor.

2.4.10 A Material Supplier is a person or entity retained by the Contractor to provide material or equipment for the Work.

2.4.11 Others means other contractors, material suppliers and persons at the Worksite who are not employed by the Contractor or Subcontractors.

2.4.12 The term Overhead shall mean 1) payroll costs and other compensation of Contractor employees in the Contractor's principal and branch offices; 2) general and administrative expenses of the Contractor's principal and branch offices including deductibles paid on any insurance policy, charges against the Contractor for delinquent payments, and costs related to the correction of defective work; and 3) the Contractor's capital expenses, including interest on capital used for the Work.

2.4.13 Owner is the person or entity identified in Article 1, and includes the Owner's Representative.

2.4.14 The Project, as identified in Article 1, is the building, facility or other improvements for which the Contractor is to perform Work under this Agreement. It may also include construction by the Owner or Others.

2.4.15 The Schedule of the Work is the document prepared by the Contractor that specifies the dates on which the Contractor plans to begin and complete various parts of the Work, including dates on which information and approvals are required from the Owner.

2.4.16 A Subcontractor is a person or entity retained by the Contractor as an independent contractor to provide the labor, materials, equipment or services necessary to complete a specific portion of the Work. The term Subcontractor does not include the Architect/Engineer or Others.

2.4.17 Substantial Completion of the Work, or of a designated portion, occurs on the date when the Work is sufficiently complete in accordance with the Contract Documents so that the Owner may occupy or utilize the Project, or a designated portion, for the use for which it is intended, without unscheduled disruption. The issuance of a certificate of occupancy is not a prerequisite for Substantial Completion if the certificate of occupancy cannot be obtained due to factors beyond the Contractor's control. This date shall be confirmed by a Certificate of Substantial Completion signed by the Owner and Contractor.

2.4.18 A Sub-subcontractor is a person or entity who has an agreement with a Subcontractor to perform any portion of the Subcontractor's Work.

2.4.19 Terrorism means a violent act, or an act that is dangerous to human life, property or infrastructure, that is committed by an individual or individuals and that appears to be part of an effort to coerce a civilian population or to influence the policy or affect the conduct of any government by coercion. Terrorism includes, but is not limited to, any act certified by the United

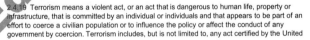

States government as an act of terrorism pursuant to the Terrorism Risk Insurance Act, as amended.

2.4.20 Work means the construction and services necessary or incidental to fulfill the Contractor's obligations for the Project in conformance with this Agreement and the other Contract Documents. The Work may refer to the whole Project or only a part of the Project if work is also being performed by the Owner or Others.

 2.4.20.1 Changed Work means work that is different from the original scope of Work; or work that changes the Contract Price or Contract Time.

 2.4.20.2 Defective Work is any portion of the Work that is not in conformance with the Contract Documents, as more fully described in Paragraphs 3.5 and 3.8.

2.4.21 Worksite means the geographical area at the location of the Project as identified in Article 1 where the Work is to be performed.

ARTICLE 3

CONTRACTOR'S RESPONSIBILITIES

3.1 GENERAL RESPONSIBILITIES

 3.1.1 The Contractor shall provide all labor, materials, equipment and services necessary to complete the Work, all of which shall be provided in full accord with and reasonably inferable from the Contract Documents as being necessary to produce the indicated results.

 3.1.2 The Contractor shall be responsible for the supervision and coordination of the Work, including the construction means, methods, techniques, sequences and procedures utilized, unless the Contract Documents give other specific instructions. In such case, the Contractor shall not be liable to the Owner for damages resulting from compliance with such instructions unless the Contractor recognized and failed to timely report to the Owner any error, inconsistency, omission or unsafe practice that it discovered in the specified construction means, methods, techniques, sequences or procedures.

 3.1.3 The Contractor shall perform Work only within locations allowed by the Contract Documents, applicable permits and applicable local law.

3.2 COOPERATION WITH WORK OF OWNER AND OTHERS

 3.2.1 The Owner may perform work at the Worksite directly or by Others. Any agreements with Others to perform construction or operations related to the Project shall include provisions pertaining to insurance, indemnification, waiver of subrogation, coordination, interference, cleanup and safety which are substantively the same as the corresponding provisions of this Agreement.

 3.2.2 In the event that the Owner elects to perform work at the Worksite directly or by Others, the Contractor and the Owner shall coordinate the activities of all forces at the Worksite and agree upon fair and reasonable schedules and operational procedures for Worksite activities. The Owner shall require each separate contractor to cooperate with the Contractor and assist with the coordination of activities and the review of construction schedules and operations. The Contract Price and Contract Time shall be equitably adjusted, as mutually agreed by the Parties, for changes made necessary by the coordination of construction activities, and the Schedule of the Work shall be revised accordingly. The Contractor, Owner and Others shall adhere to the revised construction schedule until it may subsequently be revised.

5

3.2.3 With regard to the work of the Owner and Others, the Contractor shall (a) proceed with the Work in a manner which does not hinder, delay or interfere with the work of the Owner or Others or cause the work of the Owner or Others to become defective, (b) afford the Owner or Others reasonable access for introduction and storage of their materials and equipment and performance of their activities, and (c) coordinate the Contractor's construction and operations with theirs as required by this Paragraph 3.2.

3.2.4 Before proceeding with any portion of the Work affected by the construction or operations of the Owner or Others, the Contractor shall give the Owner prompt written notification of any defects the Contractor discovers in their work which will prevent the proper execution of the Work. The Contractor's obligations in this Paragraph do not create a responsibility for the work of the Owner or Others, but are for the purpose of facilitating the Work. If the Contractor does not notify the Owner of patent defects interfering with the performance of the Work, the Contractor acknowledges that the work of the Owner or Others is not defective and is acceptable for the proper execution of the Work. Following receipt of written notice from the Contractor of defects, the Owner shall promptly inform the Contractor what action, if any, the Contractor shall take with regard to the defects.

3.3 RESPONSIBILITY FOR PERFORMANCE

3.3.1 In order to facilitate its responsibilities for completion of the Work in accordance with and as reasonably inferable from the Contract Documents, prior to commencing the Work the Contractor shall examine and compare the drawings and specifications with information furnished by the Owner pursuant to Paragraph 4.3, relevant field measurements made by the Contractor and any visible conditions at the Worksite affecting the Work.

3.3.2 If in the course of the performance of the obligations in Subparagraph 3.3.1 the Contractor discovers any errors, omissions or inconsistencies in the Contract Documents, the Contractor shall promptly report them to the Owner. It is recognized, however, that the Contractor is not acting in the capacity of a licensed design professional, and that the Contractor's examination is to facilitate construction and does not create an affirmative responsibility to detect errors, omissions or inconsistencies or to ascertain compliance with applicable laws, building codes or regulations. Following receipt of written notice from the Contractor of defects, the Owner shall promptly inform the Contractor what action, if any, the Contractor shall take with regard to the defects.

3.3.3 The Contractor shall have no liability for errors, omissions or inconsistencies discovered under Subparagraphs 3.3.1 and 3.3.2 unless the Contractor knowingly fails to report a recognized problem to the Owner.

3.3.4 The Contractor may be entitled to additional costs or time because of clarifications or instructions arising out of the Contractor's reports described in the three preceding Subparagraphs.

3.4 CONSTRUCTION PERSONNEL AND SUPERVISION

3.4.1 The Contractor shall provide competent supervision for the performance of the Work. Before commencing the Work, Contractor shall notify Owner in writing of the name and qualifications of its proposed superintendent(s) and project manager so Owner may review the individual's qualifications. If, for reasonable cause, the Owner refuses to approve the individual, or withdraws its approval after once giving it, Contractor shall name a different superintendent or project manager for Owner's review. Any disapproved superintendent shall not perform in that capacity thereafter at the Worksite.

3.4.2 The Contractor shall be responsible to the Owner for acts or omissions of parties or entities performing portions of the Work for or on behalf of the Contractor or any of its Subcontractors.

3.4.3 The Contractor shall permit only qualified persons to perform the Work. The Contractor shall enforce safety procedures, strict discipline and good order among persons performing the Work. If the Owner determines that a particular person does not follow safety procedures, or is unfit or unskilled for the assigned work, the Contractor shall immediately reassign the person upon receipt of the Owner's written notice to do so.

3.4.4 CONTRACTOR'S REPRESENTATIVE The Contractor's authorized representative is _____. The Contractor's Representative shall possess full authority to receive instructions from the Owner and to act on those instructions. The Contractor shall notify the Owner in writing of a change in the designation of the Contractor's Representative.

3.5 WORKMANSHIP The Work shall be executed in accordance with the Contract Documents in a workmanlike manner. All materials used in the Work shall be furnished in sufficient quantities to facilitate the proper and expeditious execution of the Work and shall be new except such materials as may be expressly provided in the Contract Documents to be otherwise.

3.6 MATERIALS FURNISHED BY THE OWNER OR OTHERS In the event the Work includes installation of materials or equipment furnished by the Owner or Others, it shall be the responsibility of the Contractor to examine the items so provided and thereupon handle, store and install the items, unless otherwise provided in the Contract Documents, with such skill and care as to provide a satisfactory and proper installation. Loss or damage due to acts or omissions of the Contractor shall be the responsibility of the Contractor and may be deducted from any amounts due or to become due the Contractor. Any defects discovered in such materials or equipment shall be reported at once to the Owner. Following receipt of written notice from the Contractor of defects, the Owner shall promptly inform the Contractor what action, if any, the Contractor shall take with regard to the defects.

3.7 TESTS AND INSPECTIONS

3.7.1 The Contractor shall schedule all required tests, approvals and inspections of the Work or portions thereof at appropriate times so as not to delay the progress of the Work or other work related to the Project. The Contractor shall give proper notice to all required parties of such tests, approvals and inspections. If feasible, the Owner and Others may timely observe the tests at the normal place of testing. Except as provided in Subparagraph 3.7.3, the Owner shall bear all expenses associated with tests, inspections and approvals required by the Contract Documents, which, unless otherwise agreed to, shall be conducted by an independent testing laboratory or entity retained by the Owner. Unless otherwise required by the Contract Documents, required certificates of testing, approval or inspection shall be secured by the Contractor and promptly delivered to the Owner.

3.7.2 If the Owner or appropriate authorities determine that tests, inspections or approvals in addition to those required by the Contract Documents will be necessary, the Contractor shall arrange for the procedures and give timely notice to the Owner and Others who may observe the procedures. Costs of the additional tests, inspections or approvals are at the Owner's expense except as provided in Subparagraph 3.7.3.

3.7.3 If the procedures described in Subparagraphs 3.7.1 and 3.7.2 indicate that portions of the Work fail to comply with the Contract Documents, the Contractor shall be responsible for costs of correction and retesting.

3.8 WARRANTY

3.8.1 The Contractor warrants that all materials and equipment shall be new unless otherwise specified, of good quality, in conformance with the Contract Documents, and free from defective

7

workmanship and materials. At the Owner's request, the Contractor shall furnish satisfactory evidence of the quality and type of materials and equipment furnished. The Contractor further warrants that the Work shall be free from material defects not intrinsic in the design or materials required in the Contract Documents. The Contractor's warranty does not include remedies for defects or damages caused by normal wear and tear during normal usage, use for a purpose for which the Project was not intended, improper or insufficient maintenance, modifications performed by the Owner or Others, or abuse. The Contractor's warranty pursuant to this Paragraph 3.8 shall commence on the Date of Substantial Completion.

3.8.2 The Contractor shall obtain from its Subcontractors and Material Suppliers any special or extended warranties required by the Contract Documents. All such warranties shall be listed in an attached Exhibit to this Agreement. Contractor's liability for such warranties shall be limited to the one-year correction period referred to in Paragraph 3.9. After that period Contractor shall assign them to the Owner and provide reasonable assistance to the Owner in enforcing the obligations of Subcontractors or Material Suppliers.

3.9 CORRECTION OF WORK WITHIN ONE YEAR

3.9.1 If, prior to Substantial Completion and within one year after the date of Substantial Completion of the Work, any Defective Work is found, the Owner shall promptly notify the Contractor in writing. Unless the Owner provides written acceptance of the condition, the Contractor shall promptly correct the Defective Work at its own cost and time and bear the expense of additional services required for correction of any Defective Work for which it is responsible. If within the one-year correction period the Owner discovers and does not promptly notify the Contractor or give the Contractor an opportunity to test or correct Defective Work as reasonably requested by the Contractor, the Owner waives the Contractor's obligation to correct that Defective Work as well as the Owner's right to claim a breach of the warranty with respect to that Defective Work.

3.9.2 With respect to any portion of Work first performed after Substantial Completion, the one-year correction period shall be extended by the period of time between Substantial Completion and the actual performance of the later Work. Correction periods shall not be extended by corrective work performed by the Contractor.

3.9.3 If the Contractor fails to correct Defective Work within a reasonable time after receipt of written notice from the Owner prior to final payment, the Owner may correct it in accordance with the Owner's right to carry out the Work in Paragraph 11.2. In such case, an appropriate Change Order shall be issued deducting the cost of correcting such deficiencies from payments then or thereafter due the Contractor. If payments then or thereafter due Contractor are not sufficient to cover such amounts, the Contractor shall pay the difference to the Owner.

3.9.4 If after the one-year correction period but before the applicable limitation period the Owner discovers any Defective Work, the Owner shall, unless the Defective Work requires emergency correction, promptly notify the Contractor. If the Contractor elects to correct the Work, it shall provide written notice of such intent within fourteen (14) Days of its receipt of notice from the Owner. The Contractor shall complete the correction of Work within a mutually agreed timeframe. If the Contractor does not elect to correct the Work, the Owner may have the Work corrected by itself or Others and charge the Contractor for the reasonable cost of the correction. Owner shall provide Contractor with an accounting of correction costs it incurs.

3.9.5 If the Contractor's correction or removal of Defective Work causes damage to or destroys other completed or partially completed Work or existing buildings, the Contractor shall be responsible for the cost of correcting the destroyed or damaged property.

3.9.6 The one-year period for correction of Defective Work does not constitute a limitation period with respect to the enforcement of the Contractor's other obligations under the Contract Documents.

3.9.7 Prior to final payment, at the Owner's option and with the Contractor's agreement, the Owner may elect to accept Defective Work rather than require its removal and correction. In such case the Contract Price shall be equitably adjusted for any diminution in the value of the Project caused by such Defective Work.

3.10 CORRECTION OF COVERED WORK

3.10.1 On request of the Owner, Work that has been covered without a requirement that it be inspected prior to being covered may be uncovered for the Owner's inspection. The Owner shall pay for the costs of uncovering and replacement if the Work proves to be in conformance with the Contract Documents, or if the defective condition was caused by the Owner or Others. If the uncovered Work proves to be defective, the Contractor shall pay the costs of uncovering and replacement.

3.10.2 If contrary to specific requirements in the Contract Documents or contrary to a specific request from the Owner, a portion of the Work is covered, the Owner, by written request, may require the Contractor to uncover the Work for the Owner's observation. In this circumstance the Work shall be replaced at the Contractor's expense and with no adjustment to the Contract Time.

3.11 SAFETY OF PERSONS AND PROPERTY

3.11.1 SAFETY PRECAUTIONS AND PROGRAMS The Contractor shall have overall responsibility for safety precautions and programs in the performance of the Work. While this Paragraph 3.11 establishes the responsibility for safety between the Owner and Contractor, it does not relieve Subcontractors of their responsibility for the safety of persons or property in the performance of their work, nor for compliance with the provisions of applicable laws and regulations.

3.11.2 The Contractor shall seek to avoid injury, loss or damage to persons or property by taking reasonable steps to protect:

3.11.2.1 its employees and other persons at the Worksite;

3.11.2.2 materials and equipment stored at onsite or offsite locations for use in the Work; and

3.11.2.3 property located at the site and adjacent to Work areas, whether or not the property is part of the Work.

3.11.3 CONTRACTOR'S SAFETY REPRESENTATIVE The Contractor's Worksite Safety Representative is _____, who shall act as the Contractor's authorized safety representative with a duty to prevent accidents in accordance with Subparagraph 3.11.2. If no individual is identified in this Paragraph 3.11, the authorized safety representative shall be the Contractor's Representative. The Contractor shall report immediately in writing to the Owner all recordable accidents and injuries occurring at the Worksite. When the Contractor is required to file an accident report with a public authority, the Contractor shall furnish a copy of the report to the Owner.

3.11.4 The Contractor shall provide the Owner with copies of all notices required of the Contractor by law or regulation. The Contractor's safety program shall comply with the requirements of governmental and quasi-governmental authorities having jurisdiction.

3.11.5 Damage or loss not insured under property insurance which may arise from the Work, to the

9

extent caused by the negligent acts or omissions of the Contractor, or anyone for whose acts the Contractor may be liable, shall be promptly remedied by the Contractor.

3.11.6 If the Owner deems any part of the Work or Worksite unsafe, the Owner, without assuming responsibility for the Contractor's safety program, may require the Contractor to stop performance of the Work or take corrective measures satisfactory to the Owner, or both. If the Contractor does not adopt corrective measures, the Owner may perform them and deduct their cost from the Contract Price. The Contractor agrees to make no claim for damages, for an increase in the Contract Price or for a change in the Contract Time based on the Contractor's compliance with the Owner's reasonable request.

3.12 EMERGENCIES

3.12.1 In an emergency, the Contractor shall act in a reasonable manner to prevent personal injury or property damage. Any change in the Contract Price or Contract Time resulting from the actions of the Contractor in an emergency situation shall be determined as provided in Article 8.

3.13 HAZARDOUS MATERIALS

3.13.1 A Hazardous Material is any substance or material identified now or in the future as hazardous under any federal, state or local law or regulation, or any other substance or material that may be considered hazardous or otherwise subject to statutory or regulatory requirement governing handling, disposal or cleanup. The Contractor shall not be obligated to commence or continue work until any Hazardous Material discovered at the Worksite has been removed, rendered or determined to be harmless by the Owner as certified by an independent testing laboratory and approved by the appropriate government agency.

3.13.2 If after the commencement of the Work Hazardous Material is discovered at the Worksite, the Contractor shall be entitled to immediately stop Work in the affected area. The Contractor shall report the condition to the Owner, the Architect/Engineer, and, if required, the government agency with jurisdiction.

3.13.3 The Contractor shall not be required to perform any Work relating to or in the area of Hazardous Material without written mutual agreement.

3.13.4 The Owner shall be responsible for retaining an independent testing laboratory to determine the nature of the material encountered and whether the material requires corrective measures or remedial action. Such measures shall be the sole responsibility of the Owner, and shall be performed in a manner minimizing any adverse effects upon the Work. The Contractor shall resume Work in the area affected by any Hazardous Material only upon written agreement between the Parties after the Hazardous Material has been removed or rendered harmless and only after approval, if necessary, of the governmental agency with jurisdiction.

3.13.5 If the Contractor incurs additional costs or is delayed due to the presence or remediation of Hazardous Material, the Contractor shall be entitled to an equitable adjustment in the Contract Price or the Contract Time.

3.13.6 To the extent not caused by the negligent acts or omissions of the Contractor, its Subcontractors and Sub-subcontractors, and the agents, officers, directors and employees of each of them, the Owner shall defend, indemnify and hold harmless the Contractor, its Subcontractors and Sub-subcontractors, and the agents, officers, directors and employees of each of them, from and against all claims, damages, losses, costs and expenses, including but not limited to reasonable attorneys' fees, costs and expenses incurred in connection with any dispute resolution

10

process, to the extent permitted pursuant to Paragraph 6.6, arising out of or relating to the performance of the Work in any area affected by Hazardous Material.

3.13.7 MATERIALS BROUGHT TO THE WORKSITE

3.13.7.1 Material Safety Data (MSD) sheets as required by law and pertaining to materials or substances used or consumed in the performance of the Work, whether obtained by the Contractor, Subcontractors, the Owner or Others, shall be maintained at the Worksite by the Contractor and made available to the Owner, Subcontractors and Others.

3.13.7.2 The Contractor shall be responsible for the proper delivery, handling, application, storage, removal and disposal of all materials and substances brought to the Worksite by the Contractor in accordance with the Contract Documents and used or consumed in the performance of the Work.

3.13.7.3 To the extent caused by the negligent acts or omissions of the Contractor, its agents, officers, directors and employees, the Contractor shall indemnify and hold harmless the Owner, its agents, officers, directors and employees, from and against any and all claims, damages, losses, costs and expenses, including but not limited to attorneys' fees, costs and expenses incurred in connection with any dispute resolution procedure, arising out of or relating to the delivery, handling, application, storage, removal and disposal of all materials and substances brought to the Worksite by the Contractor in accordance with the Contract Documents.

3.13.8 The terms of this Paragraph 3.13 shall survive the completion of the Work or any termination of this Agreement.

3.14 SUBMITTALS

3.14.1 The Contractor shall submit to the Owner, and, if directed, to its Architect/Engineer, for review and approval all shop drawings, samples, product data and similar submittals required by the Contract Documents. Submittals may be submitted in electronic form if required in accordance with ConsensusDOCS 200.2 and Subparagraph 4.6.1. The Contractor shall be responsible to the Owner for the accuracy and conformity of its submittals to the Contract Documents. The Contractor shall prepare and deliver its submittals to the Owner in a manner consistent with the Schedule of the Work and in such time and sequence so as not to delay the performance of the Work or the work of the Owner and Others. When the Contractor delivers its submittals to the Owner, the Contractor shall identify in writing for each submittal all changes, deviations or substitutions from the requirements of the Contract Documents. The review and approval of any Contractor submittal shall not be deemed to authorize changes, deviations or substitutions from the requirements of the Contract Documents unless express written approval is obtained from the Owner specifically authorizing such deviation, substitution or change. To the extent a change, deviation or substitution causes an impact to the Contract Price or Contract Time, such approval shall be promptly memorialized in a Change Order. Further, the Owner shall not make any change, deviation or substitution through the submittal process without specifically identifying and authorizing such deviation to the Contractor. In the event that the Contract Documents do not contain submittal requirements pertaining to the Work, the Contractor agrees upon request to submit in a timely fashion to the Owner for review and approval any shop drawings, samples, product data, manufacturers' literature or similar submittals as may reasonably be required by the Owner.

3.14.2 The Owner shall be responsible for review and approval of submittals with reasonable promptness to avoid causing delay.

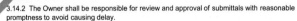

11

3.14.3 The Contractor shall perform all Work strictly in accordance with approved submittals. Approval of shop drawings is not authorization to Contractor to perform Changed Work, unless the procedures of Article 8 are followed. Approval does not relieve the Contractor from responsibility for Defective Work resulting from errors or omissions of any kind on the approved Shop Drawings.

3.14.4 Record copies of the following, incorporating field changes and selections made during construction, shall be maintained at the Project site and available to the Owner upon request: drawings, specifications, addenda, Change Order and other modifications, and required submittals including product data, samples and shop drawings.

3.14.5 No substitutions shall be made in the Work unless permitted in the Contract Documents and then only after the Contractor obtains approvals required under the Contract Documents for substitutions. All such substitutions shall be promptly memorialized in a Change Order no later than seven (7) Days following approval by the Owner and, if applicable, provide for an adjustment in the Contract Price or Contract Time.

3.14.6 The Contractor shall prepare and submit to the Owner

_____ final marked-up as-built drawings,

or

_____ updated electronic data, in accordance with ConsensusDOCS 200.2 and Paragraph 4.6.1,

or

_____ such documentation as defined by the Parties by attachment to this Agreement,

in general documenting how the various elements of the Work were actually constructed or installed.

3.15 PROFESSIONAL SERVICES The Contractor may be required to procure professional services in order to carry out its responsibilities for construction means, methods, techniques, sequences and procedures for such services specifically called for by the Contract Documents. The Contractor shall obtain these professional services and any design certifications required from licensed design professionals. All drawings, specifications, calculations, certifications and submittals prepared by such design professionals shall bear the signature and seal of such design professionals and the Owner and the Architect/Engineer shall be entitled to rely upon the adequacy, accuracy and completeness of such design services. If professional services are specifically required by the Contract Documents, the Owner shall indicate all required performance and design criteria. The Contractor shall not be responsible for the adequacy of such performance and design criteria. The Contractor shall not be required to provide such services in violation of existing laws, rules and regulations in the jurisdiction where the Project is located.

3.16 WORKSITE CONDITIONS

3.16.1 WORKSITE VISIT The Contractor acknowledges that it has visited, or has had the opportunity to visit, the Worksite to visually inspect the general and local conditions which could affect the Work.

3.16.2 CONCEALED OR UNKNOWN SITE CONDITIONS If the conditions at the Worksite are (a) subsurface or other physical conditions which are materially different from those indicated in the Contract Documents, or (b) unusual or unknown physical conditions which are materially different from conditions ordinarily encountered and generally recognized as inherent in Work provided for in the Contract Documents, the Contractor shall stop Work and give immediate written notice of the

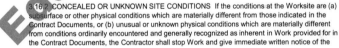

12

condition to the Owner and the Architect/Engineer. The Contractor shall not be required to perform any work relating to the unknown condition without the written mutual agreement of the Parties. Any change in the Contract Price or the Contract Time as a result of the unknown condition shall be determined as provided in Article 8. The Contractor shall provide the Owner with written notice of any claim as a result of unknown conditions within the time period set forth in Paragraph 8.4.

3.17 PERMITS AND TAXES

3.17.1 Contractor shall give public authorities all notices required by law and, except for permits and fees which are the responsibility of the Owner pursuant to Paragraph 4.4, shall obtain and pay for all necessary permits, licenses and renewals pertaining to the Work. Contractor shall provide to Owner copies of all notices, permits, licenses and renewals required under this Agreement.

3.17.2 Contractor shall pay all applicable taxes legally enacted when bids are received or negotiations concluded for the Work provided by the Contractor.

3.17.3 The Contract Price or Contract Time shall be equitably adjusted by Change Order for additional costs resulting from any changes in laws, ordinances, rules and regulations enacted after the date of this Agreement, including increased taxes.

3.17.4 If in accordance with the Owner's direction, the Contractor claims an exemption for taxes, the Owner shall indemnify and hold the Contractor harmless from any liability, penalty, interest, fine, tax assessment, attorneys' fees or other expense or cost incurred by the Contractor as a result of any such action.

3.18 CUTTING, FITTING AND PATCHING

3.18.1 The Contractor shall perform cutting, fitting and patching necessary to coordinate the various parts of the Work and to prepare its Work for the work of the Owner or Others.

3.18.2 Cutting, patching or altering the work of the Owner or Others shall be done with the prior written approval of the Owner. Such approval shall not be unreasonably withheld.

3.19 CLEANING UP

3.19.1 The Contractor shall regularly remove debris and waste materials at the Worksite resulting from the Work. Prior to discontinuing Work in an area, the Contractor shall clean the area and remove all rubbish and its construction equipment, tools, machinery, waste and surplus materials. The Contractor shall minimize and confine dust and debris resulting from construction activities. At the completion of the Work, the Contractor shall remove from the Worksite all construction equipment, tools, surplus materials, waste materials and debris.

3.19.2 If the Contractor fails to commence compliance with cleanup duties within two (2) business Days after written notification from the Owner of non-compliance, the Owner may implement appropriate cleanup measures without further notice and the cost shall be deducted from any amounts due or to become due the Contractor in the next payment period.

3.20 ACCESS TO WORK The Contractor shall facilitate the access of the Owner, Architect/Engineer and Others to Work in progress.

3.21 CONFIDENTIALITY Unless compelled by law, a governmental agency or authority, an order of a court of competent jurisdiction, or a validly issued subpoena, the Contractor shall treat as confidential and not disclose to third persons, except Subcontractors, Sub-subcontractors and Material Suppliers as is necessary for the performance of the Work, or use for its own benefit, any of the Owner's confidential

---

information, know-how, discoveries, production methods and the like that may be disclosed to the Contractor or which the Contractor may acquire in connection with the Work. The Owner shall treat as confidential information all of the Contractor's estimating systems and historical and parameter cost data that may be disclosed to the Owner in connection with the performance of this Agreement. The Owner and the Contractor shall each specify those items to be treated as confidential and shall mark them as "Confidential." In the event of a legal compulsion or other order seeking disclosure of any Confidential Information, the Contractor or Owner, as the case may be, shall promptly notify the other party to permit that party's legal objection, if necessary.

ARTICLE 4

OWNER'S RESPONSIBILITIES

4.1 INFORMATION AND SERVICES Any information or services to be provided by the Owner shall be provided in a timely manner so as not to delay the Work.

4.2 FINANCIAL INFORMATION Prior to commencement of the Work and thereafter at the written request of the Contractor, the Owner shall provide the Contractor with evidence of Project financing. Evidence of such financing shall be a condition precedent to the Contractor's commencing or continuing the Work. The Contractor shall be notified prior to any material change in Project financing.

4.3 WORKSITE INFORMATION Except to the extent that the Contractor knows of any inaccuracy, the Contractor is entitled to rely on Worksite information furnished by the Owner pursuant to this Paragraph 4.3. To the extent the Owner has obtained, or is required elsewhere in the Contract Documents to obtain, the following Worksite information, the Owner shall provide at the Owner's expense and with reasonable promptness:

4.3.1 information describing the physical characteristics of the site, including surveys, site evaluations, legal descriptions, data or drawings depicting existing conditions, subsurface conditions and environmental studies, reports and investigations;

4.3.2 tests, inspections and other reports dealing with environmental matters, Hazardous Material and other existing conditions, including structural, mechanical and chemical tests, required by the Contract Documents or by law; and

4.3.3 any other information or services requested in writing by the Contractor which are relevant to the Contractor's performance of the Work and under the Owner's control.

The information required by Paragraph 4.3 shall be provided in reasonable detail. Legal descriptions shall include easements, title restrictions, boundaries, and zoning restrictions. Worksite descriptions shall include existing buildings and other construction and all other pertinent site conditions. Adjacent property descriptions shall include structures, streets, sidewalks, alleys, and other features relevant to the Work. Utility details shall include available services, lines at the Worksite and adjacent thereto and connection points. The information shall include public and private information, subsurface information, grades, contours, and elevations, drainage data, exact locations and dimensions, and benchmarks that can be used by the Contractor in laying out the Work.

4.4 BUILDING PERMIT, FEES AND APPROVALS Except for those permits and fees related to the Work which are the responsibility of the Contractor pursuant to Subparagraph 3.17.1, the Owner shall secure and pay for all other permits, approvals, easements, assessments and fees required for the development, construction, use or occupancy of permanent structures or for permanent changes in existing facilities, including the building permit.

4.5 MECHANICS AND CONSTRUCTION LIEN INFORMATION Within seven (7) Days after receiving the Contractor's written request, the Owner shall provide the Contractor with the information necessary to give notice of or enforce mechanics lien rights and, where applicable, stop notices. This information shall include the Owner's interest in the real property on which the Project is located and the record legal title.

4.6 CONTRACT DOCUMENTS Unless otherwise specified, Owner shall provide _____ (_____) hard copies of the Contract Documents to the Contractor without cost.

> 4.6.1 DIGITIZED DOCUMENTS If the Owner requires that the Owner, Architect/Engineer and Contractor exchange documents and data in electronic or digital form, prior to any such exchange, the Owner, Architect/Engineer and Contractor shall agree on a written protocol governing all exchanges in ConsensusDOCS 200.2 or a separate Agreement, which, at a minimum, shall specify (1) the definition of documents and data to be accepted in electronic or digital form or to be transmitted electronically or digitally; (2) management and coordination responsibilities; (3) necessary equipment, software and services; (4) acceptable formats, transmission methods and verification procedures; (5) methods for maintaining version control; (6) privacy and security requirements; and (7) storage and retrieval requirements. Except as otherwise agreed to by the Parties in writing, the Parties shall each bear their own costs as identified in the protocol. In the absence of a written protocol, use of documents and data in electronic or digital form shall be at the sole risk of the recipient.

4.7 OWNER'S REPRESENTATIVE The Owner's authorized representative is _____ The representative shall be fully acquainted with the Project, and shall have authority to bind the Owner in all matters requiring the Owner's approval, authorization or written notice. If the Owner changes its representative or the representative's authority as listed above, the Owner shall immediately notify the Contractor in writing.

4.8 OWNER'S CUTTING AND PATCHING Cutting, patching or altering the Work by the Owner or Others shall be done with the prior written approval of the Contractor, which approval shall not be unreasonably withheld.

4.9 OWNER'S RIGHT TO CLEAN UP In case of a dispute between the Contractor and Others with regard to respective responsibilities for cleaning up at the Worksite, the Owner may implement appropriate cleanup measures after two (2) business Days' notice and allocate the cost among those responsible during the following pay period.

4.10 COST OF CORRECTING DAMAGED OR DESTROYED WORK With regard to damage or loss attributable to the acts or omissions of the Owner or Others and not to the Contractor, the Owner may either (a) promptly remedy the damage or loss or (b) accept the damage or loss. If the Contractor incurs additional costs or is delayed due to such loss or damage, the Contractor shall be entitled to an equitable adjustment in the Contract Price or Contract Time.

ARTICLE 5

SUBCONTRACTS

5.1 SUBCONTRACTORS The Work not performed by the Contractor with its own forces shall be performed by Subcontractors.

5.2 AWARD OF SUBCONTRACTS AND OTHER CONTRACTS FOR PORTIONS OF THE WORK

> 5.2.1 As soon after the award of this Agreement as possible, the Contractor shall provide the Owner and if directed, the Architect/Engineer with a written list of the proposed Subcontractors and

15

significant material suppliers. If the Owner has a reasonable objection to any proposed Subcontractor or material supplier, the Owner shall notify the Contractor in writing. Failure to promptly object shall constitute acceptance.

5.2.2 If the Owner has reasonably and promptly objected as provided in Subparagraph 5.2.1, the Contractor shall not contract with the proposed Subcontractor or material supplier, and the Contractor shall propose another acceptable Subcontractor to the Owner. To the extent the substitution results in an increase or decrease in the Contract Price or Contract Time, an appropriate Change Order shall be issued as provided in Article 8.

5.3 BINDING OF SUBCONTRACTORS AND MATERIAL SUPPLIERS The Contractor agrees to bind every Subcontractor and Material Supplier (and require every Subcontractor to so bind its subcontractors and material suppliers) to all the provisions of this Agreement and the Contract Documents as they apply to the Subcontractor's and Material Supplier's portions of the Work.

5.4 LABOR RELATIONS (Insert here any conditions, obligations or requirements relative to labor relations and their effect on the Project. Legal counsel is recommended.)

5.5 CONTINGENT ASSIGNMENT OF SUBCONTRACTS

5.5.1 If this Agreement is terminated, each subcontract agreement shall be assigned by the Contractor to the Owner, subject to the prior rights of any surety, provided that:

5.5.1.1 this Agreement is terminated by the Owner pursuant to Paragraphs 11.3 or 11.4; and

5.5.1.2 the Owner accepts such assignment after termination by notifying the Subcontractor and Contractor in writing, and assumes all rights and obligations of the Contractor pursuant to each subcontract agreement.

5.5.2 If the Owner accepts such an assignment, and the Work has been suspended for more than thirty (30) consecutive Days, following termination, if appropriate, the Subcontractor's compensation shall be equitably adjusted as a result of the suspension.

ARTICLE 6
CONTRACT TIME

6.1 PERFORMANCE OF THE WORK

6.1.1 DATE OF COMMENCEMENT The Date of Commencement is the date of this Agreement as first written in Article 1 unless otherwise set forth below: (Insert here any special provisions concerning notices to proceed and the Date of Commencement):

6.1.2 TIME Substantial Completion of the Work shall be achieved in _____ (_____) Days from the Date of Commencement. Unless otherwise specified in the Certificate of Substantial Completion, the Contractor shall achieve Final Completion within _____ (_____) Days after the date of Substantial Completion, subject to adjustments as provided for in the Contract Documents.

6.1.3 Time limits stated above are of the essence of this Agreement.

6.1.4 Unless instructed by the Owner in writing, the Contractor shall not knowingly commence the Work before the effective date of insurance that is required to be provided by the Contractor and

Owner.

6.2 SCHEDULE OF THE WORK

6.2.1 Before submitting the first application for payment, the Contractor shall submit to the Owner, and if directed, its Architect/Engineer, a Schedule of the Work that shall show the dates on which the Contractor plans to commence and complete various parts of the Work, including dates on which information and approvals are required from the Owner. On the Owner's written approval of the Schedule of the Work, the Contractor shall comply with it unless directed by the Owner to do otherwise or the Contractor is otherwise entitled to an adjustment in the Contract Time. The Contractor shall update the Schedule of the Work on a monthly basis or at appropriate intervals as required by the conditions of the Work and the Project.

6.2.2 The Owner may determine the sequence in which the Work shall be performed, provided it does not unreasonably interfere with the Schedule of the Work. The Owner may require the Contractor to make reasonable changes in the sequence at any time during the performance of the Work in order to facilitate the performance of work by the Owner or Others. To the extent such changes increase Contractor's time and costs the Contract Price and Contract Time shall be equitably adjusted.

6.3 DELAYS AND EXTENSIONS OF TIME

6.3.1 If the Contractor is delayed at any time in the commencement or progress of the Work by any cause beyond the control of the Contractor, the Contractor shall be entitled to an equitable extension of the Contract Time. Examples of causes beyond the control of the Contractor include, but are not limited to, the following: acts or omissions of the Owner, the Architect/Engineer or Others; changes in the Work or the sequencing of the Work ordered by the Owner, or arising from decisions of the Owner that impact the time of performance of the Work; transportation delays not reasonably foreseeable; labor disputes not involving the Contractor; general labor disputes impacting the Project but not specifically related to the Worksite; fire; terrorism; epidemics; adverse governmental actions, unavoidable accidents or circumstances; adverse weather conditions not reasonably anticipated; encountering Hazardous Materials; concealed or unknown conditions; delay authorized by the Owner pending dispute resolution; and suspension by the Owner under Paragraph 11.1. The Contractor shall submit any requests for equitable extensions of Contract Time in accordance with the provisions of Article 8.

6.3.2 In addition, if the Contractor incurs additional costs as a result of a delay that is caused by acts or omissions of the Owner, the Architect/Engineer or Others, changes in the Work or the sequencing of the Work ordered by the Owner, or arising from decisions of the Owner that impact the time of performance of the Work, encountering Hazardous Materials, or concealed or unknown conditions, delay authorized by the Owner pending dispute resolution or suspension by the Owner under Paragraph 11.1, the Contractor shall be entitled to an equitable adjustment in the Contract Price subject to Paragraph 6.6.

6.3.3 NOTICE OF DELAYS In the event delays to the Work are encountered for any reason, the Contractor shall provide prompt written notice to the Owner of the cause of such delays after Contractor first recognizes the delay. The Owner and Contractor agree to undertake reasonable steps to mitigate the effect of such delays.

6.4 NOTICE OF DELAY CLAIMS If the Contractor requests an equitable extension of Contract Time or an equitable adjustment in Contract Price as a result of a delay described in Subparagraphs 6.3.1 and 6.3.2, the Contractor shall give the Owner written notice of the claim in accordance with Paragraph 8.4. If

---

17

the Contractor causes delay in the completion of the Work, the Owner shall be entitled to recover its additional costs subject to Paragraph 6.6. The Owner shall process any such claim against the Contractor in accordance with Article 8.

6.5 LIQUIDATED DAMAGES

6.5.1 SUBSTANTIAL COMPLETION The Owner and the Contractor agree that this Agreement _____ shall/ _____ shall not (indicate one) provide for the imposition of liquidated damages based on the Date of Substantial Completion.

6.5.1.1 The Contractor understands that if the Date of Substantial Completion established by this Agreement, as may be amended by subsequent Change Order, is not attained, the Owner will suffer damages which are difficult to determine and accurately specify. The Contractor agrees that if the Date of Substantial Completion is not attained the Contractor shall pay the Owner _____ Dollars ($_____) as liquidated damages and not as a penalty for each Days that Substantial Completion extends beyond the Date of Substantial Completion. The liquidated damages provided herein shall be in lieu of all liability for any and all extra costs, losses, expenses, claims, penalties and any other damages of whatsoever nature incurred by the Owner which are occasioned by any delay in achieving the Date of Substantial Completion.

6.5.2 FINAL COMPLETION The Owner and the Contractor agree that this Agreement _____ shall/ _____ shall not (indicate one) provide for the imposition of liquidated damages based on the Date of Final Completion.

6.5.2.1 The Contractor understands that if the Date of Final Completion established by this Agreement, as may be amended by subsequent Change Order is not attained, the Owner will suffer damages which are difficult to determine and accurately specify. The Contractor agrees that if the Date of Final Completion is not attained the Contractor shall pay the Owner _____ Dollars ($_____) as liquidated damages and not as a penalty for each Days that Final Completion extends beyond the Date of Final Completion. The liquidated damages provided herein shall be in lieu of all liability for any and all extra costs, losses, expenses, claims, penalties and any other damages of whatsoever nature incurred by the Owner which are occasioned by any delay in achieving the Date of Final Completion.

6.5.3 OTHER LIQUIDATED DAMAGES The Owner and the Contractor may agree upon the imposition of liquidated damages based on other project milestones or performance requirements. Such agreement shall be included as an exhibit to this Agreement.

6.6 LIMITED MUTUAL WAIVER OF CONSEQUENTIAL DAMAGES Except for damages mutually agreed upon by the Parties as liquidated damages in Paragraph 6.5 and excluding losses covered by insurance required by the Contract Documents, the Owner and the Contractor agree to waive all claims against each other for any consequential damages that may arise out of or relate to this Agreement, except for those specific items of damages excluded from this waiver as mutually agreed upon by the Parties and identified below. The Owner agrees to waive damages including but not limited to the Owner's loss of use of the Project, any rental expenses incurred, loss of income, profit or financing related to the Project, as well as the loss of business, loss of financing, principal office overhead and expenses, loss of profits not related to this Project, loss of reputation, or insolvency. The Contractor agrees to waive damages including but not limited to loss of business, loss of financing, principal office overhead and expenses, loss of profits not related to this Project, loss of bonding capacity, loss of reputation, or insolvency. The following items of damages are excluded from this mutual waiver:

18

6.6.1 The provisions of this Paragraph shall also apply to the termination of this Agreement and shall survive such termination. The Owner and the Contractor shall require similar waivers in contracts with Subcontractors and Others retained for the project.

ARTICLE 7

CONTRACT PRICE

7.1 LUMP SUM As full compensation for performance by the Contractor of the Work in conformance with the Contract Documents, the Owner shall pay the Contractor the lump sum price of _____ Dollars ($_____). The lump sum price is hereinafter referred to as the Contract Price, which shall be subject to increase or decrease as provided in Article 8.

7.2 ALLOWANCES

7.2.1 All allowances stated in the Contract Documents shall be included in the Contract Price. While the Owner may direct the amounts of, and particular material suppliers or subcontractors for, specific allowance items, if the Contractor reasonably objects to a material supplier or subcontractor, it shall not be required to contract with them. The Owner shall select allowance items in a timely manner so as not to delay the Work.

7.2.2 Allowances shall include the costs of materials, supplies and equipment delivered to the Worksite, less applicable trade discounts and including requisite taxes, unloading and handling at the Worksite, and labor and installation, unless specifically stated otherwise. The Contractor's Overhead and profit for the allowances shall be included in the Contract Price, but not in the allowances. The Contract Price shall be adjusted by Change Order to reflect the actual costs when they are greater than or less than the allowances.

ARTICLE 8

CHANGES

Changes in the Work that are within the general scope of this Agreement shall be accomplished, without invalidating this Agreement, by Change Order and Interim Directed Change.

8.1 CHANGE ORDER

8.1.1 The Contractor may request or the Owner may order changes in the Work or the timing or sequencing of the Work that impacts the Contract Price or the Contract Time. All such changes in the Work that affect Contract Time or Contract Price shall be formalized in a Change Order. Any such requests for a change in the Contract Price or the Contract Time shall be processed in accordance with this Article 8.

8.1.2 The Owner and the Contractor shall negotiate in good faith an appropriate adjustment to the Contract Price or the Contract Time and shall conclude these negotiations as expeditiously as possible. Acceptance of the Change Order and any adjustment in the Contract Price or Contract Time shall not be unreasonably withheld.

8.2 INTERIM DIRECTED CHANGE

8.2.1 The Owner may issue a written Interim Directed Change directing a change in the Work prior to reaching agreement with the Contractor on the adjustment, if any, in the Contract Price or the Contract Time.

19

8.2.2 The Owner and the Contractor shall negotiate expeditiously and in good faith for appropriate adjustments, as applicable, to the Contract Price or the Contract Time arising out of an Interim Directed Change. As the Changed Work is performed, the Contractor shall submit its costs for such work with its application for payment beginning with the next application for payment within thirty (30) Days of the issuance of the Interim Directed Change. If there is a dispute as to the cost to the Owner, the Owner shall pay the Contractor fifty percent (50%) of its estimated cost to perform the work. In such event, the Parties reserve their rights as to the disputed amount, subject to the requirements of Article 12.

8.2.3 When the Owner and the Contractor agree upon the adjustment in the Contract Price or the Contract Time, for a change in the Work directed by an Interim Directed Change, such agreement shall be the subject of a Change Order. The Change Order shall include all outstanding Interim Directed Changes on which the Owner and Contractor have reached agreement on Contract Price or Contract Time issued since the last Change Order.

8.3 DETERMINATION OF COST

8.3.1 An increase or decrease in the Contract Price or the Contract Time resulting from a change in the Work shall be determined by one or more of the following methods:

8.3.1.1 unit prices set forth in this Agreement or as subsequently agreed;

8.3.1.2 a mutually accepted, itemized lump sum;

8.3.1.3 costs calculated on a basis agreed upon by the Owner and Contractor plus _____ % Overhead and _____ % profit, or

8.3.1.4 if an increase or decrease cannot be agreed to as set forth in Clauses .1 through .3 above, and the Owner issues an Interim Directed Change, the cost of the change in the Work shall be determined by the reasonable actual expense and savings of the performance of the Work resulting from the change. If there is a net increase in the Contract Price, the Contractor's Overhead and profit shall be adjusted accordingly. In case of a net decrease in the Contract Price, the Contractor's Overhead and profit shall not be adjusted unless ten percent (10%) or more of the Project is deleted. The Contractor shall maintain a documented, itemized accounting evidencing the expenses and savings.

8.3.2 If unit prices are set forth in the Contract Documents or are subsequently agreed to by the Parties, but the character or quantity of such unit items as originally contemplated is so different in a proposed Change Order that the original unit prices will cause substantial inequity to the Owner or the Contractor, such unit prices shall be equitably adjusted.

8.3.3 If the Owner and the Contractor disagree as to whether work required by the Owner is within the scope of the Work, the Contractor shall furnish the Owner with an estimate of the costs to perform the disputed work in accordance with the Owner's interpretations. If the Owner issues a written order for the Contractor to proceed, the Contractor shall perform the disputed work and the Owner shall pay the Contractor fifty percent (50%) of its estimated cost to perform the work. In such event, both Parties reserve their rights as to whether the work was within the scope of the Work, subject to the requirements of Article 12. The Owner's payment does not prejudice its right to be reimbursed should it be determined that the disputed work was within the scope of Work. The Contractor's receipt of payment for the disputed work does not prejudice its right to receive full payment for the disputed work should it be determined that the disputed work is not within the scope of the Work.

8.4 CLAIMS FOR ADDITIONAL COST OR TIME Except as provided in Subparagraph 6.3.2 and Paragraph 6.4 for any claim for an increase in the Contract Price or the Contract Time, the Contractor shall give the Owner written notice of the claim within fourteen (14) Days after the occurrence giving rise to the claim or within fourteen (14) Days after the Contractor first recognizes the condition giving rise to the claim, whichever is later. Except in an emergency, notice shall be given before proceeding with the Work. Thereafter, the Contractor shall submit written documentation of its claim, including appropriate supporting documentation, within twenty-one (21) Days after giving notice, unless the Parties mutually agree upon a longer period of time. The Owner shall respond in writing denying or approving the Contractor's claim no later than fourteen (14) Days after receipt of the Contractor's claim. Any change in the Contract Price or the Contract Time resulting from such claim shall be authorized by Change Order.

ARTICLE 9

PAYMENT

9.1 SCHEDULE OF VALUES Within twenty-one (21) Days from the date of execution of this Agreement, the Contractor shall prepare and submit to the Owner, and if directed, the Architect/Engineer, a schedule of values apportioned to the various divisions or phases of the Work. Each line item contained in the schedule of values shall be assigned a value such that the total of all items shall equal the Contract Price.

9.2 PROGRESS PAYMENTS

9.2.1 APPLICATIONS The Contractor shall submit to the Owner and the Architect/Engineer a monthly application for payment no later than the _____ Day of the calendar month for the preceding thirty (30) Days. Contractor's applications for payment shall be itemized and supported by the Contractor's schedule of values and any other substantiating data as required by this Agreement. Payment applications shall include payment requests on account of properly authorized Change Orders or Interim Directed Change. The Owner shall pay the amount otherwise due on any payment application, as certified by the Architect/Engineer, no later than twenty (20) Days after the Contractor has submitted a complete and accurate payment application, or such shorter time period as required by applicable state statute. The Owner may deduct from any progress payment amounts as may be retained pursuant to Subparagraph 9.2.4.

9.2.2 STORED MATERIALS AND EQUIPMENT Unless otherwise provided in the Contract Documents, applications for payment may include materials and equipment not yet incorporated into the Work but delivered to and suitably stored onsite or offsite including applicable insurance, storage and costs incurred transporting the materials to an offsite storage facility. Approval of payment applications for stored materials and equipment stored offsite shall be conditioned on submission by the Contractor of bills of sale and proof of required insurance, or such other procedures satisfactory to the Owner to establish the proper valuation of the stored materials and equipment, the Owner's title to such materials and equipment, and to otherwise protect the Owner's interests therein, including transportation to the site.

9.2.3 LIEN WAIVERS AND LIENS

9.2.3.1 PARTIAL LIEN WAIVERS AND AFFIDAVITS If required by the Owner, as a prerequisite for payment, the Contractor shall provide partial lien and claim waivers in the amount of the application for payment and affidavits from its Subcontractors, and Material Suppliers for the completed Work. Such waivers shall be conditional upon payment. In no event shall the Contractor be required to sign an unconditional waiver of lien or claim, either partial or final, prior to receiving payment or in an amount in excess of what it has been paid.

21

9.2.3.2 RESPONSIBILITY FOR LIENS If Owner has made payments in the time required by this Article 9, the Contractor shall, within thirty (30) Days after filing, cause the removal of any liens filed against the premises or public improvement fund by any party or parties performing labor or services or supplying materials in connection with the Work. If the Contractor fails to take such action on a lien, the Owner may cause the lien to be removed at the Contractor's expense, including bond costs and reasonable attorneys' fees. This Clause shall not apply if there is a dispute pursuant to Article 12 relating to the subject matter of the lien.

9.2.4 RETAINAGE From each progress payment made prior to Substantial Completion the Owner may retain _____ percent (_____ %) of the amount otherwise due after deduction of any amounts as provided in Paragraph 9.3, and in no event shall such percentage exceed any applicable statutory requirements. If the Owner chooses to use this retainage provision:

9.2.4.1 after the Work is fifty percent (50%) complete, the Owner shall withhold no additional retainage and shall pay the Contractor the full amount of what is due on account of progress payments;

9.2.4.2 the Owner may, in its sole discretion, reduce the amount to be retained at any time;

9.2.4.3 the Owner may release retainage on that portion of the Work a Subcontractor has completed in whole or in part, and which the Owner has accepted.

In lieu of retainage, the Contractor may furnish a retention bond or other security interest, acceptable to the Owner, to be held by the Owner.

9.3 ADJUSTMENT OF CONTRACTOR'S PAYMENT APPLICATION The Owner may adjust or reject a payment application or nullify a previously approved payment application, in whole or in part, as may reasonably be necessary to protect the Owner from loss or damage based upon the following, to the extent that the Contractor is responsible therefor under this Agreement:

9.3.1 the Contractor's repeated failure to perform the Work as required by the Contract Documents;

9.3.2 loss or damage arising out of or relating to this Agreement and caused by the Contractor to the Owner or to Others to whom the Owner may be liable;

9.3.3 the Contractor's failure to properly pay Subcontractors and Material Suppliers following receipt of such payment from the Owner;

9.3.4 rejected, nonconforming or defective Work not corrected in a timely fashion;

9.3.5 reasonable evidence of delay in performance of the Work such that the Work will not be completed within the Contract Time, and

9.3.6 reasonable evidence demonstrating that the unpaid balance of the Contract Price is insufficient to fund the cost to complete the Work.

9.3.7 third party claims involving the Contractor or reasonable evidence demonstrating that third party claims are likely to be filed unless and until the Contractor furnishes the Owner with adequate security in the form of a surety bond, letter of credit or other collateral or commitment which are sufficient to discharge such claims if established.

No later than seven (7) Days after receipt of an application for payment, the Owner shall give written notice to the Contractor, at the time of disapproving or nullifying all or part of an application for payment, stating its specific reasons for such disapproval or nullification, and the remedial actions to be taken by the Contractor in order to receive payment. When the above reasons for disapproving or nullifying an

22

application for payment are removed, payment will be promptly made for the amount previously withheld.

9.4 ACCEPTANCE OF WORK Neither the Owner's payment of progress payments nor its partial or full use or occupancy of the Project constitutes acceptance of Work not complying with the Contract Documents.

9.5 PAYMENT DELAY If for any reason not the fault of the Contractor the Contractor does not receive a progress payment from the Owner within seven (7) Days after the time such payment is due, as defined in Subparagraph 9.2.1, then the Contractor, upon giving seven (7) Days' written notice to the Owner, and without prejudice to and in addition to any other legal remedies, may stop Work until payment of the full amount owing to the Contractor has been received, including interest from the date payment was due in accordance with Paragraph 9.9. The Contract Price and Contract Time shall be equitably adjusted by a Change Order for reasonable cost and delay resulting from shutdown, delay and start-up.

9.6 SUBSTANTIAL COMPLETION

9.6.1 The Contractor shall notify the Owner and, if directed, its Architect/Engineer when it considers Substantial Completion of the Work or a designated portion to have been achieved. The Owner, with the assistance of its Architect/Engineer, shall promptly conduct an inspection to determine whether the Work or designated portion can be occupied or utilized for its intended use by the Owner without excessive interference in completing any remaining unfinished Work by the Contractor. If the Owner determines that the Work or designated portion has not reached Substantial Completion, the Owner shall promptly compile a list of items to be completed or corrected so the Owner may occupy or utilize the Work or designated portion for its intended use. The Contractor shall promptly complete all items on the list.

9.6.2 When Substantial Completion of the Work or a designated portion is achieved, the Contractor shall prepare a Certificate of Substantial Completion that shall establish the date of Substantial Completion, and the respective responsibilities of the Owner and Contractor for interim items such as security, maintenance, utilities, insurance and damage to the Work. In the absence of a clear delineation of responsibilities, the Owner shall assume all responsibilities for items such as security, maintenance, utilities, insurance, and damage to the Work. The certificate shall also list the items to be completed or corrected, and establish the time for their completion or correction. The Certificate of Substantial Completion shall be submitted by the Contractor to the Owner for written acceptance of responsibilities assigned in the Certificate.

9.6.3 Unless otherwise provided in the Certificate of Substantial Completion, warranties required by the Contract Documents shall commence on the date of Substantial Completion of the Work or a designated portion.

9.6.4 Upon acceptance by the Owner of the Certificate of Substantial Completion, the Owner shall pay to the Contractor the remaining retainage held by the Owner for the Work described in the Certificate of Substantial Completion less a sum equal to two hundred percent (200%) of the estimated cost of completing or correcting remaining items on that part of the Work, as agreed to by the Owner and Contractor as necessary to achieve final completion. Uncompleted items shall be completed by the Contractor in a mutually agreed upon timeframe. The Owner shall pay the Contractor monthly the amount retained for unfinished items as each item is completed.

9.7 PARTIAL OCCUPANCY OR USE

9.7.1 The Owner may occupy or use completed or partially completed portions of the Work when (a) the portion of the Work is designated in a Certificate of Substantial Completion, (b) appropriate insurer(s) consent to the occupancy or use, and (c) appropriate public authorities authorize the

IMPORTANT: A vertical line in the margin indicates a change has been made to the original text. Prior to signing, recipients may wish to request from the party producing the document a "redlined" version indicating changes to the original text. Consultation with legal and insurance counsel and careful review of the entire document are strongly encouraged.
ConsensusDOCS 200 • STANDARD AGREEMENT AND GENERAL CONDITIONS BETWEEN OWNER AND CONTRACTOR (Where the Contract Price is a Lump Sum) Copyright © 2007, ConsensusDOCS LLC. YOU ARE ALLOWED TO USE THIS DOCUMENT FOR ONE CONTRACT ONLY. YOU MAY MAKE 9 COPIES OF THE COMPLETED DOCUMENT FOR DISTRIBUTION TO THE CONTRACT'S PARTIES. ANY OTHER USES, INCLUDING COPYING THE FORM DOCUMENT, ARE STRICTLY PROHIBITED.

occupancy or use. Such partial occupancy or use shall constitute Substantial Completion of that portion of the Work.

9.8 FINAL COMPLETION AND FINAL PAYMENT

9.8.1 Upon notification from the Contractor that the Work is complete and ready for final inspection and acceptance, the Owner with the assistance of its Architect/Engineer shall promptly conduct an inspection to determine if the Work has been completed and is acceptable under the Contract Documents.

9.8.2 When Final Completion has been achieved, the Contractor shall prepare for the Owner's acceptance a final application for payment stating that to the best of the Contractor's knowledge, and based on the Owner's inspections, the Work has reached Final Completion in accordance with the Contract Documents.

9.8.3 Final payment of the balance of the Contract Price shall be made to the Contractor within twenty (20) Days after the Contractor has submitted a complete and accurate application for final payment, including submissions required under Subparagraph 9.8.4, and a Certificate of Final Completion has been executed by the Owner and the Contractor.

9.8.4 Final payment shall be due on the Contractor's submission of the following to the Owner:

9.8.4.1 an affidavit declaring any indebtedness connected with the Work, e.g. payrolls or invoices for materials or equipment, to have been paid, satisfied or to be paid with the proceeds of final payment, so as not to encumber the Owner's property;

9.8.4.2 as-built drawings, manuals, copies of warranties and all other close-out documents required by the Contract Documents;

9.8.4.3 release of any liens, conditioned on final payment being received;

9.8.4.4 consent of any surety; and

9.8.4.5 any outstanding known and unreported accidents or injuries experienced by the Contractor or its Subcontractors at the Worksite.

9.8.5 If, after Substantial Completion of the Work, the Final Completion of a portion of the Work is materially delayed through no fault of the Contractor, the Owner shall pay the balance due for portion(s) of the Work fully completed and accepted. If the remaining contract balance for Work not fully completed and accepted is less than the retained amount prior to payment, the Contractor shall submit to the Owner, and, if directed, the Architect/Engineer, the written consent of any surety to payment of the balance due for portions of the Work that are fully completed and accepted. Such payment shall not constitute a waiver of claims, but otherwise shall be governed by these final payment provisions.

9.8.6 OWNER RESERVATION OF CLAIMS Claims not reserved in writing by the Owner with the making of final payment shall be waived except for claims relating to liens or similar encumbrances, warranties, Defective Work and latent defects.

9.8.7 CONTRACTOR ACCEPTANCE OF FINAL PAYMENT Unless the Contractor provides written identification of unsettled claims known to the Contractor at the time of making application for final payment, acceptance of final payment constitutes a waiver of such claims.

9.9 LATE PAYMENT Payments due but unpaid shall bear interest from the date payment is due at the statutory rate at the place of the Project.

ARTICLE 10

INDEMNITY, INSURANCE, WAIVERS AND BONDS

10.1 INDEMNITY

10.1.1 To the fullest extent permitted by law, the Contractor shall indemnify and hold harmless the Owner, the Owner's officers, directors, members, consultants, agents and employees, the Architect/Engineer and Others (the Indemnitees) from all claims for bodily injury and property damage, other than to the Work itself and other property insured under Subparagraph 10.3.1 including reasonable attorneys' fees, costs and expenses, that may arise from the performance of the Work, but only to the extent caused by the negligent acts or omissions of the Contractor, Subcontractors or anyone employed directly or indirectly by any of them or by anyone for whose acts any of them may be liable. The Contractor shall be entitled to reimbursement of any defense costs paid above Contractor's percentage of liability for the underlying claim to the extent provided for under Subparagraph 10.1.2.

10.1.2 To the fullest extent permitted by law, the Owner shall indemnify and hold harmless the Contractor, its officers, directors, members, consultants, agents, and employees, Subcontractors or anyone employed directly or indirectly by any of them or anyone for whose acts any of them may be liable from all claims for bodily injury and property damage, other than property insured under Subparagraph 10.3.1, including reasonable attorneys' fees, costs and expenses, that may arise from the performance of work by Owner, Architect/Engineer or Others, but only to the extent caused by the negligent acts or omissions of the Owner, Architect/Engineer or Others. The Owner shall be entitled to reimbursement of any defense costs paid above Owner's percentage of liability for the underlying claim to the extent provided for under Subparagraph 10.1.1.

10.1.3 NO LIMITATION ON LIABILITY In any and all claims against the Indemnitees by any employee of the Contractor, anyone directly or indirectly employed by the Contractor or anyone for whose acts the Contractor may be liable, the indemnification obligation shall not be limited in any way by any limitation on the amount or type of damages, compensation or benefits payable by or for the Contractor under Workers' Compensation acts, disability benefit acts or other employment benefit acts.

10.2 INSURANCE

10.2.1 Prior to the start of the Work, the Contractor shall procure and maintain in force Workers' Compensation Insurance, Employers' Liability Insurance, Business Automobile Liability Insurance, and Commercial General Liability Insurance (CGL). The CGL policy shall include coverage for liability arising from premises, operations, independent contractors, products-completed operations, personal injury and advertising injury, contractual liability, and broad form property damage. The Contractor's Employers' Liability, Business Automobile Liability, and Commercial General Liability policies, as required in this Subparagraph 10.2.1, shall be written with at least the following limits of liability:

    10.2.1.1 Employers' Liability Insurance

        a. $_____

            Bodily Injury by Accident

            Each Accident

        b. $_____

Bodily Injury by Disease

Policy Limit

c. $ _____

Bodily Injury by Disease

Each Employee

10.2.1.2 Business Automobile Liability Insurance

a. $ _____

Each Accident

10.2.1.3 Commercial General Liability Insurance

a. $ _____

Each Occurrence

b. $ _____

General Aggregate

c. $ _____

Products/Completed

Operations Aggregate

d. $ _____

Personal and Advertising

Injury Limit

10.2.2 Employers' Liability, Business Automobile Liability and Commercial General Liability coverage required under Subparagraph 10.2.1 may be arranged under a single policy for the full limits required or by a combination of underlying policies with the balance provided by Excess or Umbrella Liability policies.

10.2.3 The Contractor shall maintain in effect all insurance coverage required under Subparagraph 10.2.1 with insurance companies lawfully authorized to do business in the jurisdiction in which the Project is located. If the Contractor fails to obtain or maintain any insurance coverage required under this Agreement, the Owner may purchase such coverage and charge the expense to the Contractor, or terminate this Agreement.

10.2.4 The policies of insurance required under Subparagraph 10.2.1 shall contain a provision that the coverage afforded under the policies shall not be cancelled or allowed to expire until at least thirty (30) Days' prior written notice has been given to the Owner. The Contractor shall maintain completed operations liability insurance for one year after acceptance of the Work, Substantial Completion of the Project, or to the time required by the Contract Documents, whichever is longer. Prior to commencement of the Work, Contractor shall furnish the Owner with certificates evidencing the required coverage.

10.3 PROPERTY INSURANCE

26

10.3.1 Before the start of Work, the Owner shall obtain and maintain a Builder's Risk Policy upon the entire Project for the full cost of replacement at the time of loss. This insurance shall also name the Contractor, Subcontractors, Sub-subcontractors, Material Suppliers and Architect/Engineer as insureds. This insurance shall be written as a Builder's Risk Policy or equivalent form to cover all risks of physical loss except those specifically excluded by the policy, and shall insure at least against the perils of fire, lightning, explosion, windstorm, hail, smoke, aircraft (except aircraft, including helicopter, operated by or on behalf of Contractor) and vehicles, riot and civil commotion, theft, vandalism, malicious mischief, debris removal, flood, earthquake, earth movement, water damage, wind damage, testing if applicable, collapse however caused, and damage resulting from defective design, workmanship or material, and material or equipment stored offsite, onsite or in transit. The Owner shall be solely responsible for any deductible amounts or coinsurance penalties. This policy shall provide for a waiver of subrogation in favor of the Contractor, Subcontractors, Sub-subcontractors, Material Suppliers and Architect/Engineer. This insurance shall remain in effect until final payment has been made or until no person or entity other than the Owner has an insurable interest in the property to be covered by this insurance, whichever is sooner. Partial occupancy or use of the Work shall not commence until the Owner has secured the consent of the insurance company or companies providing the coverage required in this Subparagraph 10.3.1. Prior to commencement of the Work, the Owner shall provide a copy of the property policy or policies obtained in compliance with this Subparagraph 10.3.1.

10.3.2 If the Owner does not intend to purchase the property insurance required by this Agreement, including all of the coverages and deductibles described herein, the Owner shall give written notice to the Contractor and the Architect/Engineer before the Work is commenced. The Contractor may then provide insurance to protect its interests and the interests of the Subcontractors and Sub-subcontractors, including the coverage of deductibles. The cost of this insurance shall be charged to the Owner in a Change Order. The Owner shall be responsible for all of Contractor's costs reasonably attributed to the Owner's failure or neglect in purchasing or maintaining the coverage described above.

10.3.2.1 If the Owner does not obtain insurance to cover the risk of physical loss resulting from Terrorism, the Owner shall give written notice to the Contractor before the Work commences. The Contractor may then provide insurance to protect its interests and the interests of the Subcontractors and Sub-subcontractors against such risk of loss, including the coverage of deductibles. The cost of this insurance shall be charged to the Owner in a Change Order.

10.3.3 Owner and Contractor waive all rights against each other and their respective employees, agents, contractors, subcontractors and sub-subcontractors, and the Architect/Engineer for damages caused by risks covered by the property insurance except such rights as they may have to the proceeds of the insurance and such rights as the Contractor may have for the failure of the Owner to obtain and maintain property insurance in compliance with Subparagraph 10.3.1.

10.3.4 To the extent of the limits of Contractor's Commercial General Liability Insurance specified in Subparagraph 10.2.1 or _____ Dollars ($_____), whichever is more, the Contractor shall indemnify and hold harmless the Owner against any and all liability, claims, demands, damages, losses and expenses, including attorneys' fees, in connection with or arising out of any damage or alleged damage to any of Owner's existing adjacent property that may arise from the performance of the Work, to the extent caused by the negligent acts or omissions of the Contractor, Subcontractor or anyone employed directly or indirectly by any of them or by anyone for whose acts any of them may be liable.

27

10.3.5 RISK OF LOSS Except to the extent a loss is covered by applicable insurance, risk of loss or damage to the Work shall be upon the Contractor until the Date of Substantial Completion, unless otherwise agreed to by the Parties.

10.4 OWNER'S INSURANCE

10.4.1 BUSINESS INCOME INSURANCE The Owner may procure and maintain insurance against loss of use of the Owner's property caused by fire or other casualty loss.

10.4.2 OWNER'S LIABILITY INSURANCE The Owner shall either self-insure or obtain and maintain its own liability insurance for protection against claims arising out of the performance of this Agreement, including without limitation, loss of use and claims, losses and expenses arising out of the Owner's errors or omissions.

10.5 ADDITIONAL LIABILITY COVERAGE

10.5.1 The Owner _____ shall/_____ shall not (indicate one) require Contractor to purchase and maintain liability coverage, primary to Owner's coverage under Subparagraph 10.4.2.

10.5.2 If required by Subparagraph 10.5.1, the additional liability coverage required of the Contractor shall be

(Designate required coverage(s)):

_____ .1 Additional Insured. Owner shall be named as an additional insured on Contractor's Commercial General Liability Insurance specified for operations and completed operations, but only with respect to liability for bodily injury, property damage or personal and advertising injury to the extent caused by the negligent acts or omissions of Contractor, or those acting on Contractor's behalf, in the performance of Contractor's Work for Owner at the Worksite.

_____ .2 OCP. Contractor shall provide an Owners' and Contractors' Protective Liability Insurance ("OCP") policy with limits equal to the limits on Commercial General Liability Insurance specified, or limits as otherwise required by Owner.

Any documented additional cost in the form of a surcharge associated with procuring the additional liability coverage in accordance with this Subparagraph shall be paid by the Owner directly or the costs may be reimbursed by Owner to Contractor by increasing the Contract Price to correspond to the actual cost required to purchase and maintain the additional liability coverage. Prior to commencement of the Work, Contractor shall obtain and furnish to the Owner a certificate evidencing that the additional liability coverages have been procured.

10.6 ROYALTIES, PATENTS AND COPYRIGHTS The Contractor shall pay all royalties and license fees which may be due on the inclusion of any patented or copyrighted materials, methods or systems selected by the Contractor and incorporated in the Work. The Contractor shall defend, indemnify and hold the Owner harmless from all suits or claims for infringement of any patent rights or copyrights arising out of such selection. The Owner agrees to defend, indemnify and hold the Contractor harmless from any suits or claims of infringement of any patent rights or copyrights arising out of any patented or copyrighted materials, methods or systems specified by the Owner and Architect/Engineer.

10.7 BONDS

10.7.1 Performance and Payment Bonds

(Mark one only)

are _____ / are not _____

required of the Contractor. Such bonds shall be issued by a surety admitted in the State in which the Project is located and must be acceptable to the Owner. Owner's acceptance shall not be withheld without reasonable cause. The penal sum of the Payment Bond and of the Performance Bond shall each be 100% of the original Contract Price. Any increase in the Contract Price that exceeds 10% in the aggregate shall require a rider to the Bonds increasing penal sums accordingly. Up to such 10% amount, the penal sum of the bond shall remain equal to 100% of the Contract Price. The Contractor shall endeavor to keep its surety advised of changes potentially impacting the Contract Time and Contract Price, though the Contractor shall require that its surety waives any requirement to be notified of any alteration or extension of time. The Contractor's Payment Bond for the Project, if any, shall be made available by the Owner for review and copying by the Subcontractor.

10.8 PROFESSIONAL LIABILITY INSURANCE To the extent the Contractor is required to procure design services under this Agreement, in accordance with Paragraph 3.15, the Contractor shall require the designers to obtain professional liability insurance for claims arising from the negligent performance of professional services under this Agreement, with a company reasonably satisfactory to the Owner, including coverage for all professional liability caused by any of the Designer's(s') consultants, written for not less than $_____ per claim and in the aggregate with the deductible not to exceed $_____. The deductible shall be paid by the Designer.

ARTICLE 11

SUSPENSION, NOTICE TO CURE AND TERMINATION OF THE AGREEMENT

11.1 SUSPENSION BY OWNER FOR CONVENIENCE

11.1.1 OWNER SUSPENSION Should the Owner order the Contractor in writing to suspend, delay, or interrupt the performance of the Work for such period of time as may be determined to be appropriate for the convenience of the Owner and not due to any act or omission of the Contractor or any person or entity for whose acts or omissions the Contractor may be liable, then the Contractor shall immediately suspend, delay or interrupt that portion of the Work as ordered by the Owner. The Contract Price and the Contract Time shall be equitably adjusted by Change Order for the cost and delay resulting from any such suspension.

11.1.2 Any action taken by the Owner that is permitted by any other provision of the Contract Documents and that results in a suspension of part or all of the Work does not constitute a suspension of Work under this Paragraph 11.1.

11.2 NOTICE TO CURE A DEFAULT If the Contractor persistently refuses or fails to supply enough properly skilled workers, proper materials, or equipment, to maintain the approved Schedule of the Work in accordance with Article 6, or fails to make prompt payment to its workers, Subcontractors or Material Suppliers, disregards laws, ordinances, rules, regulations or orders of any public authority having jurisdiction, or is otherwise guilty of a material breach of a provision of this Agreement, the Contractor may be deemed in default. If the Contractor fails within seven (7) Days after receipt of written notification to commence and continue satisfactory correction of such default with diligence and promptness, then the Owner shall give the Contractor a second notice to correct the default within a three (3) Days period. If the Contractor fails to promptly commence and continue satisfactory correction of the default following receipt of such second notice, the Owner without prejudice to any other rights or remedies may:

11.2.1 supply workers and materials, equipment and other facilities as the Owner deems necessary for the satisfactory correction of the default, and charge the cost to the Contractor, who shall be

---

liable for the payment of same including reasonable Overhead, profit and attorneys' fees;

11.2.2 contract with Others to perform such part of the Work as the Owner determines shall provide the most expeditious correction of the default, and charge the cost to the Contractor;

11.2.3 withhold payment due the Contractor in accordance with Paragraph 9.3; and

11.2.4 in the event of an emergency affecting the safety of persons or property, immediately commence and continue satisfactory correction of such default as provided in Subparagraphs 11.2.1 and 11.2.2 without first giving written notice to the Contractor, but shall give prompt written notice of such action to the Contractor following commencement of the action.

11.3 OWNER'S RIGHT TO TERMINATE FOR DEFAULT

11.3.1 TERMINATION BY OWNER FOR DEFAULT If, within seven (7) Days of receipt of a notice to cure pursuant to Paragraph 11.2, the Contractor fails to commence and satisfactorily continue correction of the default set forth in the notice to cure, the Owner may notify the Contractor that it intends to terminate this Agreement for default absent appropriate corrective action within fourteen (14) additional Days. After the expiration of the additional fourteen (14) Days period, the Owner may terminate this Agreement by written notice absent appropriate corrective action. Termination for default is in addition to any other remedies available to Owner under Paragraph 11.2. If the Owner's cost arising out of the Contractor's failure to cure, including the cost of completing the Work and reasonable attorneys' fees, exceeds the unpaid Contract Price, the Contractor shall be liable to the Owner for such excess costs. If the Owner's costs are less than the unpaid Contract Price, the Owner shall pay the difference to the Contractor. In the event the Owner exercises its rights under this Paragraph 11.3, upon the request of the Contractor the Owner shall furnish to the Contractor a detailed accounting of the cost incurred by the Owner.

11.3.2 USE OF CONTRACTOR'S MATERIALS, SUPPLIES AND EQUIPMENT If the Owner or Others perform work under this Paragraph 11.3, the Owner shall have the right to take and use any materials, supplies and equipment belonging to the Contractor and located at the Worksite for the purpose of completing any remaining Work. Immediately upon completion of the Work, any remaining materials, supplies or equipment not consumed or incorporated in the Work shall be returned to the Contractor in substantially the same condition as when they were taken, reasonable wear and tear excepted.

11.3.3 If the Contractor files a petition under the Bankruptcy Code, this Agreement shall terminate if the Contractor or the Contractor's trustee rejects the Agreement or, if there has been a default, the Contractor is unable to give adequate assurance that the Contractor will perform as required by this Agreement or otherwise is unable to comply with the requirements for assuming this Agreement under the applicable provisions of the Bankruptcy Code.

11.3.4 The Owner shall make reasonable efforts to mitigate damages arising from Contractor default, and shall promptly invoice the Contractor for all amounts due pursuant to Paragraphs 11.2 and 11.3.

11.4 TERMINATION BY OWNER FOR CONVENIENCE

11.4.1 Upon written notice to the Contractor, the Owner may, without cause, terminate this Agreement. The Contractor shall immediately stop the Work, follow the Owner's instructions regarding shutdown and termination procedures, and strive to minimize any further costs.

11.4.2 If the Owner terminates this Agreement pursuant to this Paragraph 11.4, the Contractor shall be paid:

11.4.2.1 for the Work performed to date including overhead and profit;

11.4.2.2 for all demobilization costs and costs incurred as a result of the termination but not including overhead or profit on work not performed;

11.4.2.3 and shall receive a premium as set forth in a schedule below. (Insert here the schedule agreed to by the Parties.)

11.4.3 If the Owner terminates this Agreement pursuant to Paragraphs 11.3 or 11.4, the Contractor shall:

11.4.3.1 execute and deliver to the Owner all papers and take all action required to assign, transfer and vest in the Owner the rights of the Contractor to all materials, supplies and equipment for which payment has or will be made in accordance with the Contract Documents and all Subcontracts, orders and commitments which have been made in accordance with the Contract Documents;

11.4.3.2 exert reasonable effort to reduce to a minimum the Owner's liability for subcontracts, orders and commitments that have not been fulfilled at the time of the termination;

11.4.3.3 cancel any Subcontracts, orders and commitments as the Owner directs; and

11.4.3.4 sell at prices approved by the Owner any materials, supplies and equipment as the Owner directs, with all proceeds paid or credited to the Owner.

11.5 CONTRACTOR'S RIGHT TO TERMINATE

11.5.1 Upon seven (7) Days' written notice to the Owner, the Contractor may terminate this Agreement if the Work has been stopped for a thirty (30) Days period through no fault of the Contractor for any of the following reasons:

11.5.1.1 under court order or order of other governmental authorities having jurisdiction;

11.5.1.2 as a result of the declaration of a national emergency or other governmental act during which, through no act or fault of the Contractor, materials are not available; or

11.5.1.3 suspension by Owner for convenience pursuant to Paragraph 11.1

11.5.2 In addition, upon seven (7) Days' written notice to the Owner, the Contractor may terminate the Agreement if the Owner:

11.5.2.1 fails to furnish reasonable evidence pursuant to Paragraph 4.2 that sufficient funds are available and committed for Project financing, or

11.5.2.2 assigns this Agreement over the Contractor's reasonable objection, or

11.5.2.3 fails to pay the Contractor in accordance with this Agreement and the Contractor has complied with the notice provisions of Paragraph 9.5, or

11.5.2.4 otherwise materially breaches this Agreement.

11.5.3 Upon termination by the Contractor in accordance with Paragraph 11.5, the Contractor shall be entitled to recover from the Owner payment for all Work executed and for any proven loss, cost or expense in connection with the Work, including all demobilization costs plus reasonable overhead and profit on Work not performed.

31

11.6 OBLIGATIONS ARISING BEFORE TERMINATION Even after termination pursuant to Article 11, the provisions of this Agreement still apply to any Work performed, payments made, events occurring, costs charged or incurred or obligations arising before the termination date.

ARTICLE 12

DISPUTE MITIGATION AND RESOLUTION

12.1 WORK CONTINUANCE AND PAYMENT Unless otherwise agreed in writing, the Contractor shall continue the Work and maintain the Schedule of the Work during any dispute mitigation or resolution proceedings. If the Contractor continues to perform, the Owner shall continue to make payments in accordance with this Agreement.

12.2 DIRECT DISCUSSIONS If the Parties cannot reach resolution on a matter relating to or arising out of the Agreement, the Parties shall endeavor to reach resolution through good faith direct discussions between the Parties' representatives, who shall possess the necessary authority to resolve such matter and who shall record the date of first discussions. If the Parties' representatives are not able to resolve such matter within five (5) business Days of the date of first discussion, the Parties' representatives shall immediately inform senior executives of the Parties in writing that resolution was not effected. Upon receipt of such notice, senior executives of the Parties shall meet within five (5) business Days to endeavor to reach resolution. If the dispute remains unresolved after fifteen (15) Days from the date of first discussion, the Parties shall submit such matter to the dispute mitigation and dispute resolution procedures selected herein.

12.3 MITIGATION If the Parties select one of the dispute mitigation procedures provided in this Paragraph 12.3, disputes remaining unresolved after direct discussions shall be directed to the selected mitigation procedure. The dispute mitigation procedure shall result in a nonbinding finding on the matter, which may be introduced as evidence at a subsequent binding adjudication of the matter, as designated in Paragraph 12.5. The Parties agree that the dispute mitigation procedure shall be

(Designate only one):

_____ Project Neutral

_____ Dispute Review Board

12.3.1 MITIGATION PROCEDURES The Project Neutral/Dispute Review Board shall be mutually selected and appointed by the Parties and shall execute a retainer agreement with the Parties establishing the scope of the Project Neutral/Dispute Review Board's responsibilities. The costs and expenses of the Project Neutral/Dispute Review Board shall be shared equally by the Parties. The Project Neutral/Dispute Review Board shall be available to either Party, upon request, throughout the course of the Project, and shall make regular visits to the Project so as to maintain an up-to-date understanding of the Project progress and issues and to enable the Project Neutral/Dispute Review Board to address matters in dispute between the Parties promptly and knowledgeably. The Project Neutral/Dispute Review Board shall issue nonbinding findings within five (5) business Days of referral of the matter to the Project Neutral, unless good cause is shown.

12.3.2 If the matter remains unresolved following the issuance of the nonbinding finding by the mitigation procedure or if the Project Neutral/Dispute Review Board fails to issue nonbinding findings within five (5) business Days of the referral, the Parties shall submit the matter to the binding dispute resolution procedure designated in Paragraph 12.5.

12.4 MEDIATION If direct discussions pursuant to Paragraph 12.2 do not result in resolution of the matter and no dispute mitigation procedure is selected under Paragraph 12.3, the Parties shall endeavor to resolve the matter by mediation through the current Construction Industry Mediation Rules of the American Arbitration Association, or the Parties may mutually agree to select another set of mediation rules. The administration of the mediation shall be as mutually agreed by the Parties. The mediation shall be convened within thirty (30) business Days of the matter first being discussed and shall conclude within forty-five (45) business Days of the matter first being discussed. Either Party may terminate the mediation at any time after the first session, but the decision to terminate shall be delivered in person by the terminating Party to the non-terminating Party and to the mediator. The costs of the mediation shall be shared equally by the Parties.

12.5 BINDING DISPUTE RESOLUTION If the matter is unresolved after submission of the matter to a mitigation procedure or to mediation, the Parties shall submit the matter to the binding dispute resolution procedure designated herein.

(Designate only one:)

_____ Arbitration using the current Construction Industry Arbitration Rules of the American Arbitration Association or the Parties may mutually agree to select another set of arbitration rules. The administration of the arbitration shall be as mutually agreed by the Parties.

_____ Litigation in either the state or federal court having jurisdiction of the matter in the location of the Project.

12.5.1 The costs of any binding dispute resolution procedures shall be borne by the non-prevailing Party, as determined by the adjudicator of the dispute.

12.5.2 VENUE The venue of any binding dispute resolution procedure shall be the location of the Project, unless the Parties agree on a mutually convenient location.

12.6 MULTIPARTY PROCEEDING All parties necessary to resolve a matter shall be parties to the same dispute resolution procedure. Appropriate provisions shall be included in all other contracts relating to the Work to provide for the joinder or consolidation of such dispute resolution procedures.

12.7 LIEN RIGHTS Nothing in this Article 12 shall limit any rights or remedies not expressly waived by the Contractor that the Contractor may have under lien laws.

ARTICLE 13

MISCELLANEOUS PROVISIONS

13.1 ASSIGNMENT Neither the Owner nor the Contractor shall assign their interest in this Agreement without the written consent of the other except as to the assignment of proceeds. The terms and conditions of this Agreement shall be binding upon both Parties, their partners, successors, assigns and legal representatives. Neither Party to this Agreement shall assign the Agreement as a whole without written consent of the other except that the Owner may assign the Agreement to a wholly owned subsidiary of Owner when Owner has fully indemnified Contractor or to an institutional lender providing construction financing for the Project as long as the assignment is no less favorable to the Contractor than this Agreement. In the event of such assignment, the Contractor shall execute any consents reasonably required. In such event, the wholly-owned subsidiary or lender shall assume the Owner's rights and obligations under the Contract Documents. If either Party attempts to make such an assignment, that Party shall nevertheless remain legally responsible for all obligations under this Agreement, unless otherwise agreed by the other Party.

13.2 GOVERNING LAW This Agreement shall be governed by the law in effect at the location of the

Project.

13.3 SEVERABILITY The partial or complete invalidity of any one or more provisions of this Agreement shall not affect the validity or continuing force and effect of any other provision.

13.4 NO WAIVER OF PERFORMANCE The failure of either Party to insist, in any one or more instances, on the performance of any of the terms, covenants or conditions of this Agreement, or to exercise any of its rights, shall not be construed as a waiver or relinquishment of such term, covenant, condition or right with respect to further performance or any other term, covenant, condition or right.

13.5 TITLES AND GROUPINGS The titles given to the articles of this Agreement are for ease of reference only and shall not be relied upon or cited for any other purpose. The grouping of the articles in this Agreement and of the Owner's specifications under the various headings is solely for the purpose of convenient organization and in no event shall the grouping of provisions, the use of paragraphs or the use of headings be construed to limit or alter the meaning of any provisions.

13.6 JOINT DRAFTING The Parties expressly agree that this Agreement was jointly drafted, and that both had opportunity to negotiate its terms and to obtain the assistance of counsel in reviewing its terms prior to execution. Therefore, this Agreement shall be construed neither against nor in favor of either Party, but shall be construed in a neutral manner.

13.7 RIGHTS AND REMEDIES The Parties' rights, liabilities, responsibilities and remedies with respect to this Agreement, whether in contract, tort, negligence or otherwise, shall be exclusively those expressly set forth in this Agreement.

13.8 OTHER PROVISIONS

ARTICLE 14

CONTRACT DOCUMENTS

14.1 The Contract Documents in existence at the time of execution of this Agreement are as follows:

14.2 INTERPRETATION OF CONTRACT DOCUMENTS

14.2.1 The drawings and specifications are complementary. If Work is shown only on one but not on the other, the Contractor shall perform the Work as though fully described on both consistent with the Contract Documents and reasonably inferable from them as being necessary to produce the indicated results.

14.2.2 In case of conflicts between the drawings and specifications, the specifications shall govern. In any case of omissions or errors in figures, drawings or specifications, the Contractor shall immediately submit the matter to the Owner for clarification. The Owner's clarifications are final and binding on all Parties, subject to an equitable adjustment in Contract Time or Price pursuant to Articles 6 and 7 or dispute resolution in accordance with Article 12.

14.2.3 Where figures are given, they shall be preferred to scaled dimensions.

14.2.4 Any terms that have well-known technical or trade meanings, unless otherwise specifically defined in this Agreement, shall be interpreted in accordance with their well-known meanings.

14.2.5 In case of any inconsistency, conflict or ambiguity among the Contract Documents, the

34

documents shall govern in the following order: (a) Change Orders and written amendments to this Agreement; (b) this Agreement; (c) subject to Subparagraph 14.2.2 the drawings (large scale governing over small scale), specifications and addenda issued prior to the execution of this Agreement; (d) approved submittals; (e) information furnished by the Owner pursuant to Paragraph 4.3; (f) other documents listed in this Agreement. Among categories of documents having the same order of precedence, the term or provision that includes the latest date shall control. Information identified in one Contract Document and not identified in another shall not be considered a conflict or inconsistency.

This Agreement is entered into as of the date entered in Article 1.

ATTEST: ....

OWNER: _____

BY: ....

PRINT NAME _____

PRINT TITLE _____

ATTEST: ....

CONTRACTOR: _____

BY: ....

PRINT NAME _____

PRINT TITLE _____

35

ConsensusDOCS is proudly endorsed by the following

CONSENSUSDOCS 750

STANDARD FORM OF AGREEMENT BETWEEN CONTRACTOR AND SUBCONTRACTOR

This document was developed through a collaborative effort of entities representing a wide cross-section of the construction industry. The organizations endorsing this document believe it represents a fair and reasonable consensus among the collaborating parties of allocation of risk and responsibilities in an effort to appropriately balance the critical interests and concerns of all project participants.

These endorsing organizations recognize and understand that users of this document must review and adapt this document to meet their particular needs, the specific requirements of the project, and applicable laws. Users are encouraged to consult legal, insurance and surety advisors before modifying or completing this document. Further information on this document and the perspectives of endorsing organizations is available in the ConsensusDOCS Guidebook.

TABLE OF ARTICLES

1

7. CHANGES IN THE SUBCONTRACT WORK

8. PAYMENT

9. INDEMNITY, INSURANCE AND WAIVER OF SUBROGATION

10. CONTRACTOR'S RIGHT TO PERFORM SUBCONTRACTOR'S RESPONSIBILITIES AND TERMINATION OF AGREEMENT

11. DISPUTE RESOLUTION

12. MISCELLANEOUS PROVISIONS

13. EXISTING SUBCONTRACT DOCUMENTS

This Agreement has important legal and insurance consequences. Consultations with an attorney and with insurance and surety consultants are encouraged with respect to its completion or modification. Notes indicate where information is to be inserted to complete this Agreement.

ARTICLE 1
AGREEMENT

This Agreement is made this _____ Day of _____ in the year _____, by and between the

CONTRACTOR

and the

SUBCONTRACTOR

for services in connection with the

SUBCONTRACT WORK

for the following

PROJECT

whose

OWNER is

The ARCHITECT/ENGINEER for the Project is

Notice to the Parties shall be given at the above addresses.

2

ARTICLE 2

SCOPE OF WORK

2.1 SUBCONTRACT WORK   The Contractor contracts with the Subcontractor as an independent contractor to provide all labor, materials, equipment and services necessary or incidental to complete the work for the project described in Article 1 and as may be set forth in further detail in Exhibit A, in accordance with, and reasonably inferable from, that which is indicated in the Subcontract Documents, and consistent with the Progress Schedule, as may change from time to time. The Subcontractor shall perform the Subcontract Work under the general direction of the Contractor and in accordance with the Subcontract Documents.

2.2 CONTRACTOR'S WORK   The Contractor's Work is the construction and services required of the Contractor to fulfill its obligations pursuant to its agreement with the Owner (the Work). The Subcontract Work is a portion of the Contractor's Work.

2.2.1   The Contractor and the Subcontractor shall perform their obligations with integrity, ensuring at a minimum that:

2.2.1.1   Conflicts of interest shall be avoided or disclosed promptly to the other Party; and

2.2.1.2   The Contractor and the Subcontractor warrant that they have not and shall not pay nor receive any contingent fees or gratuities to or from the other Party, including their agents, officers and employees, Subcontractors or others for whom they may be liable, to secure preferential treatment.

2.3 SUBCONTRACT DOCUMENTS   The Subcontract Documents include this Agreement, the Owner-Contractor agreement, special conditions, general conditions, specifications, drawings, addenda, Subcontract Change Orders, approved submittals, amendments and any pending and exercised alternates. The Contractor shall provide to the Subcontractor, prior to the execution of this Agreement, copies of the existing Subcontract Documents to which the Subcontractor will be bound. The Subcontractor similarly shall provide copies of applicable portions of the Subcontract Documents to its proposed subcontractors and suppliers. Nothing shall prohibit the Subcontractor from obtaining copies of the Subcontract Documents from the Contractor at any time after the Subcontract Agreement is executed. The Subcontract Documents existing at the time of the execution of this Agreement are listed in Article 13.

2.3.1   ELECTRONIC DOCUMENTS   If the Owner requires that the Owner, Architect/Engineer, Contractor and Subcontractors exchange documents and data in electronic or digital form, prior to any such exchange, the Owner, Architect/Engineer and Contractor shall agree in ConsensusDOCS 200.2 or a written protocol governing all exchanges, which, at a minimum, shall specify: (1) the definition of documents and data to be accepted in electronic or digital form or to be transmitted electronically or digitally; (2) management and coordination responsibilities; (3) necessary equipment, software and services; (4) acceptable formats, transmission methods and verification procedures; (5) methods for maintaining version control; (6) privacy and security requirements; and (7) storage and retrieval requirements. The Subcontractor shall provide whatever input is needed to assist the Contractor in developing the protocol and shall be bound by the requirements of the written protocol. Except as otherwise agreed to by the Parties in writing, the Parties shall each bear their own costs as identified in the protocol. In the absence of a written protocol, use of documents and data in electronic or digital form shall be at the sole risk of the recipient.

2.4 CONFLICTS   In the event of a conflict between this Agreement and the other Subcontract Documents, this Agreement shall govern.

3

2.5 EXTENT OF AGREEMENT   Nothing in this Agreement shall be construed to create a contractual relationship between persons or entities other than the Contractor and Subcontractor. This Agreement is solely for the benefit of the Parties, represents the entire and integrated agreement between the Parties, and supersedes all prior negotiations, representations, or agreements, either written or oral.

2.6 DEFINITIONS

2.6.1 Wherever the term Progress Schedule is used in this Agreement, it shall be read as Project Schedule when that term is used in the Subcontract Documents.

2.6.2 Whenever the term Change Order is used in this Agreement, it shall be read as Change Document when that term is used in the Subcontract Documents.

2.6.3 Unless otherwise indicated, the term Day shall mean calendar day.

ARTICLE 3

SUBCONTRACTOR'S RESPONSIBILITIES

3.1 OBLIGATIONS   The Contractor and Subcontractor are hereby mutually bound by the terms of this Agreement. To the extent the terms of the Owner-Contractor agreement apply to the Subcontract Work, then the Contractor hereby assumes toward the Subcontractor all the obligations, rights, duties, and redress that the Owner under the prime contract assumes toward the Contractor. In an identical way, the Subcontractor hereby assumes toward the Contractor all the same obligations, rights, duties, and redress that the Contractor assumes toward the Owner and Architect/Engineer under the prime contract. In the event of an inconsistency among the documents, the specific terms of this Agreement shall govern.

3.2 RESPONSIBILITIES   The Subcontractor agrees to furnish its diligent efforts and judgment in the performance of the Subcontract Work and to cooperate with the Contractor so that the Contractor may fulfill its obligations to the Owner. The Subcontractor shall furnish all of the labor, materials, equipment, and services, including but not limited to, competent supervision, shop drawings, samples, tools, and scaffolding as are necessary for the proper performance of the Subcontract Work. The Subcontractor shall provide the Contractor a list of its proposed subcontractors and suppliers, and be responsible for taking field dimensions, providing tests, obtaining required permits related to the Subcontract Work and affidavits, ordering of materials and all other actions as required to meet the Progress Schedule.

3.3 INCONSISTENCIES AND OMISSIONS   The Subcontractor shall make a careful analysis and comparison of the drawings, specifications, other Subcontract Documents and information furnished by the Owner relative to the Subcontract Work. Such analysis and comparison shall be solely for the purpose of facilitating the Subcontract Work and not for the discovery of errors, inconsistencies or omissions in the Subcontract Documents nor for ascertaining if the Subcontract Documents are in accordance with applicable laws, statutes, ordinances, building codes, rules or regulations. Should the Subcontractor discover any errors, inconsistencies or omissions in the Subcontract Documents, the Subcontractor shall report such discoveries to the Contractor in writing within three (3) Days. Upon receipt of notice, the Contractor shall instruct the Subcontractor as to the measures to be taken, and the Subcontractor shall comply with the Contractor's instructions. If the Subcontractor performs work knowing it to be contrary to any applicable laws, statutes, ordinances, building codes, rules or regulations without notice to the Contractor and advance approval by appropriate authorities, including the Contractor, the Subcontractor shall assume appropriate responsibility for such work and shall bear all associated costs, charges, fees and expenses necessarily incurred to remedy the violation. Nothing in this paragraph shall relieve the Subcontractor of responsibility for its own errors, inconsistencies and omissions.

3.4 SITE VISITATION   Prior to performing any portion of the Subcontract Work, the Subcontractor shall

4

conduct a visual inspection of the Project site to become generally familiar with local conditions and to correlate site observations with the Subcontract Documents. If the Subcontractor discovers any discrepancies between its site observations and the Subcontract Documents, such discrepancies shall be promptly reported to the Contractor.

3.5 INCREASED COSTS OR TIME The Subcontractor may assert a Claim as provided in Article 7 if Contractor's clarifications or instructions in responses to requests for information are believed to require additional time or cost. If the Subcontractor fails to perform the reviews and comparisons required in Paragraphs 3.3 and 3.4, above, to the extent the Contractor is held liable to the Owner because of the Subcontractor's failure, the Subcontractor shall pay the costs and damages to the Contractor that would have been avoided if the Subcontractor had performed those obligations.

3.6 COMMUNICATIONS Unless otherwise provided in the Subcontract Documents and except for emergencies, Subcontractor shall direct all communications related to the Project to the Contractor.

3.7 SUBMITTALS

3.7.1 The Subcontractor promptly shall submit for approval to the Contractor all shop drawings, samples, product data, manufacturers' literature and similar submittals required by the Subcontract Documents. Submittals shall be submitted in electronic form if required in accordance with Subparagraph 2.3.1. The Subcontractor shall be responsible to the Contractor for the accuracy and conformity of its submittals to the Subcontract Documents. The Subcontractor shall prepare and deliver its submittals to the Contractor in a manner consistent with the Progress Schedule and in such time and sequence so as not to delay the Contractor or others in the performance of the Work. The approval of any Subcontractor submittal shall not be deemed to authorize deviations, substitutions or changes in the requirements of the Subcontract Documents unless express written approval is obtained from the Contractor and Owner authorizing such deviation, substitution or change. Such approval shall be promptly memorialized in a Subcontract Change Order with in seven (7) Days following approval by the Contractor and, if applicable, provide for an adjustment in the Subcontract Amount or Subcontract Time. In the event that the Subcontract Documents do not contain submittal requirements pertaining to the Subcontract Work, the Subcontractor agrees upon request to submit in a timely fashion to the Contractor for approval any shop drawings, samples, product data, manufacturers' literature or similar submittals as may reasonably be required by the Contractor, Owner or Architect/Engineer.

3.7.2 The Contractor, Owner, and Architect/Engineer are entitled to rely on the adequacy, accuracy and completeness of any professional certifications required by the Subcontract Documents concerning the performance criteria of systems, equipment or materials, including all relevant calculations and any governing performance requirements.

3.8 DESIGN DELEGATION

3.8.1 If the Subcontract Documents (1) specifically require the Subcontractor to procure design services and (2) specify all design and performance criteria, the Subcontractor shall provide those design services necessary to satisfactorily complete the Subcontract Work. Design services provided by the Subcontractor shall be procured from licensed design professionals retained by the Subcontractor as permitted by the law of the place where the Project is located (the Designer). The Designer's signature and seal shall appear on all drawings, calculations, specifications, certifications, Shop Drawings and other submittals prepared by the Designer. Shop Drawings and other submittals related to the Subcontract Work designed or certified by the Designer, if prepared by others, shall bear the Subcontractor's and the Designer's written approvals when submitted to the Contractor. The Contractor shall be entitled to rely upon the adequacy, accuracy and completeness

---

5

of the services, certifications or approvals performed by the Designer.

3.8.2 If the Designer is an independent professional, the design services shall be procured pursuant to a separate agreement between the Subcontractor and the Designer. The Subcontractor-Designer agreement shall not provide for any limitation of liability, except to the extent that consequential damages are waived pursuant to Subparagraph 5.4.1, or exclusion from participation in the multiparty proceedings requirement of Paragraph 11.4. The Designer(s) is (are) _____. The Subcontractor shall notify the Contractor in writing if it intends to change the Designer. The Subcontractor shall be responsible for conformance of its design with the information given and the design concept expressed in the Subcontract Documents. The Subcontractor shall not be responsible for the adequacy of the performance or design criteria required by the Subcontract Documents.

3.8.3 The Subcontractor shall not be required to provide design services in violation of any applicable law.

3.9 TEMPORARY SERVICES The Subcontractor's and Contractor's respective responsibilities for temporary services are set forth in Exhibit _____.

3.10 COORDINATION The Subcontractor shall:

3.10.1 cooperate with the Contractor and all others whose work may interface with the Subcontract Work;

3.10.2 specifically note and immediately advise the Contractor of any such interface with the Subcontract Work; and

3.10.3 participate in the preparation of coordination drawings and work schedules in areas of congestion.

3.11 SUBCONTRACTOR'S REPRESENTATIVE The Subcontractor shall designate a person, subject to Contractor's approval, who shall be the Subcontractor's authorized representative. This representative shall be the only person to whom the Contractor shall issue instructions, orders or directions, except in an emergency. The Subcontractor's representative is _____ who is agreed to by the Contractor.

3.12 TESTS AND INSPECTIONS The Subcontractor shall schedule all required tests, approvals and inspections of the Subcontract Work at appropriate times so as not to delay the progress of the work. The Subcontractor shall give proper written notice to all required Parties of such tests, approvals and inspections. Except as otherwise provided in the Subcontract Documents the Subcontractor shall bear all expenses associated with tests, inspections and approvals required of the Subcontractor by the Subcontract Documents which, unless otherwise agreed to, shall be conducted by an independent testing laboratory or entity approved by the Contractor and Owner. Required certificates of testing, approval or inspection shall, unless otherwise required by the Subcontract Documents, be secured by the Subcontractor and promptly delivered to the Contractor.

3.13 CLEANUP

3.13.1 The Subcontractor shall at all times during its performance of the Subcontract Work keep the Work site clean and free from debris resulting from the Subcontract Work. Prior to discontinuing the Subcontract Work in an area, the Subcontractor shall clean the area and remove all its rubbish and its construction equipment, tools, machinery, waste and surplus materials. Subcontractor shall make provisions to minimize and confine dust and debris resulting from its construction activities. The Subcontractor shall not be held responsible for unclean conditions caused by others.

3.13.2 If the Subcontractor fails to commence compliance with cleanup duties within two (2) business Days after written notification from the Contractor of non-compliance, the Contractor may implement appropriate cleanup measures without further notice and the cost thereof shall be deducted from any amounts due or to become due the Subcontractor in the next payment period.

3.14 SAFETY

3.14.1 The Subcontractor is required to perform the Subcontract Work in a safe and reasonable manner. The Subcontractor shall seek to avoid injury, loss or damage to persons or property by taking reasonable steps to protect:

3.14.1.1 Employees and other persons at the site;

3.14.1.2 Materials and equipment stored at the site or at off-site locations for use in performance of the Subcontract Work; and

3.14.1.3 All property and structures located at the site and adjacent to work areas, whether or not said property or structures are part of the Project or involved in the Work.

3.14.2 The Subcontractor shall give all required notices and comply with all applicable rules, regulations, orders and other lawful requirements established to prevent injury, loss or damage to persons or property.

3.14.3 The Subcontractor shall implement appropriate safety measures pertaining to the Subcontract Work and the Project, including establishing safety rules, posting appropriate warnings and notices, erecting safety barriers, and establishing proper notice procedures to protect persons and property at the site and adjacent to the site from injury, loss or damage.

3.14.4 The Subcontractor shall exercise extreme care in carrying out any of the Subcontract Work which involves explosive or other dangerous methods of construction or hazardous procedures, materials or equipment. The Subcontractor shall use properly qualified individuals or entities to carry out the Subcontract Work in a safe and reasonable manner so as to reduce the risk of bodily injury or property damage.

3.14.5 Damage or loss not insured under property insurance and to the extent caused by the negligent acts or omissions of the Subcontractor, or anyone for whose acts the Subcontractor may be liable, shall be promptly remedied by the Subcontractor. Damage or loss to the extent caused by the negligent acts or omissions of the Contractor, or anyone for whose acts the Contractor may be liable, shall be promptly remedied by the Contractor.

3.14.6 The Subcontractor is required to designate an individual at the site in the employ of the Subcontractor who shall act as the Subcontractor's designated safety representative with a duty to prevent accidents. Unless otherwise identified by the Subcontractor in writing to the Contractor, the designated safety representative shall be the Subcontractor's project superintendent. Such safety representative shall attend site safety meetings as requested by the Contractor.

3.14.7 The Subcontractor has an affirmative duty not to overload the structures or conditions at the site and shall take reasonable steps not to load any part of the structures, or site so as to give rise to an unsafe condition or create an unreasonable risk of bodily injury or property damage. The Subcontractor shall have the right to request, in writing, from the Contractor loading information concerning the structures at the site.

3.14.8 The Subcontractor shall give prompt written notice to the Contractor of any accident involving bodily injury requiring a physician's care, any property damage exceeding Five Hundred Dollars

($500.00) in value, or any failure that could have resulted in serious bodily injury, whether or not such an injury was sustained.

3.14.9 Prevention of accidents at the site is the responsibility of the Contractor, Subcontractor, and all other subcontractors, persons and entities at the site. Establishment of a safety program by the Contractor shall not relieve the Subcontractor or other Parties of their safety responsibilities. The Subcontractor shall establish its own safety program implementing safety measures, policies and standards conforming to those required or recommended by governmental and quasi-governmental authorities having jurisdiction and by the Contractor and Owner, including, but not limited to, requirements imposed by the Subcontract Documents. The Subcontractor shall comply with the reasonable recommendations of insurance companies having an interest in the Project, and shall stop any part of the Subcontract Work which the Contractor deems unsafe until corrective measures satisfactory to the Contractor shall have been taken. The Contractor's failure to stop the Subcontractor's unsafe practices shall not relieve the Subcontractor of the responsibility therefor. The Subcontractor shall notify the Contractor immediately following a reportable incident under applicable rules, regulations, orders and other lawful requirements, and promptly confirm the notice in writing. A detailed written report shall be furnished if requested by the Contractor. To the fullest extent permitted by law, each Party to this Agreement shall indemnify the other party from and against fines or penalties imposed as a result of safety violations, but only to the extent that such fines or penalties are caused by its failure to comply with applicable safety requirements. This indemnification obligation does not extend to additional or increased fines that result from repeated or willful violations not caused by the Subcontractor's failure to comply with applicable rules, regulations, orders and other lawful requirements.

3.15 PROTECTION OF THE WORK The Subcontractor shall take necessary precautions to properly protect the Subcontract Work and the work of others from damage caused by the Subcontractor's operations. Should the Subcontractor cause damage to the Work or property of the Owner, the Contractor or others, the Subcontractor shall promptly remedy such damage to the satisfaction of the Contractor, or the Contractor may, after forty-eight (48) hours written notice to the Subcontractor, remedy the damage and deduct its cost from any amounts due or to become due the Subcontractor, unless such costs are recovered under applicable property insurance.

3.16 PERMITS, FEES, LICENSES AND TAXES The Subcontractor shall give timely notices to authorities pertaining to the Subcontract Work, and shall be responsible for all permits, fees, licenses, assessments, inspections, testing and taxes necessary to complete the Subcontract Work in accordance with the Subcontract Documents. To the extent reimbursement is obtained by the Contractor from the Owner under the Owner-Contractor agreement, the Subcontractor shall be compensated for additional costs resulting from taxes enacted after the date of this Agreement.

3.17 ASSIGNMENT OF SUBCONTRACT WORK The Subcontractor shall neither assign the whole nor any part of the Subcontract Work without prior written approval of the Contractor.

3.18 HAZARDOUS MATERIALS To the extent that the Contractor has rights or obligations under the Owner-Contractor agreement or by law regarding hazardous materials as defined by the Subcontract Document within the scope of the Subcontract Work, the Subcontractor shall have the same rights or obligations.

3.19 MATERIAL SAFETY DATA (MSD) SHEETS The Subcontractor shall submit to the Contractor all Material Safety Data Sheets required by law for materials or substances necessary for the performance of the Subcontract Work. MSD sheets obtained by the Contractor from other subcontractors or sources shall be made available to the Subcontractor by the Contractor.

3.20 LAYOUT RESPONSIBILITY AND LEVELS   The Contractor shall establish principal axis lines of the building and site, and benchmarks. The Subcontractor shall lay out and be strictly responsible for the accuracy of the Subcontract Work and for any loss or damage to the Contractor or others by reason of the Subcontractor's failure to lay out or perform Subcontract Work correctly. The Subcontractor shall exercise prudence so that the actual final conditions and details shall result in alignment of finish surfaces.

3.21 WARRANTIES   The Subcontractor warrants that all materials and equipment shall be new unless otherwise specified, of good quality, in conformance with the Subcontract Documents, and free from defective workmanship and materials. The Subcontractor further warrants that the Work shall be free from material defects not intrinsic in the design or materials required in the Subcontract Documents. The Subcontractor's warranty does not include remedies for defects or damages caused by normal wear and tear during normal usage, use for a purpose for which the Project was not intended, improper or insufficient maintenance, modifications performed by Others, or abuse. The Subcontractor's warranties shall commence on the date of Substantial Completion of the Work or a designated portion.

3.22 UNCOVERING/CORRECTION OF SUBCONTRACT WORK

   3.22.1 UNCOVERING OF SUBCONTRACT WORK

      3.22.1.1 If required in writing by the Contractor, the Subcontractor must uncover any portion of the Subcontract Work which has been covered by the Subcontractor in violation of the Subcontract Documents or contrary to a directive issued to the Subcontractor by the Contractor. Upon receipt of a written directive from the Contractor, the Subcontractor shall uncover such work for the Contractor's or Owner's inspection and restore the uncovered Subcontract Work to its original condition at the Subcontractor's time and expense.

      3.22.1.2 The Contractor may direct the Subcontractor to uncover portions of the Subcontract Work for inspection by the Owner or Contractor at any time. The Subcontractor is required to uncover such work whether or not the Contractor or Owner had requested to inspect the Subcontract Work prior to it being covered. Except as provided in Subparagraph 3.22.1.1, this Agreement shall be adjusted by change order for the cost and time of uncovering and restoring any work which is uncovered for inspection and proves to be installed in accordance with the Subcontract Documents, provided the Contractor had not previously instructed the Subcontractor to leave the work uncovered. If the Subcontractor uncovers work pursuant to a directive issued by the Contractor, and such work upon inspection does not comply with the Subcontract Documents, the Subcontractor shall be responsible for all costs and time of uncovering, correcting and restoring the work so as to make it conform to the Subcontract Documents. If the Contractor or some other entity for which the Subcontractor is not responsible caused the nonconforming condition, the Contractor shall be required to adjust this Agreement by change order for all such costs and time.

   3.22.2 CORRECTION OF WORK

      3.22.2.1 If the Architect/Engineer or Contractor rejects the Subcontract Work or the Subcontract Work is not in conformance with the Subcontract Documents, the Subcontractor shall promptly correct the Subcontract Work whether it had been fabricated, installed or completed. The Subcontractor shall be responsible for the costs of correcting such Subcontract Work, any additional testing, inspections, and compensation for services and expenses of the Architect/Engineer and Contractor made necessary by the defective Subcontract Work.

      3.22.2.2 In addition to the Subcontractor's obligations under Paragraph 3.21, the

Subcontractor agrees to promptly correct, after receipt of a written notice from the Contractor, all Subcontract Work performed under this Agreement which proves to be defective in workmanship or materials within a period of one year from the date of Substantial Completion of the Subcontract Work or for a longer period of time as may be required by specific warranties in the Subcontract Documents. Substantial Completion of the Subcontract Work, or of a designated portion, occurs on the date when construction is sufficiently complete in accordance with the Subcontract Documents so that the Owner can occupy or utilize the Project, or a designated portion, for the use for which it is intended. If, during the one-year period, the Contractor fails to provide the Subcontractor with prompt written notice of the discovery of defective or nonconforming Subcontract Work, the Contractor shall neither have the right to require the Subcontractor to correct such Subcontract Work nor the right to make claim for breach of warranty. If the Subcontractor fails to correct defective or nonconforming Subcontract Work within a reasonable time after receipt of notice from the Contractor, the Contractor may correct such Subcontract Work pursuant to Subparagraph 10.1.1.

3.22.3 The Subcontractor's correction of Subcontract Work pursuant to this Paragraph 3.22 shall not extend the one-year period for the correction of Subcontract Work, but if Subcontract Work is first performed after Substantial Completion, the one-year period for corrections shall be extended by the time period after Substantial Completion and the performance of that portion of Subcontract Work. The Subcontractor's obligation to correct Subcontract Work within one year as described in this Paragraph 3.22 does not limit the enforcement of Subcontractor's other obligations with regard to the Agreement and the Subcontract Documents.

3.22.4 If the Subcontractor's correction or removal of Subcontract Work destroys or damages completed or partially completed work of the Owner, the Contractor or any separate contractors or subcontractors, the Subcontractor shall be responsible for the reasonable cost of correcting such destroyed or damaged property.

3.22.5 If portions of Subcontract Work which do not conform with the requirements of the Subcontract Documents are neither corrected by the Subcontractor nor accepted by the Contractor, the Subcontractor shall remove such Subcontract Work from the Project site if so directed by the Contractor.

3.23 MATERIALS OR EQUIPMENT FURNISHED BY OTHERS In the event the scope of the Subcontract Work includes installation of materials or equipment furnished by others, it shall be the responsibility of the Subcontractor to exercise proper care in receiving, handling, storing and installing such items, unless otherwise provided in the Subcontract Documents. The Subcontractor shall examine the items provided and report to the Contractor in writing any items it may discover that do not conform to requirements of the Subcontract Documents. The Subcontractor shall not proceed to install non-conforming items without further instructions from the Contractor. Loss or damage due to acts or omissions of the Subcontractor shall, upon two (2) business Days written notice to the Subcontractor be deducted from any amounts due or to become due the Subcontractor.

3.24 SUBSTITUTIONS No substitutions shall be made in the Subcontract Work unless permitted in the Subcontract Documents, and only upon the Subcontractor first receiving all approvals required under the Subcontract Documents for substitutions.

3.25 USE OF CONTRACTOR'S EQUIPMENT The Subcontractor, its agents, employees, subcontractors or suppliers shall use the Contractor's equipment only with the express written permission of the Contractor's designated representative and in accordance with the Contractor's terms and conditions for such use. If the Subcontractor or any of its agents, employees, subcontractors or suppliers utilize any

10

of the Contractor's equipment, including machinery, tools, scaffolding, hoists, lifts or similar items owned, leased or under the control of the Contractor, the Subcontractor shall indemnify and be liable to the Contractor as provided in Article 9 for any loss or damage (including bodily injury or death) which may arise from such use, except to the extent that such loss or damage is caused by the negligence of the Contractor's employees operating the Contractor's equipment.

3.26 WORK FOR OTHERS   Until final completion of the Subcontract Work, the Subcontractor agrees not to perform any work directly for the Owner or any tenants, or deal directly with the Owner's representatives in connection with the Subcontract Work, unless otherwise approved in writing by the Contractor.

3.27 SYSTEMS AND EQUIPMENT STARTUP   With the assistance of the Owner's maintenance personnel and the Contractor, the Subcontractor shall direct the check-out and operation of systems and equipment for readiness, and assist in their initial startup and the testing of the Subcontract Work.

3.28 COMPLIANCE WITH LAWS   The Subcontractor agrees to be bound by, and at its own costs comply with, all federal, state and local laws, ordinances and regulations (the Laws) applicable to the Subcontract Work, including but not limited to, equal employment opportunity, minority business enterprise, women's business enterprise, disadvantaged business enterprise, safety and all other Laws with which the Contractor must comply. The Subcontractor shall be liable to the Contractor and the Owner for all loss, cost and expense attributable to any acts of commission or omission by the Subcontractor, its employees and agents resulting from the failure to comply with Laws, including, but not limited to, any fines, penalties or corrective measures, except as provided in Subparagraph 3.14.9.

3.29 CONFIDENTIALITY   To the extent the Owner-Contractor agreement provides for the confidentiality of any of the Owner's proprietary or otherwise confidential information disclosed in connection with the performance of this Agreement, the Subcontractor is equally bound by the Owner's confidentiality requirements.

3.30 ROYALTIES, PATENTS AND COPYRIGHTS   The Subcontractor shall pay all royalties and license fees which may be due on the inclusion of any patented or copyrighted materials, methods or systems selected by the Subcontractor and incorporated in the Subcontract Work. The Subcontractor shall defend, indemnify and hold the Contractor and Owner harmless from all suits or claims for infringement of any patent rights or copyrights arising out of such selection. The Subcontractor shall be liable for all loss, including all costs, expenses, and attorneys' fees, but shall not be responsible for such defense or loss when a particular design, process or product of a particular manufacturer or manufacturers is required by the Subcontract Documents. However, if the Subcontractor has reason to believe that a particular design, process or product required by the Subcontract Documents is an infringement of a patent, the Subcontractor shall promptly furnish such information to the Contractor or be responsible to the Contractor and Owner for any loss sustained as a result.

3.31 LABOR RELATIONS   (Insert here any conditions, obligations or requirements relative to labor relations and their effect on the project. Legal counsel is recommended.)

ARTICLE 4

CONTRACTOR'S RESPONSIBILITIES

4.1 CONTRACTOR'S REPRESENTATIVE   The Contractor shall designate a person who shall be the Contractor's authorized representative. The Contractor's representative shall be the only person the Subcontractor shall look to for instructions, orders or directions, except in an emergency. The Contractor's

11

IMPORTANT: A vertical line in the margin indicates a change has been made to the original text. Prior to signing, recipients may wish to request from the party producing the document a "redlined" version indicating changes to the original text. Consultation with legal and insurance counsel and careful review of the entire document are strongly encouraged.
ConsensusDOCS 750 • STANDARD FORM OF AGREEMENT BETWEEN CONTRACTOR AND SUBCONTRACTOR Copyright © 2007, ConsensusDOCS LLC; revised May 2009. YOU ARE ALLOWED TO USE THIS DOCUMENT FOR ONE CONTRACT ONLY. YOU MAY MAKE 9 COPIES OF THE COMPLETED DOCUMENT FOR DISTRIBUTION TO THE CONTRACT'S PARTIES. ANY OTHER USES, INCLUDING COPYING THE FORM DOCUMENT, ARE STRICTLY PROHIBITED.

Representative is _____.

4.2 OWNER'S ABILITY TO PAY

4.2.1 The Subcontractor shall have the right upon request to receive from the Contractor such information as the Contractor has obtained relative to the Owner's financial ability to pay for the Work, including any subsequent material variation in such information. The Contractor, however, does not warrant the accuracy or completeness of the information provided by the Owner.

4.2.2 If the Subcontractor does not receive the information referenced in Subparagraph 4.2.1 with regard to the Owner's ability to pay for the Work as required by the Contract Documents, the Subcontractor may request the information from the Owner or the Owner's lender.

4.3 CONTRACTOR APPLICATION FOR PAYMENT   Upon request, the Contractor shall give the Subcontractor a copy of the most current Contractor application for payment reflecting the amounts approved or paid by the Owner for the Subcontract Work performed to date.

4.4 INFORMATION OR SERVICES   The Subcontractor is entitled to request through the Contractor any information or services relevant to the performance of the Subcontract Work which is under the Owner's control. The Subcontractor also is entitled to request through the Contractor any information necessary to give notice of or enforce mechanics lien rights and, where applicable, stop notices. This information shall include the Owner's interest in the real property on which the Project is located and the recorded legal title. To the extent the Contractor receives such information and services, the Contractor shall provide them to the Subcontractor. The Contractor, however, does not warrant the accuracy or completeness of the information provided by the Owner. To the extent the Owner provides any warranty of Owner provided information, the Contractor agrees to permit the Subcontractor to prosecute a claim in the name of the Contractor for the use and benefit of the Subcontractor, pursuant to Subparagraph 5.3.2.

4.5 STORAGE AREAS   The Contractor shall allocate adequate storage areas, if available, for the Subcontractor's materials and equipment during the course of the Subcontract Work. Unless otherwise agreed upon, the Contractor shall reimburse the Subcontractor for the additional costs of having to relocate such storage areas at the direction of the Contractor.

4.6 TIMELY COMMUNICATIONS   The Contractor shall transmit to the Subcontractor, with reasonable promptness, all submittals, transmittals, and written approvals relative to the Subcontract Work. Unless otherwise specified in the Subcontract Documents, communications by and with the Subcontractor's subcontractors, materialmen and suppliers shall be through the Subcontractor.

4.7 USE OF SUBCONTRACTOR'S EQUIPMENT   The Contractor, its agents, employees or suppliers shall use the Subcontractor's equipment only with the express written permission of the Subcontractor's designated representative and in accordance with the Subcontractor's terms and conditions for such use. If the Contractor or any of its agents, employees or suppliers utilize any of the Subcontractor's equipment, including machinery, tools, scaffolding, hoists, lifts or similar items owned, leased or under the control of the Subcontractor, the Contractor shall indemnify and be liable to the Subcontractor as provided in Article 9 for any loss or damage (including bodily injury or death) which may arise from such use, except to the extent that such loss or damage is caused by the negligence of the Subcontractor's employees operating the Subcontractor's equipment.

ARTICLE 5

PROGRESS SCHEDULE

5.1 TIME IS OF THE ESSENCE   Time is of the essence for both Parties. They mutually agree to see to the performance of their respective obligations so that the entire Project may be completed in accordance with the Subcontract Documents and particularly the Progress Schedule as set forth in Exhibit _____.

5.2 SCHEDULE OBLIGATIONS The Subcontractor shall provide the Contractor with any scheduling information proposed by the Subcontractor for the Subcontract Work. In consultation with the Subcontractor, the Contractor shall prepare the schedule for performance of the Work (the Progress Schedule) and shall revise and update such schedule, as necessary, as the Work progresses. Both the Contractor and the Subcontractor shall be bound by the Progress Schedule. The Progress Schedule and all subsequent changes and additional details shall be submitted to the Subcontractor promptly and reasonably in advance of the required performance. The Contractor shall have the right to determine and, if necessary, change the time, order and priority in which the various portions of the Work shall be performed and all other matters relative to the Subcontract Work. To the extent such changes increase Subcontractor's time and costs, the Subcontract Amount and Subcontract Time shall be equitably adjusted.

5.3 DELAYS AND EXTENSIONS OF TIME

5.3.1 OWNER CAUSED DELAY Subject to Subparagraph 5.3.2, if the commencement or progress of the Subcontract Work is delayed without the fault or responsibility of the Subcontractor, the time for the Subcontract Work shall be extended by Subcontract Change Order and the Subcontract Price equitably adjusted to the extent obtained by the Contractor under the Subcontract Documents, and the Progress Schedule shall be revised accordingly.

5.3.2 CLAIMS RELATING TO OWNER The Subcontractor agrees to initiate all claims for which the Owner is or may be liable in the manner and within the time limits provided in the Subcontract Documents for like claims by the Contractor upon the Owner and in sufficient time for the Contractor to initiate such claims against the Owner in accordance with the Subcontract Documents. At the Subcontractor's request and expense to the extent agreed upon in writing, the Contractor agrees to permit the Subcontractor to prosecute a claim in the name of the Contractor for the use and benefit of the Subcontractor in the manner provided in the Subcontract Documents for like claims by the Contractor upon the Owner.

5.3.3 CONTRACTOR CAUSED DELAY Nothing in this Article shall preclude the Subcontractor's recovery of delay damages caused by the Contractor to the extent not otherwise precluded by this Agreement.

5.3.4 CLAIMS RELATING TO CONTRACTOR The Subcontractor shall give the Contractor written notice of all claims not included in Subparagraph 5.3.2 within fourteen (14) Days of the Subcontractor's knowledge of the facts giving rise to the event for which claim is made. Thereafter, the Subcontractor shall submit written documentation of its claim, including appropriate supporting documentation, within twenty-one (21) Days after giving notice, unless the Parties agree upon a longer period of time. The Contractor shall respond in writing denying or approving, in whole or in part the Subcontractor's claim no later than fourteen (14) Days after receipt of the Subcontractor's documentation of claim. All unresolved claims, disputes and other matters in question between the Contractor and the Subcontractor not relating to claims included in Subparagraph 5.3.2 shall be resolved in the manner provided in Article 11.

5.4 LIMITED MUTUAL WAIVER OF CONSEQUENTIAL DAMAGES

5.4.1 Except for damages provided for by the Subcontract Documents as liquidated damages and excluding losses covered by insurance required by the Subcontract Documents, the Contractor and

Subcontractor waive claims against each other for consequential damages arising out of or relating to this Agreement, to the same extent the Owner-Contractor agreement furnished to the Subcontractor in accordance with Paragraph 2.3 provides for a mutual waiver of consequential damages by the Owner and Contractor, including to the extent provided in the Owner-Contractor agreement, damages for loss of business, loss of financing, principal office overhead and expenses, loss of profits not related to this Project, loss of bonding capacity, loss of reputation, or insolvency. Similarly, the Subcontractor shall obtain in another agreement from its Sub-Subcontractors mutual waivers of consequential damages that correspond to the Subcontractor's waiver of consequential damages herein. To the extent applicable, this mutual waiver applies to consequential damages due to termination by the Contractor or the Owner in accordance with this Agreement or the Owner-Contractor agreement. The provisions of this Article shall also apply to and survive termination of this Agreement.

5.5 LIQUIDATED DAMAGES

5.5.1 If the Subcontract Documents furnished to the Subcontractor in accordance with Paragraph 2.3 provide for liquidated damages or other damages for delay beyond the completion date set forth in the Subcontract Documents that are not specifically addressed as a liquidated damage item in this Agreement, and such damages are assessed, the Contractor may assess a share of the damages against the Subcontractor in proportion to the Subcontractor's share of the responsibility for the damages. However, the amount of such assessment shall not exceed the amount assessed against the Contractor. This Paragraph shall not limit the Subcontractor's liability to the Contractor for the Contractor's actual damages caused by the Subcontractor.

5.5.2 To the extent the Owner-Contractor Agreement provides for a mutual waiver of consequential damages by the Owner and the Contractor, damages for which the Contractor is liable to the Owner including those related to Paragraph 9.1 are not consequential damages for the purpose of this waiver. Similarly, to the extent the Subcontractor-Sub-Subcontractor agreement provides for a mutual waiver of consequential damages by the Owner and the Contractor, damages for which the Subcontractor is liable to lower-tiered parties due to the fault of the Owner or Contractor are not consequential damages for the purpose of this waiver.

ARTICLE 6

SUBCONTRACT AMOUNT

As full compensation for performance of this Agreement, Contractor agrees to pay Subcontractor in current funds for the satisfactory performance of the Subcontract Work subject to all applicable provisions of the Subcontract:

(a) the fixed-price of _____ Dollars
($_____) subject to additions and deductions as provided for in the Subcontract Documents; or

(b) alternates and unit prices in accordance with the attached schedule of Alternates and Unit Prices and estimated quantities, which is incorporated by reference and identified as Exhibit _____; or

(c) time and material rates and prices in accordance with the attached Schedule of Labor and Material Costs which is incorporated by reference and identified as Exhibit _____.

The fixed-price, unit prices or time and material rates and prices are referred to as the Subcontract Amount.

ARTICLE 7

CHANGES IN THE SUBCONTRACT WORK

7.1 SUBCONTRACT CHANGE ORDERS When the Contractor orders in writing, the Subcontractor, without nullifying this Agreement, shall make any and all changes in the Subcontract Work which are within the general scope of this Agreement. Any adjustment in the Subcontract Amount or Subcontract Time shall be authorized by a Subcontract Change Order. No adjustments shall be made for any changes performed by the Subcontractor that have not been ordered by the Contractor. A Subcontract Change Order is a written instrument prepared by the Contractor and signed by the Subcontractor stating their agreement upon the change in the Subcontract Work.

7.2 CONSTRUCTION CHANGE DIRECTIVES To the extent that the Subcontract Documents provide for Construction Change Directives in the absence of agreement on the terms of a Subcontract Change Order, the Subcontractor shall promptly comply with the Construction Change Directive and be entitled to apply for interim payment if the Subcontract Documents so provide.

7.3 UNKNOWN CONDITIONS If in the performance of the Subcontract Work the Subcontractor finds latent, concealed or subsurface physical conditions which differ materially from those indicated in the Subcontract Documents or unknown physical conditions of an unusual nature, which differ materially from those ordinarily found to exist, and not generally recognized as inherent in the kind of work provided for in this Agreement, the Subcontract Amount or the Progress Schedule shall be equitably adjusted by a Subcontract Change Order within a reasonable time after the conditions are first observed. The adjustment which the Subcontractor may receive shall be limited to the adjustment the Contractor receives from the Owner on behalf of the Subcontractor, or as otherwise provided under Subparagraph 5.3.2.

7.4 ADJUSTMENTS IN SUBCONTRACT AMOUNT If a Subcontract Change Order requires an adjustment in the Subcontract Amount, the adjustment shall be established by one of the following methods:

7.4.1 mutual acceptance of an itemized lump sum;

7.4.2 unit prices as indicated in the Subcontract Documents or as subsequently agreed to by the Parties; or

7.4.3 costs determined in a manner acceptable to the Parties and a mutually acceptable fixed or percentage fee; or

7.4.4 another method provided in the Subcontract Documents.

7.5 SUBSTANTIATION OF ADJUSTMENT If the Subcontractor does not respond promptly or disputes the method of adjustment, the method and the adjustment shall be determined by the Contractor on the basis of reasonable expenditures and savings of those performing the Work attributable to the change, including, in the case of an increase in the Subcontract Amount, an allowance for overhead and profit of the percentage provided in Paragraph 7.6, or if none is provided as mutually agreed upon by the Parties. The Subcontractor may contest the reasonableness of any adjustment determined by the Contractor. The Subcontractor shall maintain for the Contractor's review and approval an appropriately itemized and substantiated accounting of the following items attributable to the Subcontract Change Order:

7.5.1 labor costs, including Social Security, health, welfare, retirement and other fringe benefits as normally required, and state workers' compensation insurance;

7.5.2 costs of materials, supplies and equipment, whether incorporated in the Subcontract Work or consumed, including transportation costs;

7.5.3 costs of renting machinery and equipment other than hand tools;

7.5.4 costs of bond and insurance premiums, permit fees and taxes attributable to the change; and

7.5.5 costs of additional supervision and field office personnel services necessitated by the change.

7.6 Adjustments shall be based on net change in Subcontractor's reasonable cost of performing the changed Subcontract Work plus, in case of a net increase in cost, an agreed upon sum for overhead and profit not to exceed _____ percent (_____ %).

7.7 NO OBLIGATION TO PERFORM The Subcontractor shall not perform changes in the Subcontract Work until a Subcontract Change Order has been executed or written instructions have been issued in accordance with Paragraphs 7.2 and 7.9.

7.8 EMERGENCIES In an emergency affecting the safety of persons or property, the Subcontractor shall act, at its discretion, to prevent threatened damage, injury or loss. Any change in the Subcontract Amount or the Progress Schedule on account of emergency work shall be determined as provided in this Article.

7.9 INCIDENTAL CHANGES The Contractor may direct the Subcontractor to perform incidental changes in the Subcontract Work which do not involve adjustments in the Subcontract Amount or Subcontract Time. Incidental changes shall be consistent with the scope and intent of the Subcontract Documents. The Contractor shall initiate an incidental change in the Subcontract Work by issuing a written order to the Subcontractor. Such written notice shall be carried out promptly and are binding on the Parties.

**ARTICLE 8
PAYMENT**

8.1 SCHEDULE OF VALUES As a condition to payment, the Subcontractor shall provide a schedule of values satisfactory to the Contractor not more than fifteen (15) Days from the date of execution of this Agreement.

8.2 PROGRESS PAYMENTS

8.2.1 APPLICATIONS The Subcontractor's applications for payment shall be itemized and supported by substantiating data as required by the Subcontract Documents. If the Subcontractor is obligated to provide design services pursuant to Paragraph 3.8, Subcontractor's applications for payment shall show the Designer's fee and expenses as a separate cost item. The Subcontractor's application shall be notarized if required and if allowed under the Subcontract Documents may include properly authorized Subcontract Construction Change Directives. The Subcontractor's progress payment application for the Subcontract Work performed in the preceding payment period shall be submitted for approval of the Contractor in accordance with the schedule of values if required and Subparagraphs 8.2.2, 8.2.3, and 8.2.4. The Contractor shall incorporate the approved amount of the Subcontractor's progress payment application into the Contractor's payment application to the Owner for the same period and submit it to the Owner in a timely fashion. The Contractor shall immediately notify the Subcontractor of any changes in the amount requested on behalf of the Subcontractor.

8.2.2 RETAINAGE The rate of retainage shall be _____ percent (_____ %), which is equal to the percentage retained from the Contractor's payment by the Owner for the Subcontract Work. If the Subcontract Work is satisfactory and the Subcontract Documents provide

16

for reduction of retainage at a specified percentage of completion, the Subcontractor's retainage shall also be reduced when the Subcontract Work has attained the same percentage of completion and the Contractor's retainage for the Subcontract Work has been so reduced by the Owner.

8.2.3 TIME OF APPLICATION The Subcontractor shall submit progress payment applications to the Contractor no later than the _____ Day of each payment period for the Subcontract Work performed up to and including the _____ Day of the payment period indicating work completed and, to the extent allowed under Subparagraph 8.2.4, materials suitably stored during the preceding payment period.

8.2.4 STORED MATERIALS Unless otherwise provided in the Subcontract Documents, applications for payment may include materials and equipment not yet incorporated in the Subcontract Work but delivered to and suitably stored on-site or off-site including applicable insurance, storage and costs incurred transporting the materials to an off-site storage facility. Approval of payment applications for such stored items on or off the site shall be conditioned upon submission by the Subcontractor of bills of sale and required insurance or such other procedures satisfactory to the Owner and Contractor to establish the Owner's title to such materials and equipment, or otherwise to protect the Owner's and Contractor's interest including transportation to the site.

8.2.5 TIME OF PAYMENT Progress payments to the Subcontractor for satisfactory performance of the Subcontract Work shall be made no later than seven (7) Days after receipt by the Contractor of payment from the Owner for the Subcontract Work. If payment from the Owner for such Subcontract Work is not received by the Contractor, through no fault of the Subcontractor, the Contractor will make payment to the Subcontractor within a reasonable time for the Subcontract Work satisfactorily performed.

8.2.6 PAYMENT DELAY If the Contractor has received payment from the Owner and if for any reason not the fault of the Subcontractor, the Subcontractor does not receive a progress payment from the Contractor within seven (7) Days after the date such payment is due, as defined in Subparagraph 8.2.5, or, if the Contractor has failed to pay the Subcontractor within a reasonable time for the Subcontract Work satisfactorily performed, the Subcontractor, upon giving seven (7) Days' written notice to the Contractor, and without prejudice to and in addition to any other legal remedies, may stop work until payment of the full amount owing to the Subcontractor has been received. The Subcontract Amount and Time shall be adjusted by the amount of the Subcontractor's reasonable and verified cost of shutdown, delay, and startup, which shall be effected by an appropriate Subcontractor Change Order.

8.2.7 PAYMENTS WITHHELD The Contractor may reject a Subcontractor payment application in whole or in part or withhold amounts from a previously approved Subcontractor payment application, as may reasonably be necessary to protect the Contractor from loss or damage for which the Contractor may be liable and without incurring an obligation for late payment interest based upon:

8.2.7.1 the Subcontractor's repeated failure to perform the Subcontract Work as required by this Agreement;

8.2.7.2 loss or damage arising out of or relating to this Agreement and caused by the Subcontractor to the Owner, Contractor or others to whom the Contractor may be liable;

8.2.7.3 the Subcontractor's failure to properly pay for labor, materials, equipment or supplies furnished in connection with the Subcontract Work;

8.2.7.4 rejected, nonconforming or defective Subcontract Work which has not been corrected

---

17

in a timely fashion;

8.2.7.5 reasonable evidence of delay in performance of the Subcontract Work such that the Work will not be completed within the Subcontract Time, and that the unpaid balance of the Subcontract Amount is not sufficient to offset the liquidated damages or actual damages that may be sustained by the Contractor as a result of the anticipated delay caused by the Subcontractor;

8.2.7.6 reasonable evidence demonstrating that the unpaid balance of the Subcontract Amount is insufficient to cover the cost to complete the Subcontract Work;

8.2.7.7 third party claims involving the Subcontractor or reasonable evidence demonstrating that third party claims are likely to be filed unless and until the Subcontractor furnishes the Contractor with adequate security in the form of a surety bond, letter of credit or other collateral or commitment which are sufficient to discharge such claims if established.

No later than seven (7) Days after receipt of an application for payment, the Contractor shall give written notice to the Subcontractor, at the time of disapproving or nullifying all or part of an application for payment, stating its specific reasons for such disapproval or nullification, and the remedial actions to be taken by the Subcontractor in order to receive payment. When the above reasons for disapproving or nullifying an application for payment are removed, payment will be promptly made for the amount previously withheld.

8.3 FINAL PAYMENT

8.3.1 APPLICATION Upon acceptance of the Subcontract Work by the Owner and the Contractor and receipt from the Subcontractor of evidence of fulfillment of the Subcontractor's obligations in accordance with the Subcontract Documents and Subparagraph 8.3.2, the Contractor shall incorporate the Subcontractor's application for final payment into the Contractor's next application for payment to the Owner without delay, or notify the Subcontractor if there is a delay and the reasons therefor.

8.3.2 REQUIREMENTS Before the Contractor shall be required to incorporate the Subcontractor's application for final payment into the Contractor's next application for payment, the Subcontractor shall submit to the Contractor:

8.3.2.1 an affidavit that all payrolls, bills for materials and equipment, and other indebtedness connected with the Subcontract Work for which the Owner or its property or the Contractor or the Contractor's surety might in any way be liable, have been paid or otherwise satisfied;

8.3.2.2 consent of surety to final payment, if required;

8.3.2.3 satisfaction of required closeout procedures;

8.3.2.4 other data, if required by the Contractor or Owner, such as receipts, releases, and waivers of liens to the extent and in such form as may be required by the Subcontract Documents;

8.3.2.5 written warranties, equipment manuals, startup and testing required in Paragraph 3.28; and

8.3.2.6 as-built drawings if required by the Subcontract Documents.

8.3.3 TIME OF PAYMENT Final payment of the balance due of the Subcontract Amount shall be made to the Subcontractor within seven (7) Days after receipt by the Contractor of final payment from the Owner for such Subcontract Work.

8.3.4 FINAL PAYMENT DELAY   If the Owner or its designated agent does not issue a certificate for final payment or the Contractor does not receive such payment for any cause which is not the fault of the Subcontractor, the Contractor shall promptly inform the Subcontractor in writing. The Contractor shall also diligently pursue, with the assistance of the Subcontractor, the prompt release by the Owner of the final payment due for the Subcontract Work. At the Subcontractor's request and expense, to the extent agreed upon in writing, the Contractor shall institute reasonable legal remedies to mitigate the damages and pursue payment of the Subcontractor's final payment including interest. If final payment from the Owner for such Subcontract Work is not received by the Contractor, through no fault of the Subcontractor, the Contractor will make payment to the Subcontractor within a reasonable time.

8.3.5 WAIVER OF CLAIMS   Final payment shall constitute a waiver of all claims by the Subcontractor relating to the Subcontract Work, but shall in no way relieve the Subcontractor of liability for the obligations assumed under Paragraphs 3.21 and 3.22, or for faulty or defective work or services discovered after final payment, nor relieve the Contractor for claims made in writing by the Subcontractor as required by the Subcontract Documents prior to its application for final payment as unsettled at the time of such payment.

8.4 LATE PAYMENT INTEREST   Progress payments or final payment due and unpaid under this Agreement, as defined in Subparagraphs 8.2.5, 8.3.3 and 8.3.4, shall bear interest from the date payment is due at the prevailing Statutory rate at the place of the Project. However, if the Owner fails to timely pay the Contractor as required under the Owner-Contractor agreement through no fault or neglect of the Contractor, and the Contractor fails to timely pay the Subcontractor as a result of such nonpayment, the Contractor's obligation to pay the Subcontractor interest on corresponding payments due and unpaid under this Agreement shall be extinguished by the Contractor promptly paying to the Subcontractor the Subcontractor's proportionate share of the interest, if any, received by the Contractor from the Owner on such late payments.

8.5 CONTINUING OBLIGATIONS   Provided the Contractor is making payments on or has made payments to the Subcontractor in accordance with the terms of this Agreement, the Subcontractor shall reimburse the Contractor for any costs and expenses for any claim, obligation or lien asserted before or after final payment is made that arises from the performance of the Subcontract Work. The Subcontractor shall reimburse the Contractor for costs and expenses including attorneys' fees and costs and expenses incurred by the Contractor in satisfying, discharging or defending against any such claims, obligation or lien including any action brought or judgment recovered. In the event that any applicable law, statute, regulation or bond requires the Subcontractor to take any action prior to the expiration of the reasonable time for payment referenced in Subparagraph 8.2.5 in order to preserve or protect the Subcontractor's rights, if any, with respect to mechanic's lien or bond claims, then the Subcontractor may take that action prior to the expiration of the reasonable time for payment and such action will not create the reimbursement obligation recited above nor be in violation of this Agreement or considered premature for purposes of preserving and protecting the Subcontractor's rights.

8.6 PAYMENT USE RESTRICTION   Payments received by the Subcontractor shall be used to satisfy the indebtedness owed by the Subcontractor to any person furnishing labor or materials, or both, for use in performing the Subcontract Work through the most current period applicable to progress payments received from the Contractor before it is used for any other purpose. In the same manner, payments received by the Contractor from the Owner for the Subcontract Work shall be dedicated to payment to the Subcontractor. This provision shall bear on this Agreement only, and is not for the benefit of third parties. Moreover, it shall not be construed by the Parties to this Agreement or third parties to require that dedicated sums of money or payments be deposited in separate accounts, or that there be other restrictions on commingling of funds. Neither shall these mutual covenants be construed to create any fiduciary duty on the Subcontractor or Contractor, nor create any tort cause of action or liability for breach of trust, punitive damages, or other equitable remedy or liability for alleged breach.

8.7 PAYMENT USE VERIFICATION   If the Contractor has reason to believe that the Subcontractor is not complying with the payment terms of this Agreement, the Contractor shall have the right to contact the Subcontractor's subcontractors and suppliers to ascertain whether they are being paid by the Subcontractor in accordance with this Agreement.

8.8 PARTIAL LIEN WAIVERS AND AFFIDAVITS   As a prerequisite for payments, the Subcontractor shall provide, in a form satisfactory to the Owner and Contractor, partial lien or claim waivers in the amount of the application for payment and affidavits covering its subcontractors and suppliers for completed Subcontract Work. Such waivers may be conditional upon payment. In no event shall Contractor require the Subcontractor to provide an unconditional waiver of lien or claim, either partial or final, prior to receiving payment or in an amount in excess of what it has been paid.

8.9 SUBCONTRACTOR PAYMENT FAILURE   Upon payment by the Contractor, the Subcontractor shall promptly pay its subcontractors and suppliers the amounts to which they are entitled. In the event the Contractor has reason to believe that labor, material or other obligations incurred in the performance of the Subcontract Work are not being paid, the Contractor may give written notice of a potential claim or lien to the Subcontractor and may take any steps deemed necessary to assure that progress payments are utilized to pay such obligations, including but not limited to the issuance of joint checks. If upon receipt of notice, the Subcontractor does not (a) supply evidence to the satisfaction of the Contractor that the moneys owing have been paid; or (b) post a bond indemnifying the Owner, the Contractor, the Contractor's surety, if any, and the premises from a claim or lien, the Contractor shall have the right to withhold from any payments due or to become due to the Subcontractor a reasonable amount to protect the Contractor from any and all loss, damage or expense including attorneys' fees that may arise out of or relate to any such claim or lien.

8.10 SUBCONTRACTOR ASSIGNMENT OF PAYMENTS   The Subcontractor shall not assign any moneys due or to become due under this Agreement, without the written consent of the Contractor, unless the assignment is intended to create a new security interest within the scope of Article 9 of the Uniform Commercial Code. Should the Subcontractor assign all or any part of any moneys due or to become due under this Agreement to create a new security interest or for any other purpose, the instrument of assignment shall contain a clause to the effect that the assignee's right in and to any money due or to become due to the Subcontractor shall be subject to the claims of all persons, firms and corporations for services rendered or materials supplied for the performance of the Subcontract Work.

8.11 PAYMENT NOT ACCEPTANCE   Payment to the Subcontractor does not constitute or imply acceptance of any portion of the Subcontract Work.

ARTICLE 9

INDEMNITY, INSURANCE AND WAIVER OF SUBROGATION

9.1 INDEMNITY

9.1.1 INDEMNITY   To the fullest extent permitted by law, the Subcontractor shall indemnify and hold harmless the Contractor, Architect/Engineer, the Owner and their agents, consultants and employees (the Indemnitees) from all claims for bodily injury and property damage other than to the Work itself that may arise from the performance of the Subcontract Work, including reasonable attorneys' fees, costs and expenses, that arise from the performance of the Work, but only to the extent caused by the negligent acts or omissions of the Subcontractor, the Subcontractor's Sub-Subcontractors or anyone employed directly or indirectly by any of them or by anyone for whose acts any of them may be liable. The Subcontractor shall be entitled to reimbursement of any defense cost paid above Subcontractor's percentage of liability for the underlying claim to the extent attributable to the negligent acts or omissions of the Indemnitees.

9.1.2 NO LIMITATION ON LIABILITY In any and all claims against the Indemnitees by any employee of the Subcontractor, anyone directly or indirectly employed by the Subcontractor or anyone for whose acts the Subcontractor may be liable, the indemnification obligation shall not be limited in any way by any limitation on the amount or type of damages, compensation or benefits payable by or for the Subcontractor under workers' compensation acts, disability benefit acts or other employee benefit acts.

9.2 INSURANCE

9.2.1 SUBCONTRACTOR'S INSURANCE Before commencing the Subcontract Work, and as a condition of payment, the Subcontractor shall purchase and maintain insurance that will protect it from the claims arising out of its operations under this Agreement, whether the operations are by the Subcontractor, or any of its consultants or subcontractors or anyone directly or indirectly employed by any of them, or by anyone for whose acts any of them may be liable.

9.2.2 MINIMUM LIMITS OF LIABILITY The Subcontractor shall procure and maintain with insurance companies licensed in a the jurisdiction in which the Project is located and acceptable to the Contractor, which acceptance shall not be unreasonably withheld, at least the limits of liability as set forth in Exhibit _____.

9.2.3 PROFESSIONAL LIABILITY INSURANCE

9.2.3.1 PROFESSIONAL LIABILITY INSURANCE The Subcontractor shall require the Designer(s) to maintain Professional Liability Insurance with a company reasonably satisfactory to the Contractor, including contractual liability insurance against the liability assumed in Paragraph 3.8, and including coverage for any professional liability caused by any of the Designer's(s') consultants. Said insurance shall have specific minimum limits as set forth below:

    Limit of $ _____ per claim.

    General Aggregate of $ ____ for the subcontract services rendered.

The Professional Liability Insurance shall contain prior acts coverage sufficient to cover all subcontract services rendered by the Designer. Said insurance shall be continued in effect with an extended period of _____ years following final payment to the Designer.

Such insurance shall have a maximum deductible amount of $_____ per occurrence. The deductible shall be paid by the Subcontractor or Designer.

9.2.3.2 The Subcontractor shall require the Designer to furnish to the Subcontractor and Contractor, before the Designer commences its services, a copy of its professional liability policy evidencing the coverages required in this Paragraph. No policy shall be cancelled or modified without thirty (30) Days' prior written notice to the Subcontractor and Contractor.

9.2.4 NUMBER OF POLICIES Commercial General Liability Insurance and other liability insurance may be arranged under a single policy for the full limits required or by a combination of underlying

21

policies with the balance provided by an Excess or Umbrella Liability Policy.

9.2.5 CANCELLATION, RENEWAL AND MODIFICATION The Subcontractor shall maintain in effect all insurance coverages required under this Agreement at the Subcontractor's sole expense and with insurance companies acceptable to the Contractor, which acceptance shall not be unreasonably withheld. The policies shall contain a provision that coverage will not be cancelled or not be renewed until at least thirty (30) Days' prior written notice has been given to the Contractor. Certificates of insurance showing required coverage to be in force pursuant to Subparagraph 9.2.2 shall be filed with the Contractor prior to commencement of the Subcontract Work. In the event the Subcontractor fails to obtain or maintain any insurance coverage required under this Agreement, the Contractor may purchase such coverage as desired for the Contractor's benefit and charge the expense to the Subcontractor, or terminate this Agreement.

9.2.6 CONTINUATION OF COVERAGE The Subcontractor shall continue to carry Completed Operations Liability Insurance for at least one year after either ninety (90) Days following Substantial Completion of the Work or final payment to the Contractor, whichever is earlier. Prior to commencement of the Work, Subcontractor shall furnish the Contractor with certificates evidencing the required coverages.

9.2.7 PROPERTY INSURANCE

9.2.7.1 Upon written request of the Subcontractor, the Contractor shall provide the Subcontractor with a copy of the Builder's Risk Policy of insurance or any other property or equipment insurance in force for the Project and procured by the Owner or Contractor. The Contractor shall advise the Subcontractor if a Builder's Risk Policy of insurance is not in force.

9.2.7.2 If the Owner or Contractor has not purchased property insurance reasonably satisfactory to the Subcontractor, the Subcontractor may procure such insurance as will protect the interests of the Subcontractor, its subcontractors and their subcontractors in the Subcontract Work. The cost of this insurance shall be charged to the Contractor in a Change Order.

9.2.7.3 If not covered under the Builder's Risk Policy of insurance or any other property or equipment insurance required by the Subcontract Documents, the Subcontractor shall procure and maintain at the Subcontractor's own expense property and equipment insurance for the Subcontract Work including portions of the Subcontract Work stored off the site or in transit, when such portions of the Subcontract Work are to be included in an application for payment under Article 8.

9.2.8 WAIVER OF SUBROGATION

9.2.8.1 The Contractor and Subcontractor waive all rights against each other, the Owner and the Architect/Engineer, and any of their respective consultants, subcontractors, and sub-subcontractors, agents and employees, for damages caused by perils to the extent covered by the proceeds of the insurance provided in Subparagraph 9.2.7, except such rights as they may have to the insurance proceeds. The Subcontractor shall require similar waivers from its subcontractors.

9.2.9 ENDORSEMENT If the policies of insurance referred to in this Article require an endorsement to provide for continued coverage where there is a waiver of subrogation, the owners of such policies will cause them to be so endorsed.

9.2.10 CONTRACTOR'S LIABILITY INSURANCE The Contractor shall obtain and maintain its own

IMPORTANT: A vertical line in the margin indicates a change has been made to the original text. Prior to signing, recipients may wish to request from the party producing the document a "redlined" version indicating changes to the original text. Consultation with legal and insurance counsel and careful review of the entire document are strongly encouraged.

ConsensusDOCS 750 • STANDARD FORM OF AGREEMENT BETWEEN CONTRACTOR AND SUBCONTRACTOR Copyright © 2007, ConsensusDOCS LLC; revised May 2009. YOU ARE ALLOWED TO USE THIS DOCUMENT FOR ONE CONTRACT ONLY. YOU MAY MAKE 9 COPIES OF THE COMPLETED DOCUMENT FOR DISTRIBUTION TO THE CONTRACT'S PARTIES. ANY OTHER USES, INCLUDING COPYING THE FORM DOCUMENT, ARE STRICTLY PROHIBITED.

liability insurance for protection against claims arising out of the performance of this Agreement, including without limitation, loss of use and claims, losses and expenses arising out of the Contractor's errors or omissions.

9.2.11 ADDITIONAL LIABILITY COVERAGE Contractor _____ shall/_____ shall not (indicate one) require Subcontractor to purchase and maintain liability coverage, primary to Contractor's coverage under Subparagraph 9.2.10.

9.2.11.1 If required by Subparagraph 9.2.11, the additional liability coverage required of the Subcontractor shall be:

[Designate Required Coverage(s)]

_____.1 ADDITIONAL INSURED. Contractor shall be named as an additional insured on Subcontractor's Commercial General Liability Insurance specified, for operations and completed operations, but only with respect to liability for bodily injury, property damage or personal and advertising injury to the extent caused by the negligent acts or omissions of Subcontractor, or those acting on Subcontractor's behalf, in the performance of Subcontract Work for Contractor at the Project site.

_____.2 OCP. Subcontractor shall provide an Owners' and Contractors' Protective Liability Insurance ("OCP") policy with limits equal to the limits on Commercial General Liability Insurance specified, or limits as otherwise required by Contractor.

Any documented additional cost in the form of a surcharge associated with procuring the additional liability coverage in accordance with this Subparagraph shall be paid by the Contractor directly or the costs may be reimbursed by Contractor to Subcontractor by increasing the Subcontract Amount to correspond to the actual cost required to purchase and maintain the additional liability coverage. Prior to commencement of the Subcontract Work, Subcontractor shall obtain and furnish to the Contractor a certificate evidencing that the additional liability coverages have been procured.

9.2.12 RISK OF LOSS Except to the extent a loss is covered by applicable insurance, risk of loss or damage to the Subcontract Work shall be upon the Subcontractor until the Date of Substantial Completion, unless otherwise agreed to by the Parties.

9.3 BONDS

9.3.1 The Subcontractor _____ shall/_____ shall not furnish to the Contractor, as the named Obligee, appropriate surety bonds to secure the faithful performance of the Subcontract Work and to satisfy all Subcontractor payment obligations related to Subcontract Work. Such bonds shall be issued by a surety admitted in the State in which the Project is located and shall be acceptable to the Contractor. Contractor's acceptance shall not be withheld without reasonable cause.

9.3.2 If a performance or payment bond, or both, are required of the Subcontractor under this Agreement, the bonds shall be in a form and by a surety acceptable to the Contractor, and in the full amount of the Subcontract Amount, unless otherwise specified. Contractor's acceptance shall not be withheld without reasonable cause.

9.3.3 The Subcontractor shall be reimbursed, without retainage, for the cost of any required performance or payment bonds simultaneously with the first progress payment. The reimbursement amount for the Subcontractor bonds shall be _____ percent (_____%) of the Subcontract Amount, which sum is included in the Subcontract Amount.

If acceptable to the Contractor, the Subcontractor may in lieu of retainage, furnish a retention bond or other security interest, acceptable to the Contractor, to be held by the Contractor.

9.3.4 In the event the Subcontractor shall fail to promptly provide any required bonds, the Contractor may terminate this Agreement and enter into a subcontract for the balance of the Subcontract Work with another subcontractor. All Contractor costs and expenses incurred by the Contractor as a result of said termination shall be paid by the Subcontractor.

9.3.5 PAYMENT BOND REVIEW The Contractor _____ has/_____ has not provided the Owner a payment bond. The Contractor's payment bond for the Project, if any, shall be made available by the Contractor for review and copying by the Subcontractor.

ARTICLE 10

CONTRACTOR'S RIGHT TO PERFORM SUBCONTRACTOR'S RESPONSIBILITIES AND TERMINATION OF AGREEMENT

10.1 FAILURE OF PERFORMANCE

10.1.1 NOTICE TO CURE If the Subcontractor refuses or fails to supply enough properly qualified workers, proper materials, or maintain the Progress Schedule, or fails to make prompt payment to its workers, subcontractors or suppliers, or disregards laws, ordinances, rules, regulations or orders of any public authority having jurisdiction, or otherwise is guilty of a material breach of a provision of this Agreement, the Subcontractor shall be deemed in default of this Agreement. If the Subcontractor fails within three (3) business Days after written notification to commence and continue satisfactory correction of the default with diligence and promptness, then the Contractor without prejudice to any other rights or remedies, shall have the right to any or all of the following remedies:

10.1.1.1 supply workers, materials, equipment and facilities as the Contractor deems necessary for the completion of the Subcontract Work or any part which the Subcontractor has failed to complete or perform after written notification, and charge the cost, including reasonable overhead, profit, attorneys' fees, costs and expenses to the Subcontractor;

10.1.1.2 contract with one or more additional contractors to perform such part of the Subcontract Work as the Contractor determines will provide the most expeditious completion of the Work, and charge the cost to the Subcontractor as provided under Clause 10.1.1.1; or

10.1.1.3 withhold any payments due or to become due the Subcontractor pending corrective action in amounts sufficient to cover losses and compel performance to the extent required by and to the satisfaction of the Contractor.

In the event of an emergency affecting the safety of persons or property, the Contractor may proceed as above without notice, but the Contractor shall give the Subcontractor notice promptly after the fact as a precondition of cost recovery.

10.1.2 TERMINATION BY CONTRACTOR If the Subcontractor fails to commence and satisfactorily continue correction of a default within three (3) business Days after written notification issued under Subparagraph 10.1.1, then the Contractor may, in lieu of or in addition to the remedies provided for in Subparagraph 10.1.1, issue a second written notification, to the Subcontractor and its surety, if any. Such notice shall state that if the Subcontractor fails to commence and continue correction of a default within seven (7) Days of the written notification, the Agreement will be deemed terminated. A written notice of termination shall be issued by the Contractor to the Subcontractor at

IMPORTANT: A vertical line in the margin indicates a change has been made to the original text. Prior to signing, recipients may wish to request from the party producing the document a "redlined" version indicating changes to the original text. Consultation with legal and insurance counsel and careful review of the entire document are strongly encouraged.

ConsensusDOCS 750 • STANDARD FORM OF AGREEMENT BETWEEN CONTRACTOR AND SUBCONTRACTOR Copyright © 2007, ConsensusDOCS LLC; revised May 2009. YOU ARE ALLOWED TO USE THIS DOCUMENT FOR ONE CONTRACT ONLY. YOU MAY MAKE 9 COPIES OF THE COMPLETED DOCUMENT FOR DISTRIBUTION TO THE CONTRACT'S PARTIES. ANY OTHER USES, INCLUDING COPYING THE FORM DOCUMENT, ARE STRICTLY PROHIBITED.

the time the Subcontractor is terminated. The Contractor may furnish those materials, equipment or employ such workers or subcontractors as the Contractor deems necessary to maintain the orderly progress of the Work. All costs incurred by the Contractor in performing the Subcontract Work, including reasonable overhead, profit and attorneys' fees, costs and expenses, shall be deducted from any moneys due or to become due the Subcontractor. The Subcontractor shall be liable for the payment of any amount by which such expense may exceed the unpaid balance of the Subcontract Amount. At the Subcontractor's request, the Contractor shall provide a detailed accounting of the costs to finish the Subcontract Work.

10.1.3 USE OF SUBCONTRACTOR'S EQUIPMENT   If the Contractor performs work under this Article, either directly or through other subcontractors, the Contractor or other subcontractors shall have the right to take and use any materials, implements, equipment, appliances or tools furnished by, or belonging to the Subcontractor and located at the Project site for the purpose of completing any remaining Subcontract Work. Immediately upon completion of the Subcontract Work, any remaining materials, implements, equipment, appliances or tools not consumed or incorporated in performance of the Subcontract Work, and furnished by, belonging to, or delivered to the Project by or on behalf of the Subcontractor, shall be returned to the Subcontractor in substantially the same condition as when they were taken, normal wear and tear excepted.

10.2. BANKRUPTCY

10.2.1 TERMINATION ABSENT CURE   If the Subcontractor files a petition under the Bankruptcy Code, this Agreement shall terminate if the Subcontractor or the Subcontractor's trustee rejects the Agreement or, if there has been a default, the Subcontractor is unable to give adequate assurance that the Subcontractor will perform as required by this Agreement or otherwise is unable to comply with the requirements for assuming this Agreement under the applicable provisions of the Bankruptcy Code.

10.2.2 INTERIM REMEDIES   If the Subcontractor is not performing in accordance with the Progress Schedule at the time a petition in bankruptcy is filed, or at any subsequent time, the Contractor, while awaiting the decision of the Subcontractor or its trustee to reject or to assume this Agreement and provide adequate assurance of its ability to perform, may avail itself of such remedies under this Article as are reasonably necessary to maintain the Progress Schedule. The Contractor may offset against any sums due or to become due the Subcontractor all costs incurred in pursuing any of the remedies provided including, but not limited to, reasonable overhead, profit and attorneys' fees. The Subcontractor shall be liable for the payment of any amount by which costs incurred may exceed the unpaid balance of the Subcontract Amount.

10.3 SUSPENSION BY OWNER FOR CONVENIENCE   Should the Owner suspend the Work or any part which includes the Subcontract Work for the convenience of the Owner and such suspension is not due to any act or omission of the Contractor, or any other person or entity for whose acts or omissions the Contractor may be liable, the Contractor shall notify the Subcontractor in writing and upon receiving notification the Subcontractor shall immediately suspend the Subcontract Work. To the extent provided for under the Owner-Contractor Agreement and to the extent the Contractor recovers such on the Subcontractor's behalf, the Contract Price and the Contract Time shall be equitably adjusted by Change Order for the cost and delay resulting from any such suspension. The Contractor agrees to cooperate with the Subcontractor, at the Subcontractor's expense, in the prosecution of any Subcontractor claim arising out of an Owner suspension and to permit the Subcontractor to prosecute the claim, in the name of the Contractor, for the use and benefit of the Subcontractor.

10.4 TERMINATION BY OWNER   Should the Owner terminate its contract with the Contractor or any

25

part which includes the Subcontract Work, the Contractor shall notify the Subcontractor in writing within three (3) business Days of the termination and upon written notification, this Agreement shall be terminated and the Subcontractor shall immediately stop the Subcontract Work, follow all of Contractor's instructions, and mitigate all costs. In the event of Owner termination, the Contractor's liability to the Subcontractor shall be limited to the extent of the Contractor's recovery on the Subcontractor's behalf under the Subcontract Documents, except as otherwise provided in this Agreement. The Contractor agrees to cooperate with the Subcontractor, at the Subcontractor's expense, in the prosecution of any Subcontractor claim arising out of the Owner termination and to permit the Subcontractor to prosecute the claim, in the name of the Contractor, for the use and benefit of the Subcontractor, or assign the claim to the Subcontractor. In the event Owner terminates Contractor for cause, through no fault of the Subcontractor, Subcontractor shall be entitled to recover from the Contractor its reasonable costs arising from the termination of this Agreement, including overhead and profit on Work not performed.

10.5 CONTINGENT ASSIGNMENT OF THIS AGREEMENT The Contractor's contingent assignment of this Agreement to the Owner, as provided in the Owner-Contractor agreement, is effective when the Owner has terminated the Owner-Contractor agreement for cause and has accepted the assignment by notifying the Subcontractor in writing. This contingent assignment is subject to the prior rights of a surety that may be obligated under the Contractor's bond, if any. Subcontractor consents to such assignment and agrees to be bound to the assignee by the terms of this Agreement, provided that the assignee fulfills the obligations of the Contractor.

10.6 SUSPENSION BY CONTRACTOR The Contractor may order the Subcontractor in writing to suspend all or any part of the Subcontract Work for such period of time as may be determined to be appropriate for the convenience of the Contractor. Phased Work or interruptions of the Subcontract Work for short periods of time shall not be considered a suspension. The Subcontractor, after receipt of the Contractor's order, shall notify the Contractor in writing in sufficient time to permit the Contractor to provide timely notice to the Owner in accordance with the Owner-Contractor agreement of the effect of such order upon the Subcontract Work. The Subcontract Amount or Subcontract Time shall be adjusted by Subcontract Change Order for any increase in the time or cost of performance of this Agreement caused by such suspension. No claim under this Paragraph shall be allowed for any costs incurred more than fourteen (14) Days prior to the Subcontractor's notice to the Contractor. Neither the Subcontract Amount nor the Progress Schedule shall be adjusted for any suspension, to the extent that performance would have been suspended, due in whole or in part to the fault or negligence of the Subcontractor or by a cause for which Subcontractor would have been responsible. The Subcontract Amount shall not be adjusted for any suspension to the extent that performance would have been suspended by a cause for which the Subcontractor would have been entitled only to a time extension under this Agreement.

10.7 WRONGFUL EXERCISE If the Contractor wrongfully exercises any option under this Article, the Contractor shall be liable to the Subcontractor solely for the reasonable value of Subcontract Work performed by the Subcontractor prior to the Contractor's wrongful action, including reasonable overhead and profit on the Subcontract Work performed, less prior payments made, together with reasonable overhead and profit on the Subcontract Work not executed, and other reasonable costs incurred by reason of such action.

10.8 TERMINATION BY SUBCONTRACTOR If the Subcontract Work has been stopped for thirty (30) Days because the Subcontractor has not received progress payments or has been abandoned or suspended for an unreasonable period of time not due to the fault or neglect of the Subcontractor, then the Subcontractor may terminate this Agreement upon giving the Contractor seven (7) Days' written notice. Upon such termination, Subcontractor shall be entitled to recover from the Contractor payment for all Subcontract Work satisfactorily performed but not yet paid for, including reasonable overhead, profit and

26

attorneys' fees, costs and expenses. However, if the Owner has not paid the Contractor for the satisfactory performance of the Subcontract Work through no fault or neglect of the Contractor, and the Subcontractor terminates this Agreement under this Article because it has not received corresponding progress payments, the Subcontractor shall be entitled to recover from the Contractor, within a reasonable period of time following termination, payment for all Work executed and for any proven loss, cost or expense in connection with the Work, including all demobilization costs plus reasonable overhead and profit on Work not performed. The Contractor's liability for any other damages claimed by the Subcontractor under such circumstances shall be extinguished by the Contractor pursuing said damages and claims against the Owner, on the Subcontractor's behalf, in the manner provided for in Subparagraphs 10.3 and 10.4 of this Agreement.

ARTICLE 11

DISPUTE RESOLUTION

11.1 WORK CONTINUATION AND PAYMENT Unless otherwise agreed in writing, the Subcontractor shall continue the Subcontract Work and maintain the Progress Schedule during any dispute mitigation or resolution proceedings. If the Subcontractor continues to perform, the Contractor shall continue to make payments in accordance with this Agreement.

11.2 NO LIMITATION OF RIGHTS OR REMEDIES Nothing in this Article shall limit any rights or remedies not expressly waived by the Subcontractor which the Subcontractor may have under lien laws or payment bonds.

11.3 MULTIPARTY PROCEEDING The Parties agree that all parties necessary to resolve a claim shall be parties to the same dispute resolution proceeding. To the extent disputes between the Contractor and Subcontractor involve in whole or in part disputes between the Contractor and the Owner, disputes between the Subcontractor and the Contractor shall be decided by the same tribunal and in the same forum as disputes between the Contractor and the Owner.

11.4 DISPUTES BETWEEN CONTRACTOR AND SUBCONTRACTOR In the event that the provisions for resolution of disputes between the Contractor and the Owner contained in the Subcontract Documents do not permit consolidation or joinder with disputes of third parties, such as the Subcontractor, or if such dispute is only between the Contractor and Subcontractor, then the Parties shall submit the dispute to the dispute resolution procedures set forth in Paragraph 11.5.

11.5 CONTRACTOR-SUBCONTRACTOR DISPUTE RESOLUTION

11.5.1 DIRECT DISCUSSIONS If the Parties cannot reach resolution on a matter relating to or arising out of the Agreement, the Parties shall endeavor to reach resolution through good faith direct discussions between the Parties' representatives, who shall possess the necessary authority to resolve such matter and who shall record the date of first discussions. If the Parties' representatives are not able to resolve such matter within seven (7) Days, the Parties' representatives shall immediately inform senior executives of the Parties in writing that resolution was not affected. Upon receipt of such notice, the senior executives of the Parties shall meet within seven (7) Days to endeavor to reach resolution. If the matter remains unresolved after fifteen (15) Days from the date of first discussion, the Parties shall submit such matter to the dispute resolution procedures selected in Article 11.

11.5.2 MEDIATION If direct discussions pursuant to Subparagraph 11.6.1 do not result in resolution of the matter, the Parties shall endeavor to resolve the matter by mediation through the current Construction Industry Mediation Rules of the American Arbitration Association, or the Parties

IMPORTANT: A vertical line in the margin indicates a change has been made to the original text. Prior to signing, recipients may wish to request from the party producing the document a "redlined" version indicating changes to the original text. Consultation with legal and insurance counsel and careful review of the entire document are strongly encouraged.
ConsensusDOCS 750 • STANDARD FORM OF AGREEMENT BETWEEN CONTRACTOR AND SUBCONTRACTOR Copyright © 2007, ConsensusDOCS LLC; revised May 2009. YOU ARE ALLOWED TO USE THIS DOCUMENT FOR ONE CONTRACT ONLY. YOU MAY MAKE 9 COPIES OF THE COMPLETED DOCUMENT FOR DISTRIBUTION TO THE CONTRACT'S PARTIES. ANY OTHER USES, INCLUDING COPYING THE FORM DOCUMENT, ARE STRICTLY PROHIBITED.

may mutually agree to select another set of mediation rules. The administration of the mediation shall be as mutually agreed by the Parties. The mediation shall be convened within thirty (30) working Days of the matter first being discussed and shall conclude within forty-five (45) working Days of the matter being first discussed. Either Party may terminate the mediation at any time after the first session, but the decision to terminate shall be delivered in person by the terminating Party to the non-terminating Party and to the mediator. The costs of the mediation shall by shared equally by the Parties.

11.5.3 BINDING DISPUTE RESOLUTION If the matter is unresolved after submission of the matter to a mitigation procedure or mediation, the Parties shall submit the matter to the binding dispute resolution procedure selected herein: (Designate only one)

_____ Arbitration using the current Construction Industry Arbitration Rules of the American Arbitration Association or the Parties may mutually agree to select another set of arbitration rules. The administration of the arbitration shall be as mutually agreed by the Parties.

_____ Litigation in either the state or federal court having jurisdiction of the matter in the location of the Project.

11.6 COST OF DISPUTE RESOLUTION The costs of any binding dispute resolution procedure shall be borne by the non-prevailing Party, as determined by the adjudicator of the dispute.

11.7 VENUE The venue for any binding dispute resolution proceeding shall be the location of the Project unless the Parties agree on a mutually convenient location.

ARTICLE 12

MISCELLANEOUS PROVISIONS

12.1 GOVERNING LAW   This Agreement shall be governed by the law in effect at the location of the Project.

12.2 SEVERABILITY   The partial or complete invalidity of any one or more provisions of this Agreement shall not affect the validity or continuing force and effect of any other provision.

12.3 NO WAIVER OF PERFORMANCE   The failure of either Party to insist, in any one or more instances, upon the performance of any of the terms, covenants or conditions of this Agreement, or to exercise any of its rights, shall not be construed as a waiver or relinquishment of term, covenant, condition or right with respect to further performance.

12.4 TITLES   The titles given to the Articles and Paragraphs of this Agreement are for ease of reference only and shall not be relied upon or cited for any other purpose.

12.5 OTHER PROVISIONS AND DOCUMENTS   Other provisions and documents applicable to the Subcontract Work are set forth in Exhibit _____.

12.6 JOINT DRAFTING The Parties expressly agree that this Agreement was jointly drafted, and that they both had opportunity to negotiate its terms and to obtain the assistance of counsel in reviewing its terms prior to execution. Therefore, this Agreement shall be construed neither against nor in favor of either Party, but shall be construed in a neutral manner.

ARTICLE 13

EXISTING SUBCONTRACT DOCUMENTS

13.1 INTERPRETATION OF SUBCONTRACT DOCUMENTS

13.1.1 The drawings and specifications are complementary. If Work is shown only on one but not on the other, the Subcontractor shall perform the Subcontract Work as though fully described on both consistent with the Subcontract Documents and reasonably inferable from them as being necessary to produce the indicated results.

13.1.2 In case of conflicts between the drawings and specifications, the specifications shall govern. In any case of omissions or errors in figures, drawings or specifications, the Subcontractor shall immediately submit the matter to the Contractor for clarification by the Owner. The Owner's clarifications are final and binding on all Parties, subject to an equitable adjustment in Subcontract Time or Price pursuant to Articles 5 and 6 or dispute resolution in accordance with Article 11.

13.1.3 Where figures are given, they shall be preferred to scaled dimensions.

13.1.4 Any terms that have well-known technical or trade meanings, unless otherwise specifically defined in this Agreement, shall be interpreted in accordance with their well-known meanings.

13.1.5 In case of any inconsistency, conflict or ambiguity among the Subcontract Documents, the documents shall govern in the following order: (a) Change Orders and written amendments to this Agreement; (b) this Agreement; (c) subject to Subparagraph 13.1.2 the drawings (large scale governing over small scale), specifications and addenda issued prior to the execution of this Agreement; (d) approved submittals; (e) information furnished by the Owner pursuant to Paragraph 4.5; (f) other documents listed in this Agreement. Among categories of documents having the same order of precedence, the term or provision that includes the latest date shall control. Information identified in one Contract Document and not identified in another shall not be considered a conflict or inconsistency.

As defined in Paragraph 2.3, the following Exhibits are a part of this Agreement.

EXHIBIT _____ The Subcontract Work, _____ pages.

EXHIBIT _____ The Drawings, Specifications, General and other conditions, addenda and other information. (Attach a complete listing by title, date and number of pages.)

EXHIBIT _____ Progress Schedule, _____ pages.

EXHIBIT _____ Alternates and Unit Prices, include dates when alternates and unit prices no longer apply, _____ pages.

EXHIBIT _____ Temporary Services, stating specific responsibilities of the Subcontractor, and Contractor _____ pages.

EXHIBIT _____ Temporary Services, stating specific responsibilities of the Subcontractor, _____ pages.

EXHIBIT _____ Insurance Provisions, _____ pages.

EXHIBIT _____ Other Provisions and Documents, _____ pages.

This Agreement is entered into as of the date entered in Article 1.

CONTRACTOR _____

BY: ....

NOT FOR FURTHER REPRODUCTION
TO ORDER DOCUMENT, VISIT WWW.CONSENSUSDOCS.ORG

PRINT NAME: _____

PRINT TITLE: _____

ATTEST ....

SUBCONTRACTOR: _____

BY: ....

PRINT NAME: _____

PRINT TITLE: _____

ATTEST ....

GLOSSARY

acts of God   events causing a project delay that are not caused by the owner or the contractor. Such delays often result in adjustments to the contract duration but not to the contract amount.

actual acceleration   an increase in the pace of construction as a result of a specific directive from the owner.

addenda   formal changes or clarifications issued to all identified bidders by the owner or the owner's representative during the bidding period.

additional work   construction work that was not recognized at the contract award but that must be performed in order to deliver a project as planned. A structural member may be designed to occupy the same space as air ducts. This will require additional work.

adverse weather   weather conditions not anticipated in a particular location for a particular time of the year that impede construction progress.

advertisement   a public announcement to solicit bids for a construction project.

agency agreement   an arrangement between a principal and an agent by which the agent agrees to perform certain tasks for the principal. The principal is bound by the actions of the agent.

agent   a party who acts for another party and binds that party by those acts.

alternates   an itemization of selected items of work for which bidders are asked to provide prices that will add to or subtract from the base bid. These priced items will give the owner greater flexibility in choosing items to add to or delete from the contract.

alternative dispute resolution technique (ADR)   a means used for settling conflicts by means of an alternative procedure to formal litigation.

arbitration   a well-established alternative to litigation in which the conflict is resolved by an impartial third party or an impartial panel of selected individuals.

as-built drawings   also called as-builts; project drawings that show all data concerning the actual in-place locations of all construction items, including any items that differ from what was shown in the original drawings.

balanced bid   a unit price bid that accurately reflects the actual anticipated price of each item of work to be performed.

bid bond   a surety instrument that guarantees to the owner that the bearer, if awarded the contract, will enter into a binding contract and provide all required bonds.

bid peddling the effort by a bidder, usually a subcontractor, on a project to determine the relative standing of a quoted bid. If the bid is not the lowest bid, the bidder may reassess the amount originally quoted and submit a lower bid before the deadline for bid submittal.

bid shopping see *prebid bid shopping* and *postbid bid shopping.*

bilateral contract an agreement made through the mutual promises of the contracting parties.

board of directors an elected group of individuals who are assigned the responsibility of managing a corporation. They act as agents for the stockholders and are accountable to them.

boilerplate the general conditions that outline the roles of the parties to a construction agreement and provide guidance concerning procedures to follow under varying circumstances.

bonding capacity the maximum amount of uncompleted construction work that a contractor can have under contract, above which no bonds will be provided.

brokerage a situation in which the general contractor subcontracts all the work on a project.

builder's risk insurance construction insurance that provides coverage specifically for a project that is under construction. Although this is normally considered to be fire insurance, other types of losses are also generally covered.

calendar days the time unit that may be used to define the duration of construction.

cardinal change a change order that is of such magnitude that the original scope of the project is altered to an extent that constitutes a new contract.

caveat emptor (let the buyer beware) a defense, no longer valid, in which it is assumed that a product must be accepted with whatever flaws exist at the time of purchase.

certificate of insurance a written document that serves as evidence that a particular insurance policy is in force.

change order a directive, usually authorized in writing by the owner, to alter or modify some aspect of a project. Such a directive is generally accompanied by an adjustment to the contract amount and/or the contract duration.

closed specification a specification that is expressly restrictive in stating that only one or two products will satisfy the quality requirements, or is implied when performance is so narrowly prescribed that only one or two products will satisfy the requirement.

code of ethics the written standards of behavior adopted by a profession.

completion date the time of construction; stated not in terms of a specified duration, but as a specified date by which construction must be complete.

compliance officer the title of OSHA employees who are responsible for enforcing OSHA regulations through site inspections.

complimentary bid a bid that is not prepared in earnest, but is presented to appear to be a serious bid. Such bids are usually generated through collaborative efforts with another bidder who does submit a serious bid.

condemnation the exercise of eminent domain to seize private property.

consideration an essential ingredient to a contract that implies something of value, commonly a stated sum of money.

construction schedule a detailed network analysis or bar chart of a construction project showing the sequence and duration of activities required to construct a project.

constructive acceleration an increase in the pace of construction that is not a result of a directive but is done by inference. Denial of a legitimate request for a time extension, such as for an excusable delay or a change order, would be an example.

constructor the party, also called the prime or general contractor, who has primary responsibility for the construction of a project.

contractor-caused delays construction delays caused by or under the control of the contractor. No contract adjustments are associated with such delays.

contributory negligence careless acts of an injured person that accompany the careless acts or physical conditions under the control of a second party.

cost-plus contract a contract in which the contractor is reimbursed for specified incurred costs, with an additional allowance provided for overhead and profit.

counteroffer the rejection of an offer followed by another proposal. This proposal forms a new offer. A counteroffer by a party changes the roles of the negotiating parties.

critical activity an activity in a construction schedule that must be completed in the time allotted for its completion or the project duration will be increased.

dedication public permission granted by the owner of property for the public to use a given parcel of land for a specified use. This use cannot be denied as long as consistent use is made as specified.

design-build method an arrangement by which the owner lets a single contract for both the design and the construction of a project; also known as design-construct or turnkey construction.

designer the party responsible for translating the concept of a project to a document that can be used as a guide for its construction.

design specifications "how to" specifications that state exactly what the contractor is to do in order to satisfy a quality requirement.

differing site conditions physical conditions on a site that differ from what was shown in the bidding documents or from what would reasonably be expected.

disputes review board a panel of experts selected on a project to render decisions on disputes brought to it for consideration.

dividends monetary return made by insurance companies to clients who have kept losses to an acceptably low level. Also, profits of corporations shared with stockholders.

dual gates two entrances established on a construction project where both union and nonunion workers will be employed. One entrance will be established as the "union workers only" entrance and the other entrance can be used by any worker. A nonunion worker passing through the union gate will result in "contamination" of the gate and can result in a labor dispute.

easement a restricted use of private land granted to another party. The restricted use may be in the form of the right to cross a parcel of land to gain access to another; the right to install, maintain, and monitor a gas line; and so forth.

eichleay formula a means by which delay reimbursement can be made to the contractor to compensate for home office overhead. This is a controversial approach that is not accepted universally.

eminent domain the right of the federal government, state government, or another public agency to take possession of private property and appropriate it for public use.

estoppel the legitimate, though implied, formation of a contract as evident through the actions of the parties involved. When one party places a reliance on the other party based on that party's actions, the second party cannot subsequently deny that an agreement exists.

exclusions specific items stated in an insurance policy for which no coverage is provided.

exculpatory provisions contract clauses that shift liability from one of the contracting parties to the other. In the absence of such a clause, the shift in risk will not occur.

excusable delays delays for which time extensions are granted. Such delays typically include acts of God and owner-caused delays.

executed contract an agreement in which both parties have fully performed in accordance with the contract terms.

executory contract an agreement in which one or both parties to the agreement have not yet performed.

experience modification rating a factor, unique to a company, that reflects the past claims history of that company. This factor is used to increase or decrease the basic insurance premium charges.

express contract an agreement in which the terms of the agreement, whether verbal or written, are clear, concise, explicit, and definite.

express warranty the specific statement in a contract that a warranty is to be effective on a project.

extra work see *change order.*

field change an authorized directive, usually by the owner's representative on a project, to alter or modify some aspect of the project. Normally, such a directive is not accompanied by any change in the contract amount or duration.

final completion the status of a project when all punch list items have been satisfactorily addressed and the owner officially accepts the project. At this point, the retainage can be released to the contractor.

fixed-price contract see *lump sum contract.*

force account arrangement in which the services of a contractor are reinbursed on a cost-plus basis; also referred to as a time and materials contract.

front-loading a plan initiated by a contractor in which a disproportionate share of the payments to be made to the contractor is made early in the life of a project. This is a means of getting the owner to finance a greater share of the project.

general conditions see *boilerplate.*

general contract method a common procedure in which the owner of a project contracts with a single firm, often called a prime contractor, for its construction. This firm may contract with specialty contractors for portions of the work.

holding company a firm that has a dominant interest in one or more other companies so as to be able, through its voting power, to prescribe the management policies of those companies.

holidays days on which working days are not assessed against the contract duration. Such holidays must be clearly noted in the contract documents.

housing starts the number of new homes placed under construction in a stipulated period of time, generally one month. This figure is used as an indication of the strength of the U.S. economy.

implied contract an agreement in which the terms of the agreement are not clearly stated but are established through inference and deduction.

implied warranty the general interpretation in the courts that a warranty exists despite the fact that there is no express warranty.

inclusions specific items, as stated in an insurance policy, for which coverage is specifically provided.

indemnification an exculpatory provision in which one party to an agreement agrees to hold the other party harmless. That is, one party assumes the liability that would normally fall on another party.

independent contractor a contractor hired to produce a product without being specifically supervised or constrained by specific methods and means of performance.

instructions to bidders guidelines or rules enumerated for bidders on a project concerning proper procedures for bid submittal and relevant project information.

joint venture a company formed by two or more companies in which the sole objective is typically to construct a single project.

letter of credit a demand instrument issued by a bank that guarantees the availability of a specified amount of funds to be paid to the owner in the event of contractor default.

lien a claim placed on real property.

limited partner a partner who contributes to a partnership and shares in the profits and losses, but provides no services and has no vote in matters of management.

limits of liability   the maximum coverage for which a specific liability policy is written.

liquidated damages   a specified sum of money that is charged against a contractor for each day that project completion is delayed. This amount is assumed to accurately reflect the anticipated costs of late completion.

litigation   a means of resolving disputes in the judicial system in which a formal claim is filed and the disputing parties typically obtain the services of lawyers.

loss ratio   a quotient that reflects the amount of funds expended by an insurance company for the claims of a client, divided by the amount of premiums paid by that client.

lump sum contract   a contract in which the contractor agrees to construct a project for a specified sum of money.

mechanic's lien   a right created by law that permits a worker to place a claim on land on which improvements have been made.

mediation   an alternative to litigation in which knowledgeable individuals' talents are used to get the disputing parties to agree on a compromise resolution to the conflict.

meeting of the minds   a basic ingredient in contracts in which the contracting parties agree on the basic meaning and legal implications of a contract.

minitrial   an alternative to litigation in which one or more individuals are asked to hear a case and render a decision. The rules for hearing such cases may be established on an individual basis.

morals   generally accepted standards of social behavior for a community or society.

no-damage-for-delay clause   a contract provision that is exculpatory in nature and is intended to bar the contractor from receiving monetary compensation of any kind for any delays that may occur on a project, regardless of the source of the delay.

notice to proceed   a means of notifying the contractor about the decision to award the contract and of specifying when the contract time will start.

obligee   a party to whom a duty is promised by the surety.

offer   occurs when one person signifies to another a willingness to enter into a binding contract on certain specified terms.

offer and acceptance   one of the basic ingredients of a contract. After an offer is made, a contract can become binding only if the other party agrees to be bound by the contract terms and accepts the offer.

open specification   an open or nonrestrictive specification in which a wide variety (three or more) of products are considered suitable.

owner   the party with the overall responsibility for a project beginning from inception and ending with the project sale or occupancy.

owner-caused delays   construction delays directly attributable to the action or lack of action of the owner. Generally the contract duration will be extended for such delays, and the contract amount may be adjusted.

owner-controlled insurance program (OCIP)   see *wrap-up insurance*.

partnering   voluntary arrangement whereby the parties on a construction project agree to work together as a team with a common objective and to resolve disputes at the lowest managerial level.

partnership   an association of two or more persons to carry on a business. Each person acts as an agent for the other partners.

payment bond   a surety instrument guaranteeing the bearer's payments to suppliers, laborers, and subcontractors.

penalty provision   a specified sum of money that is charged against a contractor for each day that project completion is delayed. The amount is greater than the amount of the liquidated damages because a portion of it is considered a punitive assessment. To be valid, such a provision must also compensate or reward the contractor for early completion.

performance bond a surety instrument in which the faithful performance of a contractor is guaranteed up to the face value of the bond.

performance specification a descriptive requirement that states the results or the performance of the finished product rather than the specific methods and means used to construct the product.

periodic payments see *progress payment.*

postbid bid shopping the efforts of a general contractor to get subcontractors to lower their bids after the general contractor has been declared the low bidder.

prebid bid shopping see *bid peddling.*

preconstruction conference a meeting between the owner, the general contractor, and the major subcontractors that takes place after the contract is awarded but before construction begins. This conference addresses various matters deemed to be of importance to the project.

premium a payment made to an insurance company in exchange for a specific type of coverage.

prescription the legal transfer of private property to a public agency or another private citizen that can occur when that property has been subjected to adverse use for a stipulated number of years.

professional construction management method a method in which the owner hires a construction management firm to perform professional services and represent the owner during the design and construction phases.

progress payment a payment to the contractor that compensates the contractor for work performed up to a given date. Such payments generally are made monthly and are reduced by a given percentage that is retained by the owner until final completion.

progress schedule an updated construction schedule that presents an accurate portrayal of the work accomplished up to a given date.

proprietary specification a closed specification that names a product made by a particular manufacturer.

proprietorship a firm owned by an individual.

punch list a list developed at the time of substantial completion that itemizes all remaining work tasks that must be performed before a project reaches final completion.

quantum meruit term meaning "as much as deserved"; an approach used to determine equitable compensation to be awarded under given conditions. It is implied by law that when one party benefits as a result of the materials or labor provided by another that there should be no unjust enrichment. If the parties do not have a contract, the party providing the labor and/or materials is to receive fair compensation.

ratification the approval of an unsanctioned act after it has taken place.

real property land and all items physically attached to it, such as buildings, fences, utilities, and walls.

reference specification a performance specification that defines the acceptable product in terms of its ability to satisfy a standard developed by one of several standards organizations.

regular bid a bid that conforms to the standards outlined in the instructions to bidders. Only such bids can be considered on public works projects to assure all bidders of fair and equal treatment.

reserve a given sum of money that an insurance company sets aside to cover anticipated claims costs on a case that is not closed.

retainage a stated percentage of the progress payment request that is withheld by the owner. This amount is generally used as an incentive for the contractor to complete the project in an expedient manner. It is generally returned to the contractor after final completion.

self-insurance   an arrangement by which a firm acts as its own insurance company. Specific criteria must be satisfied for a company to qualify for self-insurance.

self-performance   a mechanism by which no contracts are awarded for a construction project. The owner's own workers or employees are solely responsible for the construction effort.

separate contracts method   an arrangement by which the owner lets contracts directly to specialty contractors for various portions of the work.

site investigation   an inspection of a proposed construction site during the bidding phase. Such an inspection is often contractually required to verify that the contractor is familiar with the site.

specialty contractors   firms with skills in specific areas of construction work. Such firms are typically involved in construction projects as subcontractors.

standard specifications   a compilation of general conditions, technical specifications, and other requirements that an agency uses on numerous construction projects. Such documents are prepared by state departments of transportation and many major municipalities.

stockholders   the owners of a corporation, also known as shareholders.

subcontractor   a speciality contractor who enters into an agreement with a general contractor. The subcontractor has no contractual agreement with the owner. See *specialty contractors.*

submittals   information concerning products to be incorporated in a construction project that must be approved by the owner before they are used. This information may include samples, calculations, performance tests, and manufacturer's literature.

subrogation   an insurance term stating that when the insurance company pays for a particular loss to an injured party, the insurance company gains the injured party's right to sue the third party that caused the injury.

substantial completion   a designation of when a project is sufficiently finished to be occupied by the owner. The duration of the project is measured against substantial completion to determine when the last periodic payment can be made.

supplementary conditions   modifications or additions to the general conditions that address issues that were omitted in the general conditions, or are specific requirements for a particular project.

surety   a firm that guarantees or vouches for the performance or indebtedness of another party.

suspension of work   a halt in the construction process for any of several reasons. Work is presumably resumed after the delay.

technical specifications   a document that provides the qualitative requirements for a project in terms of materials, equipment, work performed, and so forth.

termination   the cancellation of a contract before construction is complete. May be for the owner's convenience or for contractor default.

time extensions   modifications to a construction contract in which the project duration is increased. Such modifications are often granted for excusable delays or for change orders that increase the project duration.

tort   a wrong committed by a person against a second party as a result of an act or the failure to act when that person had a duty to the second party.

turnkey construction   see *design-build method.*

ultra vires contracts   agreements made by corporations for which the proper authorization was not made. Such agreements must often be ratified in order to be binding.

umbrella coverage   an insurance policy that is generally used to provide coverage in excess of the limits of coverage of another insurance policy.

unbalanced bid   a unit price bid in which the pricing of the various items of work does not reflect the actual anticipated costs, but redistributes those costs to serve a specific objective of the bidder.

unilateral contract   a contract in which only one of the contracting parties makes a promise. The other party exchanges something other than a promise, commonly performance.

unit price contract   a contract in which payment is based on a contractor's quoted price per unit of work performed and the owner's measurement of the total number of such units installed.

unjust enrichment   a circumstance under which one party benefits at the expense of another without equitable compensation being given to the party that provided the labor and/or materials.

value engineering   a critical examination of construction contract documents performed for the owner to determine whether modifications can be made to decrease the delivered cost, reduce maintenance costs, simplify construction, reduce disputes, and the like.

warranty   certification that a certain aspect of a project is of the quality it was promised to be. In construction, such assurances are generally provided for one year from substantial completion.

warranty period   the duration for which a warranty is in effect, typically one year on construction projects.

winter exclusion period   a block of time, often consisting of several winter months, during which no working days are assessed against a contractor. Such exclusion periods are most common in states with harsh winter weather conditions.

workers' compensation   insurance coverage for the employees of a firm during their employment.

working days   the time unit used to define the duration of a construction project. These days typically consist of work days except for holidays. Allowances are often granted to extend the project duration if adverse weather is encountered.

wrap-up insurance   insurance for a project that is obtained entirely by the owner.

zoning   restrictions placed on land usage to assure orderly growth and development in a municipal area.

INDEX

Page numbers followed by f indicate figures